AF899952

Two-Dimensional-Materials-Based Membranes

Two-Dimensional-Materials-Based Membranes

Preparation, Characterization, and Applications

Edited by Gongping Liu and Wanqin Jin

WILEY-VCH

Editors

Prof. Gongping Liu
Nanjing Tech University
College of Chemical Engineering
30 Puzhu Road (S)
211816 Nanjing
China

Prof. Wanqin Jin
Nanjing Tech University
College of Chemical Engineering
30 Puzhu Road (S)
211816 Nanjing
China

Cover Image: © LAGUNA DESIGN/Getty Images

All books published by **WILEY-VCH** are carefully produced. Nevertheless, authors, editors, and publisher do not warrant the information contained in these books, including this book, to be free of errors. Readers are advised to keep in mind that statements, data, illustrations, procedural details or other items may inadvertently be inaccurate.

Library of Congress Card No.: applied for

British Library Cataloguing-in-Publication Data
A catalogue record for this book is available from the British Library.

Bibliographic information published by the Deutsche Nationalbibliothek
The Deutsche Nationalbibliothek lists this publication in the Deutsche Nationalbibliografie; detailed bibliographic data are available on the Internet at <http://dnb.d-nb.de>.

© 2022 WILEY-VCH GmbH, Boschstraße 12, 69469 Weinheim, Germany

All rights reserved (including those of translation into other languages). No part of this book may be reproduced in any form – by photoprinting, microfilm, or any other means – nor transmitted or translated into a machine language without written permission from the publishers. Registered names, trademarks, etc. used in this book, even when not specifically marked as such, are not to be considered unprotected by law.

Print ISBN: 978-3-527-34848-0
ePDF ISBN: 978-3-527-82983-5
ePub ISBN: 978-3-527-82984-2
oBook ISBN: 978-3-527-82985-9

Typesetting Straive, Chennai, India
Printing and Binding CPI Group (UK) Ltd, Croydon, CR0 4YY

C109793_260722

Contents

Preface

Synthetic membranes have been used in various processes, from water treatment, gas purification to energy conversion. The conventional membrane processes, including microfiltration, ultrafiltration, and reverse osmosis, have been commercialized and widely used in food, chemical, and medical industries. However, it remains challenging to realize the precise separation of the molecules/ions with subnanometer sizes, such as saline water, organic solvent solution, and gas mixtures. As we know, there is a general trade-off between permeability and selectivity for polymeric membranes. Although inorganic membranes could achieve both high permeability and selectivity, the high cost of membrane fabrication restricts their development. The key lies in the design and fabrication of high-performance (high permeability, selectivity, and stability) membranes for efficient molecular separation.

Two-dimensional (2D) materials (e.g. nanosheets of graphene family materials, zeolites, metal–organic frameworks, or layered double hydroxides) have become excellent nanobuilding blocks to develop next-generation separation membranes featuring unique nanopores and/or nanochannels. Recently, 2D materials-based membranes have been appearing as a very hot topic both in theoretical and experimental studies and have attracted a huge surge of interest. Important new results demonstrated that the fascinating 2D-material membranes showed extraordinary permeation properties, opening the door to ultrafast and highly selective membranes for water purification, desalination, gas separation, and bioseparation. Up to now, books lacked an instructive view on the important contemporary topic "2D materials membranes." There is an urgent need for a book that brings alive the new results, and current significances and challenges concerning 2D-material membranes. In this book, we aim to give comprehensive information on design, fabrication, and application of 2D-material membranes that are used for molecular separation.

Various membrane structures based on different 2D materials are involved in the book, including porous nanosheet membranes, 2D laminar membranes, and mixed-matrix membranes (MMMs) with 2D materials serving as nanofillers. Fabrication strategies, physicochemical properties, morphologies, host–guest interactions, transport characteristics, and mechanisms of 2D-material membranes will be thoroughly discussed based on advanced characterization techniques and theoretical calculations. Meanwhile, the book contains several typical applications of 2D-material membranes in fields, such as water desalination, ion separation, solvent dehydration, and gas separation. In addition, milestones in the large-scale fabrication of 2D-material membranes for wide implementation are briefly introduced. Finally, we conclude this book with an overview of remaining challenges and new opportunities that have opened for 2D-material membranes in molecular separation.

At last, we would like to express our appreciation and gratitude to the chapter authors: Dr. Kumar Varoon Agrawal from École Polytechnique Fédérale de Lausanne (EPFL), Dr. Xiwang Zhang from Monash University, Dr. Hongwei Zhu from Tsinghua University, Dr. Dan Zhao from National University of Singapore, Dr. Yi Liu from Dalian University of Technology, Dr. Fusheng Pan from Tianjin University, and Ho Bum Park from Hanyang University.

Professor in Chemical Engineering
Nanjing Tech University
Nanjing, China
10 January 2022

Gongping Liu
Wanqin Jin

1

Introduction

Gongping Liu and Wanqin Jin

Nanjing Tech University, College of Chemical Engineering, State Key Laboratory of Materials-Oriented Chemical Engineering, 30 Puzhu South Road, Nanjing, 211816, P. R. China

Two-dimensional (2D) materials have been emerging stars in condensed matter physics, materials science, and chemistry since the successful exfoliation of graphene by Novoselov and Geim in 2004 [1–3]. The atomic thickness and micrometer lateral dimensions endow 2D materials with great potential in membrane separation. Synthetic membranes are used widely in many separation processes, from industrial-scale processes, such as removing salt from seawater and separating atmospheric gases, to smaller-scale processes in chemical synthesis and purification [4, 5]. The membranes function by forming a selective barrier between the two phases, restricting the movement of some molecules while letting others penetrate. On one hand, the membranes assembled from ultrathin nanosheets can minimize the transport resistance and maximize the mass-transfer rate. On the other hand, the intrinsic or artificial nanopores and interlayer galleries can provide excellent molecular sieving properties [6, 7]. Hence, 2D materials, including graphene-family, zeolites, metal–organic frameworks (MOFs), covalent–organic frameworks (COFs), metal carbides and nitrides (MXenes), and layered double hydroxides (LDHs), have been demonstrated as excellent building blocks for high-performance membranes [8, 9].

2D materials are either porous or nonporous based on their atomic structure. The intrinsic nanopores in zeolites and MOFs or the drilled nanopores in graphene can provide molecular/ionic transport pathway. In contrast, the nonporous nanosheets (e.g. graphene oxide and MXene) must be assembled into laminates with interlayer channels for mass transfer, which is driven by external forces, such as pressure difference, centrifugal force, and molecular interaction [10]. Hence, according to the difference in membrane structures, 2D-material-based membranes can be categorized into three types—(i) porous nanosheet membranes, (ii) laminar membranes, and (iii) 2D-material-based mixed-matrix membranes (MMMs). The fabrication methods and unique properties of these membranes will be detailedly discussed in the following chapters. By the precise construction and manipulation of the in-plane nanopores/slits and interlayer channels, 2D-material-based membranes

Two-Dimensional-Materials-Based Membranes: Preparation, Characterization, and Applications, First Edition. Edited by Gongping Liu and Wanqin Jin.
© 2022 WILEY-VCH GmbH. Published 2022 by WILEY-VCH GmbH.

have exhibited outstanding molecular separation properties in various membrane processes, such as ultrafiltration, nanofiltration, reverse osmosis, forward osmosis, pervaporation, and gas separation [9].

With the rapid development of 2D-material-based membranes, a comprehensive summary of the breakthroughs and milestones for these membranes is urgently needed to guide newcomers and inspire innovations in this field. The following chapters summarize the recent progress of 2D-material-based membranes for molecular separation, focusing on membrane preparation, characterization, and application. Chapter 2 introduces the fabrication methods for 2D materials and membranes. Chapter 3 presents the development of porous graphene-based nanosheet membranes. Chapters 4–6 describe the design and application of graphene-based membranes for water separation, ions separation, and pervaporation, respectively. Chapter 7 shows the development of 2D-material membranes for gas separation. Chapter 8 introduces the advance in 2D-LDHs membranes. Chapter 9 demonstrates the design of MXene and other emerging 2D-material membranes. Chapter 10 summarizes the milestones for 2D-material mixed-matrix membranes. Chapter 11 reveals the transport mechanism of 2D-material membranes. Finally, Chapter 12 gives the conclusions and perspectives of this research topic.

References

1 Geim, A.K. and Grigorieva, I.V. (2013). Van der Waals heterostructures. *Nature* 499 (7459): 419–425.
2 Koltonow, A.R. and Huang, J. (2016). Two-dimensional nanofluidics. *Science* 351 (6280): 1395–1396.
3 Shen, J., Liu, G., Han, Y. et al. (2021). Artificial channels for confined mass transport at the sub-nanometre scale. *Nat. Rev. Mater.* 6 (4): 294–312.
4 Gin, D.L. and Noble, R.D. (2011). Designing the next generation of chemical separation membranes. *Science* 332 (6030): 674–676.
5 Koros, W.J. and Zhang, C. (2017). Materials for next-generation molecularly selective synthetic membranes. *Nat. Mater.* 16 (3): 289–297.
6 Cheng, L., Liu, G., and Jin, W. (2019). Recent progress in two-dimensional-material membranes for gas separation. *Acta Phys. Chim. Sin.* 35: 351090–351098.
7 Liu, G., Jin, W., and Xu, N. (2015). Graphene-based membranes. *Chem. Soc. Rev.* 44 (15): 5016–5030.
8 Liu, G., Jin, W., and Xu, N. (2016). Two-dimensional-material membranes: a new family of high-performance separation membranes. *Angew. Chem. Int. Ed.* 55 (43): 13384–13397.
9 Cheng, L., Liu, G., Zhao, J. et al. (2021). Two-dimensional-material membranes: manipulating the transport pathway for molecular separation. *Accounts Mater. Res.* 2 (2): 114–128.
10 Wang, S., Yang, L., He, G. et al. (2020). Two-dimensional nanochannel membranes for molecular and ionic separations. *Chem. Soc. Rev.* 49 (4): 1071–1089.

2

Fabrication Methods for 2D Membranes

Long Cheng, Gongping Liu, and Wanqin Jin

Nanjing Tech University, College of Chemical Engineering, State Key Laboratory of Materials-Oriented Chemical Engineering, 30 Puzhu South Road, Nanjing, 211816, P. R. China

2.1 Introduction

Two-dimensional-material membranes have demonstrated great potential in molecular separation. The synthesis of high-quality nanosheets with integrated structure is a prerequisite for fabricating defect-free two-dimensional-material membranes. The synthetic methods of nanosheets aimed at membrane application include mechanical-force exfoliation, ion-intercalation exfoliation, oxidation-assisted exfoliation, selective-etching method, chemical vapor deposition (CVD), hydro/solvothermal synthesis, and interfacial synthesis. All of these methods can be divided into two categories, top-down and bottom-up methods [1], which are discussed in detail. After the nanosheets are successfully synthesized, novel membrane fabrication methods should be proposed for tuning the in-plane pores and constructing interlayer channels. There are three main types of 2D-material membranes—(i) nanosheet membranes; (ii) laminar membranes; and (iii) mixed-matrix membranes (MMMs). The first type consists of a monolayer or a few layers of 2D material with intrinsically uniformly sized pores (e.g. zeolite, metal-organic frameworks (MOFs)) or drilled nanopores (graphene) for molecular sieving. The second type, laminar membrane, is composed of orderly stacked nanosheets with tunable interlayer channels by various assembly strategies. The size and functional groups of the interlayer channels play an important role in selective permeation of small molecules. Thirdly, the addition of 2D fillers into polymer matrix can combine the unique transport property of nanosheets and the excellent processability of a polymer, further enhancing separation performance and even mechanical strength of pristine polymer membrane. Owing to the difference in membrane structures, various fabrication methods have been proposed. In this chapter, we aim to summarize the advances in design and fabrication of nanosheets and 2D-material membranes. We focus on the discussion of different approaches for assembling the nanosheets, and the prominent effect of tuning the nanostructures

Two-Dimensional-Materials-Based Membranes: Preparation, Characterization, and Applications, First Edition. Edited by Gongping Liu and Wanqin Jin.
© 2022 WILEY-VCH GmbH. Published 2022 by WILEY-VCH GmbH.

and physicochemical properties of transport channels is highlighted. Opportunities and challenges concerning the development of fabrication methods for 2D-material membranes are discussed as well.

2.2 Synthesis of Nanosheets

The synthetic methods of 2D nanosheets can be categorized into top-down and bottom-up methods. The top-down method, including mechanical-force exfoliation, ion-intercalation exfoliation, oxidation-assisted exfoliation, and selective-etching method, relies on the exfoliation of bulk crystals into 2D nanosheets. Conversely, the bottom-up method refers to the construction of 2D frameworks from basic units. CVD, hydro/solvothermal synthesis, and interfacial synthesis are well-established bottom-up methods. All the aforementioned methods are introduced and discussed, respectively.

2.2.1 Top-Down Method

2.2.1.1 Mechanical-Force Exfoliation

The early mechanical-force exfoliation method is the micromechanical cleavage technique [2], namely "Scotch-tape" method. The micromechanical force can weaken the van der Waals interaction between the layers of bulk crystals, which is beneficial for the isolation of neighboring layers, and single- or few-layer nanosheets can be obtained after several times of cleavage. The sizes of the as-prepared nanosheets can reach a few to tens of micrometers, and this nondestructive technique can contribute to clean surfaces and excellent crystal quality, making them good candidates for investigating the intrinsic physical properties. Nevertheless, the low yield of this method restricts its practical application. There are more easily scalable strategies, such as sonication, shear force, and ball-milling exfoliations. For the sonication method, the layered crystals were firstly dispersed in a proper solvent, followed by a sonication treatment (Figure 2.1) [3]. To realize efficient exfoliation, the surface energy of the chosen solvent should match that of the bulk crystal. Coleman and coworkers pioneered the liquid exfoliation of graphite into graphene in organic solvents, which is a scalable method to produce high-quality graphene nanosheets [4]. They further extended this method for the synthesis of other 2D nanosheets, such as MoS_2, WS_2, and *h*-BN [5]. Moreover, the addition of polymers or surfactants can tune the surface tension of the aqueous solution to match the surface energy of bulk crystals. For instance, Coleman's group demonstrated that the introduction of sodium dodecylbenzenesulfonate (SDBS) or sodium cholate (SC) into the aqueous solution can largely improve the exfoliation efficiency, contributing to high yield of graphene nanosheets [6, 7]. In addition, the control of sonication parameters, such as ultrasonic powder, sonication time, sonication temperature, and the shape of vessels, can tune the properties (e.g. concentration, lateral size, and thickness) of as-synthesized nanosheets [8].

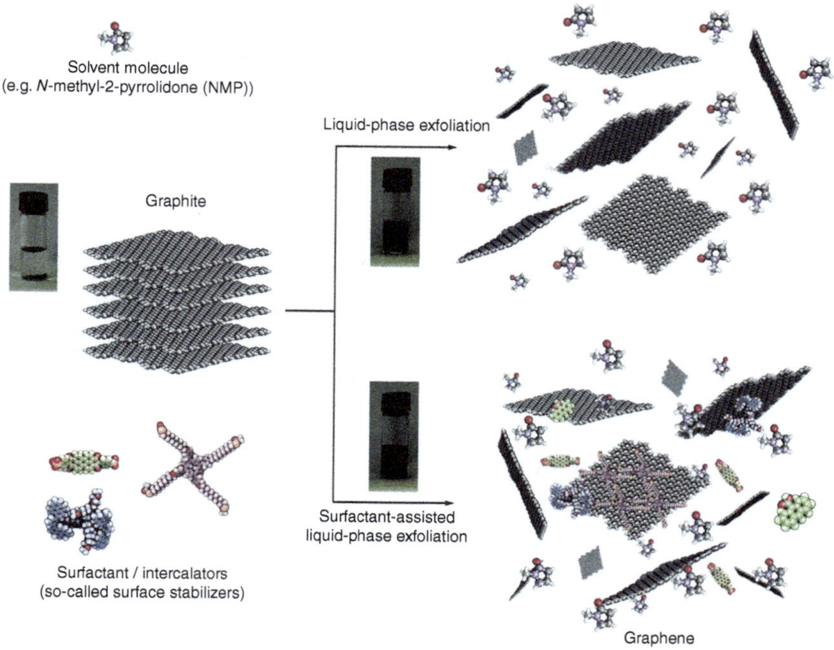

Figure 2.1 Schematic representation of the sonication exfoliation process of graphite in the absence (top right) and presence (bottom right) of surfactant molecules. Source: Ciesielski and Samori [3]/with permission from Royal Society of Chemistry.

Another mechanical force for efficient exfoliation is the shear force, which can be generated from the high-shear rates or the ball-milling impact. Coleman and coworkers employed a high-shear rotor–stator mixer to provide high-shear rates in liquid, facilitating the exfoliation of bulk graphite [9]. They successfully produced graphene nanosheets in 1-methyl-2-pyrrolidinone (NMP) solvent in this shear-force-assisted method. Besides the rotor–stator mixer, a commercially available kitchen blender was proved to generate sufficient shear rates for fabrication of various nanosheets, including graphene and MoS_2 [10, 11], which implies the potential of industrial rotating blade stirred-tank reactors in large-scale graphene production. Ball-milling method was also employed as the shear-force-assisted exfoliation. The rotating speed should be carefully controlled to avoid the in-plane defects caused by shock stress and ensure the dominant shear stress. Chen and coworkers demonstrated the exfoliation of graphite in *N,N*-dimethylformamide (DMF) by the ball-milling method [12]. After high-speed centrifugation, they obtained single- and few-layer graphene nanosheets with a thickness around 0.8–1.8 nm. The monolayer fraction could be further increased by optimizing the milling conditions (e.g. the diameters of milling balls, the milling time, and the concentration of graphite). Yang and coworkers proposed a soft-physical strategy combining the ball-milling and sonication processes to exfoliate layered $Zn_2(bim)_4$ (bim = benzimidazole) crystals [13]. The mixture of methanol and propanol is proved to be the most appropriate for the exfoliation—small methanol molecules can enter the interlayer galleries to promote the exfoliation, and propanol helps to stabilize the exfoliated nanosheets by adsorbing on the nanosheets with its hydrophobic alkane tails. Therefore, the selection of proper solvents is critical for successful exfoliation, which is universally applicable to almost all exfoliation methods. In general, the mechanical force exfoliation can contribute to high yield and massive production of nanosheets, but the sonication or ball milling may cause some defects on the nanosheets, affecting the separation performance of assembled membranes.

2.2.1.2 Ion-Intercalation Exfoliation

The ion-intercalation is accomplished by intercalating cation ions (e.g. Li^+, Na^+, and K^+) into the interlayer galleries of bulk crystals, which would weaken the var der Waals interaction between neighboring layers and expand the interlayer space for easier cleavage. In 1980s, Morrison and coworkers employed *n*-butyl lithium, as the intercalator, to synthesize MoS_2 nanosheets [14]. The bulk MoS_2 was first soaked in hexane containing *n*-butyl lithium for at least two days to intercalate Li into the MoS_2. Then the obtained Li-intercalated compound was transferred into water, followed by sonication to assist in the exfoliation. The reaction between intercalated Li and water continuously generated hydrogen gas and further facilitated the exfoliation process. In addition, Loh and coworkers proposed a two-step chemical ion-intercalation method to synthesize large-sized nanosheets (Figure 2.2) [15]. N_2H_4 was first intercalated into the bulk crystals and could expand the volume of the crystals after its decomposition. After that, the product was secondly intercalated by metal naphthalenide (metal = Li, Na, K) to form intercalated compounds. In this way, the yield of MoS_2 reached 90%, and the lateral size was up to 400 μm^2. It is worth pointing out that organometallic compounds used in this method are

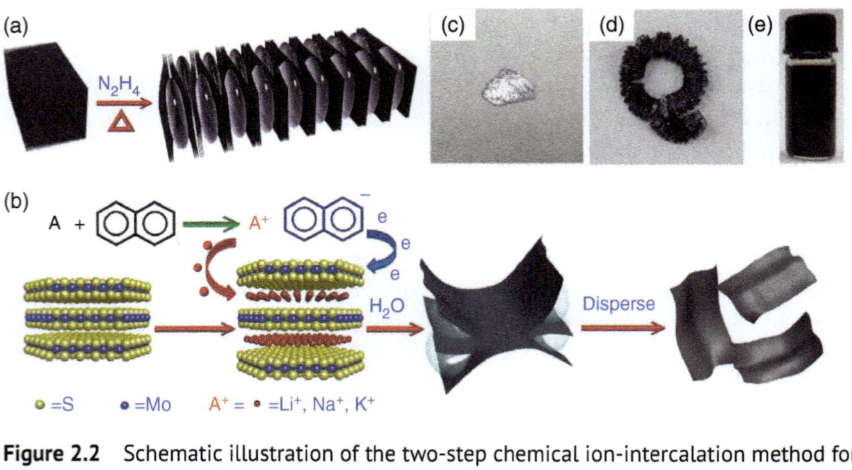

Figure 2.2 Schematic illustration of the two-step chemical ion-intercalation method for the preparation of MoS_2 nanosheets. (a) Bulk MoS_2 was pre-exfoliated by the decomposition products of N_2H_4. (b) Pre-exfoliated MoS_2 was secondly intercalated by metal naphthalenide, followed by exfoliation in water. Photographs of (c) bulk single-crystal MoS_2, (d) pre-exfoliated MoS_2, and (e) Na-exfoliated single-layer MoS_2 dispersion in water. Source: Zheng et al. [15]/with permission from Springer Nature.

highly explosive and very sensitive to moisture and oxygen, so the experiment needs to be operated in a glovebox with extreme caution. Bearing this in mind, Zheng and coworkers demonstrated that some safer inorganic salts, including NaCl and $CuCl_2$, can serve as intercalators for efficient exfoliation [16, 17]. The graphite was dispersed in saturated NaCl or $CuCl_2$ solution, which was heated at 100 °C to evaporate the water and obtain the cation-intercalated graphite. The intercalated compound was further sonicated in organic solvent (e.g. DMF, ethanol, and NMP) to obtain graphene nanosheets. Moreover, they extended this method for the synthesis of other nanosheets, including MoS_2, $MoSe_2$, WS_2, and WSe_2 [17], demonstrating the universality of the safer ion-intercalation exfoliation.

2.2.1.3 Oxidation-Assisted Exfoliation

Oxidation-assisted exfoliation is a well-established method for the synthesis of graphene oxide (GO). This method was firstly proposed by Brodie in 1859, and further developed by Staudenmaier and Hummers [18], forming the three primary synthetic routes. The exfoliation is based on the oxidation of graphite into graphite oxide by strong oxidizing agents. The generated oxygen-containing functional groups can greatly expand the interlayer space of layered graphite and weaken the van der Waals interaction between neighboring nanosheets. After the oxidation treatment, the subsequent sonication can further transform the graphite oxide into single-layer GO nanosheets. Both Brodie and Staudenmaier methods use potassium chlorate and nitric acid as the oxidizing agent, and the most commonly used Hummers' method employed the potassium permanganate and concentrated sulfuric acid to ensure sufficient oxidation. The advantage of this method lies in the high yield of single-layer GO nanosheets. However, the oxidation process can cause severe explosions at elevated temperatures, and a huge amount of wastewater containing acid and metal ions can be produced and lead to environmental pollution. Therefore, a greener and safer exfoliation method based on electrochemical oxidation process was proposed for the synthesis of GO nanosheets. Ren and coworkers reported a two-step electrochemical method to fabricate GO nanosheets with a high level of oxygen content (Figure 2.3) [19]. In the first intercalation step, the commercial graphite paper serving as the anode was immersed in 98 wt% H_2SO_4 at a low voltage of 1.6 V, resulting in the blue graphite intercalation compound. The second step (rapid oxidation process) required a suitable concentration of H_2SO_4, normally in the range of 40–60 wt%, to achieve highly oxidized GO with C/O ratio of 1.5–1.8, surpassing that of classic chemical oxidation routes (around 2.2) [20]. In addition, the as-synthesized GO nanosheets exhibited good dispersibility and nanoscale thickness, which are similar to those synthesized by Hummers' method. They attributed the extensive graphite oxidation to the abundant radical intermediates (e.g. $^*OH$, $^*O$, and $^*OOH$) from the electrolytic water splitting. In this way, the oxidation of graphene lattice was only within a few seconds, which is much faster than the Hummers' method. The electrochemical oxidation method combined the advantages of safety, environmental friendliness, ultrafast synthesis, and easiness to scale up, which paves the way for industrial production and application of GO sheets at a low cost.

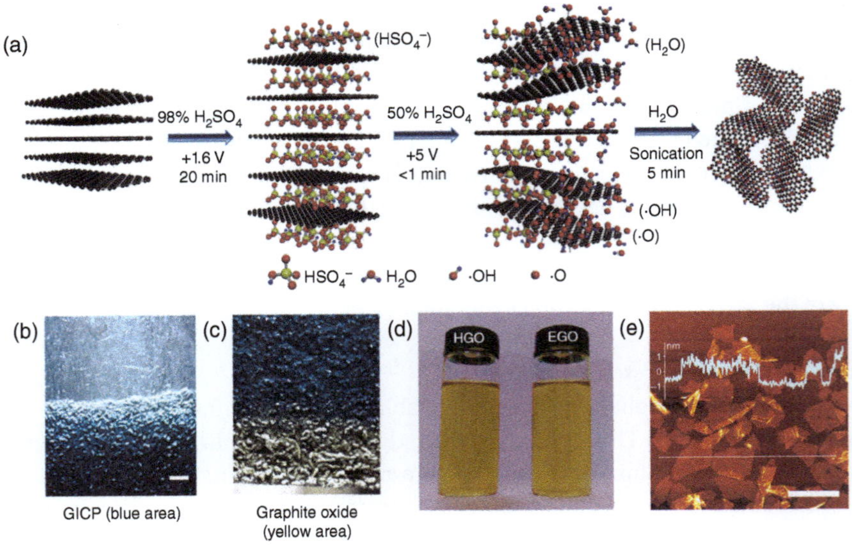

Figure 2.3 Synthesis of GO nanosheets by water electrolytic oxidation. (a) Schematic illustration of the fabrication process of electrochemically synthesized GO (EGO). Source: (a) Pei et al. [19]/Springer Nature/CC BY 4.0. (b) Photograph of the graphite intercalation compound paper (GICP) (blue area) after the reaction with 98 wt% H_2SO_4 at 1.6 V for 20 minutes. (c) Photograph of GO (yellow area) resulting from the water electrolytic oxidation of GICP in 50 wt% H_2SO_4 at 5 V for 30 seconds. (d) Comparison of EGO with GO prepared by Hummers' method (HGO) in water (1 mg ml^{-1}). (e) AFM images of EGO nanosheets. Source: Pei et al. [19]/with permission from Springer Nature.

2.2.1.4 Selective-Etching Method

In 2011, Gogotsi and coworkers proposed the selective-etching method for the exfoliation of bulk MAX phases into the nanosheets of transition metal carbides and/or carbonitrides (MXenes) [21]. The term MAX phases reflect the chemical composition of $M_{n+1}AX_n$, where $n = 1$, 2, or 3 (M_2AX, M_3AX_2, or M_4AX_3), "M" is an early transition metal, "A" is an A group (mostly groups 13 and 14) element, and "X" is C and/or N [22]. In the MAX phases, the M layers are nearly closed packed with the X atoms filling the octahedral sites, and the $M_{n+1}X_n$ layers are interleaved with layers of A atoms. The M—X bond has a mixed covalent/metallic/ionic character, while the M—A bond is metallic. Compared with the weak van der Waals interactions in graphite and transition metal dichalcogenides (TMDs), the metallic bonds are too strong to be broken by the traditional mechanical forces. To solve this problem, Gogotsi and coworkers reported the selective etching of the Al from Ti_3AlC_2 by introducing the hydrogen fluoride (HF) (Figure 2.4) [22]. Owing to the differences in character and relative strengths of the Ti—Al bond compared with the Ti—C bond, the Al layers can be selectively etched without disrupting the Ti—C bond. The etched intermediate product was further sonicated to obtain Ti_3C_2 nanosheets. MXenes prepared by this method possess numerous terminations (e.g. O, –OH, and –F) on the surface, which are favorable for stabilizing them in solution and the assembly into membranes. It is worth pointing out that the etchant HF is

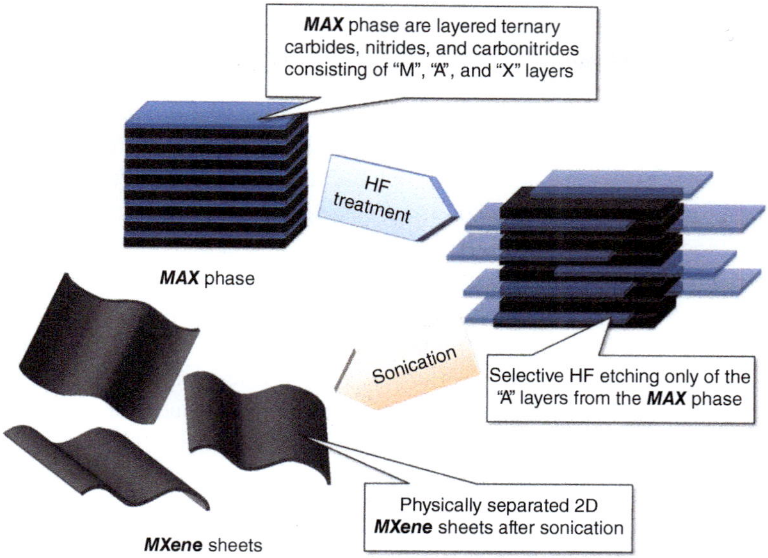

Figure 2.4 Schematic illustration of the synthesis process of MXenes from MAX phases by selective-etching method. Source: Naguib et al. [22]. Reproduced with permission. Copyright 2012, American Chemical Society.

highly corrosive and dangerous. Therefore, Gogotsi's group developed the mixture of fluoride salts and HCl or H_2SO_4 as the alternative to the corrosive HF [23], which improved the safety of the selective-etching method. Moreover, more kinds of MXenes, such as Ti_2C, Nb_2C, V_2C, and Nb_4C_3, have been successfully synthesized by this method [24], demonstrating its great potential for wider application. The limitation is that the method is not applicable to the synthesis of other nanosheets, such as graphene or TMDs.

2.2.2 Bottom-Up Method

2.2.2.1 Chemical Vapor Deposition

The first industrial application of CVD technique was dated back to 1897 when de Lodyguine reduced tungsten hexachloride with hydrogen to deposit tungsten onto the carbon filament of lamps [25]. Then this technique was further employed for the production of high-purity materials, including Ti, Ta, Zr, and Si. Over the past decade, CVD has been considered as an efficient technique for the massive production of 2D nanosheets. Typically, the substrates are placed in a furnace, and gas/vapor precursors are cycled through the chamber to react and/or decompose on the surface of substrates. Sometimes the catalysts, in the shape of a film, are needed in the reaction process. Then the nanosheets can be obtained and further transferred to the desired substrates. Somani and coworkers pioneered the synthesis of graphene films from camphor pyrolysis on Ni substrates via the CVD method [26]. Despite the film thickness of about 30 layers, this work demonstrated the feasibility of the CVD method for the fabrication of graphene. Following this work, many efforts have been devoted to optimizing the experimental conditions, such as the precursors, substrates, catalysts, temperature, and atmospheres, which are critical to the structural features of the obtained graphene products. Pollard et al. demonstrated the CVD growth of single-layer graphene on the Ni film evaporated on the SiO_2/Si substrate [27]. It is worth noting that the Ni film served as the substrate, which supported the growing graphene and the catalyst facilitating the nucleation of precursors at the same time. In addition, Li et al. synthesized large-area graphene film with lateral size of centimeters and single-layer rate over 95% by CVD method on copper foil substrate [28]. Methane and hydrogen were chosen as gas sources, and the substrate was removed by the treatment with iron nitrite solution to obtain freestanding graphene films. Similarly, the CVD growth method can be applied for the fabrication of other nanosheets, including *h*-BN [28, 29], MoS_2 [30], and so on. Despite the great potential of CVD technique in the massive production of nanosheets with high crystal quality and purity, the synthetic conditions, such as high temperatures (~1000 °C) and inert atmosphere, lead to high production costs and restrict its industrial application.

2.2.2.2 Hydro/Solvothermal Synthesis

The hydro/solvothermal synthesis, in which water or organic solvent serves as the reaction medium in a sealed vessel at high temperature and pressure, has been a popular strategy for preparing nanosheets, such as 2D metal oxides, metal chalcogenides, and MOFs [31]. During the synthesis, the key issue lies in how to selectively block the growth of nanomaterials along one direction to form desired nanosheets. The introduction of structure-directing agents is demonstrated to be an effective strategy. Tsapatsis and coworkers proposed the bottom-up synthesis of MFI nanosheets via the hydrothermal growth of MFI seeds with the assistance of bis-1,5(tripropyl ammonium) pentamethylene diiodide (dC5) [32]. The dC5 was proven to direct the growth of plate-like MFI with the thin crystal dimension along the *b* axis, and the straight micropores aligned along this direction, endowing the 2.5-unit-cell thick (5 nm) MFI nanosheets with great potential in membrane separation. The hydro/solvothermal synthesis method is a scalable method for the high-yield production of ultrathin nanosheets, but the product is quite sensitive to the synthesis parameters, such as the concentration of precursors, surfactant and solvent type, and reaction temperature/time, thus influencing the reproducibility of each batch [1].

2.2.2.3 Interfacial Synthesis

The interfacial synthesis is another bottom-up method for the fabrication of 2D nanosheets. During the synthesis process, the monomers react at the confined interfaces to guarantee a low thickness. The diffusion rates of the monomers toward the interface should be carefully controlled, which determines the structural integrity of the as-prepared nanosheets. Gascon and coworkers proposed a three-layer synthesis strategy for the preparation of MOF nanosheets [33]. The system consisted of three liquid layers composed of mixtures of DMF and a suitable miscible co-solvent in appropriate ratios, which were vertically arranged according to their different densities. The top and bottom layers contained $Cu(NO_3)_2$ and 1,4-benzenedicarboxylic acid (BDCA), respectively, while the intermediate region acted as a highly diluted medium for the growth of MOF nanosheets. The obtained nanosheets exhibited lateral size of 0.5–4 μm and thickness in the range of 5–25 nm. Moreover, the method can be extended to a variety of MOFs by changing the metal nodes or organic linkers, demonstrating the versatility of the three-layer synthesis strategy. Besides MOF nanosheets, covalent organic framework (COF) nanosheets can be prepared by the interfacial synthesis method as well. Payamyar et al. successfully synthesized COF nanosheets at the water/air interface [34]. They designed an amphiphilic monomer that can preorganize on the water surface and realize face-to-face stacking of the anthracene units. As a result, the monomers were connected via the photochemical anthracene [4+4]-cycloaddition dimerization to construct COF nanosheets. Although the interfacial synthesis method can produce nanosheets with integrated structure, the yield is relatively low and cannot meet the demand for industrial-scale production.

2.3 Membrane Structures and Fabrication Methods

According to the differences in membrane structures, two-dimensional-material membranes can be categorized into three main types, including nanosheet membranes, laminar membranes, and MMMs, and the fabrication methods for each structure differ from one another. In this section, we focus on various strategies for membrane fabrication and introduce the latest advances in different types of membranes. Importantly, the effect of constructing and tuning the in-plane pores or interlayer channels on the membrane structure and separation performance is systematically summarized. In addition, the pros and cons of the fabrication methods are briefly compared. In the last part, some hybrid membranes beyond the mentioned types are discussed as well.

2.3.1 Two-Dimensional-Material Nanosheet Membranes

2.3.1.1 Zeolite Membrane

Zeolites are microporous materials based on crystalline silica with pore sizes within several angstroms to nanometers. The thickness of 2D zeolite nanosheets can be reduced as thin as the dimension of unit cells, which enables the nanosheets to be promising material for fabricating ultrathin membranes with superior separation performance. Tsapatsis and coworkers have been devoted to the fabrication of high-quality zeolite nanosheets and molecular-sieving membranes. In 2011, they reported the synthesis of highly crystalline MFI nanosheets via the melt-blending method combined with sonication [35]. The obtained nanosheets were ~3.4 nm, only 1.5-unit-cell thick along the *b* axis. Then they employed filtration method to deposit nanosheets on the porous support to form the nanosheet coating, which was oriented since the short dimension of nanosheets was perpendicular to the support surface. Nevertheless, nanometer-sized pinhole defects existed in the deposits, which seriously affected the separation performance. The secondary solvothermal growth method was proposed to eliminate the defects, and the preferred orientation could be maintained by the introduction of appropriate structure-directing agents. The thickness of as-prepared ultrathin zeolite membranes can be reduced to ~100 nm, which was 10 times thinner than typical zeolite membranes [36]. Importantly, the membrane showed excellent molecular-sieving property and high-separation performance for xylene isomer mixtures. Although the zeolite nanosheets synthesized in this method were of high purity, the yield was low (~0.01% w/w nanosheets in the dispersion) and the lateral size of nanosheets was relatively small. Bearing this in mind, they further demonstrated the direct synthesis of zeolite nanosheets via hydrothermal growth of MFI seeds [32], which was previously discussed in Section 2.2.2. The as-synthesized nanosheets combined the lateral dimension of conventional MFI crystals (1.5–2.0 μm) with nanometer-scale thickness. The large lateral dimensions were favor for lowering the density of grain boundaries in the membranes, which would lead to nonselective transport pathways and reduce the sieving property. After gel-free secondary growth, a dense layer with thickness ranging between 250 nm and 1 μm was formed on the porous support. The successful synthesis of high-quality zeolite nanosheets endowed the membrane with excellent *p*-xylene/*o*-xylene separation performance (*p*-xylene permeance: 0.56×10^{-6} mol Pa^{-1} m^{-2} s^{-1}, separation factor: 2000 at 150 °C). The membrane throughput can be further enhanced if the membrane thickness can be reduced to 10 nm. Furthermore, more zeolite crystals with layered structures should be discovered for improving the diversity of zeolite nanosheets. The early advance in 2D zeolite membranes lays the foundation for the fast development of 2D-material membranes.

2.3.1.2 MOF Membrane

Owing to the abundant porous structure and functional surface groups, MOFs can serve as building blocks for molecular-sieve membranes. A large amount of MOF crystals with layered structures can be exfoliated into ultrathin nanosheets, demonstrating the great potential of 2D MOF membranes. However, it is challenging to retain the morphological and structural integrity of the synthesized nanosheets, which is a critical factor for fabricating defect-free MOF membranes. Yang and coworkers demonstrated the fabrication of ultrathin MOF nanosheets and high-performance molecular-sieve membranes [13]. The $Zn_2(bim)_4$ nanosheets were exfoliated via a soft-physical strategy, which was discussed in Section 2.2.1. Filtration method is commonly used for the preparation of nanosheet membranes. Nevertheless, they found this method could lead to the restacking of nanosheets back to ordered pristine structures, which would partially or totally block the molecular-sieve pores. A hot-drop coating method was proposed to realize the disordered stacking of nanosheets. In a typical preparation, a certain amount of nanosheet suspension was deposited dropwise onto the surface of the porous support, which was heated on a heating plate. As a result, the MOF nanosheets were uniformly deposited on the support to form a continuous layer. The texture of the underlying support was distinguishable, indicating an ultrathin MOF layer. In addition, the detection of aluminum signal from Al_2O_3 support by X-ray photoelectron spectroscopy (XPS) proved that the membrane was only several nanometers thick. Moreover, the structure–performance relationship was investigated as well. The low-angle hump and (002) reflection corresponded to the expanded stacking and ordered restacking of nanosheets, respectively, which would block the pathway for H_2 and reduced the separation performance. The ultrathin MOF membrane, consisting of disordered nanosheets, exhibited excellent H_2/CO_2 separation performance with H_2 permeance of up to 2700 GPU and selectivity reaching 291. Wang et al. proposed the freeze–thaw approach to exfoliate the bulk MAMS-1 (mesh adjustable molecular sieve, Ni_8(5-bbdc)$_6$(μ-OH)$_4$, 5-bbdc = 5-*tert*-butyl-1,3-benzenedicarboxylate) crystals into ~4 nm nanosheets [37]. Then the MAMS-1 nanosheets were assembled into molecular-sieve membranes via the hot-drop casting method. The as-prepared membranes possessed pore openings parallel to gas concentration gradient, which allowed high gas flux and high selectivity at the same time. In addition, they demonstrated the reversed thermos-switchable property of the gas transport pathways. The H_2/CO_2 selectivity exhibited a temperature-dependent behavior, with the maximum of 245 at 20 °C and the minimum of c. 5 at 100 °C. Based on *in situ* variable temperature powder X-ray diffraction (PXRD) and molecular dynamics simulation, this phenomenon was attributed to the flexibility of the MAMS-1 nanosheets. Hence, the MOF membrane with reversed thermo-switchable molecular-sieving property could have potential in the temperature-related gas separation process.

2.3.1.3 Porous Graphene Membrane

The perfect single-layer graphene sheet is impermeable to gases as small as helium because the electron density of its aromatic rings is too substantial to allow the transport of atoms and molecules through these rings [38]. Therefore, researchers were motivated to drill holes in graphene sheets to fabricate porous graphene membranes, which hold the potential to achieve superior separation performance due to their single-atom thinness. At first, theoretical studies on porous graphene membranes were reported. Jiang et al. designed pores functionalized with different atoms on the graphene nanosheet and investigated the gas separation performance of the porous graphene membrane by first-principles density-functional theory calculations [39]. Cohen-Tanugi and Grossman reported the desalination performance of single-layer graphene with nanometer-scale pores based on classical molecular dynamics [40]. They found that the pore diameter of the porous nanosheet determined the salt rejection performance and the hydrophilic groups, such as hydroxyl groups, bonded to the pore edges contributed to prominently enhanced water flux. In addition, the water permeability of the porous graphene was several orders of magnitude higher than the conventional reverse osmosis membrane, which exhibited the great potential of porous graphene in water purification.

Apart from the theoretical calculations, experimental studies on drilling pores in graphene sheets were reported. The ultraviolet-induced oxidative etching method was firstly proposed for creating pores in graphene membranes [41]. After the introduction of nanopores, the pressurized blister test combined with mechanical resonance was employed to monitor the molecular transport through nanopores. The as-fabricated membrane exhibited a gas permeation rate comparable to that of computational modeling and molecular-sieving property, making an important step toward the realization of macroscopic porous graphene membranes. Celebi et al. reported the physically perforated double-layer graphene by the focused ion beam [42]. The perforated graphene membrane possessed lots of nanopores with narrowly distributed diameters ranging from <10 nm to 1 μm, endowing the membrane with orders-of-magnitude enhancement in gas, water, and water vapor permeances. Moreover, the O_2 plasma-etching method was employed to fabricate porous graphene membranes for water desalination [43]. The membrane with finely tuned nanopores showed fast water transport and a salt rejection rate of nearly 100%, proving the feasibility of porous graphene membrane for desalination. To create nanopores with angstrom precision at a high pore density, Zhao et al. developed synergistic defect nucleation and pore-expansion strategy by combining O_2 plasma and O_3 treatment [44]. Raman spectra were employed to detect the evolution of defects and porosity of the porous graphene nanosheet. As a result, they achieved a high density (c. 2.1×10^{12} cm^{-2}) of H_2-sieving pores and record-high H_2/CH_4 and H_2/C_3H_8 separation performance. More importantly, this pore-etching strategy is highly scalable and will contribute to the rapid development of single-layer graphene-based membranes.

Although many strategies have been proposed to make pores on the graphene nanosheet, the problem of structural integrity of the atomic-thin membrane will restrict its practical application. Yuan and coworkers reported the fabrication of large-area graphene-nanomesh/single-walled carbon nanotube (GNM/SWNT) hybrid membrane with excellent mechanical strength [45]. In a typical process, single-layer graphene was prepared on Cu foil via the CVD method. A layer of SWNT membrane consisting of interconnected SWNTs was transferred on top of the graphene, and a freestanding membrane with SWNT-supported graphene was obtained by etching the Cu foil. Then a layer of mesoporous SiO_2 film with perpendicular mesoporous channels was created on the graphene surface, serving as the porous template. O_2 plasma was employed to drill pores in the graphene to form the graphene nanomesh, and the meso-SiO_2 layer was removed by hydrofluoric acid etching. As a result, the GNM/SWNT hybrid membrane was obtained, in which the mechanically strong SWNT webs formed strong π–π interaction with the supported GNM and acted as a microscopic framework to support the GNM, ensuring the structural integrity of the atomically thin GNM over macroscopic scale. Moreover, the high-density subnanometer pores in the GNM allow the fast transport of water molecules, while blocking solute ions/molecules to enable selective separation. The as-fabricated ultrathin GNM/SWNT hybrid membrane show high water permeance and a high rejection ratio for salt ions or organic molecules.

In all, the porous graphene membrane has demonstrated its huge potential in the process, such as water purification and gas separation, but the technical issue of creating high-density nanopores with narrowly distributed diameter in the membrane area of m^2 level remains to be solved.

2.3.2 Two-Dimensional-Material Laminar Membranes

2.3.2.1 Assembly Strategies of Laminates

For the laminar membrane assembled from 2D nanosheets, the selection of proper assembly strategy determines the microstructure of membranes and thus the transport behavior. For GO membranes prepared by self-assembly, the intrinsic repulsive force between the carboxyl group of neighboring nanosheets will cause the disordered stacking of nanosheets and the generation of nonselective defects [8], which greatly influence the separation performance of laminar membranes. Hence, various driving forces, including pressure difference, shear force, centrifugal force, and electric-field force, are employed to regulate the assembly of 2D nanosheets and reduce the interlayer defect.

To construct continuous GO laminates with few defects, vacuum filtration method was employed to surpass the repulsive force between neighboring nanosheets. Yu and coworkers fabricated ultrathin (1.8–18 nm) GO membranes on the anodic aluminum oxide (AAO) support via filtration for hydrogen purification [46]. The membrane thickness could be easily tuned by controlling the deposition amount of GO nanosheets. They observed that the reduced GO membrane with narrowed interlayer spacing showed similar gas permeance to the pristine membrane, suggesting the selective structural defects within GO flakes instead of the interlayer spacing determined the gas transport behavior. The as-prepared GO membrane exhibited ultrahigh H_2/CO_2 selectivity of 3400 and H_2/N_2 selectivity of 900. Moreover, our group firstly proposed a vacuum suction method for depositing GO nanosheets on the ceramic hollow fiber (Figure 2.5) [47], which possesses high-packing density and is beneficial to the scale-up fabrication of GO membranes. Owing to the preferential water-sorption ability and fast water diffusivity through the GO layer, the as-fabricated GO membrane showed excellent pervaporation dehydration performance for the dimethyl carbonate (DMC)/water mixture. Similarly, based on the driving force of pressure difference, pressurized filtration method was employed to prepare the freestanding GO membrane (Figure 2.6a) [48]. The filtration pressure was proved to influence the packing density of GO nanosheets and thus the interlayer spacing of GO laminates. The freestanding GO membrane was applied in the pervaporation separation of water/ethanol mixture, but the binary-component selectivity was much lower than the ideal water/ethanol selectivity, which was attributed to the swelling of GO laminates in aqueous solution. The strategy for stabilizing GO membranes is discussed in Section 2.3.2.2.

Besides the pressure difference, centrifugal force was employed to assist the assembly of GO laminates. Kim et al. reported the fabrication of few-layered GO membranes via the spin–coating method [52]. They employed two different coating methods to prepare GO membranes. In method one, they contacted the surface of the support to the air–liquid interface of a GO solution, followed by spin coating. For comparison, they directly spin-coated the GO solution on the support in method two. As a result, two GO membranes exhibited similar surface morphology but quite different gas transport behavior. The repulsive edge-to-edge

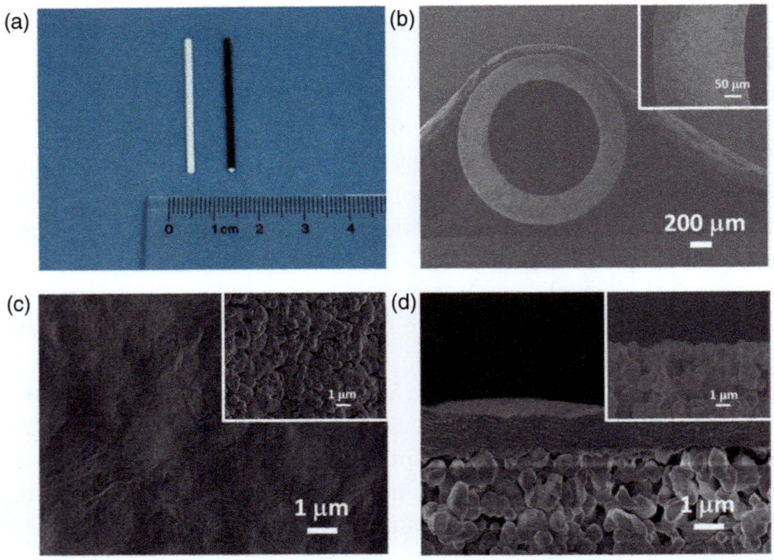

Figure 2.5 GO membranes fabricated by vacuum suction method. (a) Digital pictures of blank hollow fiber (white) and GO membrane (black). SEM images of (b) a ceramic hollow fiber (inset: an enlarged cross section of the ceramic hollow fiber), (c) the surface of the blank hollow fiber (inset) and the GO membrane, and (d) the cross section of the blank hollow fiber (inset) and the GO membrane. Source: Huang et al. [47]/with permission from John Wiley & Sons, Inc.

interaction from the carboxylic groups of GO nanosheets led to a heterogeneous GO deposition in method one, while the face-to-face attractive capillary forces during spin coating overwhelmed the repulsive interaction in method two, contributing to the dense stacking of GO nanosheets. The GO membrane prepared by method two formed highly interlocked structure and exhibited excellent CO_2/N_2 separation performance. In addition, we firstly proposed to fabricate GO laminar membranes by the spray–evaporation method (Figure 2.6b) [49]. Spraying is an efficient coating method with high controllability for realizing the homogeneous coating. We found that the spraying times and evaporation rate of the casting solution could be manipulated to control the microstructure of GO membrane and thus the H_2/CO_2 separation performance. It was demonstrated that the heating temperature and the ethanol/water mass ratio in the solvent of GO solution together influenced the evaporation rate. The optimized GO membrane exhibited superior H_2/CO_2 separation performance exceeding the upper bound of polymeric membranes.

To promote the industrial application of two-dimensional-material membranes, the precondition is realizing the fabrication of large-area membranes based on ultrathin nanosheets. Majumder and coworkers successfully fabricate large-area GO membranes ($13 \times 14\,cm^2$) via the shear force-assisted assembly (Figure 2.6c,d) [50]. They innovatively concentrated the GO solution to the liquid crystalline phase ($40\,mg\,ml^{-1}$) by introducing the hydrophilic polymer hydrogel beads into aqueous

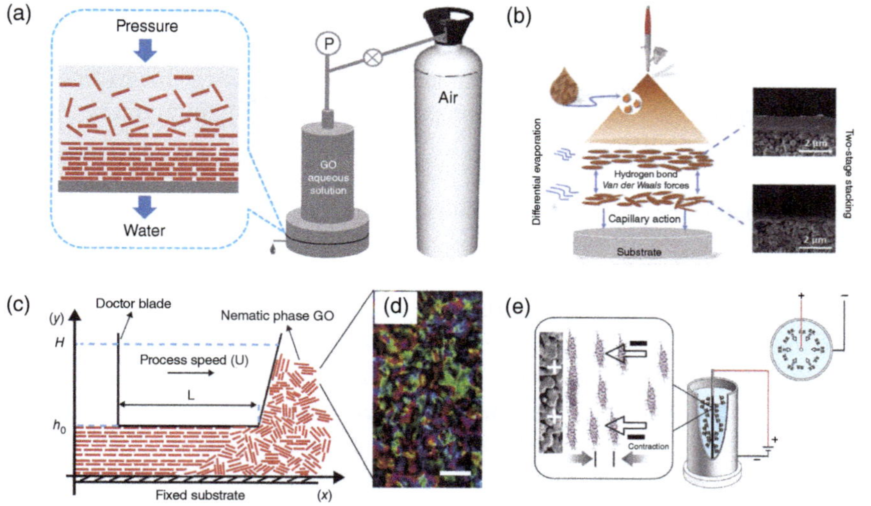

Figure 2.6 (a) Schematic of a pressurized ultrafiltration system. Source: Tang et al. [48]/ with permission of Elsevier. (b) Schematic of the assembly of GO laminates via spray-evaporation method. Source: Guan et al. [49]/with permission from Elsevier. (c) Schematic of shear-alignment processing of a nematic GO to a film, and (d) Polarized light images of fully nematic GO at 40 mg ml^{-1} (scale bar, 1 μm). Source: Akbari et al. [50]/Springer Nature/CC BY 4.0. (e) Schematic of GO electrophoresis deposition on PSSHF with a circular electric field. Source: Qi et al. [51]/with Springer Nature/CC BY 4.0.

solution. Then the nematic phase of GO was shear aligned to form continuous and highly ordered GO layer by the doctor blade. The viscosity of the GO casting solution, which was controlled by the GO concentration from 0.1 to 60 mg ml^{-1}, was demonstrated to be the dominant factor on the imposed shear stress and influenced the membrane structure. They observed that the uniformity of the cast film increased with increasing GO concentration, and the GO suspension of 40 and 60 mg ml^{-1} contributed to the best uniformity and continuity. Moreover, they employed a gravure printer to make large-area GO membranes, verifying the proficiency of the method in membrane fabrication. The reduced GO membrane, which was exposed to the hydrazine vapor, showed high retention (>90%) for organic molecules and modest (30–40%) retention of monovalent and divalent salts during the nanofiltration process.

Besides the mechanical force mentioned above, electric-field force was utilized to facilitate the assembly of GO nanosheets. Zeng and coworkers proposed to deposit ultrathin GO layer on the porous stainless steel hollow fiber (PSSHF) via a facile electrophoresis deposition (ED) method [51]. In a typical fabrication process, a circular electric field was formed between the concentric electrodes of PSSHF and stainless steel tubular container for the GO solution, ensuring a uniform driving force from every direction (Figure 2.6e). As a result, the oxygen-containing group of GO nanosheets was partially reduced by the cathode, leading to the narrowed 2D channel of GO laminates for enhanced molecular sieving. The GO membrane fabricated by the ED method exhibits a sharp cutoff between C2 (ethane and ethene) and C3 (propane and propene) and nearly complete rejections for the alcohol and ion in aqueous solutions. In all, various methods were proposed for the efficient assembly of GO nanosheets, and the assembly method directly influenced the microstructure of GO membrane and thus the separation performance in nanofiltration, pervaporation dehydration, and gas separation.

2.3.2.2 Nanostructure Controlling of Laminar Membranes

Although many methods have been proposed to assemble ultrathin nanosheets into laminar structures, it is crucial to precisely manipulate the microstructure and physicochemical property of the interlayer or in-plane channels for desired separation performance [53]. In addition, the stability issue of two-dimensional-material membranes (2DMMs) in the aqueous system should be addressed. The strategies, such as chemical reduction or crosslinking, have been employed to enhance membrane stability. Apart from the size of interlayer or in-plane channels, the surface property (e.g. hydrophilicity, charge property) was proved to influence the transport behavior of molecules/ions. Herein, the nanostructure controlling of 2DMMs has been discussed from the three aspects—(i) interlayer channels; (ii) in-plane channels; and (iii) surface properties.

Interlayer Channel Owing to the rich oxygen-containing group decorated on the GO nanosheets, GO membranes face the issue of poor structural stability in aqueous solution [54]. Chemical reduction was employed to stabilize the membrane structure, but the interlayer height of 2D channels was significantly narrowed, leading to much lower permeance. Hence, researchers proposed to enlarge the channel size by intercalating nanomaterials (e.g. carbon dots, carbon nanotubes) [55, 56] into the GO interlayer. Whereas, the uneven intercalation would disturb the pristine GO laminate and severely affect the separation efficiency. It remains challenging to boost the water permeance meanwhile without sacrificing the selectivity of GO membranes.

We proposed the *in situ* intercalation method involving the synthesis of nanoparticles (NPs)@2D materials as building blocks for constructing laminar membranes on the inner surface of ceramic tubes (Figure 2.7) [57]. In a typical process, NPs (e.g. Fe_3O_4, UiO-66, and TiO_2) uniformly *in situ* grown on the plane of GO nanosheet via the solvothermal method. Afterward, the as-synthesized NPs@GO nanosheets were deposited on the inner surface of the ceramic tube to form the separation layer by the pressurized filtration. Owing to the *in situ* intercalation of NPs in the GO laminate, the nanochannels were efficiently enlarged to maximize the water permeance, meanwhile maintaining the laminar structure to preserve the selectivity of membranes. Moreover, GO nanosheet was partially reduced, endowing the membrane with enhanced stability in water. The as-fabricated NPs@GO membranes exhibited excellent nanofiltration performance as well as good stability under high pressure and cross-flow operation. Wang and coworkers developed the sacrificed NP template method to fabricate MXene laminar membrane for water purification [58]. The positively charged $Fe(OH)_3$ NP was selected to intercalate the negatively charged MXene nanosheets, leading to expanded nanochannels for fast water transport. Subsequently, hydrochloric acid solution (HCl) was employed to remove the NPs and

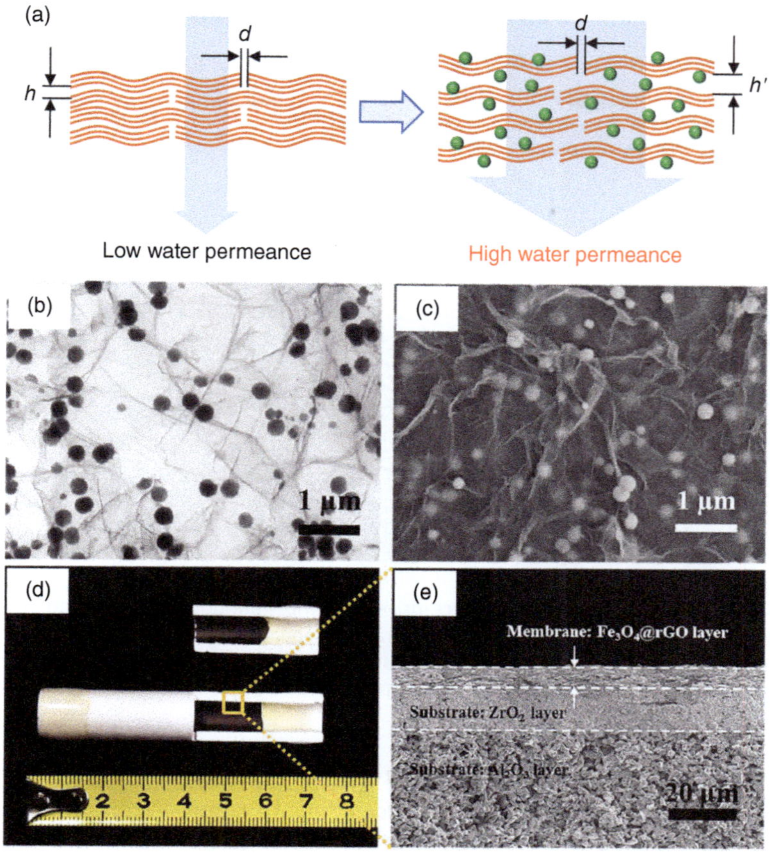

Figure 2.7 Fabrication and characterization of the Fe_3O_4@GO membrane. (a) Schematic of the enlarged interlayer channel and improved water permeance of nanoparticle intercalated GO membranes. (b) TEM image of Fe_3O_4@GO nanosheets. (c) SEM surface, (d) digital photographs (blank part of the inner surface is covered by sealing epoxy), and (e) SEM cross section of Fe_3O_4@GO membrane deposited on inner surface of a ceramic tube. Source: Zhang et al. [57]/with permission from John Wiley & Sons, Inc.

further reduce the transport resistance. As a result, the MXene membrane exhibited excellent water permeance (>1000 $l\,m^{-2}\,h^{-1}\,bar^{-1}$) and a high rejection rate (90%) for molecules with sizes larger than 2.5 nm.

The NP intercalation method was proven to efficiently enhance the water flux during the nanofiltration process. Nevertheless, it remains challenging to finely regulate the interlayer gallery of 2D laminates for precise gas separation. Bearing this in mind, we developed an external force (EF)-driven assembly method to manipulate the 2D nanochannels (Figure 2.8) [8]. The introduced EFs were classified into "outer" EFs and "inner" EFs. The "outer" EFs were applied outside the GO laminate, consisting of compressive, centrifugal, and shear forces; the "inner" EFs referred to molecular interactions between polymer chains and GO nanosheets, which were applied inside the laminate. The employed EFs contributed

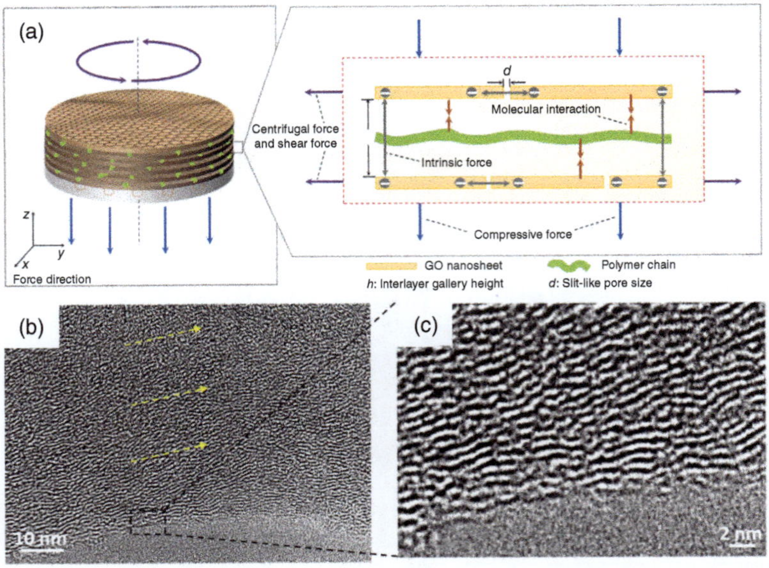

Figure 2.8 (a) Schematic of the external force-driven assembly method for fabricating 2D nanochannels. (b) and (c) TEM images of GO membranes prepared by the external force-driven assembly method. The yellow dashed arrows are eye-guiding lines indicating the orientations of GO. Source: Shen et al. [8]/with permission from American Chemical Society.

to surpass the repulsive electrostatic forces between GO nanosheets and eliminate the nonselective defects. As a result, highly ordered nanochannels were generated and the size of the interlayer gallery could be precisely manipulated for fast and selective gas transport. The as-fabricated ~0.4 nm 2D channels endowed the membrane with remarkable H_2/CO_2 and H_2/C_3H_8 separation performance. Moreover, the chemical tuning of the interlayer microenvironment was employed to manipulate the gas transport behavior in MXene membranes. We reported the fabrication of ultrathin (~20 nm) MXene membrane with well-stacked structure, which exhibited outstanding molecular-sieving property for H_2 preferential transport [59]. The pristine MXene membrane showed excellent H_2/CO_2 separation performance with H_2 permeance of 1584 GPU and H_2/CO_2 selectivity of 27. Then the chemical microenvironment of MXene nanochannels was tuned by intercalating borate and polyethylenimine (PEI) molecules into the interlayer galleries [59]. As a result, the pristine MXene membrane with H_2-selective channels was transformed to the chemically modified MXene membrane with CO_2-selective channels, which showed the CO_2 permeance of 350 GPU with a CO_2/CH_4 selectivity of 15.3.

Apart from separating the gas mixture, desalinating water is another huge challenge for membranes assembled from 2D nanosheets. It is extremely difficult to reduce the interlayer spacing to exclude small ions meanwhile maintaining the interlayer gallery against the swelling tendency of GO membranes in aqueous solution. We demonstrated the cationic control of the interlayer spacing of GO membranes with ångström precision using various cations, such as K^+, Na^+,

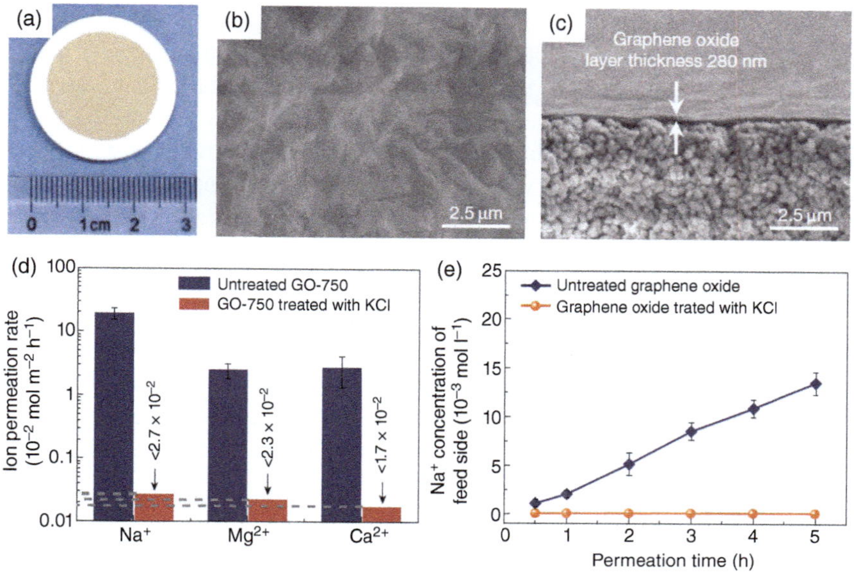

Figure 2.9 Fabrication, characterization, and ion permeation tests of GO membranes with cation-controlled interlayer spacing. (a) Digital photo, (b) surface SEM image, and (c) cross-sectional SEM image of an Al_2O_3-supported GO membrane. (d) Na^+, Mg^{2+}, and Ca^{2+} permeation rates of untreated and KCl-treated GO membranes. Dashed lines indicate the detection limits of the different cations. (e) Na^+ permeation rates of untreated GO membranes ($71.84 \pm 6.75 \times 10^{-2}$ mol m^{-2} h^{-1}) and KCl-treated GO membranes ($0.48 \pm 0.07 \times 10^{-2}$ mol m^{-2} h^{-1}) with a thickness of about 280 nm. Source: Chen et al. [60]/with permission from Springer Nature.

and Ca^{2+} (Figure 2.9) [60]. In a typical process, the freestanding GO membrane was prepared via the drop-casting method. Then the membrane was immersed in the aqueous solutions containing salts for the intercalation of cation into the interlayer gallery. Based on the strong hydrated cation–π interactions between hydrated cations and the aromatic ring, the interlayer spacing of GO membranes could be precisely controlled, with the order from widest to narrowest spacing was $MgCl_2 > LiCl > CaCl_2 >$ pure water > NaCl > KCl. Interestingly, we found K^+, Na^+, and Ca^{2+} fixed the interlayer spacing and excluded other cations that required larger interlayer spacings. Moreover, we fabricated GO membranes on the ceramic substrates and employed them for ion permeation tests to prove the controlling effect of cations. The KCl-treated GO membranes showed low Na^+, Mg^{2+}, and Ca^{2+} permeation rates below the cation detection limits, demonstrating the ion-sieving effect of the membrane with the ion rejection of more than 99% relative to untreated GO membranes (here ion rejection refers to the reduction in ion permeation rate). Meanwhile, water could still pass through the KCl-controlled membrane, showing a flux of about 0.1 l m^{-2} h^{-1}. In addition, we found that the K^+-controlled GO membrane rejected K^+ itself, which was attributed to the narrowed interlayer spacing from the distortion of the hydrated structure of K^+ inside the GO nanosheets. The density-functional theory computations show that other cations (Fe^{2+}, Co^{2+},

Cu^{2+}, Cd^{2+}, Cr^{2+}, and Pb^{2+}) have a much stronger cation–π interaction with the graphene sheet, suggesting that other ions could be used to produce a wider range of interlayer spacings for various applications, such as lithium-based batteries and supercapacitors.

As mentioned above, GO membranes faced the challenge of structural stability in aqueous solution. The undesired water-induced effects, including the redispersion and peeling of stacked GO laminates, severely influence the separation performance and hinder their practical application. To address this issue, we developed a molecular bridge strategy [61], in which an interlaminar short-chain molecular bridge and an interfacial long-chain molecular bridge together strengthened the membrane structure (Figure 2.10a). To construct the interlaminar molecular bridge, polydopamine (PDA) was employed to covalently interlock the GO laminate through condensation and nucleophilic addition reactions, which ensured the robust GO structure to resist the tendency toward swelling. In addition, based on the chemical

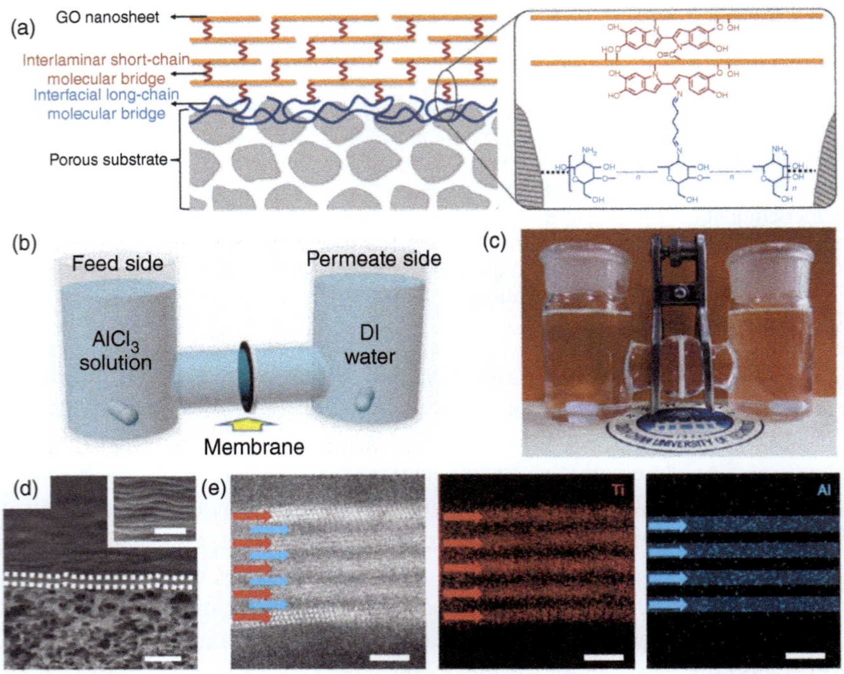

Figure 2.10 (a) Schematic of stabilizing a GO membrane through interlaminar short-chain and interfacial long-chain molecular bridges. Source: Zhang et al. [61]. Reproduced/with permission of John Wiley & Sons. (b) Schematic drawing. Source: Ding et al. [62]/with permission of Springer Nature and (c) digital photograph of the U-shaped permeation device for cation intercalation and ion permeation experiments. (d) Cross-sectional SEM image of Al^{3+}-intercalated MXene membranes supported on a PES substrate. Scale bar, 5 μm. The inset shows the lamellar cross section of membranes with higher magnification. Scale bar, 500 nm. (e) HAADF image of the cross section of Al^{3+}-intercalated MXene membranes (left). In the corresponding STEM-EDS mappings of Ti (center) and Al (right), the red and blue arrows give a guide to view the Ti and Al layers, respectively. Scale bars, 2 nm. Source: (c–e) Zhang et al. [61]/with permission from Springer Nature.

Schiff-base reaction and physical van der Waals interactions, interfacial long-chain molecular bridges were formed to connect the GO layer and substrate, efficiently enhancing the interfacial adhesion. Owing to the rational design and construction of molecular bridges, the stabilized GO membranes showed outstanding durability in harsh conditions, such as cross-flow, high-pressure, and long-term filtration. This general and scalable approach for stabilizing GO membranes makes a great step toward the practical application of GO membranes in water purification. The swelling problem also restricts the application of MXene membranes in aqueous environments. Wang and coworkers proposed the Al^{3+} intercalation method to fabricate the non-swelling MXene membranes (Figure 2.10b–e) [62]. The pristine MXene membrane was prepared via vacuum filtration, and then intercalated by Al^{3+} via a concentration diffusion method. The MXene membrane was fixed between the feed and permeate compartments of a U-shaped permeation device, which contained $AlCl_3$ solution and deionized water, respectively. As a result, the Al^{3+} diffused into the MXene laminate and connected the nanosheets by interacting with oxygen groups on the MXene surface, contributing to the non-swelling property of Al^{3+}-intercalated MXene membranes.

By the precise manipulation of the size and chemical microstructure of the interlayer gallery, the 2DMMs exhibit outstanding separation performance in water purification and gas separation. Moreover, the chemical or physical crosslinking strategy is proposed to resist the swelling tendency of the stacked nanosheets, which prominently enhances the structural stability of 2DMMs and further pushes the membrane toward practical application.

Surface Properties Based on the solution–diffusion theory, the permeant dissolves in the membrane and then diffuses through the membrane down a concentration gradient. Separation is achieved between different permeants due to the difference in the number of molecules dissolved in the membrane and the rate at which the molecules diffuse through the membrane. The dissolved amount, one of the influencing factors, is dominated by the adsorption property of the membrane material. For the manipulation of the nanostructure of 2DMMs, tuning the surface properties, such as hydrophilicity, can intensify the water adsorption ability, contributing to the sufficient utilization of fast transport channels and thus prominently enhanced water flux. Based on this conception, we proposed to fabricate a bio-inspired membrane coupling a water-capturing layer and GO laminates with water transport channels [63]. The pristine GO laminate was firstly prepared on the ceramic hollow fiber support, and then the hydrophilic chitosan (CS) layer was fabricated on the GO laminate by the vacuum suction method. XPS result demonstrated that an Al2p peak from α-Al_2O_3 hollow fiber support emerged in the spectra of CS@GO layer, which revealed that the thickness of the CS coating is less than 10 nm (the maximum probing depth of XPS in the work). The synergistic effect of highly enhanced water sorption from the hydrophilic layer and water transport channel from the GO laminate realized fast and selective water transport through the integrated membrane. The as-fabricated CS@GO membrane exhibited outstanding water-selective permeation property with an excellent water flux of over $10\,000\,g\,m^{-2}\,h^{-1}$, which surpassed the performance upper bound of state-of-art membranes for butanol dehydration.

The surface hydrophilization strategy validates that the controlling of membrane surface property will facilitate the sufficient utilization of interlayer channels and intensify the molecular transport. Similarly, we developed the strategy of manipulating the surface charge property of GO membranes to realize the controllable ion transport through the laminates [64]. Different kinds of polyelectrolyte layers were fabricated on top of the pristine GO membrane to create positive/negative charge and tunable charge densities on the membrane surface via a dip-coating method. The surface polyelectrolyte layer provides desirable interaction with charged ions to control the ionic transport, meanwhile, the underlying GO laminate with 2D capillaries offered fast water transport channels. By rationally designing the surface charge and interlayer channels, the surface-charged GO membrane showed remarkable enhancement of ion rejection with intrinsically high water permeance, exceeding the performance limits of state-of-the-art nanofiltration membranes. In all, the approach of tuning the surface properties (e.g. hydrophilicity, charge property) of GO membranes establishes a multifunctional platform that could be extended to other 2DMMs and various applications, such as biomimetic ion channels, energy storages, and conversions.

2.3.3 Two-Dimensional-Materials-Based Mixed-Matrix Membranes (MMMs)

2.3.3.1 Fabrication Methods of MMMs

Owing to the relatively low cost and the feature of easy processing, the polymer membrane dominates in the industrial application of membrane separation technology. Nevertheless, the trade-off between permeability and selectivity restricts the improvement of separation performance. With the emergence of novel 2D materials with unique transport properties, researchers attempted to incorporate them into the polymer matrix to fabricate MMMs, further enhancing the separation efficiency. We firstly proposed to fabricate GO-incorporated polyether block amide (PEBA) membranes via a simple physical mixing method (Figure 2.11a–d) [65]. Usually, the repulsive electrostatic interactions caused by the negatively charged carboxyl groups of GO nanosheets lead to nonselective defects in GO membranes. In contrast, when GO nanosheets were assembled in polymeric environment, the favorable molecular interactions, such as the hydrogen bonding between GO and the PEBA chain, would stimulate the stacking of GO nanosheets into well-defined laminates, where the interlayer spacing could provide molecular-sieving channels. Transmission electron microscopy (TEM) images confirmed the *d*-spacing of GO laminates was ~0.7 nm. After subtracting the graphene thickness of 0.35 nm, the interlayer spacing (empty space for molecular transport) was calculated to be 0.35 nm, which is in the range of the kinetic diameters of industrial gases (CO_2: 0.33 nm; N_2: 0.36 nm) and could provide selective transport channels for gas separation. By coupling the molecular-sieving and strong CO_2 adsorption property of GO laminates, the GO/PEBA MMMs exhibited extraordinarily high and stable

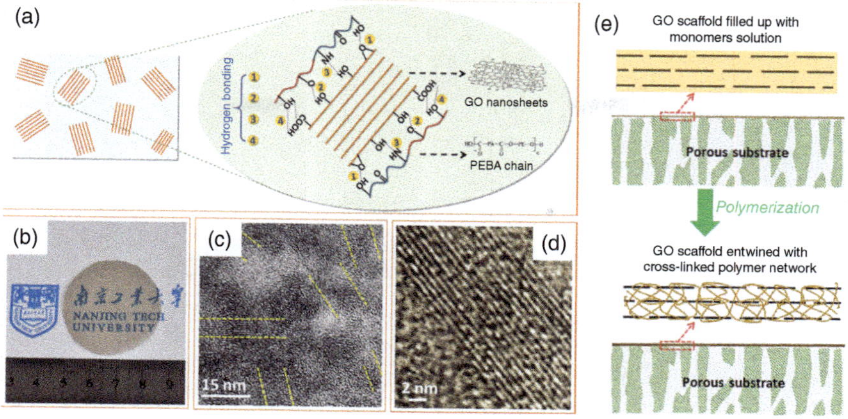

Figure 2.11 (a) Schematic of the assembly of GO nanosheets in polymeric environment based on hydrogen bonds formed between different groups on GO and the PEBA chain Shen et al. [62]/with permission of John Wiley & Sons. (b) Digital photographs of the membrane with 0.1 wt% GO [65]. (c) Overview (the yellow dashed lines are eye-guiding lines indicating the GO laminates in these regions) and (d) expanded TEM image of the cross section of GO/PEBA membrane. Source: Shen et al. [65]/with permission from John Wiley & Sons, Inc. (e) Schematic of the fabrication of a GO–polymer network composite membrane for forward osmosis desalination. Source: Kim et al. [66]/with permission of Royal Society of Chemistry.

CO_2/N_2 separation performance transcending the upper bounds of state-of-the-art membranes.

To further improve the dispersibility of GO nanosheets in the polymer matrix, Wang and coworkers developed the *in situ* polymerization strategy to fabricate GO-based MMMs for water desalination (Figure 2.11e) [66]. In a typical process, the mixed solution of GO and organic monomers was spin-coated on the nylon support to form a layer of GO scaffold filled with the monomer solution, followed by the free-radical polymerization process. As a result, a highly crosslinked polymer network-entwined GO layer (less than 40 nm) was produced, which was mechanically strong and chemically durable, while maintaining the fast water transport within GO laminates. The as-prepared GO–polymer membrane endured high chlorine exposure at 1000 ppm for 24 hours, confirming its excellent chlorine tolerance. During the forward osmosis process, the ultrathin composite membrane exhibited outstanding water flux of $25.8\,l\,m^{-1}\,h^{-1}$ along with NaCl rejection of 99.9%. This *in situ* polymerization method provides a new platform for developing 2D-material-based MMMs for water desalination.

2.3.3.2 Effect of Physicochemical Properties of 2D Fillers

Our previous work demonstrated that GO nanosheets can be engineered in the polymeric environment and assembled into molecular-sieving laminates for fast and selective CO_2 separation. However, the physicochemical properties (e.g. the lateral size of nanosheet, oxidized degree) of nanosheets could greatly influence the microstructures and thus the separation performance of the as-fabricated MMMs. We firstly investigated the effect of GO lateral size on the membrane structure

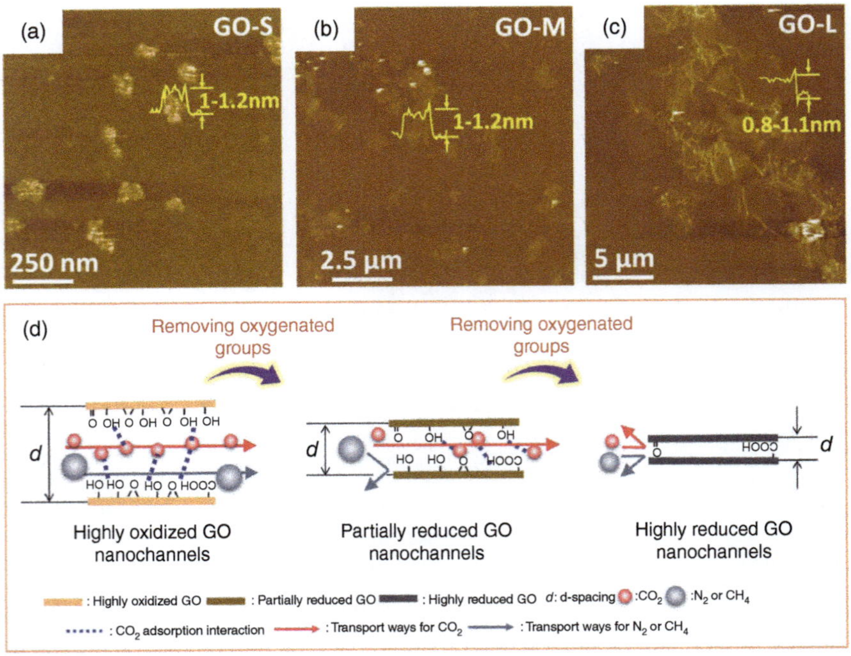

Figure 2.12 AFM images for GO sheets with lateral sizes of (a) 100–200 nm, (b) 1–2 μm, and (c) 5–10 μm. Source: Shen et al. [67]/with permission from John Wiley & Sons, Inc. (d) Schematic of the microstructures and molecular transport mechanisms of GO nanochannels with controlled oxygenated groups. Source: Shen et al. [68]. Reproduced/with permission of Royal Society of Chemistry.

and separation performance of GO/PEBA MMMs by varying the GO lateral sizes (Figure 2.12a–c) [67]. On the one hand, tuning the lateral size can change the diffusion length of the targeted molecules and thus the transport behavior. On the other hand, the mobility of polymer chains was restricted with increasing the GO lateral size, and especially large-sized GO can lead to poor dispersion properties in membrane and the gas barrier effect. Among the as-fabricated membranes, the GO/PEBA membrane loaded with 0.1 wt% medium-lateral sized (1–2 μm) GO nanosheets exhibited the highest CO_2/N_2 separation performance (CO_2 permeability: 110 Barrer, CO_2/N_2 mixed gas selectivity: 80) transcending the Robeson upper bound. Besides the effect of GO lateral size, we explored the effect of oxidized degrees of GO nanosheets on the gas transport behavior (Figure 2.12d) [68]. A facile thermal annealing method was employed to controllably remove the oxygen-containing groups for producing GO nanosheets with different oxidized degrees. The high content of oxygen-containing groups was proven to increase the dispersibility of GO in polymeric environments and induce stronger hydrogen bonding between GO and polymer chains. In addition, the size of the interlayer nanochannels and the adsorption property of the laminates were simultaneously affected with varying the oxidized degree. After systematic investigation, the GO laminate with O/C ratio of 0.55 contributed to CO_2-philic interlayer channels of

0.36 nm for the fast permeation of CO_2 molecules, endowing the membrane with excellent CO_2/N_2 separation performance for practical CO_2 capture.

2.3.4 Other Hybrid Membranes

2D nanosheets featuring large lateral size combined with atomic thickness enable them as multifunctional frameworks for growing seeding layer, supporting ultrathin polymeric layer, and reducing the pore penetration, which contributes to the generation of high-performance membranes with different configurations. Wang and coworkers reported the fabrication of ultrathin (100 nm) zeolitic imidazolate framework-8 (ZIF-8)/GO hybrid membrane by a 2D nano-hybrid seeding strategy (Figure 2.13) [69]. They grew ZIF-8 nanocrystals on both sides of 2D GO nanosheets, producing the ZIF-8/GO hybrid nanosheets. The as-prepared flexible nanosheets allowed for seeding substrates with large pores and rough surfaces, free from the substrate modification in the traditional synthesis using nanocrystal seeds. Then the seeding layer was prepared by spin coating of the ZIF-8/GO suspension on the substrate twice for uniform surface coverage. Finally, the contra-diffusion method was employed to realize the secondary growth of seeding layer into the detect-free ZIF-8/GO membrane. The seeding layer assembled from 2D hybrid nanosheets provided 2D confined space and uniform coating of ZIF-8 nanocrystals, facilitating the fast crystal intergrowth and the formation of ultrathin ZIF-8/GO layer. The resulting ultrathin membranes show excellent molecular-sieving gas separation properties with the CO_2/N_2 selectivity of 7.0. This 2D nano-hybrid seeding strategy can be readily extended to the fabrication of other defect-free and ultrathin MOF or zeolite molecular-sieving membranes for a wide range of separation applications.

Thermally rearranged (TR) polymers are an important class of microporous polymers with remarkable gas separation performance. However, it remains challenging to engineer the TR polymers into ultrathin membranes, which is highly

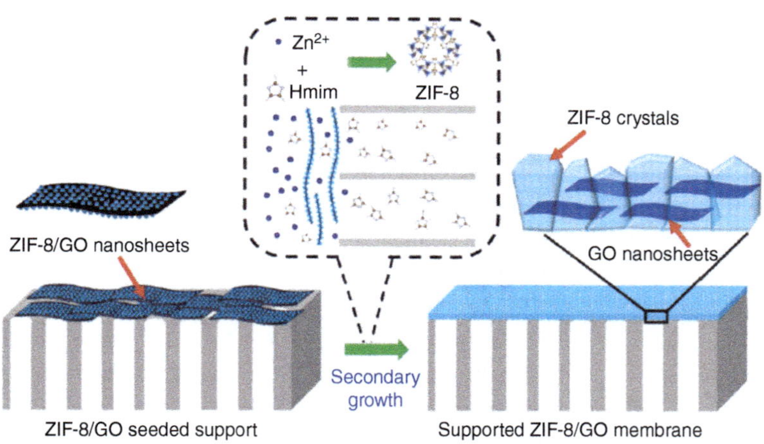

Figure 2.13 Illustration of the synthesis process of ultrathin ZIF-8/GO membrane by a 2D nano-hybrid seeding strategy. Source: Hu et al. [69]. Reproduced/with permission of John Wiley & Sons.

demanded for practical application. Wang and coworkers proposed to employ the GO nanosheet as 2D scaffold inside the TR polymer to fabricate ultrathin TR membranes [70]. During the fabrication process, the precursor solution containing GO and the polymer were spin-coated on the AAO substrate to form the continuous membrane layer. Then the GO–polymer membrane was converted into TR polymer membrane containing rGO nanosheets by thermal annealing under an argon atmosphere. The incorporation of GO scaffold into the polymer phase enabled the formation of ultrathin TR polymer layer, meanwhile enhancing the mechanical robustness of the membrane. The as-prepared membrane exhibited remarkable gas separation performance, with 482 times higher CO_2 permeance (1784 vs. 3.7 GPU) than the freestanding TR membrane at a similar CO_2/CH_4 selectivity of 32. This 2D-scaffold strategy can be employed to process other polymer materials into ultrathin membranes with prominently enhanced molecular transport properties.

During the fabrication of ultrathin polymer membranes, the pore penetration of the casting solution usually occurs and increases the transport resistance, severely affecting the separation performance. To address this problem, researchers proposed to add an intermediate gutter layer (e.g. polydimethylsiloxane [PDMS]) between the selective layer and the porous support. However, the PDMS gutter layer could account for part of the transport resistance and affect the separation efficiency. Fu and coworkers employed 2D MOF nanosheets to fabricate a highly permeable gutter with reduced gas resistance compared with the PDMS layer, contributing to the formation of ultrathin polymeric layer for efficient CO_2 capture [71]. The ultrathin (~3–4 nm) zinc(II) tetrakis(4-carboxy-phenyl)porphyrin) (ZnTCPP) MOF nanosheets were synthesized via a modified surfactant-assisted method, followed by vacuum filtration to produce the flat MOF layer on the PAN support. Then the polymer solution was coated on the PAN/MOF to fabricate flexible thin-film composite membranes (TFCMs) by spin coating. By comparing the effect of several kinds of MOF nanosheets serving as gutter layers, they summarized that the flexibility of the MOF nanosheet determined the quality of the gutter layer and thus the separation performance. Typically, the lower flexibility of nanosheets would generate larger free voids during the stacking, which permitted the polymer solution to penetrate the gutter layer and led to lower gas permeance. The as-fabricated TFCM with optimized MOF gutter layer showed high CO_2 permeance of ~2100 GPU and CO_2/N_2 ideal selectivity of ~30, satisfying the desired properties for an economically viable post-combustion CO_2 capture membrane.

Owing to the robust, ordered, and tunable porous network structures, COF nanosheets have exhibited great potential in molecular separation. However, there are only a few reports on 2D COF membranes for separation of small molecules, which is attributed to the relatively large pore size (0.8–4.7 nm) compared with the kinetic diameter of most small molecules, such as water (0.26 nm) and CO_2 (0.33 nm). In addition, the weak interlayer π–π interactions between COF nanosheets lead to the poor mechanical strength of COF membranes. Jiang and coworkers developed a mixed-dimensional assembly strategy to fabricate 1D cellulose nanofibers (CNFs)/2D COF composite membranes with molecular-sieving channels [72]. In a typical process, the CNF dispersion was added into the $TpTG_{Cl}$

dispersion with vigorous stirring, forming the mixed-dimensional $TpTG_{Cl}$@CNF nanocomposites through self-assembly. The mixture was further centrifugated and dialyzed to produce the stable dispersion of $TpTG_{Cl}$@CNF nanocomposites, which was filtrated through the PAN substrate to prepare the $TpTG_{Cl}$@CNF membranes via the vacuum-assisted method. Based on the sheltering effect of CNFs, the equivalent pore size of membranes was effectively tailored from 1 nm to 0.45 nm, prominently enhancing the molecular-sieving property. Moreover, the electrostatic interaction and the hydrogen bonds between $TpTG_{Cl}$ and CNF rendered the COF membrane high mechanical strength. As a result, the $TpTG_{Cl}$@CNF membranes exhibited a flux of 8.53 kg m^{-2} h^{-1} with a separation factor of 3876 for *n*-butanol dehydration, and high permeance of 42.8 l m^{-2} h^{-1} bar^{-1} with a rejection of 96.8% for Na_2SO_4 removal. This mixed-dimensional design can be employed to fabricate a series of various COF membranes for energy and environment-relevant separation processes.

2.4 Summary and Outlook

With the emergence of novel two-dimensional materials, they have become excellent building blocks for the fabrication of high-performance separation membranes. Owing to the characteristic of atomic thinness and large lateral size, the nanosheets can be easily assembled into ultrathin membranes down to 10 nm thickness based on various fabrication strategies. By the facile manipulation of the microstructure and chemical property of the molecular transport channels, 2DMMs exhibited outstanding separation performance exceeding the performance upper bound in many separation processes, such as nanofiltration, organic solvent dehydration, and gas separation. With regard to the further development of the fabrication methods for 2D membranes, we encourage the community to pay more attention to the following topics:

(i) Toolkit for manipulating the molecular transport channels. Based on the efficient strategies for manipulating the microstructures and chemical properties of the transport channel, we can systematically conclude the critical parameters (e.g. in-plane pore size, interlayer spacing, and mass friction of functional groups) dominating the transport behavior and categorize the strategies for tuning these parameters. Hence, a powerful toolkit, which matches the tuning strategies with the targeted parameters, can be constructed to guide the design and fabrication of 2D-material membranes for different separation processes.

(ii) Nanochannel characterization and membrane-formation mechanism. Advanced characterization techniques with atomic-scale resolution, such as positron annihilation lifetime spectra and *in situ* high-resolution transmission electron microscopy, are required to monitor the synthesis process of nanosheets and membranes, and reveal the nanochannel structures within 2D-material membranes. The influencing factors for the synthesis of nanosheets and membranes can be summarized to guide the design and fabrication of membranes. Moreover, based on the understanding of the architecture of 2D nanosheets and transport nanochannels, the membrane-formation mechanism needs to be established to optimize the membrane fabrication strategy.

(iii) Large-scale fabrication. This includes the large-scale fabrication of 2D materials and membranes and the design of modules for ultrathin membranes. The prerequisite condition for fabricating high-performance 2D-material membranes is the successful synthesis of 2D nanosheets with integrated structure. The nanosheets with micron-lateral size and atomic thickness will be beneficial to the production of defect-free membranes. Greener and safer methods for massively synthesizing 2D nanosheets need to be developed to satisfy the demand for fabrication of large-area membranes. In addition, new methods for fabricating defect-free 2D-material membranes on large scale should be developed. Owing to the 1–2 orders of magnitude lower thickness of 2D-material membranes compared with that of traditional membranes, it is imperative to design specially made membrane modules to maintain the structural integrity and thus the separation performance of 2D-material membranes.

References

1 Tan, C., Cao, X., Wu, X.-J. et al. (2017). Recent advances in ultrathin two-dimensional nanomaterials. *Chem. Rev.* 117 (9): 6225–6331.

2 Novoselov, K.S., Geim, A.K., Morozov, S.V. et al. (2004). Electric field effect in atomically thin carbon films. *Science* 306 (5696): 666–669.

3 Ciesielski, A. and Samori, P. (2014). Graphene via sonication assisted liquid-phase exfoliation. *Chem. Soc. Rev.* 43 (1): 381–398.

4 Hernandez, Y., Nicolosi, V., Lotya, M. et al. (2008). High-yield production of graphene by liquid-phase exfoliation of graphite. *Nat. Nanotechnol.* 3 (9): 563–568.

5 Coleman, J.N., Lotya, M., O'Neill, A. et al. (2011). Two-dimensional nanosheets produced by liquid exfoliation of layered materials. *Science* 331 (6017): 568–571.

6 Lotya, M., Hernandez, Y., King, P.J. et al. (2009). Liquid phase production of graphene by exfoliation of graphite in surfactant/water solutions. *J. Am. Chem. Soc.* 131 (10): 3611–3620.

7 Lotya, M., King, P.J., Khan, U. et al. (2010). High-concentration, surfactant-stabilized graphene dispersions. *ACS Nano* 4 (6): 3155–3162.

8 Shen, J., Liu, G., Huang, K. et al. (2016). Subnanometer two-dimensional graphene oxide channels for ultrafast gas sieving. *ACS Nano* 10 (3): 3398–3409.

9 Paton, K.R., Varrla, E., Backes, C. et al. (2014). Scalable production of large quantities of defect-free few-layer graphene by shear exfoliation in liquids. *Nat. Mater.* 13 (6): 624–630.

10 Yi, M. and Shen, Z. (2014). Kitchen blender for producing high-quality few-layer graphene. *Carbon* 78: 622–626.

11 Varrla, E., Backes, C., Paton, K.R. et al. (2015). Large-scale production of size-controlled MoS_2 nanosheets by shear exfoliation. *Chem. Mater.* 27 (3): 1129–1139.

12 Zhao, W., Fang, M., Wu, F. et al. (2010). Preparation of graphene by exfoliation of graphite using wet ball milling. *J. Mater. Chem.* 20 (28): 5817–5819.

13 Peng, Y., Li, Y., Ban, Y. et al. (2014). Metal-organic framework nanosheets as building blocks for molecular sieving membranes. *Science* 346 (6215): 1356–1359.

14 Joensen, P., Frindt, R., and Morrison, S.R. (1986). Single-layer mos2. *Mater. Res. Bull.* 21 (4): 457–461.

15 Zheng, J., Zhang, H., Dong, S. et al. (2014). High yield exfoliation of two-dimensional chalcogenides using sodium naphthalenide. *Nat. Commun.* 5 (1): 1–7.

16 Niu, L., Li, M., Tao, X. et al. (2013). Salt-assisted direct exfoliation of graphite into high-quality, large-size, few-layer graphene sheets. *Nanoscale* 5 (16): 7202–7208.

17 Niu, L., Li, K., Zhen, H. et al. (2014). Salt-assisted high-throughput synthesis of single-and few-layer transition metal dichalcogenides and their application in organic solar cells. *Small* 10 (22): 4651–4657.

18 Hummers, W.S. Jr. and Offeman, R.E. (1958). Preparation of graphitic oxide. *J. Am. Chem. Soc.* 80 (6): 1339–1339.

19 Pei, S., Wei, Q., Huang, K. et al. (2018). Green synthesis of graphene oxide by seconds timescale water electrolytic oxidation. *Nat. Commun.* 9 (1): 1–9.

20 Yang, S., Zhang, P., Nia, A.S. et al. (2020). Emerging 2D materials produced via electrochemistry. *Adv. Mater.* 32 (10): 1907857.

21 Naguib, M., Kurtoglu, M., Presser, V. et al. (2011). Two-dimensional nanocrystals produced by exfoliation of Ti_3AlC_2. *Adv. Mater.* 23 (37): 4248–4253.

22 Naguib, M., Mashtalir, O., Carle, J. et al. (2012). Two-dimensional transition metal carbides. *ACS Nano* 6 (2): 1322–1331.

23 Ghidiu, M., Lukatskaya, M.R., Zhao, M.-Q. et al. (2014). Conductive two-dimensional titanium carbide 'clay'with high volumetric capacitance. *Nature* 516 (7529): 78–81.

24 Naguib, M., Mochalin, V.N., Barsoum, M.W. et al. (2014). 25th anniversary article: MXenes: a new family of two-dimensional materials. *Adv. Mater.* 26 (7): 992–1005.

25 Yu, J., Li, J., Zhang, W. et al. (2015). Synthesis of high quality two-dimensional materials via chemical vapor deposition. *Chem. Sci.* 6 (12): 6705–6716.

26 Somani, P.R., Somani, S.P., and Umeno, M. (2006). Planer nano-graphenes from camphor by CVD. *Chem. Phys. Lett.* 430 (1-3): 56–59.

27 Pollard, A., Nair, R., Sabki, S. et al. (2009). Formation of monolayer graphene by annealing sacrificial nickel thin films. *J. Phys. Chem. C* 113 (38): 16565–16567.

28 Li, X., Cai, W., An, J. et al. (2009). Large-area synthesis of high-quality and uniform graphene films on copper foils. *Science* 324 (5932): 1312–1314.

29 Kim, G., Jang, A.-R., Jeong, H.Y. et al. (2013). Growth of high-crystalline, single-layer hexagonal boron nitride on recyclable platinum foil. *Nano Lett.* 13 (4): 1834–1839.

30 Zhan, Y., Liu, Z., Najmaei, S. et al. (2012). Large-area vapor-phase growth and characterization of MoS_2 atomic layers on a SiO_2 substrate. *Small* 8 (7): 966–971.

31 Wang, S., Yang, L., He, G. et al. (2020). Two-dimensional nanochannel membranes for molecular and ionic separations. *Chem. Soc. Rev.* 49 (4): 1071–1089.

32 Jeon, M.Y., Kim, D., Kumar, P. et al. (2017). Ultra-selective high-flux membranes from directly synthesized zeolite nanosheets. *Nature* 543 (7647): 690–694.

33 Rodenas, T., Luz, I., Prieto, G. et al. (2015). Metal–organic framework nanosheets in polymer composite materials for gas separation. *Nat. Mater.* 14 (1): 48–55.

34 Payamyar, P., Kaja, K., Ruiz-Vargas, C. et al. (2014). Synthesis of a covalent monolayer sheet by photochemical anthracene dimerization at the air/water interface and its mechanical characterization by AFM indentation. *Adv. Mater.* 26 (13): 2052–2058.

35 Varoon, K., Zhang, X., Elyassi, B. et al. (2011). Dispersible exfoliated zeolite nanosheets and their application as a selective membrane. *Science* 334 (6052): 72–75.

36 Liu, G., Jin, W., and Xu, N. (2016). Two-dimensional-material membranes: a new family of high-performance separation membranes. *Angew. Chem. Int. Ed.* 55 (43): 13384–13397.

37 Wang, X., Chi, C., Zhang, K. et al. (2017). Reversed thermo-switchable molecular sieving membranes composed of two-dimensional metal-organic nanosheets for gas separation. *Nat. Commun.* 8 (1): 1–10.

38 Bunch, J.S., Verbridge, S.S., Alden, J.S. et al. (2008). Impermeable atomic membranes from graphene sheets. *Nano Lett.* 8 (8): 2458–2462.

39 Jiang, D.-E., Cooper, V.R., and Dai, S. (2009). Porous graphene as the ultimate membrane for gas separation. *Nano Lett.* 9 (12): 4019–4024.

40 Cohen-Tanugi, D. and Grossman, J.C. (2012). Water desalination across nanoporous graphene. *Nano Lett.* 12 (7): 3602–3608.

41 Koenig, S.P., Wang, L., Pellegrino, J. et al. (2012). Selective molecular sieving through porous graphene. *Nat. Nanotechnol.* 7 (11): 728–732.

42 Celebi, K., Buchheim, J., Wyss, R.M. et al. (2014). Ultimate permeation across atomically thin porous graphene. *Science* 344 (6181): 289–292.

43 Surwade, S.P., Smirnov, S.N., Vlassiouk, I.V. et al. (2015). Water desalination using nanoporous single-layer graphene. *Nat. Nanotechnol.* 10 (5): 459–464.

44 Zhao, J., He, G., Huang, S. et al. (2019). Etching gas-sieving nanopores in single-layer graphene with an angstrom precision for high-performance gas mixture separation. *Sci. Adv.* 5 (1): eaav1851.

45 Yang, Y., Yang, X., Liang, L. et al. (2019). Large-area graphene-nanomesh/carbon-nanotube hybrid membranes for ionic and molecular nanofiltration. *Science* 364 (6445): 1057–1062.

46 Li, H., Song, Z., Zhang, X. et al. (2013). Ultrathin, molecular-sieving graphene oxide membranes for selective hydrogen separation. *Science* 342 (6154): 95–98.

47 Huang, K., Liu, G., Lou, Y. et al. (2014). A graphene oxide membrane with highly selective molecular separation of aqueous organic solution. *Angew. Chem. Int. Ed.* 53 (27): 6929–6932.

48 Tang, Y.P., Paul, D.R., and Chung, T.S. (2014). Free-standing graphene oxide thin films assembled by a pressurized ultrafiltration method for dehydration of ethanol. *J. Membr. Sci.* 458: 199–208.

49 Guan, K., Shen, J., Liu, G. et al. (2017). Spray-evaporation assembled graphene oxide membranes for selective hydrogen transport. *Sep. Purif. Technol.* 174: 126–135.

50 Akbari, A., Sheath, P., Martin, S.T. et al. (2016). Large-area graphene-based nanofiltration membranes by shear alignment of discotic nematic liquid crystals of graphene oxide. *Nat. Commun.* 7 (1): 1–12.

51 Qi, B., He, X., Zeng, G. et al. (2017). Strict molecular sieving over electrodeposited 2D-interspacing-narrowed graphene oxide membranes. *Nat. Commun.* 8 (1): 1–10.

52 Kim, H.W., Yoon, H.W., Yoon, S.-M. et al. (2013). Selective gas transport through few-layered graphene and graphene oxide membranes. *Science* 342 (6154): 91–95.

53 Cheng, L., Liu, G., Zhao, J. et al. (2021). Two-dimensional-material membranes: manipulating the transport pathway for molecular separation. *Accounts Mater. Res.* 2 (2): 114–128.

54 Perreault, F., De Faria, A.F., and Elimelech, M. (2015). Environmental applications of graphene-based nanomaterials. *Chem. Soc. Rev.* 44 (16): 5861–5896.

55 Wang, W., Eftekhari, E., Zhu, G. et al. (2014). Graphene oxide membranes with tunable permeability due to embedded carbon dots. *Chem. Commun.* 50 (86): 13089–13092.

56 Gao, S.J., Qin, H., Liu, P. et al. (2015). SWCNT-intercalated GO ultrathin films for ultrafast separation of molecules. *J. Mater. Chem. A* 3 (12): 6649–6654.

57 Zhang, M., Guan, K., Shen, J. et al. (2017). Nanoparticles@ rGO membrane enabling highly enhanced water permeability and structural stability with preserved selectivity. *AlChE J.* 63 (11): 5054–5063.

58 Ding, L., Wei, Y., Wang, Y. et al. (2017). A two-dimensional lamellar membrane: MXene nanosheet stacks. *Angew. Chem. Int. Ed.* 56 (7): 1825–1829.

59 Shen, J., Liu, G., Ji, Y. et al. (2018). 2D MXene nanofilms with tunable gas transport channels. *Adv. Funct. Mater.* 28 (31): 1801511.

60 Chen, L., Shi, G., Shen, J. et al. (2017). Ion sieving in graphene oxide membranes via cationic control of interlayer spacing. *Nature* 550 (7676): 380–383.

61 Zhang, M., Mao, Y., Liu, G. et al. (2019). Molecular bridges stabilize graphene oxide membranes in water. *Angew. Chem. Int. Ed.* 59 (4): 1689–1695.

62 Ding, L., Li, L., Liu, Y. et al. (2020). Effective ion sieving with $Ti_3C_2T_x$ MXene membranes for production of drinking water from seawater. *Nat. Sustain.* 3: 1–7.

63 Huang, K., Liu, G., Shen, J. et al. (2015). High-efficiency water-transport channels using the synergistic effect of a hydrophilic polymer and graphene oxide laminates. *Adv. Funct. Mater.* 25 (36): 5809–5815.

64 Zhang, M., Guan, K., Ji, Y. et al. (2019). Controllable ion transport by surface-charged graphene oxide membrane. *Nat. Commun.* 10 (1): 1253.

65 Shen, J., Liu, G., Huang, K. et al. (2015). Membranes with fast and selective gas-transport channels of laminar graphene oxide for efficient CO_2 capture. *Angew. Chem. Int. Ed.* 127 (2): 588–592.

66 Kim, S., Lin, X., Ou, R. et al. (2017). Highly crosslinked, chlorine tolerant polymer network entwined graphene oxide membrane for water desalination. *J. Mater. Chem. A* 5 (4): 1533–1540.

67 Shen, J., Zhang, M., Liu, G. et al. (2016). Size effects of graphene oxide on mixed matrix membranes for CO_2 separation. *AlChE J.* 62 (8): 2843–2852.

68 Shen, J., Zhang, M., Liu, G. et al. (2016). Facile tailoring of the two-dimensional graphene oxide channels for gas separation. *RSC Adv.* 6 (59): 54281–54285.

69 Hu, Y., Wei, J., Liang, Y. et al. (2016). Zeolitic imidazolate framework/graphene oxide hybrid nanosheets as seeds for the growth of ultrathin molecular sieving membranes. *Angew. Chem. Int. Ed.* 128 (6): 2088–2092.

70 Kim, S., Hou, J., Wang, Y. et al. (2018). Highly permeable thermally rearranged polymer composite membranes with a graphene oxide scaffold for gas separation. *J. Mater. Chem. A* 6 (17): 7668–7674.

71 Liu, M., Xie, K., Nothling, M.D. et al. (2018). Ultrathin metal–organic framework nanosheets as a gutter layer for flexible composite gas separation membranes. *ACS Nano* 12 (11): 11591–11599.

72 Yang, H., Yang, L., Wang, H. et al. (2019). Covalent organic framework membranes through a mixed-dimensional assembly for molecular separations. *Nat. Commun.* 10: 102101.

3

Nanoporous Single-Layer Graphene Membranes for Gas Separation

Luc Bondaz, Shiqi Huang, Mojtaba Rezaei, Shaoxian Li, and Kumar Varoon Agrawal

Laboratory of Advanced Separations, École Polytechnique Fédérale de Lausanne (EPFL), Institute of Chemical Sciences and Engineering (ISIC), Rue de l'Industrie 17, Case Postale 440, CH-1950 Sion, Switzerland

3.1 Introduction

The conventional gas-separation processes such as cryogenic distillation, absorption, and adsorption have been quite successful for a wide range of gas-separation applications, e.g. air fractionation, carbon capture, natural-gas sweetening, biogas processing, and olefin/paraffin separation [1]. These processes rely on thermal energy, exploiting the differences in volatilities, solubilities, and binding energies of the gas molecules. As a result, they are energy-intensive. Also, they are generally attractive only for mid-to-large-scale operations attributing to their capital-intensive setups [2, 3]. However, energy and cost-efficient routes have become increasingly important in the light of tightening environmental regulations and increasing carbon-emission tax. Membrane-based separation processes, by the virtue of non-reliance on thermal energy but compression (electrical energy), can be highly efficient for a number of separations [4–6]. Additionally, they are, by design, environment-friendly (no chemical is used and no waste is produced) [7–10]. Because of their simple and modular design, membrane processes can be conveniently retrofitted to existing infrastructure. They are also ideal for a small-scale decentralized operation (e.g. carbon capture from community-scale boilers and oxygen enrichment for hospitals).

Currently, the membranes for gas separation are dominated by polymeric materials [11]. This is mainly because polymer can be rapidly processed into a thin film composed of a gas-selective layer (skin layer) with a typical thickness of 100–1000 nm, thanks to the phase-inversion process reported by Loeb and Sourirajan in 1963 [12]. However, their performance is challenging to be improved beyond a certain limit [13], imposed by the free volume between the polymeric chains, chain stiffness, and molecular solubility [14]. This performance constraint, famously known as the Robeson upper bound, has been surpassed by the next-generation polymeric materials that are intrinsically nanoporous, e.g. polymers with intrinsic

Two-Dimensional-Materials-Based Membranes: Preparation, Characterization, and Applications, First Edition. Edited by Gongping Liu and Wanqin Jin.
© 2022 WILEY-VCH GmbH. Published 2022 by WILEY-VCH GmbH.

microporosity (PIM) [15]. However, the chain packing in nanoporous polymers tends to age, which reduces the free volume as a function of time.

To improve the performance of gas-separation membranes, nanoporous inorganic materials (e.g. zeolites, metal–organic frameworks or MOFs) have been extensively investigated either in a polycrystalline film morphology or in a mixed-matrix configuration, where crystals would be dispersed in a polymeric matrix [16–19]. However, the complex crystallization and activation route for polycrystalline zeolite membranes have become a show-stopper, except in a few cases where high chemical and thermal stability of inorganic materials are needed (membrane reactors) or where extremely high-molecular selectivities are needed (pervaporation). MOF-based membranes, especially those where the intrinsic framework flexibility issues have been addressed [20, 21], are quite promising. Mixed-matrix membranes yield attractive performance [22]; however, the realization of 100-nm-thick mixed-matrix skin layer is challenging. Another interesting class of gas separation membranes is those based on carbon molecular sieves (CMS) that have shown exceptionally high gas-pair selectivities even from the difficult to separate gas pairs, e.g. N_2 and CH_4 [23]. A key challenge with this material lies with reducing the thickness of the selective layer below 1 μm to improve the gas permeance, as the skin layer thickness in the state-of-the-art CMS membranes is ~O (10 μm). The adoption of inorganic membranes can be accelerated if the improvement in gas permeance is high enough to justify their high capital cost [24, 25]. A path forward toward this goal is to reduce the molecular diffusion path length across the selective layer of these membranes. This can increase the gas permeance and hence reduce the needed membrane area for processing a certain volume of feed mixture, and therefore, the net capital cost of the process. A very interesting outcome of this is that while high-performance membranes with high gas permeance can be much more expensive on the basis of cost of unit membrane area than the conventional gas permeation membranes (where the typical membrane element cost is ~10–50 US\$ m^{-2} and module cost is 100–250 US\$ m^{-2}) and yet the overall capital cost can be comparable or even lower attributing to the reduced needed area [26]. High-permeance membranes also allow energy-efficient process design, e.g. the use of vacuum pumps instead of compressors in the separation processes such as postcombustion capture [26, 27].

The shortest gas diffusion path across the selective layer of the membrane can be realized in the case of a truly two-dimensional (2D) layer comprised of atom-thick film with holes or pores that modulate the flow rate of the gas. With the advent of graphene, starting from its successful isolation by top-down micromechanical exfoliation in 2004 [28] to the bottom-up synthesis by the chemical vapor deposition process [29], a 2D skin layer for high-permeance gas separation is now feasible. Since pristine graphene is impermeable to gases, a two-dimensional nanopore has to be incorporated in single-layer graphene (SLG) for separating gases based on their relative size. The first theoretical [30] and experimental [31] proof-of-concept papers on gas separation from nanoporous SLG (N-SLG) membranes were reported almost a decade ago in year 2009 and 2012, respectively. Since then, a lot of progress has been made on understanding the gas transport behavior of the graphene nanopores, the etching and functionalizing chemistry of nanopores, and scaling-up aspects of graphene membranes. Herein, we review these advances for gas separation by focusing on (i) factors that influence gas transport, e.g. electron-density gap in the nanopore relative to the size of the molecules, pore-edge functionalization, pore-size distribution (PSD), and pore density, (ii) recent advances in incorporating molecular-sieving nanopores in SLG, and (iii) the scale-up of N-SLG membranes.

3.2 Gas-Separation Potential of N-SLG Membranes

SLG is made of an atom-thick lattice of carbon where the atoms are held together by a strong sp^2-hybridized covalent bond [32]. Figure 3.1a shows the structure of graphene lattice where the unit cell is composed of only two atoms with the C—C bond length of 0.144 nm which results in a strong covalent-bonded honeycomb lattice. The covalent-bonded lattice of graphene is quite strong which makes graphene thermally and chemically robust in many environments relevant to the field of molecular separation. The high thermal stability allows one to heat graphene in reducing environments.

It is possible to prepare a stable membrane skin layer based on suspended SLG because graphene possesses an outstanding mechanical property [34] with one of the highest reported Young's modulus (1 TPa) implying that it would require an extremely large force to deform graphene's lattice. The ultimate tensile strength, determining the stress at which graphene film will rupture, is also exceptionally high (130 GPa) [35]. Interestingly, Cohen-Tanugi and Grossman showed that the nanoporous graphene with porosity up to 15% still possesses a large Young's modulus (500 GPa) and large fracture stress [33]. Another unique advantage of graphene compared to traditional inorganic materials used for membranes is

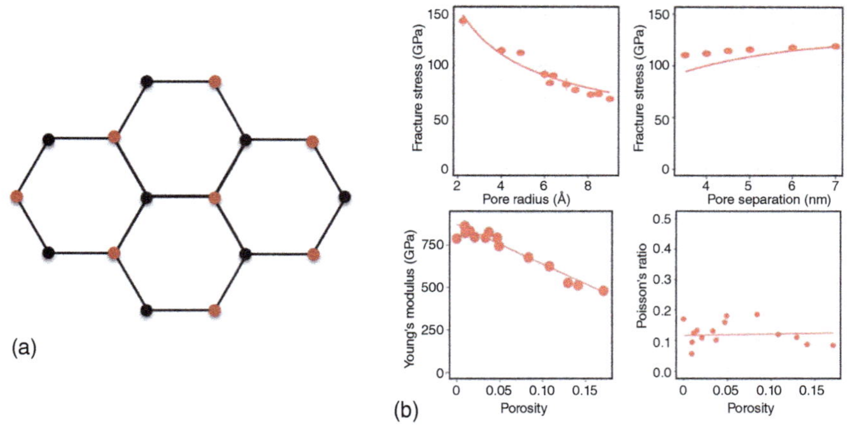

Figure 3.1 (a) Lattice structure of graphene showing the unit cell composed of two carbon atoms (red and black). Source: Kumar Varoon Agrawal. (b) Excellent mechanical properties of porous graphene reflected by high fracture stress and Young's modulus. Source: Cohen-Tanugi and Grossman [33]/with permission of American Chemical Society.

that the graphene film is extremely flexible because of its 2D lattice. This flexibility prevents crack formation in the graphene film during careful handling and membrane fabrication. This would allow one to produce graphene by roll-by-roll manufacturing methods and develop spiral-wound membranes that are certainly unique for membranes based on inorganic materials.

One of the intrinsic challenges in working with graphene as a membrane material is that there is no "free space" in the honeycomb lattice for gas diffusion even for the smallest of the gas molecules (He) [36]. This makes pristine SLG an excellent gas barrier that is impermeable to all gas molecules. Recently, Geim and coworkers demonstrated that H_2 diffusion through the honeycomb lattice is faster than He because H_2 transport is facilitated by the dissociation of molecular hydrogen at the catalytically active graphene ripples [37]. Still, H_2 transport requires that the molecule overcomes a large energy barrier of 1 eV that essentially results in an infinitesimally small H_2 permeance at room temperature. Luckily, molecular-sized carbon vacancy defects (nanopores) can be incorporated in SLG by physical and chemical etching, e.g. by knocking out carbon atoms using high-energy electron or ion beam, by gasifying the carbon atoms via oxidation, or by generating gas-permeable grain-boundary defects in SLG. The incorporation of these nanopores makes SLG very attractive for gas separation because fundamentally, PSD in graphene is tunable and, in principle, one can design an ensemble of apertures where the relative permeation rates of a given molecular pair can be optimized based on the PSD. These 2D nanopores constitute the ultimate selective layer because of the extremely small diffusion path length across the nanopore. Essentially, there exists only one transition state for diffusion, at the center of the nanopore, when the size of the molecule is comparable to that of the van der Waals gap in the nanopore [38, 39]. This leads to extremely fast molecular permeation events, only limited by the relative-size-dependent energy barrier that the molecule experiences at the center of the nanopore. In this way, one can obtain large gas permeance and attractive selectivity, as long as a narrow PSD and a moderate to a high pore density can be incorporated.

There are two ways in which gas molecules can cross graphene nanopores in a suspended SLG membrane (Figure 3.2)—(i) in the adsorbed phase or (ii) directly from the gas phase [40, 41]. In the adsorbed-phase transport, molecules first

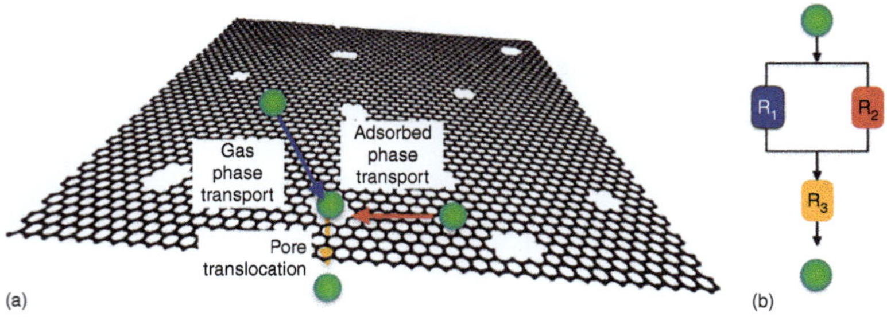

Figure 3.2 (a) Gas transport across nanopores in SLG where a schematic shows various stages of transport. Source: Kumar Varoon Agrawal. (b) A resistance model that combines the transport resistance to arrive at the overall flux. Source: Kumar Varoon Agrawal.

adsorb on the graphene lattice away from the pore, subsequently diffuse on the surface toward the pore to arrive and absorb on top of the pore. If the pore size is commensurate to that of the molecule, then the adsorbed molecule needs to overcome an energy barrier to translocate the pore. This step then becomes the rate-limiting step, and therefore, the gas flux is essentially the rate of overcoming the transition state. Yuan et al. developed the adsorbed-phase transport mechanism where the concentration on the basal plane of graphene was evaluated by computing Henry's coefficient [39]. The transition-state theory, expressed often using the Arrhenius equation to describe the rate of transition-state crossover, was used to calculate the translocation probability as a function of temperature. By assuming pore translocation as the rate-limiting step, which is expected to be the case when there is nonzero energy barrier to translocate the pore, they were able to generate an upper bound for CO_2/CH_4 separation. Interestingly, a trade-off between the permeance and selectivity was noticed. A maximum CO_2/CH_4 separation factor of 100 was computed at a high permeability coefficient (defined as permeance per pore) of 10^{-22} mol s^{-1} Pa^{-1}. On the other hand, no selectivity could be observed at higher coefficient of 10^{-21} mol s^{-1} Pa^{-1}. These models are proving to be helpful in understanding the gas transport across the nanoporous SLG membrane.

The transport of gas molecules across the nanopore can also happen directly from the gas phase. This is possible because the graphene nanopore is only an atom thick, much smaller than the typical mean free path of the gas molecules, which then permits the translocation of gases across the pore directly in the gas phase. The gas molecules can crossover the nanopore with flux determined by the rate of impingement of the molecules on top of the nanopore combined with the energy barrier to translocate the pore [42]. For small pores, the incidence angle with respect to the normal to the surface becomes important. For example, too large incidence angles will lead to a collision between the molecule and the pore edge allowing the molecule to reflect back in the feed side, preventing the pore translocation. Overall, molecular sieving can be achieved when the pore size is adequate to differentiate two gas molecules based on the difference in the corresponding energy barriers for translocation.

One can assign transport resistances, R_1, R_2, and R_3 for each event (gas-phase impingement to the pore, adsorbed-phase diffusion to the pore, and pore translocation, respectively, Figure 3.2a) and add them to get overall transport rate, analogs to electrical resistance (Figure 3.2b).

$$R = R_3 + \frac{1}{R_1^{-1} + R_2^{-1}} \tag{3.1}$$

These resistances are described by Eqs. 3.2–3.4 [41].

$$R_1^{-1} = \sqrt{\frac{\pi}{2mk_BT}}\frac{D_P^2}{4} \tag{3.2}$$

$$R_2^{-1} = \sqrt{\frac{\pi k_BT}{2m}}D_PH_{\text{sur}} \tag{3.3}$$

$$R_3^{-1} \propto \text{A}\exp\left(-\frac{E_{\text{act}}}{k_BT}\right) \tag{3.4}$$

where D_p is the effective pore diameter, k_B is the Boltzmann constant, T is the temperature, m is the mass of the molecule, H_{sur} is Henry's coefficient for the adsorption of molecules on graphene basal plane, A is the pre-exponential factor for the Arrhenius term, and E_{act} is the energy barrier that the molecule experiences for pore translocation. The proportionally constant in Eq. (3.4) depends on the concentration of gas molecules arriving at the transition state from gas phase as well as adsorbed phase.

The permeation rate, N_p, from the nanopore of interest can be estimated from Eq. (3.5):

$$N_p = \frac{\Delta P}{R} \tag{3.5}$$

where ΔP corresponds to the transmembrane pressure difference. Equations (3.1)–(3.5) allow one to model the permeance and selectivity of a gas pair from an ensemble of nanopores.

The first study on understanding the gas transport across graphene nanopores was reported in 2009 by Jiang et al., where they predicted H_2 sieving with a large H_2/CH_4 selectivity from small nanopores in graphene [30]. They studied gas transport across two different subnanometer-sized nanopores by using the first-principle density-functional theory (DFT). The first nanopore was designed by removing ten carbon atoms from the graphene lattice. This pore can be referred to as pore-10, where the number 10 indicates the number of missing carbon atoms. There are several possible geometries/isomers for pore-10 depending on the evolution of shape based on the location in lattice from where the carbon atoms are removed. In their study, Jiang et al. fully passivated pore-10 edges by H leading to an extremely small electron-density gap of 2.50 Å (Figure 3.3a). Given the kinetic diameter of H_2 is 2.89 Å and that of CH_4 is 2.80 Å, an extremely high H_2/CH_4 selectivity is expected from this kind of pore. Indeed, they reported selectivity of 10^{23} and attributed this to an extremely large energy barrier (1.6 eV) for the translocation of CH_4. For the second nanopore, they only functionalized four edge atoms of pore-10 with H, while the other four non-passivated atoms were substituted with N (Figure 3.3b). The resulting electron-density gap was reported as approximately 3 Å. This nanopore also yielded

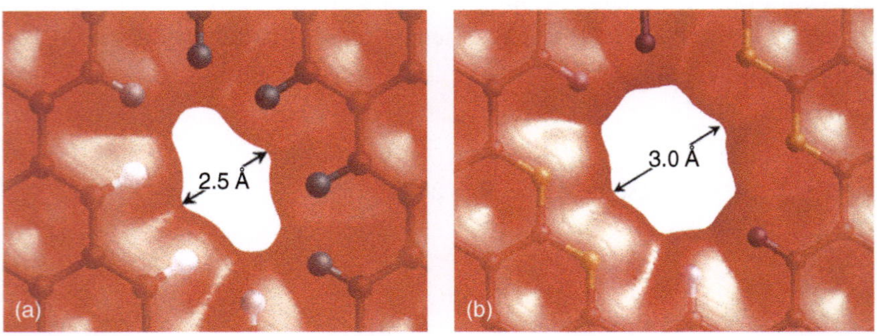

Figure 3.3 (a) All-hydrogen passivated pore-10 in SLG with an electron-density gap of 2.5 Å, and (b) nitrogen-functionalized pore-10 where electron-density gap is 3.0 Å. Source: Jiang et al. [30]/with permission of American Chemical Society.

an extremely large H_2/CH_4 selectivity of 10^8 attributing to a large energy barrier for CH_4 (0.51 eV) compared to that for H_2 (0.04 eV). With these two nanopores, the study showed that controlling the size of nanopores with respect to the kinetic diameter of gas molecules is a good strategy to obtain molecular-sieving-based separation.

Since the above study in 2009, there have been numerous studies on the gas transport across nanoporous graphene. We discuss a few of these studies highlighting the important aspects of nanopore geometry (electron-density-gap dependent on the number of missing carbon atoms), pore functionalization, and pore density.

As an early work in this field, Blankenburg et al. analyzed the gas transport across N-SLG-like porous network designed by self-assembly of the hexaiodo-substituted macrocycle cyclohexa-*m*-phenylene (CHP) [43]. This network consists of nanopores resembling H-functionalized pores made by six missing carbon atoms (pore-6, Figure 3.4a, left). These nanopores comprise of even smaller electron-density gap than that of H-functionalized pore-10 studied by Jiang et al. As a result, energy-barrier-dependent size-sieving effect was predicted for several gas molecules. The resistance to diffuse across the nanopore (Figure 3.4b, right) increased in the order of $H_2 > He > Ne > O_2 > CO_2 > NH_3 > CO > N_2 > Ar$ following the kinetic diameter of the molecules with a few exceptions that can be explained by the relative differences in the adsorption energy on the graphene basal plane (Eq. (3.3)). It is interesting to note that the energy barrier for O_2 from such nanopores is already quite high (>1 eV), where one would observe negligible O_2 permeance. Therefore, such networks would be mainly interesting for sieving of H_2 (barrier of ~0.30 eV), He (barrier of ~0.37 eV), and Ne (barrier of ~0.60 eV) from the larger gas molecules.

The above studies relied on predicting gas flux solely on the energy barrier computed by DFT. While a reasonable accuracy on molecular selectivity is expected, the calculation of gas flux can be off by several orders of magnitude because of the assumptions on the pre-exponential coefficient to the energy barrier. Several reports used molecular dynamics (MD) simulations to count gas permeation events across graphene nanopores and confirmed the high gas permeance, especially when the energy barrier from the nanopore is not too high. For example, Schrier reported extremely large CO_2 permeance (3×10^5 gas permeation units or GPU; $1\ \text{GPU} = 3.35 \times 10^{-10}\ \text{mol m}^{-2}\ \text{s}^{-1}\ \text{Pa}^{-1}$) with an attractive CO_2/N_2 selectivity of 60

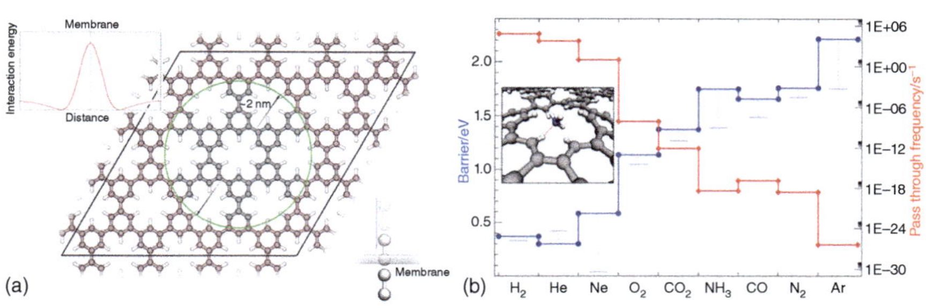

Figure 3.4 (a) A 2D nanoporous network for H_2 and He sieving. (b) Calculated energy barriers and pass through frequencies of several molecules through the 2D network. Source: Blankenburg [43]/with permission of John Wiley and Sons.

through a porous 2D network comprising of high density of pore-10 [44]. Later, Tao et al. combined DFT calculations and MD simulations to report that H_2 easily passes through H-functionalized pore-10 (energy barrier of 0.13 eV), whereas CH_4 requires a much large energy barrier (0.83 eV) [45]. The low energy barrier was reflected in a high H_2 flux (1300 mol m^{-2} s^{-1}) that translates to permeance of approximately 1 million GPU, based on the pressure used in this study. This makes H-functionalized pore-10 extremely attractive for H_2 separation from natural gas. It is important to note that these simulations were carried out with a porous lattice hosting an extremely large pore density ~O(10^{14} cm^{-2}), whereas practical N-SLG membranes are likely to have one or two orders of magnitude lower pore density attributing to difficulty in incorporating a high pore density with a narrow PSD.

Bai and coworkers highlighted the possibility of separating CH_4 from CO_2, N_2, and H_2S based on molecular sieving [46]. Using MD simulation, they studied three different pores based on the modification of pore-13 (pore made by removing 13 carbon atoms) with nitrogen and hydrogen, noted as pore 1, pore 2, and pore 3. Interestingly, they observed a significant contribution to the adsorption effects. For the bigger pore (pore 1), H_2S permeated faster than CO_2, whereas, for the smaller pore (pore 3), CO_2 permeated faster than H_2S. They reported extremely high CO_2 permeance from all three pores, above 10^5 GPU, with a pore density of 6×10^{12} cm^{-2}. Interestingly, they showed that CH_4 permeance was much slower than N_2, leading to the realization of N_2/CH_4 selectivity.

The trend of faster permeance of strongly adsorbing gases from non-sieving nanopores has been reported by several other studies. Du et al. [47] studied a number of nanopores made by missing 10–32 carbon atoms. In their case, pores were not functionalized. Their simulation indicated that beyond pore-12, N_2 permeation increased significantly compared to that of H_2. For example, H_2/N_2 selectivity was 9 for pore-11. However, bigger pores like pore-16 yielded more permeation events of N_2 than that of H_2. They report that the reason for this phenomenon is a higher van der Waals interaction between the N-SLG layer and nitrogen, leading to a formation of an adsorbed layer of N_2 on top of the nanoporous graphene lattice. Therefore, they postulated that it was easier for N_2 to find a pore by 2D diffusion. Drahushuk and Strano developed a transport model for gas permeation from graphene nanopores and reported analytical expressions for gas permeance and selectivity [48]. Their model also confirmed that while smaller nanopores sieve gases based on their size, the larger nanopores promote the transport of molecules that adsorb preferentially. For example, they reported that as one transition from nonfunctionalized pore-10 to larger pores, the H_2/N_2 selectivity goes down from over 100 to below 0.1 (Figure 3.5). It should be noted that the reverse selectivity trend (e.g. $H_2/N_2 < 1$) has not been observed in the experiments. This is mainly because direct gas-phase transport of N_2 becomes significant compared to adsorbed-phase transport for the larger pores [49], where the bombardment rate to the surface (Eq. (3.2)) is inversely proportional to the square root of the molecular weight. As a result, lighter gas molecules (H_2 as against N_2) permeate faster.

The size of a nanopore is not a unique descriptor of the pore geometry. A slight change in the pore shape can drastically change the gas permeation and selectivity.

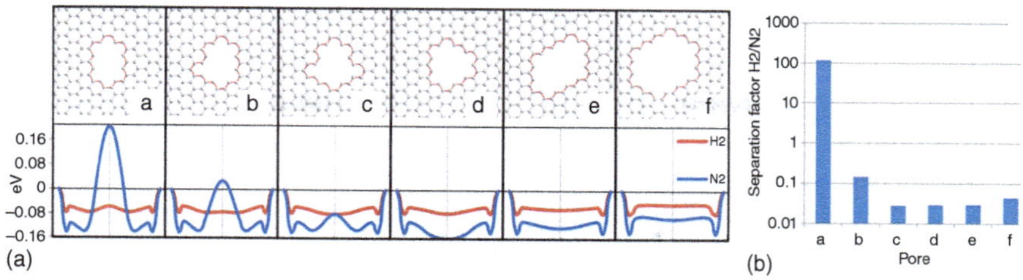

Figure 3.5 (a) Calculated adsorption energies and energy barriers of H_2 and N_2 across pore-10, pore-11, pore-12, pore-13, pore-22, and pore 32. (b) Calculated H_2/N_2 selectivities from the pores shown on the left. Source: Drahushuk and Strano [48]/with permission of American Chemical Society.

Raghavan and Gupta studied the effect of pore geometry for H_2 separation from CH_4 by focusing on the eccentricity and the jaggedness of the pore [50]. Eccentricity is a measure of how far the pore geometry is different from a circle. Jaggedness is defined as the irregularity of the pore geometry. Typically, jaggedness will increase with the number of sharp edges. They used different geometries (isomers) for the nonfunctionalized pore-18 (Figure 3.6a) and demonstrated that pore-18 isomers with high eccentricity and jaggedness can sieve H_2 from CH_4. This is remarkable because nonfunctionalized pore-18 with a low eccentricity has a relatively large electron-density gap, and H_2 sieving would not be expected from such pores. The permeances for H_2 and CH_4 are strongly dependent on the eccentricity of the pore;

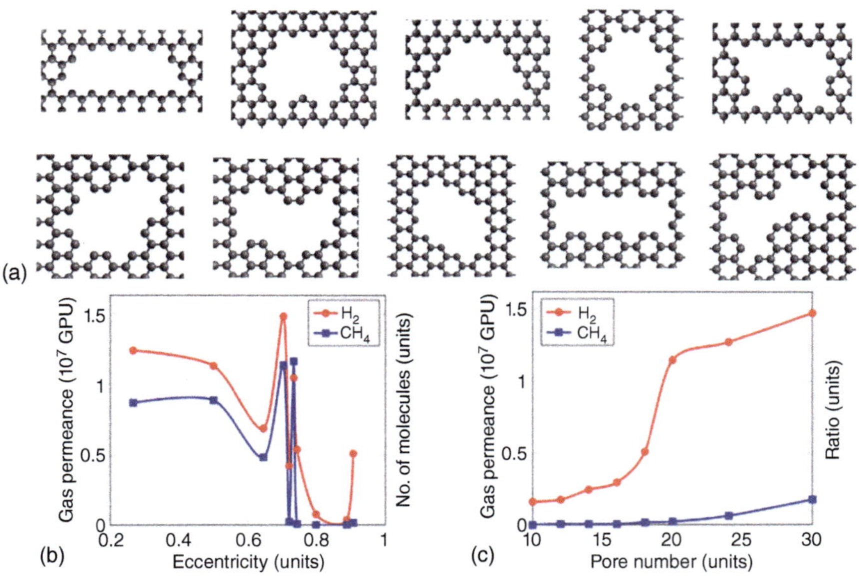

Figure 3.6 Impact of pore geometry on gas transport. (a) Various pore geometries (isomers) for pore-18. Source: Luc Bondaz. (b) H_2 and CH_4 permeances for varying degrees of eccentricity in pore-18 isomers. (c) Gas permeance from pores of varying sizes with elongated geometries. Source: Raghavan and Gupta [50]/with permission of American Chemical Society.

however, this dependency was not trivial (Figure 3.6b). A survey of elongated pores (pore-10 to pore-30) indicated that the optimal pore for H_2 sieving is an elliptical pore-20 yielding high H_2 permeance in combination with an attractive H_2/CH_4 selectivity (Figure 3.6c).

While a majority of the simulation and modeling studies assume a certain pore structure for a given number of missing carbon atoms, in reality, a graphene nanopore can have several geometric possibilities or isomers. The number of isomers increases exponentially as one increases the number of missing carbon atoms attributing to increased combinatorial possibilities of pore shape. In this case, it becomes extremely important to understand as to which pore structure would be the most probable structure that would repeat itself across N-SLG, a problem known as the isomer cataloging problem (ICP). From the point of view of membranes, this is extremely important because the pore size alone is not a sufficient indicator of molecular transport. A small change in the pore geometry can drastically impact transport properties, such as permeance and selectivity. Recently, Govind Rajan et al. reported a solution to ICP [51], and arrived at the most probable pore geometry for given pore size, i.e. number of missing carbon atoms. Starting from a single vacancy defect in graphene, they progressively increased the vacancies to check the stability and formation probability of each isomer. Generally, their model agreed with the experimental results based on transmission electron microscopy (TEM)-based structure of nanopores reported in the literature. Interestingly, they reported that the likely isomer structure is determined by the pore-formation kinetics and not the thermodynamics. It means that the probability of a given pore structure is not linked to formation energy but rather to the reaction pathway that leads to the incorporation of the pores in the graphene lattice. Recently, Huang et al. analyzed hundreds of nanopores generated by the ozone treatment of graphene and identified 49 distinct nanopore structures with missing 1 to 20 carbon atoms [52]. They also observed that several isomers occurred much more frequently and the relative population of isomers was consistent with the theoretical solution to the ICP provided by Govind Rajan et al. [51].

Gas permeance through graphene nanopores can also be tuned by pore functionalization. Using the classical MD simulations, Liu et al. confirmed the feasibility of separating CO_2 from N_2 from functionalized pore-10, where four edge carbon atoms were substituted with nitrogen, while the remaining edge carbon atoms were functionalized with hydrogen [53]. They reported a high CO_2/N_2 selectivity of 300 and CO_2 permeance of 10^5 GPU with a pore density of $4 \times 10^{12}\ cm^{-2}$. While they did not observe significant permeation events for N_2, the gas permeation rate was confirmed by DFT-derived energy-barrier calculations. This high CO_2/N_2 selectivity was attributed to electrostatic interactions between CO_2 and the nitrogen-passivated pore rim that reduced the energy barrier for CO_2 to translocate the pore. Therefore, CO_2 mass transport across the nanopore was enhanced. This study indicated that there is a rich possibility to decorate the pores with functional groups to selectively increase the adsorption of a gas molecule on graphene.

Tian et al. reported synergy between ionic liquids and graphene by designing a system where a monolayer of ionic liquid (1-ethyl-3-methylimidazolium

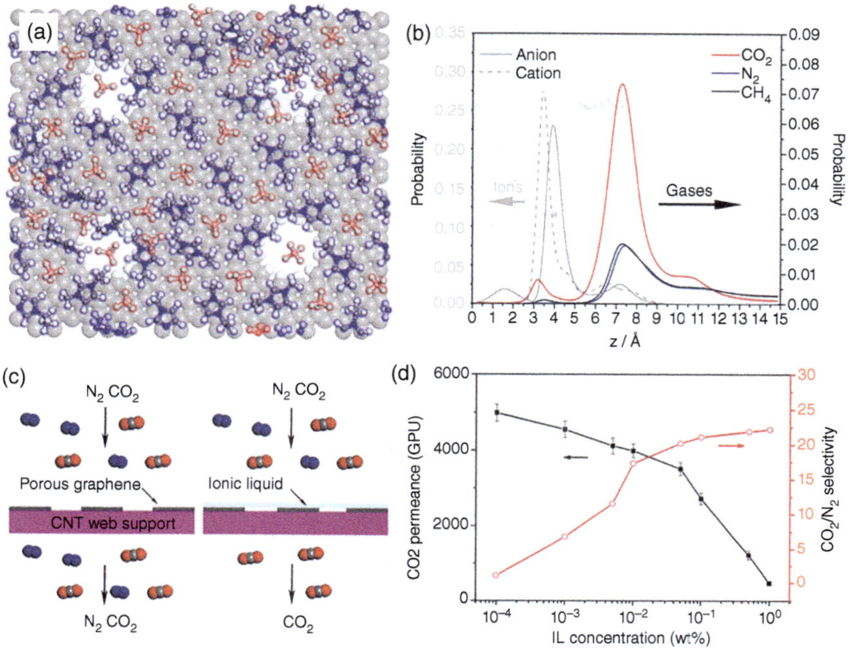

Figure 3.7 (a) A hydrogen-terminated 6-Å-sized pore covered with a monolayer of the [emim][BF_4] Color code: C, gray; H, white; N, blue; B, red; F, green. Cations and anions are highlighted with blue and red edges, respectively [54]. (b) Probability of finding ions (*y*-axis on the left) and gases (*y*-axis on the right) along the *z*-direction (perpendicular to the surface) on the feed side with graphene located at $z = 0$. Source: Tian et al. [54]/with permission of American Chemical Society. (c) Schematic showing gas transport in the presence and the absence of the ionic liquid [55] (d) CO_2 permeance and CO_2/N_2 selectivity as a function of ionic liquid concentration. Source: Guo et al. [55]/with permission of American Chemical Society.

tetrafluoroborate or [emim][BF_4]) was decorated on top of an N-SLG film hosting nonselective pores for CO_2 (Figure 3.7a) [54]. This composite film allowed selective permeation of CO_2 over N_2 and CH_4, leading to CO_2/CH_4 selectivity of ~42 and CO_2 permeance of ~10^5 GPU. This was attributed to the stronger affinity of CO_2 with the ionic liquid (Figure 3.7b) as well as the dynamic modulation of the pore size by the anions. Later, an experimental report by Guo et al. validated the concept of ionic-liquid gated N-SLG membranes [55]. Briefly, a layer of 1-butyl-3-methylimidazolium tetrafluoroborate ([Bmim] [BF_4]) was applied on a nonselective N-SLG film, and CO_2/N_2 separation performance was studied as a function of ionic liquid concentration (Figure 3.7c). They observed that CO_2 permeance reduced (4000–1200 GPU) as the concentration of ionic liquid was increased, whereas the CO_2/N_2 selectivity followed an opposite trend (18–32, Figure 3.7d).

The first proof-of-principle study on the functionalized graphene nanopore (Figure 3.8a) was reported in 2019 by He et al. where a record-high carbon capture performance was achieved by grafting CO_2-philic polymers on N-SLG [56]. Briefly,

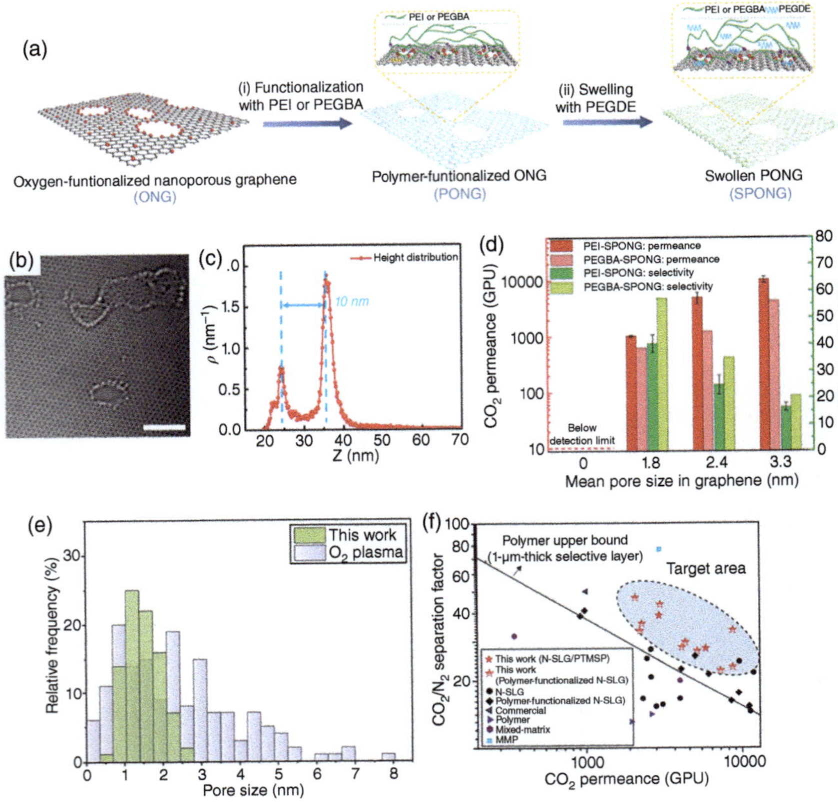

Figure 3.8 High-performance postcombustion capture membranes based on N-SLG functionalized with CO_2-philic polymers. (a) Schematic of the fabrication of the functionalized N-SLG membrane [56]. (b) Approximately 2-nm sized nanopores in SLG incorporated by O_2 plasma treatment. The scale bar in the image represents 2 nm [56]. (c) Height distribution of the grafted polymeric film showing a thickness of 10 nm [56]. (d) The evolution of CO_2/N_2 separation performance as a function of mean pore size in N-SLG [56]. Source: He et al. [56]/with permission of The Royal Society of Chemistry. (e) An improved PSD in graphene incorporated using O_3 treatment compared to that by O_2 plasma, and (f) resulting improvement in the separation performance upon using polymer functionalization. Source: Hsu et al. [57]/with permission of American Chemical Society.

a porous graphene lattice hosting a mean pore size in the range of 1.8–3.3 nm (Figure 3.8b), obtained by oxygen plasma treatment of graphene, was decorated by polyethylenimine (PEI) or poly(ethylene glycol)-bis-amine (PEGBA) such that the film thickness was between 3 and 10 nm (Figure 3.8c). The films were swollen by poly(ethylene glycol)-dimethyl-ether (PEGDE) oligomer to improve the free volume of the polymer film. The resulting film yielded CO_2 permeance of 6180 GPU with a CO_2/N_2 separation factor of 22.5 (Figure 3.8d), which is extremely attractive when compared to the state-of-the-art membranes for postcombustion capture. Later, an improved PSD in the graphene lattice (Figure 3.8e), obtained by the ozone treatment in contrast to the plasma treatment in the previous study, led to even better performance from the functionalized graphene film, where an extremely attractive separation performance was achieved (CO_2 permeance of 8730 GPU and CO_2/N_2 selectivity of 33.4, Figure 3.8f) [57].

The gas transport across N-SLG can be modulated by careful engineering of the support layer that graphene rests on. A thin dense film can be deposited on graphene similar to a masking layer in the conventional membranes. He et al. showed that masking a gas-sieving N-SLG film by a nanoporous polymer film can further improve the separation performance by completely eliminating the direct gas-phase transport through the graphene film (Figure 3.9a,b) [49]. They coated a thin film of PIM-1 on top of the N-SLG film and showed that CO_2/N_2 selectivity jumped from ~10 to over 30. By modeling the transport of CO_2 and N_2 through the ensemble of nanopores in graphene, they showed that blocking the gas-phase transport largely benefits CO_2 permeation over N_2, as the former permeates by adsorbed-phase transport by a larger extent compared to the latter (Figure 3.9c, d). They also suggested that increased compactness of the polymeric chains next to graphene can also benefit the separation performance.

The above studies demonstrate that N-SLG-based gas-separation membranes are highly promising for achieving improved gas-separation performance compared

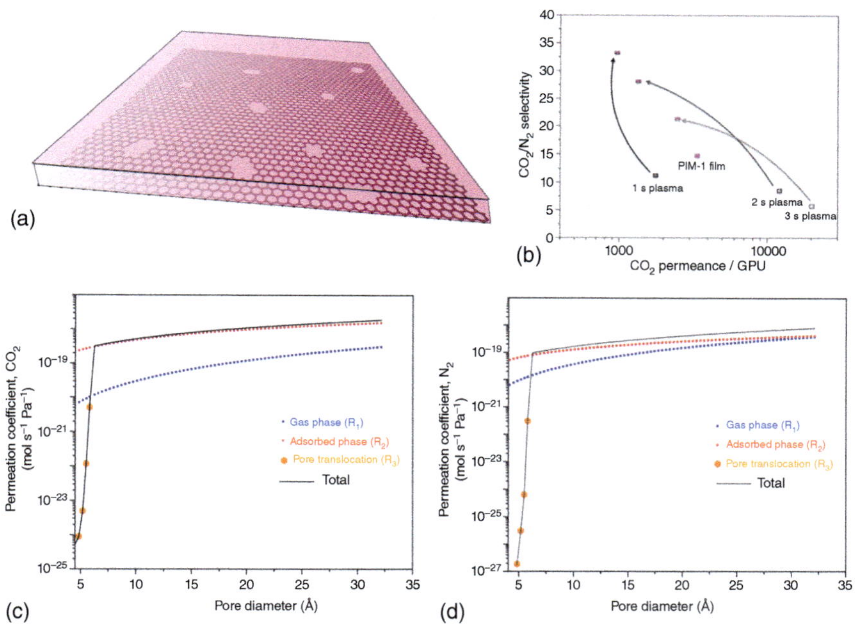

Figure 3.9 The concept of graphene pore masking to block the direct gas-phase transport for improving CO_2/N_2 separation performance. (a) Schematic of graphene nanopores masked by a polymeric film. Source: Kumar Varoon Agrawal. (b) A synergistic improvement in the selectivity of membranes when PIM-1 film is coated on top of the graphene film. Contribution of the resistances from direct gas-phase transport, adsorbed-phase transport, and pore translocation to the overall permeation coefficient of CO_2 (c) and N_2 (d). Source: He et al. [49]/with permission of John Wiley & Sons.

to the state-of-the-art. The practical implementation of high-performance N-SLG membranes is based on the incorporation of a high density of gas-sieving nanopores, and the fabrication of large-area N-SLG membranes in scalable and cost-effective manner. We discuss these topics in the next two Sections (3.3) and (3.4).

3.3 Engineering Gas-Selective Vacancy Defects

Given the gas-impermeable graphene lattice, one needs to etch molecular-sized nanopores by creating carbon vacancy defects in a controlled way. For light gas separation, this translates into incorporating vacancy defects with missing 10–16 atoms. The biggest challenge here is to achieve a moderate-to-high density (10^{10}–10^{12} cm^{-2}) of gas-permeable nanopores to achieve high permeation rates with a narrow enough PSD to realize size-sieving based separation. Otherwise, the nonselective pores would dominate the overall transport. For the nanopores operating in the gas-sieving regime, the transport rate is proportional to the energy barrier at the center of the pore (Eq. (3.4), Figure 3.10b). So, when the energy barrier is negligible, the net transport rate is about a thousand-fold higher than that of the pore where the energy barrier is moderate (0.2 eV, Figure 3.10b). For example, nonfunctionalized pore-16 is not selective for CO_2 over N_2 but pore-10 and pore-13 are, and therefore, the presence of just one pore-16 in the ensemble of thousands of pore-10 and pore-13 will ruin the selectivity. This then calls for an etching technique that is highly controllable, where the percentage of nonselective nanopores will be only a few parts per million (ppm).

The unique opportunity offered by N-SLG for gas separation, as predicted by the molecular simulations, makes it compelling to custom design and fabricate N-SLG with a suitable porous lattice structure for a given separation problem. The design of porous graphene lattice can include several elements, (i) pore density, (ii) pore

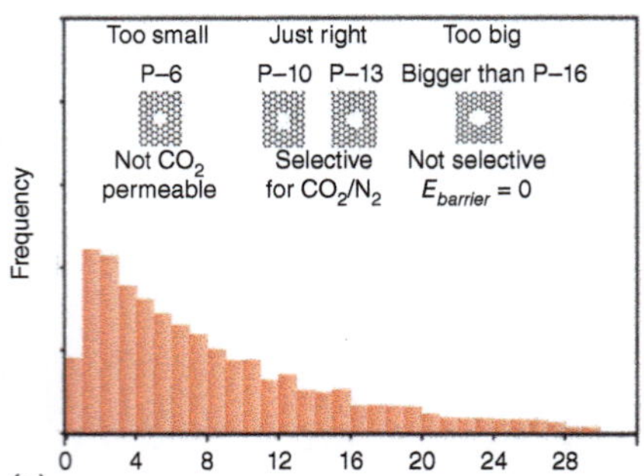

(a) (b)

$E_{barrier}$ (eV)	Relative transport rate ($1/R_3$)
0.5	10^{-5}
0.2	1
0	~2000

Figure 3.10 Illustration of complexities in selective gas transport from a pore-size distribution (PSD) of N-SLG. (a) A wide PSD consisting of pores with 1–30 missing carbon atoms is shown. Pores smaller than pore-6 do not contribute to permeation and pores bigger than pore-16 are not selective. Source: Kuang-Jung Hsu. An implication of this on the permeation properties is shown in (b) where the relative transport rates from small pore with large energy barrier (0.5 eV), and nonselective pores with no energy barrier is compared to a pore with an energy barrier of 0.2 eV.

shapes and PSD including the concentration of nonselective pores, and (iii) functionalization of pore edge. Additionally, a gas-permeable backing (mechanically reinforcing) support layer can affect the overall transport property, e.g. by blocking the direct gas transport. Among these, arguably, the most challenging aspect is the incorporation of gas-selective PSD at a moderate to a high density (10^{10}–10^{12} cm^{-2}) of gas permeable nanopores, in a scalable and cost-effective way, to make the gas-separation performance of N-SLG better than the state-of-the-art. The creation of such narrow PSD accompanying the desired pore density requires aggressive lattice etching, which is intrinsically difficult to control to limit the pore expansion. Development of etching routes, where pore nucleation and growth are decoupled, will greatly help this aspect of pore design.

Vacancy defects or nanopores can be incorporated in graphene by either bottom-up or by top-down approaches. The former involves the crystallization of graphene in a way such that lattice intrinsically comprises of gas-selective nanopores, e.g. intrinsic vacancy defects in the case of polycrystalline graphene synthesized by the chemical vapor deposition (CVD) [58–61], intrinsically porous film formed by the aryl–aryl coupling of hexaiodo-substituted macrocycle CHP [62], and graphynes [63]. The top-down route involves postsynthetic etching of the graphene lattice by physical or chemical means.

3.3.1 Bottom-Up Synthesis of N-SLG

The presence of the intrinsic vacancy defects can be beneficial for gas separation if the defect size, determined by the number of missing carbon atoms, is commensurate with the desired gas molecule [64]. High-quality CVD-derived polycrystalline SLG film, ideal for the fabrication of practical membranes, hosts a low density of intrinsic vacancy defects. These defects typically arise from either imperfect stitching of graphene grains [65] or the limited lattice etching during the high-temperature synthesis in the presence of residual oxygen in the CVD reactor [58]. The density and the size of intrinsic defects are dependent on a number of factors related to the crystallization of graphene, i.e. (i) the chemical composition of the precursor (e.g. methane or larger hydrocarbons, such as benzene), (ii) precursor concentration, (iii) the growth temperature, (iv) the orientation, roughness, carbon solubility, and activity of the catalytic substrate (Cu, Ni), and finally on the presence of residual etchant in the CVD reactor (O_2, H_2O, and CO_2) [58, 66]. For example, Kidambi et al. successfully increased the density of molecular-selective intrinsic defects in graphene, derived from CH_4 on a Cu foil, by decreasing the CVD temperature from 1000 to 900 °C [59]. The increasing density of intrinsic defects was evident from a significant rise of the *D* peak along with the appearance of the *D*′ peak. These defects could separate salt ions from the model protein molecules.

The first example of gas separation from intrinsic vacancy defects in CVD graphene was reported by Huang et al. [64]. They synthesized SLG on a Cu foil at 1000 °C and observed gas-sieving as as-synthesized graphene membranes. The carbon amorphization trajectory based on Raman spectroscopy [67] indicated that the as-synthesized SLG hosted a low density of intrinsic vacancy defects (0.025%).

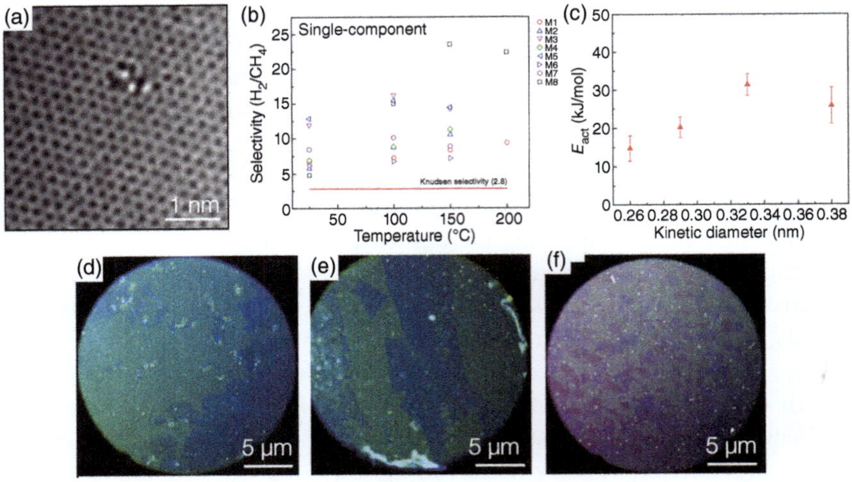

Figure 3.11 (a) HRTEM image of an intrinsic vacancy defect in as-synthesized SLG by CVD using methane as a carbon precursor, (b) corresponding H_2/CH_4 selectivities from eight separate N-SLG membranes hosting only intrinsic vacancy defects, and (c) activation energies as a function of the kinetic diameter of gases. Source: Huang et al. [64]/Springer Nature/CC BY 4.0. (d) DF TEM image from graphene synthesized at 1000 °C using benzene as growth precursor. Corresponding DF TEM images of graphene synthesized at 900 °C (e) and 825 °C (f). Source: Khan et al. [66]/with permission from Elsevier.

Interestingly, several defects were observed inside graphene grains, which are only possible if carbon lattice is etched during or after CVD (Figure 3.11a). They observed a selective hydrogen transport with H_2/CH_4 selectivity up to 25, much higher than the Knudsen selectivity (2.8) for the same gas pair (Figure 3.11b). The H_2 transport was temperature-activated with an apparent activation energy of 20.2 ± 2.7 kJ mol^{-1}, where apparent activation energy is defined as the sum of physisorption energy on top of the pore and the energy barrier to cross the nanopore (Figure 3.11c). The physisorption energy of H_2 on graphene nanopore is low (~4 kJ mol^{-1}), which makes the corresponding energy barrier close to 24 kJ mol^{-1}. This barrier is reasonably high and is close to that simulated when H_2 crosses pore-10 [30]. This study showed that despite a low density of intrinsic defects, attractive H_2 permeance of 1000 GPU could be achieved at 150 °C, which highlights the tremendous potential of N-SLG for gas separation. Later, Khan et al. reported hydrogen-sieving intrinsic defects in benzene-derived SLG on a Cu foil at a reduced CVD temperature of 900 °C [66]. The density of intrinsic vacancy defect was higher at 900 °C (5.8×10^{10} cm^{-2}) compared to that at 1000 °C (3.6×10^{10} cm^{-2}) indicating that the majority of defects were grain-boundary defects. Indeed, dark-field (DF) TEM images indicated that the grain size decreases as one reduces the growth temperature (Figure 3.11d–f). The apparent activation energy for H_2 transport from these defects was smaller (~10 kJ mol^{-1}) than that from CH_4-derived graphene reported by Huang et al. [64], indicating that these defects were relatively larger consistent with the fact that the size of intrinsic defects is expected to be a function of the size of growth precursor (benzene versus methane in these two cases).

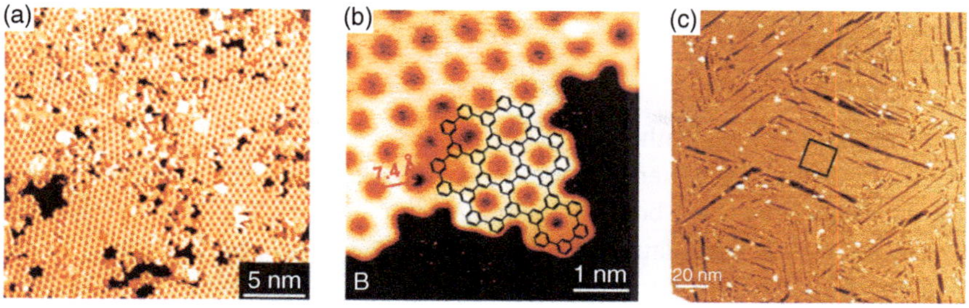

Figure 3.12 An example of the bottom-up synthesis of highly porous N-SLG. (a) STM image of the polyphenylene network on Ag (111) after polymerization at 805 K. (b) STM image of an edge of the network after polymerization at 805 K. Source: Bieri et al. [62]/with permission from Royal Society of Chemistry. (c) STM image of nanoporous graphene obtained after Ullman coupling at 200 °C on Au (111). Source: Moreno et al. [68]/with permission from AAAS.

Catalytic polymerization (coupling) of organic precursors is an emerging approach to directly synthesize porous 2D layers with precise pore topology. Bieri et al. were the first ones to demonstrate this when they studied the formation of a two-dimensional polyphenylene network by surface-promoted aryl–aryl coupling of hexaiodo-substituted macrocycle CHP on an Ag(111) substrate [62]. As a result, they could achieve a pore density exceeding $10^{14}\,cm^{-2}$, corresponding to a pore spacing of 7.4 Å (Figure 3.12a,b). This represents an upper bound of pore density that can be incorporated in SLG. Later, Moreno et al. synthesized graphene nanoribbons network by surface-catalyzed Ullman coupling of diphenyl-10,10′-dibromo-9,9′-bianthracene (DP-DBBA) on an Au(111) substrate (Figure 3.12c) [68]. The assembly consists of thermally activated reaction stages, Ullman coupling, cyclodehydrogenation, and finally dehydrogenative cross-coupling. A coupling yield reaching 100% was reported, resulting in ordered subnanometer-sized nanopores with a high pore density of $4.8\times10^{13}\,cm^{-2}$. While these reports on isoporous 2D film with high pore density are highly attractive for size-sieving, the long-range order in this route (typically 10–100 nm) needs to be significantly improved for application in membranes.

3.3.2 Postsynthetic Etching of SLG

Compared to the methods that incorporate intrinsic vacancy defects, postsynthetic etching of the graphene lattice is highly promising to fine-tune PSD for a given separation. This method, if perfected, would be able to control the pore size to the resolution of 1 missing carbon atom, and can be highly attractive for handling a wide range of molecular separation. Most of the postsynthetic etching can be categorized in two approaches—(i) physical etching marked by knocking out lattice atoms by energetic beams, and (ii) chemical etching marked by removing lattice atoms by gasification [69]. Physical etching is implemented by the collision of energetic species, such as focused electron beam or ion beams (Ar^+, Ga^+) with the lattice atoms displacing the latter from the lattice. In contrast, chemical etching is implemented by the oxidation of the carbon lattice that eventually leads to gasification event removing carbon as CO and CO_2.

3.3.2.1 Physical Etching Methods

In physical etching, high-energy electrons or ions are bombarded onto the carbon lattice that breaks the C—C bonds if the transferred energy is higher than the bond energy. A relativistic formulation of the transferred energy, T, by an electron beam with energy E to a carbon atom is given in Eq. (3.6):

$$T = E\frac{2M(E + 2m_ec^2)}{(M + m_e)^2c^2 + 2\,ME}\cos^2\gamma \tag{3.6}$$

where γ is the angle between the incident electron beam and the direction of ejection, M is the mass of carbon atom, m_e is the rest mass of electron, and c is the speed of light.

Based on Eq. (3.6), maximum energy of 15.7 eV is expected to be transferred to the carbon atom when the graphene lattice is exposed to an electron beam with an energy of 80 keV. When the energy transferred by the electron beam exceeds the energy barrier for the carbon atom to leave the graphene lattice (displacement energy or E_d), a knock-on event successfully takes place. Based on these calculations, Smith and Luzzi reported that the minimum incident electron energy needed to displace one sp^2 carbon bonded to three carbon atoms in carbon nanotubes is 86 keV [70]. Later, Kotakoski et al. showed that the knockout of carbon atoms from the pristine graphene lattice would require an electron beam with incident energy larger than 100 keV (Figure 3.13) [71]. However, much lower energy is required to knockout the edge carbon atoms mainly because they are only bonded to two carbon atoms, and as a result, E_d of edge sites (<14 eV) [72] is approximately two-third of that of the atoms in the basal plane (23 eV) [73]. Therefore, one can use an 80 keV beam to expand the pores because the energy transferred (15.7 eV) is already high enough to knockout edge atoms.

On the experimental front, one of the first demonstrations of physical etching of graphene was reported by Fischbein and Drndić [74]. They sculpted 3.5-nm-wide

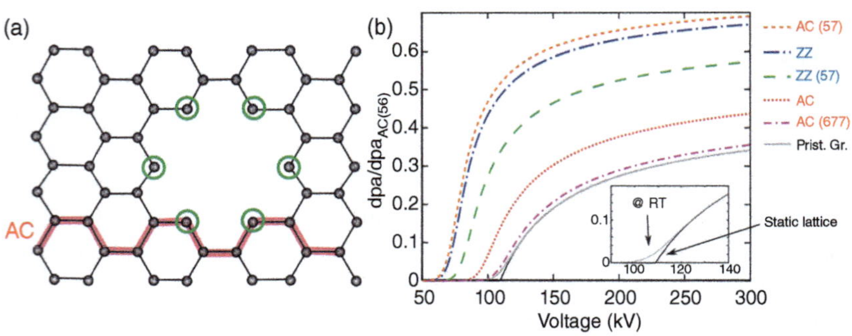

Figure 3.13 (a) Schematic of pore-6 where zigzag pore edge carbon atoms are highlighted in green [71]. (b) Relative displacement rates for various armchair (AC) and zigzag (ZZ) edge configurations. Source: Kotakoski et al. [71]/with permission of American Chemical Society.

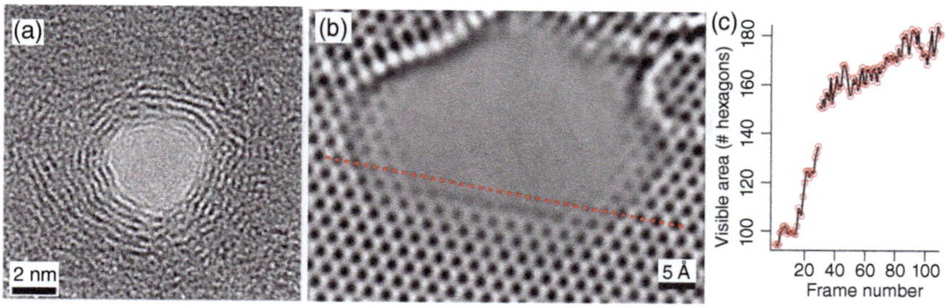

Figure 3.14 (a) TEM image of a nanopore drilled on multilayer graphene by an electron beam [74]. Source: Fischbein and Drndić [74]/with permission from AIP Publishing LLC. (b) Average intensity mapping based on 110 images taken of the nanopore under the beam. Faint atoms shown in the image represent the atoms that have been knocked out by the beam damage [75]. (c) Electron beam-led nanopore expansion rate marked by the increase in the visible area of the hole as a function of time (image frames). Source: Girit et al. [75]/with permission from American Association for the Advancement of Science – AAAS.

nanopores, slits, nanobridges, and gaps on multilayer graphene by a controlled dose of a converged 200 keV electron beam (~0.3 pA nm^{-2}, Figure 3.14a). Later, Golovchenko and coworkers drilled nanopores in graphene with a size range of 5–23 nm with a 200 keV electron beam [76]. Using this, they could study transmembrane ionic conductance as a function of pore size. Almost at the same time, both Merchant et al. [77] and Schneider et al. [78] incorporated 5 to 10 nm pores on the graphene lattice by electron beam as a high-resolution DNA sensor. By using an aberration-corrected high-resolution TEM (AC-HRTEM), capable of sub-Ångstrom resolution at 80 kV, Zettl and coworkers reported nanopore expansion dynamics (Figure 3.14b,c) [75]. They observed that zigzag (ZZ) edges are most stable compared to the armchair (AC) edges and attributed this to the fact that the latter edge type is more difficult to reconstruct by the migrating carbon atoms once an edge atom is displaced. This finding is opposite to those predicted by theory (Figure 3.13b) and could be due to the limited size of pores studied.

Focused-ion beam (FIB) shares a similar etching mechanism as that of the electron beam [72, 73]. Briefly, the energetic ions such as Ar^+, He^+, and Ga^+ collide with the carbon atoms of the graphene lattice, and knockout the carbon atoms to form vacancy defects. The minimum energy of the ions to knockout the carbon atom, E_{min}, can be estimated by the kinematical relationship assuming it is a head-on collision with the ion of mass m_i [79]:

$$E_{min} = E_d \frac{(M + m_i)^2}{4Mm_i} \tag{3.7}$$

Based on this, heavier ions, such as Ga^+, Xe^+, and Kr^+, require less energy to displace carbon atoms from the basal plane than lighter ions, such as He^+ [73]. Celebi et al. demonstrated that the sputter yield, defined as the average removal rate of the carbon atom from the lattice, using Ga^+ ion is higher than that of He^+, indicating a clear role of the size of ions on knocking out carbon atoms [80]. With a dose of $0.5–5 \times 10^{-5}$ pA nm^{-2}, nanopores with diameters between 16 and 1000 nm were incorporated in the graphene lattice. A few nanometer-sized pores could be

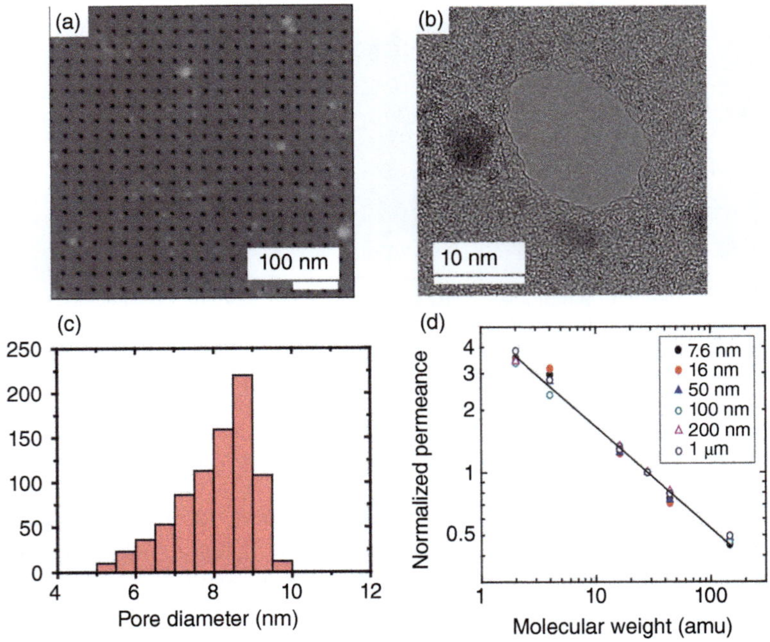

Figure 3.15 (a) An ensemble of nanopores drilled on a suspended bilayer graphene film by FIB. (b) A 15-nm-sized nanopore drilled by FIB. Source: Celebi et al. [80]/with permission from AAAS. (c) PSD from an ensemble of nanopores. (d) Gas permeance from drilled nanopores (7.6 nm to 1 μm) as a function of molecular weight. Source: Celebi et al. [80]/with permission from AAAS [65].

obtained by exposing graphene to FIB (Figure 3.15a–c). Since these pores are much bigger than the size of gas molecules but smaller than their mean free path, the gas transport through these nanopores took place in the effusive transport regime characterized by extremely large gas permeance (~100 million GPU for H_2) and gas-pair selectivity determined by the mass ratio of the molecules. They reported H_2/CO_2 selectivity of 4.0 from few-nanometer-sized pores, close to the Knudsen selectivity of 4.7 for the gas pair (Figure 3.15d). It should be noted that the FIB limits the smallest achievable pore size to about 1 nm, and it is extremely challenging to etch pores with an angstrom resolution needed for gas separation by ion beams, especially because of the fundamental difference in the E_d for pore nucleation and expansion.

To avoid a rapid etching under the high-energy beam, Russo and Golovchenko reported a decoupled pore nucleation (ion beam) and growth (electron beam) strategy [72]. Briefly, vacancy defects made of 1–2 missing atoms were incorporated by sputtering the lattice with a 3 keV Ar^+ beam at 148 K with a typical dose of 10^{13} Ar cm^{-2}. The resulting pore density was 5.1×10^{11} cm^{-2}, which corresponded to a probability of 5% for pore nucleation under these irradiation and sample conditions. They showed that a weak sputtering by a parallel 80 keV electron beam inside a TEM could controllably expand the pores, and reported a PSD centered around a pore diameter of 1.7 nm. The *in situ* ion and electron-beam etching techniques, implemented under a microscope, allow one to track the evolution of nanopore morphology during the etching process and are attractive for real-time PSD control. However, these methods are arguably challenging to optimize for drilling gas-sieving nanopores, which are only a fraction of a nanometer in size. Also, the beam treatment is difficult to implement over a large scale for practical membrane production, which limits its application to the fundamental studies.

3.3.2.2 Chemical Etching Techniques

In contrast to the physical etching techniques discussed above, the chemical etching of graphene is intrinsically scalable and has the capability to incorporate small nanopores suited for the field of gas-separation membranes. Chemical etching of the carbon lattice has large precedence in studies involving gasification of graphite, and more recently in studies on graphene oxide. Several oxidizing etchants, such as plasma, gas molecules (O_2, O_3, H_2O, CO_2, etc.), and ultraviolet (UV)/O_3 are reported in the literature for incorporating vacancy defects in graphite or graphene [81, 82]. High-temperature O_2 treatment was one of the first approaches studied for understanding the gasification kinetics of graphite. Such studies provided an early understanding of the activation energies of pore nucleation and growth. Other than the reaction with oxidizing gases, etching lattice in presence of ions and radicals (e.g. oxidative plasma, UV/O_3) has been a popular choice of etchant due to a high concentration of reactive species relatively fast reaction at room temperature [83].

The first proof of principle of gas-sieving graphene nanopores synthesized by chemical etching was provided by the pioneering study of Bunch and coworkers, who successfully introduced subnanometer nanopores in approximately 5-μm-sized micromechanically exfoliated graphene by UV/O_3 etching [31]. Gas transport was studied across bilayer and monolayer graphene microballoons by allowing the gases to pressurize the graphene-covered wafer pits. The pore size in graphene was estimated based on the kinetic diameter of the smallest molecule that did not seem to pass through the balloon. Remarkable selectivities could be obtained from bilayer graphene. H_2 and CO_2 diffused much faster than Ar, N_2, and CH_4 with selectivities exceeding 1000. Later, Karnik and coworkers reported a two-step method, decoupling pore nucleation and growth for incorporating subnanometer-sized nanopores in SLG on a centimeter-scale area [84]. Briefly, SLG was exposed to Ga^+ ion bombardment to introduce small vacancy defects. Then, the defects were enlarged by exposure to the oxidative solution of 1.875 mM $KMnO_4$ in 6.25% H_2SO_4. After a short etching time, an average pore size of approximately 0.40 nm and pore density of 3×10^{12} cm^{-2} could be obtained (Figure 3.16). Both nucleation and growth steps were required to grow sieving selective pores. Without nucleated defects, acid potassium permanganate did not etch the pristine graphene, and gallium ions bombardment was optimized to induce only small vacancy defects. Although gas-sieving was not achieved by this study, the decoupled nucleation and growth strategy can prove highly effective to realize a narrow PSD with a high pore density by generating a high density of nuclei and then switching to a controlled pore expansion chemistry. Below, we review the literature on several chemical etchants but mainly focus on gaseous etchants that have the scalability advantage

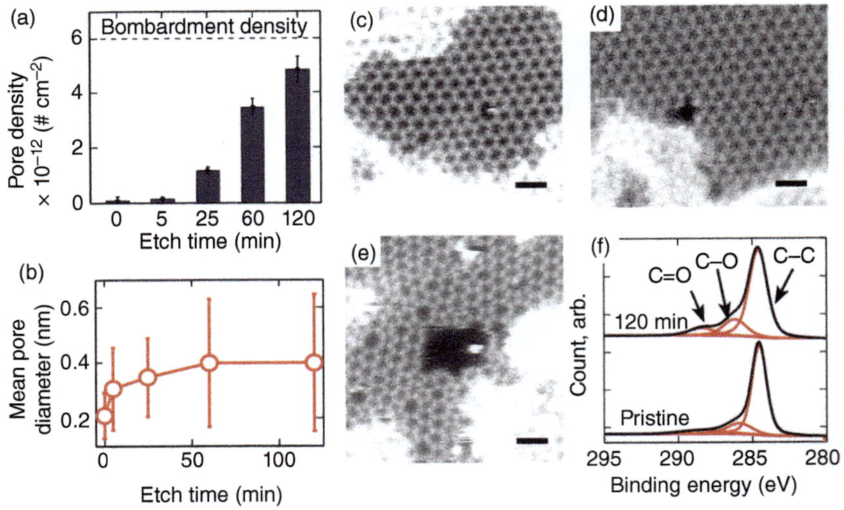

Figure 3.16 A graphene etching strategy based on nucleation by ion bombardment and expansion by chemical etching. (a) Pore density as a function of time exposure to the ion beam. (b) Mean pore size in graphene as a function of exposure time to the oxidant. Scanning transmission electron microscopy (STEM) images of nanopores formed by this etching strategy (c)–(e). X-ray photoelectron spectroscopy (XPS) comparing the pristine graphene and etched graphene showing the presence of oxygen functional groups in the etched graphene. Source: O'Hern et al. [84]/with permission from American Chemical Society.

attributing to the fact that the concentration of the gas and the temperature of the reactor can be controlled uniformly over a desired/large volume.

Plasma Treatment Air/oxygen plasma treatment has been extensively studied to etch graphene with subnanometer-sized pores attributing to the fact that such plasma consists of a high concentration of oxidative radicals that can etch graphene lattice in a few seconds at room temperature [85, 86]. Important reaction in O_2 plasma include [87, 88] (i) ionization of oxygen molecule with energetic electron ($O_2 + e^- \rightarrow O_2^+ + 2e^-$), (ii) formation of negative ion ($O_2 + e^- \rightarrow O^- + O$), and (iii) recombination reaction ($O_2^+ + O^- \rightarrow O_2 + O$). The energy of these ions, when higher than E_d, can also knockout carbon atom from the pristine lattice, and therefore, physical etching may also take place in addition to chemical etching (see Eq. (3.6) and related discussion). The high energy of oxygen ions and radicals can overcome the energy barrier for chemisorption of oxygen-functional groups on the graphitic lattice. This is likely the reason behind the rapid reaction at room temperature. The atomic oxygen is attached to the C—C bond, forming a bridge configuration (epoxy group) on the graphene surface [89]. Vacancy defects are created from the gasification of oxidized domains as CO or CO_2 but are likely promoted by ion bombardment [90, 91]. The significant role of O atoms in etching the pristine lattice and expanding the pore edges cannot be ruled out.[92, 93].

The rapid etching of graphene in oxygen plasma makes it challenging to control the etching precisely. Surwade et al. managed to generate nanometer-sized pores (~1 nm) on 5-μm-monolayer graphene by optimizing the plasma dosage [94]. Duan and coworkers overcame the rapid expansion through plasma treatment by dividing graphene lattice into nanometer-sized domains and coating a block copolymer film hosting 5–15-nm-sized pore windows [95]. More recently, they adapted this method and used mesoporous SiO_2 as a template, to etch subnanometer-sized pores (0.63 nm) with high pore density (10^{12} cm^{-2}) for water purification [96]. Jang et al. investigated a combination of Ga^+ ion and oxygen plasma treatment to decouple nucleation and growth, where oxygen plasma was used to expand the pores [97]. Using this approach, they could achieve an attractive density (1.6×10^{13} cm^{-2}) of subnanometer-sized pores that resulted in promising water transport. Xie et al. used a synergetic two-step plasma method to drill pores in bilayer graphene [98]. Firstly, few-second O_2 plasma treatment was used to nucleate defects at the surface of graphene. Subsequently, H_2 plasma at 400 °C was used to enlarge the nucleated defects. Interestingly, they reported that hydrogen plasma only expands the nucleated defects and does not lead to a new nucleation event. By combining 2 s oxygen plasma with 5 s hydrogen plasma, they could incorporate 2-nm-sized nanopores with a moderate density of 10^{11} cm^{-2}. More recently, He et al. used 1–3 s O_2 plasma to incorporate H_2-sieving nanopores in graphene [49]. The etching of graphene lattice was apparent from the increase in the intensity of the *D* peak and the appearance of the *D′* peak in the Raman spectroscopy (Figure 3.17a). AC-HRTEM-based analysis of plasma-treated graphene lattice confirmed the presence of subnanometer-sized vacancy defects (Figure 3.17b).

While plasma etching is straightforward to implement in laboratory experiments, it is relatively difficult to scale-up for the membrane application. This is because (i) plasma concentration is not uniform over a large length-scale attributing to the nonuniformity in the electric field inside the chamber. The condition of the oxygen plasma could be significantly affected by various factors, including electrical

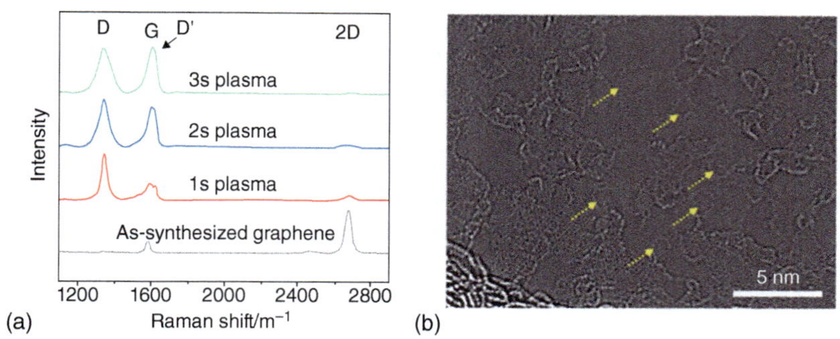

Figure 3.17 (a) Raman spectroscopy data from SLG treated with O_2 plasma for a period of —one to three seconds. (b) AC-HRTEM image of vacancy defects incorporated in the graphene lattice from 1 s plasma treatment. Source: He et al. [49]/with permission from John Wiley & Sons, Inc.

and thermal properties as well as the position of the electrodes; (ii) given the batch-to-batch nature of the process, one cannot pack a high area of graphene inside the plasma chamber; for example, one cannot roll or stack graphene inside the plasma chamber because of the spatial variation in plasma concentration. In this respect, below we review pore formation in graphene by direct exposure to oxidizing gases.

O_2 O_2-led etching of graphene has been studied in the context of nucleating pits (pores) on graphite (graphene) as well as expanding the pre-existing defects. The activation energy for pore nucleation with O_2 is reported as 200 kJ mol^{-1} (c. 2.1 eV) [99], slightly higher than the corresponding activation energy for pore expansion (127 kJ mol^{-1} or 1.3 eV). The difference in these nucleation and expansion activation energies is much lower than that in the case of beam-induced carbon displacement. This makes etching of the graphene lattice by oxidation much more attractive, especially in the context of obtaining a narrow PSD.

Liu and coworkers incorporated 20-nm-sized pores on graphene lattice by exposing it to an O_2/Ar gas flow with O_2 partial pressure of 350 torr for two hours at 450 °C [100]. These pits were also observed in the double- and triple-layer graphene (Figure 3.18a,b), confirming that the nucleation of vacancy defects on basal plane was indeed taking place in these conditions. Zettl et al. reported subnanometer pores in SLG by controlled etching in O_2 atmosphere for five hours at 200–240 °C [101]. Interestingly, structural analysis of these nanopores observed by ACHRTEM indicated that pores were created from missing 10–13 carbon atoms (Figure 3.18c,d), with pores functionalized with in-plane ether group and out-of-plane carbonyl group. Schlichting and Poulikakos reported a decoupled approach, which involved ion-beam-based nucleation and O_2-led pore expansion at 300 °C for suspended double-layer graphene [102]. The pore density was fixed by ion-beam treatment and the role of O_2 was to strictly expand the pores to allow gas transport. Exposure to O_2 for 15 minutes at 300 °C was sufficient to increase the permeance of H_2 by approximately 10-fold compared to that from nuclei created by ion beam. Using this approach, they reported H_2/CH_4 selectivity up to 9.3. Interestingly, this selectivity dropped to close to the corresponding Knudsen selectivity when the transmembrane pressure difference was increased from 0 to 1 bar. For longer etching time, they observed effusive transport (H_2/CH_4 selectivity ~ 3.8) accompanied by large H_2 permeance (10 million GPU), consistent with the pore enlargement.

Strano and coworkers studied the impact of the O_2 leak on the formation of defects during and after the CVD of SLG at 1077 °C [58]. After the synthesis of SLG, they manipulated the atmosphere of the CVD reactor by adjusting the concentration of O_2 based on controlled leak of air in the CVD reactor. Raman-based investigation of SLG exposed to O_2 in this manner showed that the number density of vacancy defects in SLG increased sharply when the concentration of O_2 in the reactor was increased. For example, by increasing the oxygen exposure from approximately 130 ppm for six to eight hours to 7% for 1 minutes at 1077 °C in H_2 atmosphere, they could increase the defect density by sixfold, from 47 ± 37 to 325 ± 173 μm^{-2}. Scanning tunneling

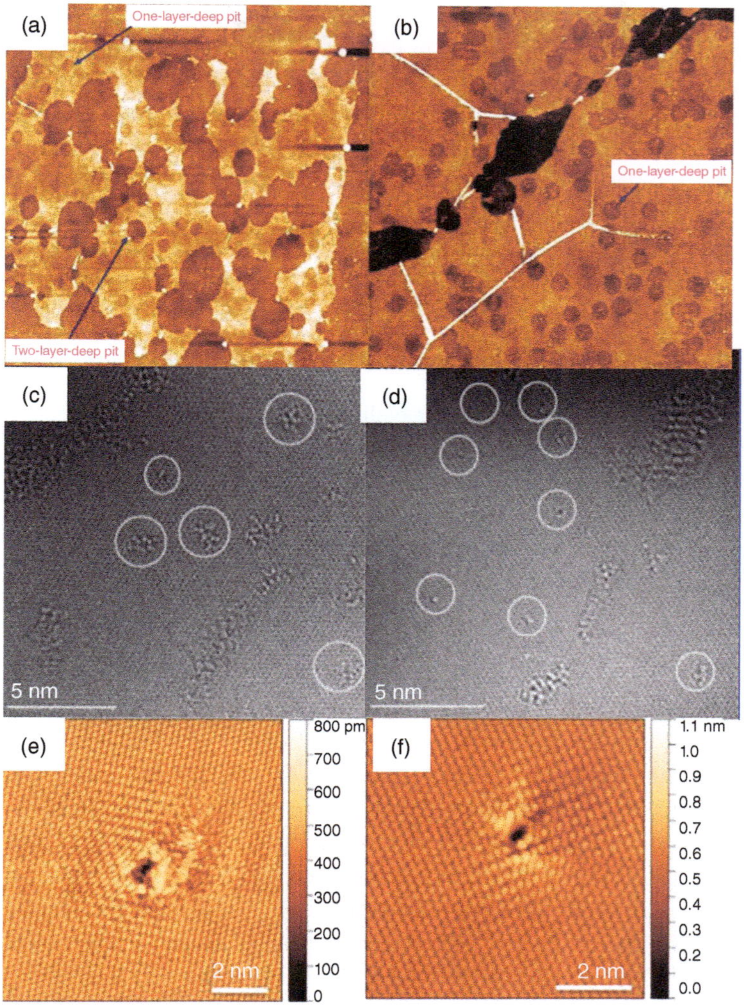

Figure 3.18 Etching of graphene by O_2. Atomic force microscopy (AFM) images of oxidized (600 °C for two hours) double-layer (a) and triple-layer graphene (c) showing the development of pits in the graphene. Source: Garaj et al. [76]/with permission from American Chemical Society. AC-HRTEM images of graphene treated with O_2 for 5 h at 240 (c) and 260 °C (d) revealing the formation of carbon vacancies. Source: Yamada et al. [101]/with permission from American Chemical Society. (e, f) STM images of vacancy defects in CVD graphene on Cu when the etching was carried out in the CVD reactor by leaking O_2 in the reactor. Source: Agrawal et al. [58]/with permission from American Chemical Society.

microscopy (STM) imaging of these defects indicated that these defects were predominately subnanometer-sized (Figure 3.18e,f), attractive for application in gas separation.

Recently, our group also demonstrated an adjustment of molecular cutoff from the N-SLG membranes using a slow expansion of nanopores by treatment with O_2 at

200 °C for a period of one to two hours [52]. Briefly, after measuring the CO_2/N_2 separation performance of N-SLG membrane, the sweep gas was switched from argon to O_2 and the temperature was increased to 200 °C. Tracking the evolution of CO_2 and N_2 in the permeate side, it became evident that the CO_2 permeation rate increased as a function of O_2 exposure time in a much more rapid way than N_2. This led to an improvement of CO_2/N_2 separation performance marked by both increase in permeance (over 10 000 GPU) as well as CO_2/N_2 selectivity (over 20). Such membranes are extremely attractive for postcombustion capture, reducing the energy consumption of capture to 1.3 MJ/kgCO_2 and capture penalty to 31.5 US$ (tonCO_2)$^{-1}$ and making N-SLG membranes competitive with the state-of-art absorption processes (capture penalty of around 50–110 US$ (tonCO_2)$^{-1}$) [26].

CO_2 CO_2 is quite interesting candidate for etching the graphene lattice because it can only expand the nanopores attributing to an extremely large energy barrier (~5 eV) for pore nucleation, which requires the formation of a single vacancy defect [103]. The majority of the experimental studies on CO_2 etching have been carried out on graphite [104–108], activated carbon [109], and charcoal [110, 111]. In its most simplified forms, the reaction of CO_2 with the carbonous materials can be written as [112]:

$$C + CO_2 \rightarrow 2CO$$

Ergun was one of the first to focus on the kinetics of the reaction between CO_2 and carbon [108]. He investigated the etching rate of graphite and activated carbon at atmospheric pressure and temperatures ranging 700–1400 °C and concluded that (i) the reaction rate is proportional to the concentration of CO_2-occupied sites, (ii) CO decreases the reaction rate by reducing the number of CO_2-occupied sites, and (iii) the release of CO from adsorbed phase to gas phase is the rate-determining step (RDS). The last claim has been debated, and other studies have indicated a possibility of a different RDS. Several studies by Yang and coworkers revealed the kinetics of graphite–CO_2 reaction [113–115]. They showed that the reaction rate follows the Langmuir–Hinshelwood model and etching is highly anisotropic along the *c*-axis of graphite. Yet, despite these and other studies on carbon gasification by CO_2, the RDS is still inconclusive. Biederman and coworkers have argued that CO_2 chemisorption is the RDS and suggested that the release of gaseous CO is much faster that makes sense based on the DFT studies (see below) [104].

Recently, a few DFT studies of carbon gasification by CO_2 were published to elucidate the mechanism of gasification and the related kinetics (Figure 3.19). Cabrera-Sanfelix was one of the first to specifically focus on the interaction of CO_2 with the defective SLG [117]. She postulated that this interaction consists of three main steps—(i) physisorption of CO_2 on top of the defect, (ii) chemisorption by lactone group formation, and (iii) desorption of gaseous CO. For etching to happen, the second CO molecule should desorb too [118–120]. Other DFT studies on this topic also agree with this analysis; however, they differ on two points—(i) the value of the energy barrier for each step and (ii) the nature of the oxygen-containing groups at the transition and the final states. Generally, the change in energy of the

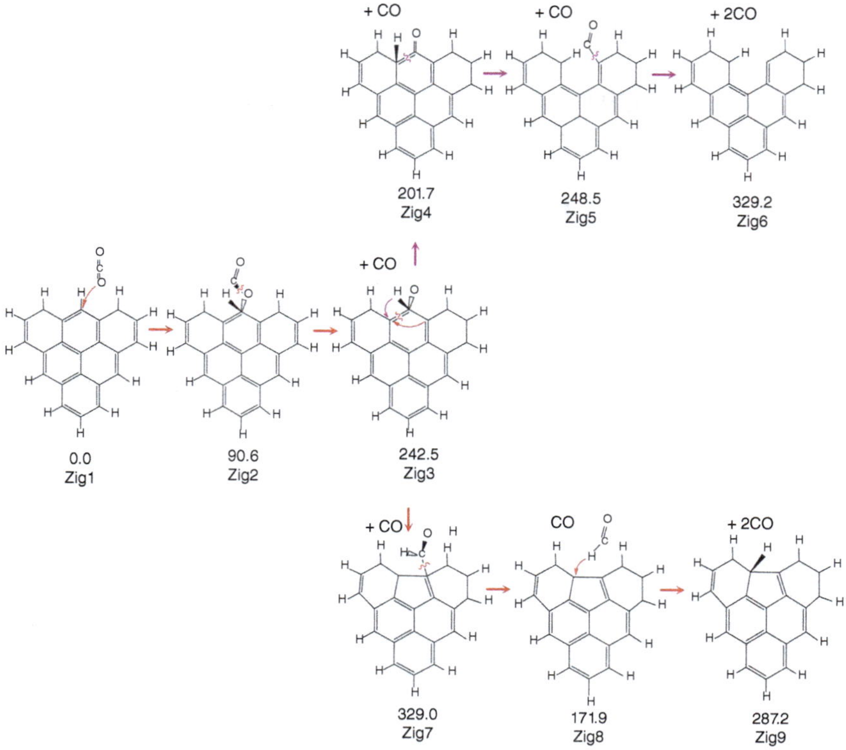

Figure 3.19 Reaction mechanism of CO_2 with the zigzag edge of graphene that ultimately leads to the gasification of edge carbon atoms as CO (the numbers written under each structure shows the potential energy of the system at that state; unit: kJ mol^{-1}). Source: Zhao et al. [116]/with permission of Elsevier.

system during the physisorption step is negligible, approximately 0.2 eV, compared to that of chemisorption [103]. Various configurations and functional groups, such as lactone, semiquinone, carbonyl, ether, furan, and epoxy, have been proposed as the structure of defected graphene at the chemisorption step, and the calculated energy barriers range from 1.7 to 3.3 eV [121, 122]. The difference in values arises from the varied choice of computational methods and nature of the defects, i.e. single vacancy, double vacancy, zigzag, and armchair edges. Desorption of the second CO molecule requires activation energy of 2.3–3.6 eV [120, 123, 124]. Alike, the difference is caused by the choice of the computational method and the structure of graphene considered at the previous step. Theoretical studies that contain all steps of the gasification process are scarce. Based on Zhao et al., the overall activation energy of gasification for the zigzag and the armchair edges are calculated to be 2.72 and 3.11 eV, respectively [116]. In comparison, the overall activation energy based on the experimental studies ranges from 1.7 to 3.7 eV [114, 125, 126]. Systematic experimental studies on the overall process of graphene etching are needed to validate the existing theoretical calculations.

It is quite difficult to etch graphene nanopores in a way to obtain extremely high pore density (~10^{13} cm^{-2}) while maintaining a narrow PSD that allows molecular differentiation. Recently, Dutta et al. simulated a unique CVD approach in the presence of CO_2 as an etchant to generate high density (3×10^{13} cm^2) of gas-selective pores [127]. Briefly, the concept dealt with a competitive etching (by CO_2) and growth (by CH_4) to arrive at a narrow PSD. A high density of elongated pores was predicted when the crystallization was modeled using the kinetic Monte Carlo simulations.

Yang et al. reported the etching behavior of graphene by gaseous CO_2 at elevated temperatures [128]. They demonstrated that CO_2 does not nucleate new defects in the graphene basal plane (up to the tested temperature of 1000 °C, which is the LPCVD process temperature), and etching initiates from the existing defects and pore edges. As a result of the presence of bond disorders and functional groups at the edges of the graphene, the chemical reactivity of the edge carbon atoms is higher than that of the perfectly bonded sp^2 carbon atoms in the basal plane [129]. Hence, the existing pores in graphene can be expanded using CO_2 without creating new pores. This characteristic of CO_2 is highly desirable as all other mentioned gaseous etchants have threshold temperatures lower than CVD conditions. For instance, both O_2 and H_2O vapor react with the graphene basal plane at temperatures below 700 °C [106, 130]. Even though Yang et al. [128] demonstrated the possibility of etching graphene by CO_2, a controllable graphene pore expansion at high temperature, i.e. CVD and near CVD temperatures, has not yet been realized. To better understand the reaction of CO_2 with graphene and realize its potential as a suitable etchant for graphene, further experimental studies are required to resolve the influencing parameters and obtain precise control over the etching kinetics.

O_3 O_3 is a good alternative to oxygen plasma where chemisorption events can take place at room temperature, thereby distinguishing O_3 from other oxidizing gases (e.g. O_2, CO_2) [64]. Traditionally, ozone has been used as an oxidant to control pollution in several contaminated effluents [131–134]. O_3 can be easily produced from O_2 by a corona discharge in a tube, a process that was developed by von Siemens in 1857 [135]. Here, an altering voltage is applied between two electrodes that lead to the formation of $\dot{O}$ ultimately forming O_3. The reaction proceeds as follows:

$$e^- + O_2 \rightarrow 2\dot{O} + e^-$$

$$\dot{O} + O_2 + M \rightarrow O_3 + M^*$$

where M is a substrate. Nowadays, inexpensive laboratory equipment can generate O_3 with concentrations of up to 20 wt%.

Theoretical studies on the reaction of O_3 with graphene are rare [136]. Existing theoretical studies on the carbon lattice predict physisorption energy in the range of 0.2–0.3 eV, which is too weak and does not explain the strong affinity of O_3 toward CNT and graphene observed by experiments at room temperature. For example, the reaction mechanism of ozone with graphene has been reported by the DFT calculations by Lee et al. [136] They predicted the energetics of O_3 physisorption and chemisorption on graphene and related energy barrier using projector augmented wave (PAW) pseudopotentials within the local-spin-density approximation (LSDA). They studied ozone physisorption as a function of its orientation on top of graphene and reported that ozone physisorbs in a form where the two terminal oxygen atoms point down toward the graphene lattice. They predicted physisorption energy of 0.25 eV with ozone subsequently chemisorbing with an energy barrier of 0.72 eV, leading to the formation of epoxy group on graphene. Based on the rapid reaction of ozone with graphene at room temperature [52], the physisorption energy should be much higher than that reported in these studies.

The epoxy groups grafted on the graphene lattice by O_3 or any other oxidative method tend to form a cluster to minimize the overall energy. In past, the clustering of the oxygen functional groups in graphene oxide has been attributed to the unzipping of graphene oxide. For example, Car and coworkers observed the occurrence of fault lines or cracks in graphene oxide (Figure 3.20a) [137]. To explain the formation of these cracks, they proposed an unzipping mechanism that is initiated by the strain generated by the coefficient alignment of the epoxy groups on the graphene lattice (Figure 3.20b). Their model explained the initial stage of the unzipping process of C—C bond underneath the epoxy groups. A subsequent study by Yang and coworkers further refined the unzipping mechanism by introducing epoxy pair or carbonyl pair into the linear epoxy chain involving both sides of the graphene layer (Figure 3.20c) [138]. According to their calculation results, for a short epoxy chain,

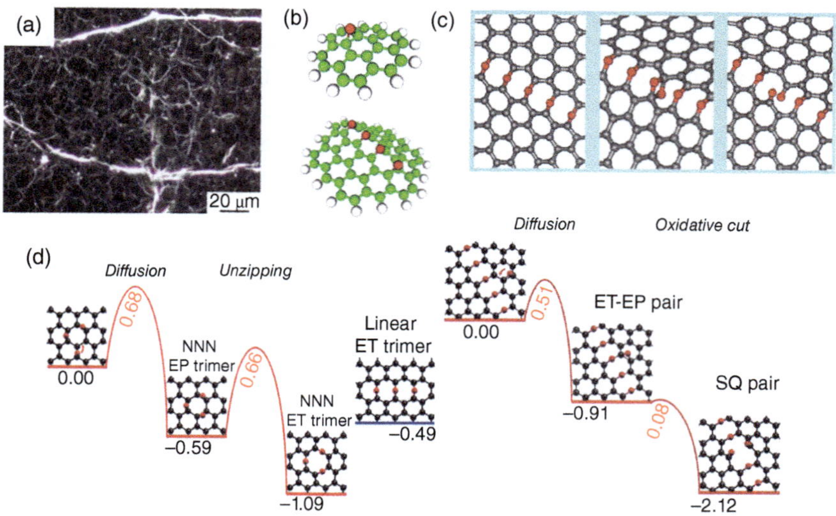

Figure 3.20 (a) Optical microscope image of cracks developing in partially oxidized graphite [137]/with permission of American Physical Society. (b) Schematic of a chain of ether groups on the graphitic lattice. Source: Li et al. [137]/with permission of American Physical Society. (c) A proposed mechanism of C—C bond cleavage in the ether chains. Source: Li et al. [138]/with permission of American Chemical Society. (d) DFT study on the formation of ether chains starting from diffusing epoxy groups which ultimately leads to bond breakage (energy reported in eV). Source: Sun and Fabris [139]/with permission of American Chemical Society.

the energy difference for forming a new epoxy pair and adding an epoxy group to extend the existed epoxy chain is approximately 0.1 eV. They concluded that new epoxy pairs could form during the growth of the existing epoxy chain. Later, Fabris and Sun proposed an integrated theory that includes epoxy and ether groups in the unzipping mechanism (Figure 3.20d) [139]. Briefly, they elucidated that the oxidized carbon network is strongly influenced by the constraint in the graphitic lattice induced by the surface functional groups. According to their model, the energetics of an oxygen adsorbate is essential for yielding the oxidative cutting of the graphene lattice (Figure 3.12f).

Based on the mechanism proposed by the DFT studies, the process of ozone etching of graphene starts with the formation of epoxy groups, followed by epoxy clusters that subsequently transform into ether clusters, and finally, the bond breakage takes place driven by the strain in the lattice, leading to the formation of two edge semiquinone groups. The gasification step then involves a further transformation of the semiquinone group into lactone and gasification into CO and CO_2 with an energy barrier of 1.1 eV. The attractive part of this process is that the energy barrier for chemisorption of O_3 and the formation of vacancy defects are close to 1 eV, and as a result, pore nucleation and pore expansion are expected to take place at a comparable rate, especially when compared to those in the physical etching routes.

On the experimental front, our group has reported ozone functionalization-based pore etching and pore-edge functionalization techniques, where the separation performance of N-SLG membranes could be optimized [52, 57, 64, 140–143]. Huang et al. prepared graphene membranes by transferring CVD graphene on a laser-drilled W foil hosting 5-μm apertures and correlated the gas permeance from the as-prepared film to the intrinsic vacancy defects in the CVD graphene [64]. Subsequently, H_2 permeance could be increased by three-fold an *in situ* treatment of the graphene membrane to O_3 in the temperature range of 80–100 °C (Figure 3.21a). This study indicated that O_3 can etch graphene at a temperature as low as 80 °C. The same study demonstrated that when O_3 is exposed to graphene near room temperature (e.g. 25 °C), there is in fact, a decrease in permeance. This can be attributed to the functionalization of the existing pore edges (likely carbonyl and ether groups) by O_3 reducing the electron-density gap in the pore. Indeed, a 1.5-fold increase in H_2/CH_4 selectivity was observed post-functionalization.

Zhao et al. developed a synergetic O_2 plasma/O_3 etching technique to successfully incorporate a high pore density in graphene for H_2/CH_4 and H_2/C_3H_8 separation, which corresponds to a resolution of 1 Å in molecular differentiation, i.e. given the difference in the kinetic diameters of H_2 (2.89 Å) and CH_4 (3.8 Å) [143]. First, 1s oxygen plasma treatment was used to nucleate defects. Subsequently, O_3 etching experiments at 60 and 150 °C were performed to enlarge the pore. Attractive H_2/CH_4 and H_2/C_3H_8 selectivities (30 and 207, respectively) combined with high H_2 permeance (~2600 GPU) could be achieved. An interesting outcome of this study was the trend in the pore density (permeance) and the concentration of nonselective pores derived from selectivity (Figure 3.21b,c). They showed that plasma treatment always leads to a tradeoff between pore density and the percentage of nonselective pores. For example, H_2/CH_4 selectivity obtained from intrinsic vacancy defects in graphene

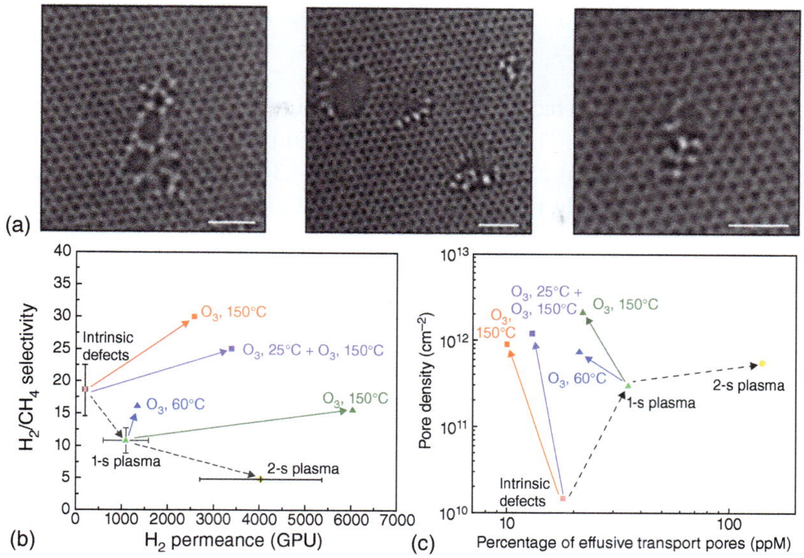

Figure 3.21 (a) HRTEM images of nanopores incorporated in graphene after O_3 etching treatment (2 min at 80 °C). The scale bar is 1 nm. Source: Huang et al. [64]/with permission from Springer Nature. (b) H_2/CH_4 selectivity plotted against H_2 permeance from various graphene membranes prepared by either the plasma etching or by the O_3 treatment or by the combination of both, and (c) corresponding evolution of pore density and the nonselective pores. Zhao, J et al. [143]/with permission of American Association for the Advancement of Science.

always reduced when new nanopores were incorporated by the plasma treatment evident from the increase in the permeance. The same decreasing trend in selectivity was observed when the plasma time was increased from 1 to 2 s. In sharp contrast, when O_3 was used to incorporate nanopores in graphene, an increase in selectivity, as well as H_2 permeance, was observed. This was true even when ozone treatment was carried out after initial plasma treatment. These trends point out to contrasting relative rates of pore nucleation and pore expansion for oxidative plasma and O_3 treatment. In the case of plasma, as discussed above, the highly energetic ions can knockout edge carbon atoms by bombardment, thereby increasing the rate of pore expansion. In contrast, during O_3 exposure, pore expansion would only take place by the gasification of the edge semiquinone groups that has a similar energy barrier to that of chemisorption of ozone (~1 eV).

Recently, our group has reported a fine control of reaction between O_3 and graphene by designing a millisecond gasification reactor that allows to limit the reaction to only a few milliseconds, while the etching can be carried out at a high temperature, e.g. between 250 and 300 °C [52, 142]. This high-temperature etching approach improves pore density given the positive energy barrier for O_3 chemisorption. A short millisecond reaction ensures that the pore expansion events are controlled. The millisecond timescale for reaction at 250 °C is evident from the energy barrier ($E_{desorption}$=1.1 eV) for CO and CO_2 desorption when one calculates the rate constant, k, for desorption as a function of temperature, T.

$$k = A\exp(-E_{desorption}/RT) \tag{3.8}$$

where A is the pre-exponential factor for the exponential relationship ($10^{13}\,s^{-1}$) and R is the universal gas constant. The above expression points to an approximate rate of 284 desorption per second at 250 °C. Therefore, pore-13 requiring 13 desorption events would form quite rapidly, in only ~50 ms at 250 °C.

The MGR is composed of a heated chamber connected to an O_3 reservoir and a purge gas stream via two synchronized millisecond leak valves (MLV) (Figure 3.22a). Graphene etching is carried out by opening the MLV for 10–200 ms, which leads to a short O_3 pulse (Figure 3.22b). The overall exposure to O_3 was controlled by varying the MLV opening time (τ), O_3 supply pressure (P_{up}), and purge delay time (t_d). The increasing level of etching of graphene (Figure 3.22d) in MGR is reflected by an increase in the defect density observed by the Raman spectroscopy (Figure 3.22c) and an increase in the H_2 permeance as a function of O_3 exposure (Figure 3.22e). The optimized conditions led to the realization of high CO_2 permeance of 2620 GPU and attractive CO_2/N_2 and CO_2/CH_4 selectivities of 27.6 and 20.0.

One can fine-tune the ozone exposure conditions to maximize pore density while minimizing the pore expansion. Recently, we showed that one can further improve the control over etching by increasing the O_3 pressure by a multi-pulsed dosing approach (Figure 3.22f) [57]. This is mainly because the nucleation rate is proportional to the concentration of the epoxy groups grafted on graphene, which in turn is proportional to the O_3 pressure, P_{O_3}. In the case of pore expansion, O_3 can directly attach to the edge of the nanopore where more than one oxygen functional group can form at the pore edge. As a result, the expansion rate constant is proportional to P^n, where $n < 1$. This study showed that when n assumes a value of 0.5 for n, the experimentally observed PSD (via AC-HRTEM) can be predicted reasonably well. The study led to important conclusions, which are as follows:

(i) Increasing P_{O_3} rather than the etching temperature is a better strategy to obtain a narrow PSD. This is because increasing the reaction temperature would proportionately accelerate the kinetics of pore nucleation as well as pore expansion, whereas increasing P_{O_3} would favor nucleation over pore expansion.
(ii) One can effectively slow down pore expansion by increasing pore density as the O_3 available for reaction would be consumed by a higher number of pores. This, of course, is valid for a regime where the availability of O_3 limits the reaction.

Based on these observations, a mathematical model was developed to predict the PSD for a given O_3 exposure profile. Several exposure profiles were screened to arrive at an optimal profile based on multi-pulsed O_3 exposure leading to attractive CO_2 permeance of 4400 ± 2070 GPU and CO_2/N_2 selectivity of 33.4 ± 7.9.

3.4 Fabrication of Large-Area N-SLG Membranes

Nowadays large area SLG can be reproducibly and routinely produced by CVD using low-cost carbon precursors, such as methane [144]. The worldwide production capacity of graphene films increased to 3.7 million m^2 in 2017 [145]. Given that high-quality N-SLG membranes would require high-quality SLG, marked by a low density of intrinsic defects and low concentration of multilayer graphene domains, it is likely that the production process of SLG would require rigorous quality control. A proof-of-concept for the production of large-area high-quality graphene in the CVD furnace has been demonstrated by rolling or stacking Cu foil in CVD furnace [146–148]. This can be further optimized for rapid membrane production. For example, in the stacked approach where the Cu foils are directly placed on top of each other using a spacer made of either graphite or a refractory metal foil (Figure 3.23) [149], or where Cu foil can be stacked without using a spacer, one can produce approximately 20 m^2 graphene film from a typically available large high-temperature furnace (8 in. in diameter, 1-m-long heating zone) in a single batch with a synthesis time of one day. The area of membranes produced from such furnaces is directly proportional to the cross-sectional area of the

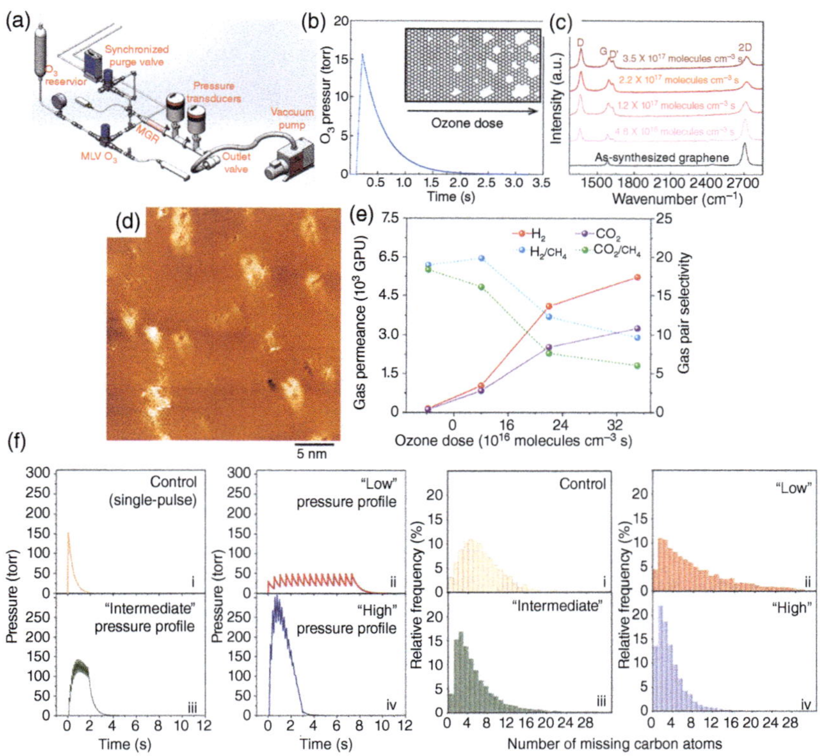

Figure 3.22 (a) Schematic of the MGR setup for etching graphene in millisecond timescale. (b) The shape profile of a single O_3 pulse in MGR. (c) The evolution of defect density in graphene as a function of O_3 dose. (d) Oxygen-functionalized pores observed by STM. (e) The evolution of gas permeance as a function of O_3 dose. Source: Huang et al. [52]/with permission from American Association for the Advancement of Science – AAAS. (f) Improvement in PSD when using multi-pulsed O_3 with larger peak pressure compared to single pulse O_3. Source: Hsu, K et al. [57]/with permission of American Chemical Society.

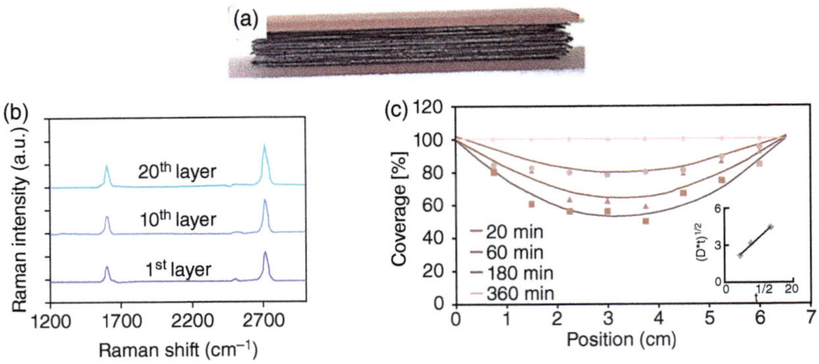

Figure 3.23 (a) Stacking of Cu foils to improve packing density of foils inside the CVD furnace. (b) Raman data from various positions of Cu foil in the stack. (c) Graphene coverage in the stack as a function of position and growth time. Source: Hsieh et al. [149]/with permission from American Chemical Society.

CVD furnace, and custom-designed industrial furnaces can produce 1000 m^2 SLG in a single batch.

The biggest challenge in the scale-up of N-SLG membrane is the crack-free transfer of SLG film from the metal foil on a porous support [150]. Even a few ppm concentration of nanoscale cracks completely ruins the gas selectivity from N-SLG membranes attributing to the fact that such cracks are nonselective and effusive flux from such pores is several orders of magnitude higher than the flux from the selective pore with a comparable area. The cracks tend to form during the transfer process because of the limited mechanical robustness of atom-thick film despite the fact that graphene is one of the strongest materials with an elastic modulus of 1 TPa. The crack formation takes place when the strain on the graphene lattice during transfer to the porous support exceeds the maximum strain (0.4%) [33, 151]. For example, a common method to transfer graphene is wet-transfer, where a mechanical reinforcement (e.g. polymer film) is coated on graphene [150]. However, when the polymer film is removed by dissolution after the transfer, the capillary forces induced by solvent drying tend to crack graphene [152]. Yet another challenge comes from the fact that even if one manages to prepare a crack-free film, the film needs to survive the pressurization needed for membrane-based gas separation. To understand the limitations of pressurization, one can refer to the generation of strain, ε, as a function of pressure difference across graphene, ΔP, and the size of the suspended domain, d, as shown in Eq. (3.9) [153]:

$$\varepsilon = \left(\frac{2}{27}\right)^{1/3}\left(\frac{d\cdot\Delta P}{\kappa(\nu)E_m t}\right)^{2/3} \tag{3.9}$$

where $\kappa(\nu)$ is a function of graphene's Poisson's ration, ν. E_m and t are Young's modulus and the thickness of graphene, respectively. This equation indicates that one needs to support graphene on the length scale of 100–500 nm, i.e. on supports with pore size of 100–500 nm for pressurization around 10 bar. Larger support pores will not prevent crack propagation in graphene upon pressurization.

To prevent cracks during transfer, several strategies have been investigated focusing on minimizing the capillary force driving the tearing of the graphene film. Celebi et al. fabricated a 2500-μm^2-sized bilayer graphene membrane by transferring two consecutive SLG layers onto a punctured SiN_X frame [80]. To avoid crack formation during the transfer, they first mechanically reinforced SLG with poly(methyl methacrylate) or PMMA, etched the Cu, and then scooped the floating SLG using another as-synthesized SLG resting on a Cu foil. This way, the second SLG layer reinforced the first SLG layer. After etching nanopores with an ion beam, they could successfully study gas transport from few-nanometer-sized nanopores operating in the effusive regime.

Boutilier et al. investigated the possibility of a defect-tolerant membrane by stacking SLG layers on a porous polycarbonate track-etched membrane (PCTEM) support as a composite graphene membrane [154]. While they could not avoid cracks in an individual layer, they noted that the flow of helium through the cracks decreases exponentially by increasing the number of stacking layers. Further, they noted that the pores of the support membrane must be smaller than the average distance between defects to isolate nonselective flow to only a small fraction of the membrane. They further discussed the relative resistance of gas transport from the porous support and the SLG film. Briefly, the support resistance should be comparable to that from the graphene film. If the resistance of the support is too high, the selectivity is expected to decrease because the nonselective pores of the support would limit the flow. Later, Boutilier et al. reported a direct-transfer approach for SLG by pressing the top surface of floating SLG without any reinforcing layer on PCTEM [152]. While the method led to a number of cracks in the film, they were able to reduce the impact of leakage from the cracks on the membrane selectivity by choosing appropriate porous support. By reducing the pore diameter of the support, the effective transport from the cracks decreased. For example, by using anodic aluminum oxide as porous support with a pore diameter of 20 nm, they reported H_2/CO_2 selectivity of ~7, exceeding the Knudsen selectivity of 4.7.

The recent reports involving a mechanical reinforcement of the SLG film using nanoporous supports that can be left behind on the graphene film have been more effective in avoiding cracks and tears [64, 96], and can evolve in a generic method to prepare high-performance N-SLG membranes. For this concept to work, the reinforcing material should have the following properties, (i) it must conformally adhere to graphene making sure that the suspended graphene domains do not exceed length scale of 100–500 nm, (ii) the layer must not limit the gas transport across graphene, and finally (iii) the layer should be chemically and thermally robust and should not limit the application of graphene for gas separation. Huang et al. demonstrated the first proof-of-concept for this approach with a novel nanoporous carbon (NPC) film-assisted transfer technique (Figure 3.24a) [64]. NPC film was prepared by a simple method, involving coating a block-copolymer (polystyrene-*co*-poly(4-vinyl pyridine) templated assembly of turanose, where turanose hydrogen bonds with the hydrophilic block. This was followed by pyrolysis at 500 °C, where the polymer is degraded and the carbon obtained from sugar forms a porous lamellar morphology with a pore opening of 20–30 nm at the top and the bottom surfaces (Figure 3.24b,c)

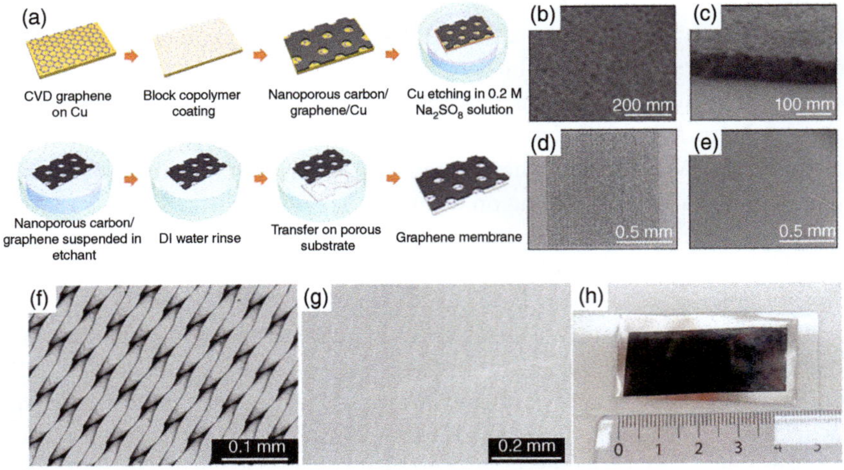

Figure 3.24 (a) Schematic of the NPC-film-based mechanical reinforcement of graphene for crack-free transfer to a perforated metal foil support. (b) Top-view SEM image of the NPC film. (c) Cross-sectional SEM image of the NPC film. (d) The perforated W foil support. (e) NPC-reinforced graphene transferred to the perforated W foil. Source: Huang et al. [64]/with permission from Springer Nature. (f) Schematic of dual carbon film (NPC and MWNT) reinforcement of graphene. (g) SEM image of low-cost metal mesh support. (h) Metal mesh support covered by crack-free graphene film using the dual carbon film reinforcement approach. (i) and (g) Picture of the centimeter-scale graphene membranes. Source: Lee et al. [140]/with permission of Elsevier.

[141]. The gas transport resistance of the NPC film was negligible (H_2 permeance up to 1 million GPU), while the carbon framework imparted the film a high chemical and thermal stability. This allowed for the fabrication of crack-free membrane on a perforated W foil with a total area of 1 mm^2 (Figure 3.24d,e). The high stability of the NPC film (crack-free up to 500 °C) allows one to study gas transport from graphene membranes to high temperatures. The low transport resistance of NPC film ensures that the gas transport from graphene is not impeded.

Recently, Lee et al. demonstrated the fabrication of centimeter-scale membranes on low-cost support by using a dual-layer reinforcement approach, where the NPC film is further mechanically reinforced by coating a film of multi-walled carbon nanotubes (MWNTs) [140]. This allowed to fabricate centimeter-scale membranes on low-cost stainless steel mesh hosting approximately 20-μm pore opening (Figure 3.24f,g), which was not possible using the standalone NPC reinforcement. The resulting centimeter-scale graphene membrane (Figure 3.24h) could be pressurized on macroporous mesh support with a pore opening of 20 μm to up to a transmembrane pressure difference of four bar and demonstrated attractive H_2/CH_4 and H_2/CO_2 selectivities of 11–23 and 5–8, respectively, higher than the corresponding Knudsen selectivities of 2.8 and 4.7, respectively.

Yang et al. succeeded in transferring SLG by using a CVD-derived single-walled carbon nanotube (SWNT) network as the mechanical reinforcement [96]. Based on the strong interaction between the SWNT and the graphene layer, a relatively large area of 0.36 cm^2 could be successfully transferred without inducing cracks. The hybrid SWCNT–graphene membrane could be pressurized for desalination application; however, its application in gas separation has not been demonstrated. Interestingly, in the study of Lee et al., just reinforcing N-SLG film by a standalone nanotube film also seemed to form crack-free films; however, gas-sieving behavior was not observed from such films indicating the presence of nanoscale cracks in the graphene film when NPC reinforcement is not used.

There have been also several reports on the transfer of graphene using a porous polymer film, where the polymer film is directly fabricated on graphene using nonsolvent-induced phase separation (NIPS) technique, a method of choice for the fabrication of polymeric membranes [155, 156]. In this case, the graphene film is supported by the bottom part of the polymeric film, where the pore size in the polymer film is typically between 100 nm to a few microns. It is desirable to reduce this pore size to a few nanometers to improve the mechanical reinforcement of graphene. Currently, this method has not been successful in preventing cracks in graphene membranes, and as a result, crack-free gas-separation membranes have not been reported by this method despite progress on optimization of such films on graphene [157, 158].

Another attractive route to prepare reinforcement of gas separation is the use of nanoporous gas-permeable polymeric film that can be left behind on the graphene film. There are several high-permeability polymer films that can be used for this, e.g. poly[1-(trimethylsilyl)-1-propyne] or PTMSP, PIMs, and Teflon. He et al. were the first to demonstrate this idea by successfully transferring SLG on macroporous support by using a thin film of PTMSP or PIM-1. An added advantage of this approach is that polymer film masks graphene nanopores and forces the gas transport to take place in adsorbed-phase by completely eliminating the direct gas-phase transport. This reduces the contribution of the nonselective pores in several cases—(i) when the gas of interest has a higher solubility in the polymer, e.g. in the case of separation of CO_2 from N_2, and (ii) when the permeance of the target gas through the polymeric layer matches that through the nanopore. In this case, the transport from the nonselective pore would be limited by the permeance of polymer film and is expected to help improve the gas-pair selectivity.

3.5 Summary and Outlook

Overall, N-SLG-based gas-separation membranes have seen rapid progress in the last decade, thanks to the improved understanding of gas transport through nanopores varying in geometry and functionality, advances in the methods to incorporate gas-sieving nanopores, and finally development of techniques that allow pressurization of suspended graphene films. Nonetheless, there are still numerous challenges in realizing the ultimate promise of N-SLG membranes for gas separation, i.e. ultrahigh permeance combined with an attractive gas-pair selectivity from membranes prepared in a scalable manner. These challenges can be overcome by:

(i) An improved understanding of the etching mechanism and evolution of PSD in graphene, which ideally results in decoupled nucleation and pore expansion routes. *In situ* studies using nondestructive characterization techniques, e.g. scanning tunneling microscope, probing events, and structures, that lead to etching will be crucial for this. Systematic studies of etching kinetics will likely give way to techniques where pore expansion can be controlled to the extent of removing one carbon atom at a time.
(ii) An improved understanding of the role of PSD in the resulting selectivity. Development of robust models that can predict gas permeance from an ensemble of nanopores without requiring computationally demanding ab initio or MD simulations will be very effective toward this goal.
(iii) The development of low-cost scalable reinforcing films that prevent cracks and tears in graphene during fabrication but also during membrane operation with large transmembrane pressure difference (tens of bars).
(iv) Improved understanding of stability of graphene nanopores under realistic operating conditions.

A better understanding of gas transport and etching mechanism will help researchers to tune the nucleation density (pore density) and growth rate (pore size) for the desired gas separation. Availability of large-area membranes will spur applied research increasing the technology readiness level of graphene membranes for a wide number of separation applications.

References

1 National Academies of Sciences and Medicine (2019). *A Research Agenda for Transforming Separation Science*. Washington, DC: National Academies Press.
2 Feron, P.H.M. (2009). The potential for improvement of the energy performance of pulverized coal fired power stations with post-combustion capture of carbon dioxide. *Energy Procedia* 1 (1): 1067–1074.
3 Washington, D.T.N.A.P. (ed.) (2019) Nationnal Academies of Sciences, *Engineering and Medecine* 2019, in *A research Agenda for Transforming Separation Science*.
4 Khalilpour, R., Mumford, K., Zhai, H. et al. (2015). Membrane-based carbon capture from flue gas: A review. *J. Clean. Prod.* 103: 286–300.
5 Yang, H., Xu, Z., Fan, M. et al. (2008). Progress in carbon dioxide separation and capture: a review. *J. Environ. Sci.* 20 (1): 14–27.
6 Mondal, M.K., Balsora, H.K., and Varshney, P. (2012). Progress and trends in CO_2 capture/separation technologies: a review. *Energy* 46 (1): 431–441.
7 Gin, D.L. and Noble, R.D. (2011). Designing the next generation of chemical separation membranes. *Science* 332 (6030): 674–676.

8 Baker, R.W. (2002). Future directions of membrane gas separation technology. *Ind. Eng. Chem. Res.* 41 (6): 1393–1411.
9 Koros, W.J. and Fleming, G.K. (1993). Membrane-based gas separation. *J. Memb. Sci.* 83: 1–80.
10 Sholl, D.S. and Lively, R.P. (2016). Seven chemical separations to change the world. *Nature* 532: 435.
11 Bernardo, P., Drioli, E., and Golemme, G. (2009). Membrane gas separation: 1 review of state of the art. *Ind. Chem. Eng.* 48 (1): 4638–4663.
12 LOEB, S. and SOURIRAJAN, S. (1963). Sea water demineralization by means of an osmotic membrane. In: *Saline Water Conversion—II*, vol. 38, 117–132 SE–9. American Chemical Society.
13 Robeson, L.M. (2008). The upper bound revisited. *J. Memb. Sci.* 320 (1–2): 390–400.
14 Freeman, B.D. (1999). Basis of permeability/selectivity tradeoff relations in polymeric gas separation membranes. *Macromolecules* 32 (2): 375–380.
15 Budd, P.M., Elabas, E.S., Ghanem, B.S. et al. (2004). Solution-processed, organophilic membrane derived from a polymer of intrinsic microporosity. *Adv. Mater.* 16 (5): 456–459.
16 Park, H.B., Kamcev, J., Robeson, L.M. et al. (2017). Maximizing the right stuff: the trade-off between membrane permeability and selectivity. *Science* 356 (6343): eaab0530.
17 Bux, H., Liang, F., Li, Y. et al. (2009). Zeolitic imidazolate framework membrane with molecular sieving properties by microwave-assisted solvothermal synthesis. *J. Am. Chem. Soc.* 131 (44): 16000–16001.
18 Zimmerman, C.M., Singh, A., and Koros, W.J. (1997). Tailoring mixed matrix composite membranes for gas separations. *J. Memb. Sci.* 137 (1–2): 145–154.
19 Kim, W.G., Lee, J.S., Bucknall, D.G. et al. (2013). Nanoporous layered silicate AMH-3/cellulose acetate nanocomposite membranes for gas separations. *J. Memb. Sci.* 441: 129–136.
20 Babu, D.J., He, G., Hao, J. et al. (2019). Restricting lattice flexibility in polycrystalline metal–organic framework membranes for carbon capture. *Adv. Mater.* 31 (28): 1900855.
21 Hao, J., Babu, D.J., Liu, Q. et al. (2021). Mechanistic study on thermally induced lattice stiffening of ZIF-8. *Chem. Mater.* 33 (11): 4035–4044.
22 Liu, G., Chernikova, V., Liu, Y. et al. (2018). Mixed matrix formulations with MOF molecular sieving for key energy-intensive separations. *Nat. Mater.* 17 (March): 1–7.
23 Zhang, C. and Koros, W.J. (2017). Ultraselective carbon molecular sieve membranes with tailored synergistic sorption selective properties. *Adv. Mater.* 1701631: 1–6.
24 Aroon, M.A., Ismail, A.F., Matsuura, T., and Montazer-Rahmati, M.M. (2010). Performance studies of mixed matrix membranes for gas separation: a review. *Sep. Purif. Technol.* 75 (3): 229–242.

25 Vu, D.Q., Koros, W.J., and Miller, S.J. (2003). Mixed matrix membranes using carbon molecular sieves: I. Preparation and experimental results. *J. Memb. Sci.* 211 (2): 311–334.

26 Micari, M., Dakhchoune, M., and Agrawal, K.V. (2021). Techno-economic assessment of postcombustion carbon capture using high-performance nanoporous single-layer graphene membranes. *J. Memb. Sci.* 624: 119103.

27 Micari, M. and Agrawal, K.V. (2022). Oxygen enrichment of air: Performance guidelines for membranes based on techno-economic assessment. *J. Memb. Sci.* 641: 119883.

28 Novoselov, K.S., Geim, A.K., Morozov, S.V. et al. (2004). Electric field effect in atomically thin carbon films. *Science* 306 (5696): 666–669.

29 Li, X., Cai, W., An, J. et al. (2009). Large-area synthesis of high-quality and uniform graphene films on copper foils. *Science* 324 (5932): 1312–1314.

30 Jiang, D., Cooper, V.R., and Dai, S. (2009). Porous graphene as the ultimate membrane for gas separation. *Nano Lett.* 9 (12): 4019–4024.

31 Koenig, S.P., Wang, L., Pellegrino, J., and Bunch, J.S. (2012). Selective molecular sieving through porous graphene. *Nat. Nanotechnol.* 7 (11): 728–732.

32 Geim, A.K. and Novoselov, K.S. (2007). The rise of graphene. *Nat. Mater.* 6 (3): 183–191.

33 Cohen-Tanugi, D. and Grossman, J.C. (2014). Mechanical strength of nanoporous graphene as a desalination membrane. *Nano Lett.* 14 (11): 6171–6178.

34 Geim, A.K. (2009). Graphene: status and prospects. *Science* 324 (5934): 1530–1534.

35 Lee, C., Wei, X., Kysar, J.W., and Hone, J. (2008). Measurement of the elastic properties and intrinsic strength of monolayer graphene. *Science* 321 (July): 385–388.

36 McEuen, P.L., Bunch, J.S., Craighead, H.G. et al. (2008). Impermeable atomic membranes from graphene sheets. *Nano Lett.* 8 (8): 2458–2462.

37 Sun, P.Z., Yang, Q., Kuang, W.J. et al. (2020). Limits on gas impermeability of graphene. *Nature* 579 (7798): 229–232.

38 Agrawal, K.V., Drahushuk, L.W., and Strano, M.S. (2016). Observation and analysis of the Coulter effect through carbon nanotube and graphene nanopores. *Philos. Trans. R. Soc. A* 374 (2060): 20150357.

39 Yuan, Z., Govind Rajan, A., Misra, R.P. et al. (2017). Mechanism and prediction of gas permeation through sub-nanometer graphene pores: comparison of theory and simulation. *ACS Nano* 11 (8): 7974–7987.

40 Sun, C., Boutilier, M.S.H., Au, H. et al. (2014). Mechanisms of molecular permeation through nanoporous graphene membranes. *Langmuir* 30 (2): 675–682.

41 Yuan, Z., Misra, R.P., Rajan, A.G. et al. (2019). Analytical prediction of gas permeation through graphene nanopores of varying sizes: understanding transitions across multiple transport regimes. *ACS Nano* 13 (10): 11809–11824.

42 Wang, L., Boutilier, M.S.H., Kidambi, P.R. et al. (2017). Fundamental transport mechanisms, fabrication and potential applications of nanoporous atomically thin membranes. *Nat. Nanotechnol.* 12 (6): 509–522.

43 Blankenburg, S., Bieri, M., Fasel, R., and M??llen, K., Pignedoli, C.A., and Passerone, D. (2010). Porous graphene as an atmospheric nanofilter. *Small* 6 (20): 2266–2271.

44 Schrier, J. (2012). Carbon dioxide separation with a two-dimensional polymer membrane. *ACS Appl. Mater. Interfaces* 4 (7): 3745–3752.

45 Tao, Y., Xue, Q., Liu, Z. et al. (2014). Tunable hydrogen separation in porous graphene membrane: first-principle and molecular dynamic simulation. *ACS Appl. Mater. Interfaces* 6: 8048–8058.

46 Sun, C., Wen, B., and Bai, B. (2015). Application of nanoporous graphene membranes in natural gas processing: molecular simulations of CH_4/CO_2, CH_4/H_2S and CH_4/N_2 separation. *Chem. Eng. Sci.* 138: 616–621.

47 Du, H., Li, J., Zhang, J. et al. (2011). Separation of hydrogen and nitrogen gases with porous graphene membrane. *J. Phys. Chem. C* 115 (47): 23261–23266.

48 Drahushuk, L.W. and Strano, M.S. (2012). Mechanisms of gas permeation through single layer graphene membranes. *Langmuir* 28 (48): 16671–16678.

49 He, G., Huang, S., Villalobos, L.F. et al. (2020). Synergistic CO_2 -sieving from polymer with intrinsic microporosity masking nanoporous single-layer graphene. *Adv. Funct. Mater.* 30 (39): 2003979.

50 Raghavan, B. and Gupta, T. (2017). H_2/CH_4 gas separation by variation in pore geometry of nanoporous graphene. *J. Phys. Chem. C* 121 (3): 1904–1909.

51 Govind Rajan, A., Silmore, K.S., Swett, J. et al. (2019). Addressing the isomer cataloguing problem for nanopores in two-dimensional materials. *Nat. Mater.* 18 (2): 129–135.

52 Huang, S., Li, S., Villalobos, L.F. et al. (2021). Millisecond lattice gasification for high-density CO_2- and O_2-sieving nanopores in single-layer graphene. *Sci. Adv.* 7 (9): eabf0116.

53 Liu, H., Dai, S., and Jiang, D. (2013). Insights into CO_2/N_2 separation through nanoporous graphene from molecular dynamics. *Nanoscale* 5 (20): 9984.

54 Tian, Z., Mahurin, S.M., Dai, S., and Jiang, D. (2017). Ion-gated gas separation through porous graphene. *Nano Lett.* 17 (3): 1802–1807.

55 Guo, W., Mahurin, S.M., Unocic, R.R. et al. (2020). Broadening the gas separation utility of monolayer nanoporous graphene membranes by an ionic liquid gating. *Nano Lett.* .

56 He, G., Huang, S., Villalobos, L.F. et al. (2019). High-permeance polymer-functionalized single-layer graphene membranes that surpass the postcombustion carbon capture target. *Energy Environ. Sci.* 12 (11): 3305–3312.

57 Hsu, K., Villalobos, L.F., Huang, S. et al. (2021). Multipulsed millisecond ozone gasification for predictable tuning of nucleation and nucleation-decoupled nanopore expansion in graphene for carbon capture. *ACS Nano* 15 (8): 13230–13239.

58 Agrawal, K.V., Benck, J.D., Yuan, Z. et al. (2017). Fabrication, pressure testing, and nanopore formation of single-layer graphene membranes. *J. Phys. Chem. C* 121 (26): 14312–14321.

59 Kidambi, P.R., Nguyen, G.D., Zhang, S. et al. (2018). Facile fabrication of large-area atomically thin membranes by direct synthesis of graphene with nanoscale porosity. *Adv. Mater.* 30 (49): 1804977.

60 Yuan, Z., He, G., Faucher, S. et al. (2021). Direct chemical vapor deposition synthesis of porous single-layer graphene membranes with high gas permeances and selectivities. *Adv. Mater.* 2104308: 1–9.

61 Villalobos, L.F., Van Goethem, C., Hsu, K.-J. et al. (2021). Bottom-up synthesis of graphene films hosting atom-thick molecular-sieving apertures. *Proc. Natl. Acad. Sci.* 118 (37): e2022201118.

62 Bieri, M., Treier, M., Cai, J. et al. (2009). Porous graphenes: two-dimensional polymer synthesis with atomic precision. *Chem. Commun. (Camb).* 45: 6919–6921.

63 Ivanovskii, A.L. (2013). Graphynes and graphdyines. *Prog. Solid State Chem.* 41 (1–2): 1–19.

64 Huang, S., Dakhchoune, M., Luo, W. et al. (2018). Single-layer graphene membranes by crack-free transfer for gas mixture separation. *Nat. Commun.* 9: 2632.

65 Huang, P.Y., Ruiz-Vargas, C.S., van der Zande, A.M. et al. (2011). Grains and grain boundaries in single-layer graphene atomic patchwork quilts. *Nature* 469 (7330): 389–392.

66 Khan, M.H., Moradi, M., Dakhchoune, M. et al. (2019). Hydrogen sieving from intrinsic defects of benzene-derived single-layer graphene. *Carbon N. Y.* 153: 458–466.

67 Cançado, L.G., Jorio, A., Ferreira, E.H.M. et al. (2011). Quantifying defects in graphene via Raman spectroscopy at different excitation energies. *Nano Lett.* 11 (8): 3190–3196.

68 Moreno, C., Vilas-Varela, M., Kretz, B. et al. (2018). Bottom-up synthesis of multifunctional nanoporous graphene. *Science* 360 (6385): 199–203.

69 Solís-Fernández, P., Paredes, J.I., Cosío, A. et al. (2010). A comparison between physically and chemically driven etching in the oxidation of graphite surfaces. *J. Colloid Interface Sci.* 344 (2): 451–459.

70 Smith, B.W. and Luzzi, D.E. (2001). Electron irradiation effects in single wall carbon nanotubes. *J. Appl. Phys.* 90 (7): 3509–3515.

71 Kotakoski, J., Santos-Cottin, D., and Krasheninnikov, A.V. (2012). Stability of graphene edges under electron beam: equilibrium energetics versus dynamic effects. *ACS Nano* 6 (1): 671–676.

72 Russo, C.J. and Golovchenko, J.A. (2012). Atom-by-atom nucleation and growth of graphene nanopores. *Proc. Natl. Acad. Sci. U. S. A.* 109 (16): 5953–5957.

73 Lehtinen, O., Kotakoski, J., Krasheninnikov, A.V. et al. (2010). Effects of ion bombardment on a two-dimensional target: atomistic simulations of graphene irradiation. *Phys. Rev. B - Condens. Matter Mater. Phys.* 81 (15): 1–4.

74 Fischbein, M.D. and Drndić, M. (2008). Electron beam nanosculpting of suspended graphene sheets. *Appl. Phys. Lett.* 93 (11): 113107-1–113107-3.

75 Girit, Ç.Ö., Meyer, J.C., Erni, R. et al. (2009). Graphene at the edge: stability and dynamics. *Science* 323 (March): 1705–1708.

76 Garaj, S., Hubbard, W., Reina, A. et al. (2010). Graphene as a subnanometre trans-electrode membrane. *Nature* 467 (7312): 190–193.

77 Merchant, C.a., Healy, K., Wanunu, M. et al. (2010). DNA translocation through graphene nanopores. *Nano Lett.* 10 (8): 2915–2921.

78 Schneider, G.F., Kowalczyk, S.W., Calado, V.E. et al. (2010). DNA translocation through graphene nanopores. *Nano Lett.* 10 (8): 3163–3167.

79 Ahlberg, P., Johansson, F.O.L., Zhang, Z.B. et al. (2016). Defect formation in graphene during low-energy ion bombardment. *APL Mater.* 4 (4): 046104-1–046104-6.

80 Celebi, K., Buchheim, J., Wyss, R.M. et al. (2014). Ultimate permeation across atomically thin porous graphene. *Science* 344 (6181): 289–292.

81 Chu, X. and Schmidt, L.D. (1992). Reactions of NO, O_2, H_2O, and CO_2 with the basal plane of graphite. *Surf. Sci.* 268 (1–3): 325–332.

82 Hughes, G., Williams, B.R., and Thomas, J.M. (1962). Etching of graphite surfaces with oxygen. *Trans. Faraday Soc.* 58: 2011–2016.

83 Hug, D., Zihlmann, S., Rehmann, M.K. et al. (2017). Anisotropic etching of graphite and graphene in a remote hydrogen plasma. *NPJ 2D Mater. Appl.* 1 (1): 1–5.

84 O'Hern, S.C., Boutilier, M.S.H., Idrobo, J.C. et al. (2014). Selective ionic transport through tunable subnanometer pores in single-layer graphene membranes. *Nano Lett.* 14 (3): 1234–1241.

85 Cuesta, A., Martínez-Alonso, A., and Tascón, J.M.D. (2001). Carbon reactivity in an oxygen plasma: a comparison with reactivity in molecular oxygen. *Carbon N. Y.* 39 (8): 1135–1146.

86 Al-Mumen, H., Rao, F., Li, W., and Dong, L. (2014). Singular sheet etching of graphene with oxygen plasma. *Nano-Micro Lett.* 6 (2): 116–124.

87 Sugiura, H., Kondo, H., Higuchi, K. et al. (2020). Reaction science of layer-by-layer thinning of graphene with oxygen neutrals at room temperature. *Carbon N. Y.* 170: 93–99.

88 Kim, J.S., Kim, G.H., Chung, T.H. et al. (2001). Characterization of an oxygen plasma by using a Langmuir probe in an inductively coupled plasma. *J. Korean Phys. Soc.* 38 (3): 259–263.

89 Paredes, J.I., Martínez-Alonso, A., and Tascón, J.M.D. (2002). Early stages of plasma oxidation of graphite: nanoscale physicochemical changes as detected by scanning probe microscopies. *Langmuir* 18 (11): 4314–4323.

90 Koizumi, K., Boero, M., Shigeta, Y., and Oshiyama, A. (2013). Atom-scale reaction pathways and free-energy landscapes in oxygen plasma etching of graphene. *J. Phys. Chem. Lett.* 4 (10): 1592–1596.

91 Seah, C.M., Vigolo, B., Chai, S.P., and Mohamed, A.R. (2016). Mechanisms of graphene fabrication through plasma-induced layer-by-layer thinning. *Carbon N. Y.* 105: 496–509.

92 Mehmood, F., Pachter, R., Lu, W., and Boeckl, J.J. (2013). Adsorption and diffusion of oxygen on single-layer graphene with topological defects. *J. Phys. Chem. C* 117 (20): 10366–10374.

93 Nicholson, K.T., Minton, T.K., and Sibener, S.J. (2005). Spatially anisotropie etching of graphite by hyperthermal atomic oxygen. *J. Phys. Chem. B* 109 (17): 8476–8480.

94 Surwade, S.P., Smirnov, S.N., Vlassiouk, I.V. et al. (2015). Water desalination using nanoporous single-layer graphene. *Nat. Nanotechnol.* 10 (5): 459–464.

95 Bai, J., Zhong, X., Jiang, S. et al. (2010). Graphene nanomesh. *Nat. Nanotechnol.* 5 (3): 190–194.

96 Yang, Y., Yang, X., Liang, L. et al. (2019). Large-area graphene-nanomesh/carbon-nanotube hybrid membranes for ionic and molecular nanofiltration. *Science* 364 (6445): 1057–1062.

97 Jang, D., Idrobo, J.C., Laoui, T., and Karnik, R. (2017). Water and solute transport governed by tunable pore size distributions in nanoporous graphene membranes. *ACS Nano* 11 (10): 10042–10052.

98 Xie, G., Yang, R., Chen, P. et al. (2014). A general route towards defect and pore engineering in graphene. *Small* 10 (11): 2280–2284.

99 Chu, X. and Schmidt, L.D. (1993). Intrinsic rates of NO_x-carbon reactions. *Ind. Eng. Chem. Res.* 32 (7): 1359–1366.

100 Liu, L., Ryu, S., Tomasik, M.R. et al. (2008). Graphene oxidation: thickness-dependent etching and strong chemical doping. *Nano Lett.* 8 (7): 1965–1970.

101 Yamada, Y., Murota, K., Fujita, R. et al. (2014). Subnanometer vacancy defects introduced on graphene by oxygen gas. *J. Am. Chem. Soc.* 136 (6): 2232–2235.

102 Schlichting, K.P. and Poulikakos, D. (2020). Selective etching of graphene membrane nanopores: from molecular sieving to extreme permeance. *ACS Appl. Mater. Interfaces* 12 (32): 36468–36477.

103 Xu, S.C., Irle, S., Musaev, D.G., and Lin, M.C. (2006). Quantum chemical prediction of reaction pathways and rate constants for dissociative adsorption of CO_x and NO_x on the graphite (0001) surface. *J. Phys. Chem. B* 110 (42): 21135–21144.

104 Biederman, D.L., Miles, A.J., Vastola, F.J., and Walker, P.L. (1976). Carbon-carbon dioxide reaction: kinetics at low pressures and hydrogen inhibition. *Carbon N. Y.* 14 (6): 351–356.

105 Chu, X. and Schmidt, L.D. (1991). Gasification of graphite studied by scanning tunneling microscopy. *Carbon N. Y.* 29 (8): 1251–1255.

106 Yang, R. (1983). Fundamental differences in the mechanisms of carbon gasification by steam and by carbon dioxide. *J. Catal.* 82 (1): 245–251.

107 Strange, J.F. and Walker, P.L. (1976). Carbon–carbon dioxide reaction: Langmuir–Hinshelwood kinetics at intermediate pressures. *Carbon N. Y.* 14 (6): 345–350.

108 Ergun, S. (1956). Kinetics of the reaction of carbon with carbon dioxide. *J. Phys. Chem.* 60 (4): 480–485.

109 Molina-Sabio, M., González, M.T., Rodriguez-Reinoso, F., and Sepúlveda-Escribano, A. (1996). Effect of steam and carbon dioxide activation in the micropore size distribution of activated carbon. *Carbon N. Y.* 34 (4): 505–509.

110 Huang, Z., Zhang, J., Zhao, Y. et al. (2010). Kinetic studies of char gasification by steam and CO_2 in the presence of H_2 and CO. *Fuel Process. Technol.* 91 (8): 843–847.

111 Turkdogan, E.T. and Vinters, J.V. (1969). Kinetics of oxidation of graphite and charcoal in carbon dioxide. *Carbon N. Y.* 7 (1): 101–117.

112 Moulijn, J.A. and Kapteijn, F. (1995). Towards a unified theory of reactions of carbon with oxygen-containing molecules. *Carbon N. Y.* 33 (8): 1155–1165.

113 Wong, C., Yang, R.T., and Halpern, B.L. (1983). The mode of attack of oxygen atoms on the basal plane of graphite. *J. Chem. Phys.* 78 (6): 3325–3328.

114 Yang, R.T. and Duan, R.Z. (1985). Kinetics and mechanism of gas-carbon reactions: conformation of etch pits, hydrogen inhibition and anisotropy in reactivity. *Carbon N. Y.* 23 (3): 325–331.

115 Yang, R., Zhang, L., Wang, Y. et al. (2010). An anisotropic etching effect in the graphene basal plane. *Adv. Mater.* 22 (36): 4014–4019.

116 Zhao, D., Liu, H., Sun, C. et al. (2018). DFT study of the catalytic effect of Na on the gasification of carbon–CO_2. *Combust. Flame* 197: 471–486.

117 Cabrera-Sanfelix, P. (2009). Adsorption and reactivity of CO_2 on defective graphene sheets. *J. Phys. Chem. A* 113 (0001): 493–498.

118 Montoya, A., Mondragón, F., and Truong, T.N. (2002). Formation of CO precursors during char gasification with O_2, CO_2 and H_2O. *Fuel Process. Technol.* 77–78 (2): 125–130.

119 Roberts, M.J., Everson, R.C., Domazetis, G. et al. (2015). Density functional theory molecular modelling and experimental particle kinetics for CO_2-char gasification. *Carbon N. Y.* 93: 295–314.

120 Zhu, Z.H., Finnerty, J., Lu, G.Q., and Yang, R.T. (2002). A comparative study of carbon gasification with O_2 and CO_2 by density functional theory calculations. *Energy and Fuels* 16 (6): 1359–1368.

121 Zhang, J., Jia, K., Lin, L. et al. (2019). Large-area synthesis of superclean graphene via selective etching of amorphous carbon with carbon dioxide. *Angew. Chemie Int. Ed.* 58 (41): 14446–14451.

122 Liu, Y. and Wilcox, J. (2011). CO_2 adsorption on carbon models of organic constituents of gas shale and coal. *Environ. Sci. Technol.* 45 (2): 809–814.

123 Montoya, A., Mondragón, F., and Truong, T.N. (2002). First-principles kinetics of CO desorption from oxygen species on carbonaceous surface. *J. Phys. Chem. A* 106 (16): 4236–4239.

124 Radovic, L.R. (2009). Active sites in graphene and the mechanism of CO_2 formation in carbon oxidation. *J. Am. Chem. Soc.* 131 (47): 17166–17175.

125 Fritz, O.W. and Hüttinger, K.J. (1993). Active sites and intrinsic rates of carbon-gas reactions-a definite confirmation with the carbon-carbon dioxide reaction. *Carbon N. Y.* 31 (6): 923–930.

126 Marchon, B., Carrazza, J., Heinemann, H., and Somorjai, G.A. (1988). TPD and XPS studies of O_2, CO_2, and H_2O adsorption on clean polycrystalline graphite. *Carbon N. Y.* 26 (4): 507–514.

127 Dutta, S., Vahdat, M.T., Rezaei, M., and Agrawal, K.V. (2019). Crystallization of gas-selective nanoporous graphene by competitive etching and growth: a modeling study. *Sci. Rep.* 9 (1): 5202.

128 Yang, X., Hu, B., Jin, Y. et al. (2017). Insight into CO_2 etching behavior for efficiently nanosizing graphene. *Adv. Mater. Interfaces* 4 (10): 1–8.

129 Wang, X., Tabakman, S.M., and Dai, H. (2008). Atomic layer deposition of metal oxides on pristine and functionalized graphene. *JACS* 130 (Figure 3): 8152–8153.

130 Hahn, J.R. (2005). Kinetic study of graphite oxidation along two lattice directions. *Carbon N. Y.* 43 (7): 1506–1511.

131 Hart, E.J. and Henglein, A. (1985). Free radical and free atom reactions in the sonolysis of aqueous iodide and formate solutions. *J. Phys. Chem.* 89 (20): 4342–4347.

132 Olson, T.M. and Barbiert, P.F. (1994). Oxidation kinetics of natural organic matter by sonolysis and ozone. *Water Res.* 28 (6): 1383–1391.

133 Chang, P.B.L. and Young, T.M. (2000). Kinetics of methyl tert-butyl ether degradation and by-product formation during UV/hydrogen peroxide water treatment. *Water Res.* 34 (8): 2233–2240.

134 Forni, L., Bahnemann, D., and Hart, E.J. (1982). Mechanism of the hydroxide ion initiated decomposition of ozone in aqueous solution. *J. Phys. Chem.* 86 (2): 255–259.

135 Siemens, W. (1857). Ueber die elektrostatische Induction und die Verzögerung des Stroms in Flaschendrähten. *Ann. Phys.* 178 (9): 66–122.

136 Lee, G., Lee, B., Kim, J., and Cho, K. (2009). Ozone adsorption on graphene: ab initio study and experimental validation. *J. Phys. Chem. C* 113 (32): 14225–14229.

137 Li, J.L., Kudin, K.N., McAllister, M.J. et al. (2006). Oxygen-driven unzipping of graphitic materials. *Phys. Rev. Lett.* 96 (17): 5–8.

138 Li, Z., Zhang, W., Luo, Y. et al. (2009). How graphene is cut upon oxidation? *J. Am. Chem. Soc.* 131 (18): 6320–6321.

139 Sun, T. and Fabris, S. (2012). Mechanisms for oxidative unzipping and cutting of graphene. *Nano Lett.* 12 (1): 17–21.

140 Lee, W., Bondaz, L., Huang, S. et al. (2021). Centimeter-scale gas-sieving nanoporous single-layer graphene membrane. *J. Memb. Sci.* 618: 118745.

141 Villalobos, L.F., Huang, S., Dakhchoune, M. et al. (2021). Rapid gas transport from nanometer-sized channels in block-copolymer templated nanoporous carbon film. *Carbon N. Y.* 173: 980–988.

142 Huang, S., Li, S., Hsu, K. et al. (2021). Systematic design of millisecond gasification reactor for the incorporation of gas-sieving nanopores in single-layer graphene. *J. Memb. Sci.* 637: 119628.

143 Zhao, J., He, G., Huang, S. et al. (2019). Etching gas-sieving nanopores in single-layer graphene with an angstrom precision for high-performance gas mixture separation. *Sci. Adv.* 5 (1): eaav1851.

144 Wang, Y., Zheng, Y., Xu, X. et al. (2011). Electrochemical delamination of CVD-grown graphene film: toward the recyclable use of copper catalyst. *ACS Nano* 5 (12): 9927–9933.

145 Lin, L., Peng, H., and Liu, Z. (2019). Synthesis challenges for graphene industry. *Nat. Mater.* 18 (6): 520–524.

146 Bae, S., Kim, H., Lee, Y. et al. (2010). Roll-to-roll production of 30-inch graphene films for transparent electrodes. *Nat. Nanotechnol.* 5 (8): 574–578.

147 Kobayashi, T., Bando, M., Kimura, N. et al. (2013). Production of a 100-m-long high-quality graphene transparent conductive film by roll-to-roll chemical vapor deposition and transfer process. *Appl. Phys. Lett.* 102 (2): 023112.

148 Polsen, E.S., McNerny, D.Q., Viswanath, B. et al. (2015). High-speed roll-to-roll manufacturing of graphene using a concentric tube CVD reactor. *Sci. Rep.* 5: 10257.

149 Hsieh, Y.P., Shih, C.H., Chiu, Y.J., and Hofmann, M. (2016). High-throughput graphene synthesis in gapless stacks. *Chem. Mater.* 28 (1): 40–43.

150 Ma, L., Ren, W., and Cheng, H. (2019). Transfer methods of graphene from metal substrates: a review. *Small Methods* 3 (7): 1900049.

151 Lee, C.-K., Hwangbo, Y., Kim, S.-M. et al. (2014). Monatomic chemical-vapor-deposited graphene membranes bridge a half-millimeter-scale gap. *ACS Nano* 8 (3): 2336–2344.

152 Boutilier, M.S.H., Jang, D., Idrobo, J.C. et al. (2017). Molecular sieving across centimeter-scale single-layer nanoporous graphene membranes. *ACS Nano* 11 (6): 5726–5736.

153 Wang, L., Drahushuk, L.W., Cantley, L. et al. (2015). Molecular valves for controlling gas phase transport made from discrete ångström-sized pores in graphene. *Nat. Nanotechnol.* 10 (9): 785–790.

154 Boutilier, M.S.H., Sun, C., O'Hern, S.C. et al. (2014). Implications of permeation through intrinsic defects in graphene on the design of defect-tolerant membranes for gas separation. *ACS Nano* 8 (1): 841–849.

155 Lalia, B.S., Kochkodan, V., Hashaikeh, R., and Hilal, N. (2013). A review on membrane fabrication: Structure, properties and performance relationship. *Desalination* 326: 77–95.

156 Sanders, D.F., Smith, Z.P., Guo, R. et al. (2013). Energy-efficient polymeric gas separation membranes for a sustainable future: a review. *Polymer (Guildf).* 54 (18): 4729–4761.

157 Qin, Y., Hu, Y., Koehler, S. et al. (2017). Ultrafast nanofiltration through large-area single-layered graphene membranes. *ACS Appl. Mater. Interfaces* 9 (11): 9239–9244.

158 Kidambi, P.R., Mariappan, D.D., Dee, N.T. et al. (2018). A scalable route to nanoporous large-area atomically thin graphene membranes by roll-to-roll chemical vapor deposition and polymer support casting. *ACS Appl. Mater. Interfaces* 10 (12): 10369–10378.

4

Graphene-Based Membranes for Water Separation

Yuan Kang and Xiwang Zhang

Monash University, Department of Chemical and Biological Engineering, Wellington Rd, Clayton, VIC 3800, Australia

4.1 Introduction

Water separation describes a series of processes where water, as the main target product, is isolated from various solutes in aqueous mixtures. It allows the collection or recovery of purified water from multiple sources, such as seawater, brackish water, and wastewater, for household, agricultural, and industrial uses [1, 2]. Among all currently available tools, membranes can efficiently realize this separation by acting as a semipermeable barrier to let water permeate while blocking unwanted solutes during processing. To maximize such differentiation in permeation, membranes are often required to have minimal thickness to yield high water throughput and appropriately sized channels to ensure low solute passage [3]. These structural requirements thus make 2D laminar membranes, particularly those composed of graphene-based materials (e.g. graphene oxide, GO), a promising candidate for water-separating applications [4, 5]. This is mostly owing to the unique layered structure of graphene-based membranes. Constructed by piling up atomically thin 2D nanosheets, graphene-based membranes can be engineered into ultrathinness to reduce water transport resistance by simply controlling layer number. More importantly, void spaces between neighboring layers will create numerous subnanoscale channels whose size falls between that of water molecules and most solutes, rendering water separation from ions and organic compounds applicable [6].

Moreover, the surge of graphene-based membranes for real-world water separations also benefits from their additional advantages [7], including (i) Scalability. To build up these membranes mainly relies on nanosheet self-stacking, and this self-assembly enables convenient membrane fabrication on large scale; (ii) Uniformity. 2D channels within the well-defined layered membranes, especially the interlayer ones, are formed with relatively uniform sizes. The narrow channel size distribution will bring about more reliable membrane performances in separating species with subtle size differences (e.g. water and hydrated ions);

Two-Dimensional-Materials-Based Membranes: Preparation, Characterization, and Applications, First Edition. Edited by Gongping Liu and Wanqin Jin.
© 2022 WILEY-VCH GmbH. Published 2022 by WILEY-VCH GmbH.

(iii) Tunability. The size and surface chemistry of these membrane channels are highly tunable, which makes their post-assembly treatments and decorations feasible to adapt into specific separations.

By virtue of the above merits, graphene-based membranes have been widely used in major water-separating processes, including reverse osmosis (RO), nanofiltration (NF), and forward osmosis (FO). Based on these previous attempts, this chapter aims to present recent progress on graphene-based membranes for water separation with its focus on two main aspects—water transport mechanisms in these membranes and how these mechanisms translate into desired separation output in three major application categories.

4.2 Water Transport Mechanisms in Graphene-Based Membranes

Water transport in 2D graphene-based laminar membranes, as in other membranes, involves the inflow of water into the membrane and its subsequent migration across membrane channels of a certain length. Therefore, the overall water permeance largely relies on its uptake at membrane entrance, traveling length inside the membrane, and flowing velocity across the channels. These key factors are immediately relevant to either membrane geometrical structure or surface chemical property, which is further influenced by working environments and the introduction of other materials. In light of this, we discuss the transport mechanisms under the mediation by (i) membrane internal geometry, (ii) surface chemistry, (iii) external environment, and (iv) guest materials.

4.2.1 Internal-Geometry-Mediated Transport

The internal geometry of graphene-based membranes refers to the configuration of various nanochannels as a result of nanosheet arrangement during membrane assembly (Figure 4.1). Horizontally, adjacent nanosheets are set apart by

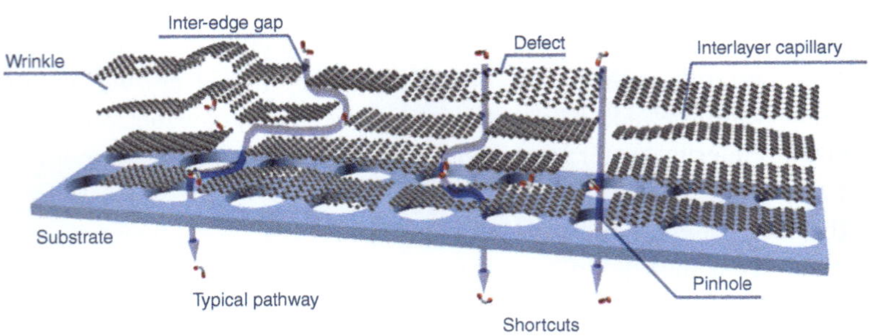

Figure 4.1 Schematic of the geometry of a typical 2D graphene-based membrane, including horizontal 2D channels, such as interlayer capillary and nanowrinkles, and vertical channels, such as inter-edge gaps, wrinkles, defects, and pinholes. They form either typical or shortcut transporting pathways, as highlighted in blue columns. Source: Yuan Kang and Xiwang Zhang.

protruding oxygenated functional groups to become interlayer capillaries. Besides, the sp^3-hybridized (functionalized) carbon atoms and topological defects, if there are any, will make membranes susceptible to corrugations and turn part of the flat capillaries into larger fuzzy wrinkles. Vertically, the edges of nanosheets in the same plane are separated, leaving inter-edge gaps to dominate flowing from one layer to another. Additionally, the intrinsic defects and pinholes may appear within or among nanosheets to cause extra cross-layer paths. When interconnected with each other, these nanochannels extending in both directions will create massive amount of water transport pathways, while their size and length will determine the uptake and traveling distance of water in a laminar membrane, respectively.

4.2.1.1 Size Effects

A typical defect-free transporting pathway comprises both inter-edge gaps and interlayer capillaries, and they are connected in succession to propagate water from the entrance to the exit of the membrane (Figure 4.1, typical pathway). To adapt into nanosized 2D channels, water molecules must isolate themselves from bulk state at energy expense [8]. From a geometrical view, such an energetic barrier is related to the size of nanochannel and limits water uptake to influence membrane permeability. Most often, the size-controlled molecular transport is believed to take place in horizontal interlayer capillaries (Figure 4.2a, left) [9]. As the most characteristic feature of GO laminates, these narrow horizontal channels are capable of opening and closing on certain conditions, allowing a tunable interlayer spacing (*d*-spacing) to dictate water intake. According to the initial work of Geim and Nair, feeding GO-based membranes with vapor can adjust *d*-spacing between 7 and 10 Å, and the enlarged channels will let more water stream in and swiftly pass the membrane in a monolayer [10]. With the relative humidity elevated from 12%, through 33% to 100%, *d*-spacing gradually can be increased from 7.5 Å, through 8.5 Å to 9.8 Å, producing water by 3.2, 4.5, and 6.5 $l\,m^{-2}\,h^{-1}\,bar^{-1}$, respectively (Figure 4.2a, middle and right) [11].

In opposition, water transport will be slowed as a result of reduced *d*-spacing, as previously observed in thermally treated GO membranes [12, 13]. This is because the protruding oxygenated functional groups on GO nanosheets, including –COOH, –OH, and –O–, are eliminated in the heat treatment (reduction procedure), causing the collapse of interlayer channels to below 4 Å. Similar trend has also been reported in UV irradiation-treated GO membranes, where channel size is reduced stepwise via a controllable photoreduction method [14]. With membrane exposure time under irradiation extending to four hours, functional groups can be gradually removed to cause *d*-spacing decrease from ~8 to ~6 Å. This accordingly suppressed water permeation, as reflected by a decline of flux from 3.7 to 1.3 $l\,m^{-2}\,h^{-1}\,bar^{-1}$ in pressure-driven filtration tests.

Besides the well-established relation between mass transport and typical interlayer capillaries, recent discoveries reveal that nanowrinkles, as a long-neglected channel, also contribute to high water throughput in graphene-based membranes due to their unique size distribution. In the majority of former studies, building

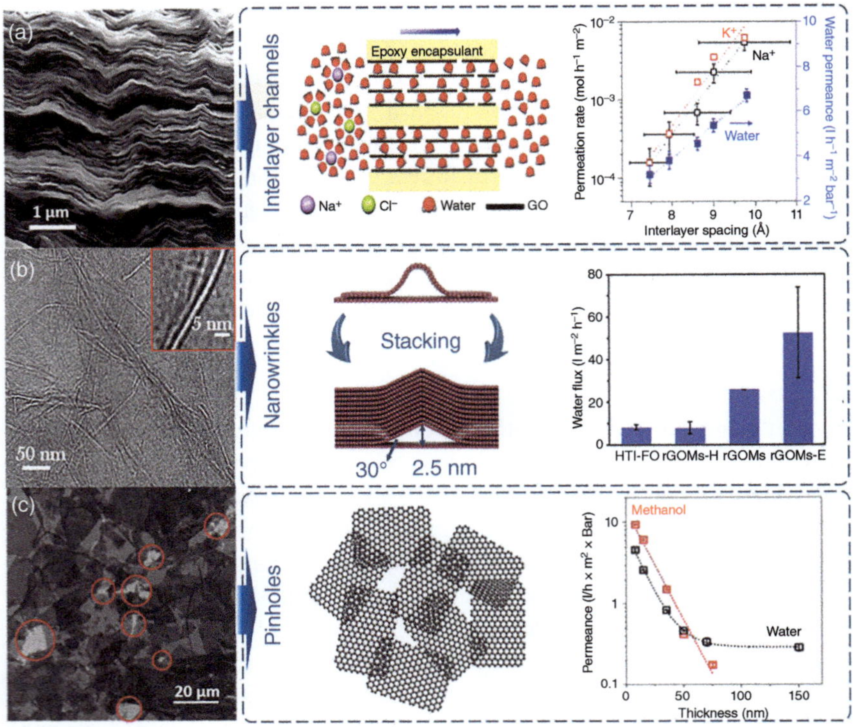

Figure 4.2 Effects of internal geometry on water transport in graphene-based laminar membranes. Row (a) from left to right: cross-section SEM image of a GO laminar membrane; mass transport through subnanoscale interlayer channels; permeation rates of water and ions through the channels of varied size in different humidity. Source: Abraham et al. [11]/with permission from Springer Nature. Row (b) from left to right: TEM image of nanowrinkles on GO laminar membrane surface; nanowrinkles formation during membrane stacking; water and salt flux in pressure-driven operation through the membranes treated by hexane, water, and ethanol. Source: Kang et al. [17]/with permission from John Wiley & Sons, INC. Row (c) from left to right: SEM image of GO laminates with imperfect stacking (pinholes); schematic of pinhole formation; solvent permeance as membrane thickness increases. Source: Yang et al. [25]/with permission from Springer Nature.

blocks are always considered flat and thus parallel to each other in prepared membranes for simple understanding. The simplification, however, drifts away from the fact that the atomic thickness of graphene-based nanosheets endows them high flexural ability to form wrinkled structures during membrane fabrication, transferring, and post-treatments [15, 16]. The work of Kang et al. found that stable nanowrinkles can be intrinsically formed throughout reduced graphene oxide (rGO) membranes and illustrate an arch-like morphology, featuring a central part (2–3 nm height), and two wedge-shaped corners (inclination angle <30°, Figure 4.2b, left and middle) [17]. Besides going through the wide center, water is also able to flow quickly via the sharp corners. This is because tremendous capillary force is induced around the narrow space and water passage can be prompted along functionality-free rGO surface. The role of these nanowrinkles as fast water tracks is also experimentally supported.

By applying solvents of different polarities, including hexane (low), water (medium), and ethanol (high), in the drying step of freshly prepared membranes, wrinkle density was found changed. With the addition of ethanol, a lot more wrinkles could be obtained in rGO membranes to achieve higher water yield (Figure 4.2b, right). Similar results were reported for hydrothermally treated graphene-based membranes [18]. By elevating treatment temperature, Qiu et al. created enlarged corrugated channels with a size ranging from 3 to 13 nm in partly reduced rGO membranes and as-obtained curved channels improved water permeation by several folds.

Likewise, the impact of inter-edge gap size on water incorporation almost follows the same rule [19]. For instance, Hu et al. examined the effect of gap width (G) on water passage in a pressure-driven process by molecular simulation [20]. It was found that wider gaps (from 1 to 8 Å) would take more water molecules (from 0 to 400×10^{-2} molecules ns^{-1} MPa^{-1}) to pass. However, to obtain deep understanding of water translocation across GO layers needs more accurate characterizations on the nanostructure of 2D laminar membranes (e.g. gap size and size distribution) in future studies.

4.2.1.2 Length Effects

Another geometrical factor to control water transport in graphene-based membranes is the average length of the transporting paths (L). It implies the total distance that water travels from membrane entrance to exit, which is inversely proportional to water permeability. In a simplified way, it can be qualitatively estimated as [10]

$$L = \frac{h}{d} \cdot l \tag{4.1}$$

where h represents membrane thickness and d represents interlayer distance. Considering the tortuous pathways in laminar membranes, the ratio of h to d (h/d) implies the turnings that water flow has to take and thus the number of individual sheets it travels by. Hence, L can be approximated if all nanosheets are assumed to be in equal length l (l is more accurately interpreted as the offset, the distance between the neighboring inter-edge of a top and bottom nanosheet) [21]. Given these geometrical variables, two common methods are adopted to make pathway length shorter for faster transport. The first is to cut down l for reduced L. A few previous studies decreased GO lateral size from microns to nanometers by strong sonication, centrifugation, and controlled suspension freezing before membrane assembly [22–24]. In one of these attempts, Vecitis and coworkers found that GO flakes sonicated in bath for ~20 minutes would have l shortened by a factor of 7, promoting membrane flux by up to 100% [14]. The second method, like practiced in other membranes, is to contain membrane thickness. An instructive message derived from Eq. (4.1) is that thinner laminar membranes, as constructed by less layers of nanosheets, will allow water to take fewer turns and the shorter pathway will pose less transporting resistance for water transport [25].

In addition to these continuous typical pathways, shortcuts may also appear in graphene-based membranes due to the existence of porous structures, especially in membranes down to below 10-nm thickness. These pores include in-plane defects incurred during the redox synthesis of GO nanosheets [26], and macrovoids (also called pinholes, Figure 4.2c, left and middle) caused by random nanosheet stacking during membrane assembly [25, 27]. Instead of going twists and turns, water molecules prefer to take a shortcut through the pores to reach next layers straightaway. This convenient cross-layer flow decreases L by a large degree and leads to ultrahigh water permeability [25, 28]. Nevertheless, the probability of water to pass through the shortcuts exponentially drops when sufficient GO layers are incrementally stacked since pores are blocked to prevent leakage. As reported by Yang et al., increasing the thickness of GO membranes from 10 to 20 nm dramatically confined the throughput from 5 to $1\,l\,m^{-2}\,h^{-1}\,bar^{-1}$ (Figure 4.2c, right), suggesting that the pore-oriented fast water transport may only take only in ultrathin membranes [25]. Moreover, the emergence of excessive porous channels in graphene-based membranes also risks compromising solute-rejecting capability due to unexpected leakage. To strike the balance of highest possible water flux and least possible solute passage thereby depends on the structure optimization of ultrathin membrane (e.g. <10 nm) via technical advancements, such as flawless nanosheet stacking, which still remains a challenge from the fabrication perspective.

4.2.2 Surface-Chemistry-Mediated Transport

In addition to membrane geometry, the surface chemistry of graphene-based membranes is also critical in mediating trans-membrane water transport. Owing to the distinct chemical nature of building blocks, resultant graphene-based (mostly GO) membranes possess both pristine regions (sp^2-bond carbon) and oxidized regions (sp^3-bond carbon) on their surface [29]. The sp^2 parts are covered by dense and delocalized π-electron cloud, while the sp^3 parts are decorated with oxygen-containing groups. The hybridized GO surface interacts with water molecules in various manners, resulting in accelerated or moderated transport at different spots in nanochannels.

For a long time, there has been a belief that water confined in GO nanochannels is able to migrate anomalously faster than it does as a classical viscous liquid. Early studies attribute this swift delivery to the synergistic roles played by the pristine and oxidized regions [30]. For one thing, the hydrophilic functional groups at nanosheet edges help water molecules to stream into the interlayer capillaries and provide them with easy access to move across different layers [31]. For another, perhaps more importantly, the in-plane hydrophobic and functional group-free sp^2 surface allow them to commence a slipping-like flowing with nearly no friction (Figure 4.3a,b) [32]. A similar alternative explanation for the ultrafast water migration is the collective motion of ice layers (Figure 4.3c) [33]. As experimentally observed under high-resolution transmission electron microscopy (HRTEM), atomic-scale ice crystal can be formed between two graphene nanosheets (chemically similar to pristine part of GO nanosheets) [34]. This is also corroborated by the calculation of Boukhvalov et al. who found that the binding energy among water molecules nanoconfined between graphene is −1.18 eV, indicative of the ice-like intermolecular hydrogen bonds. Meanwhile, much weaker interaction between pristine regions and water (−50 meV) ensures that the ice structure will not be

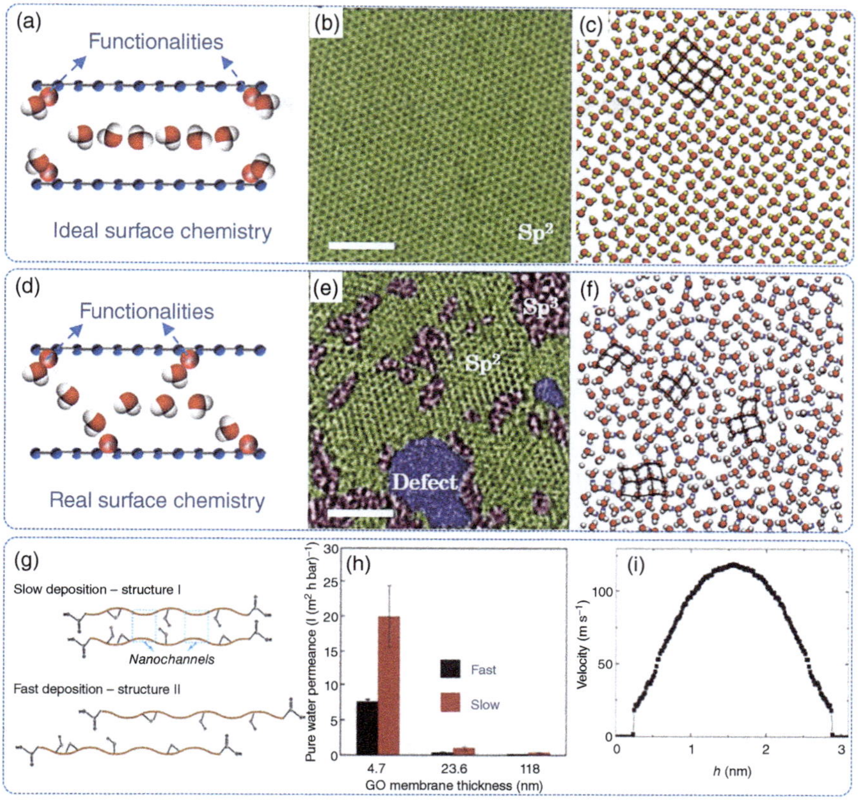

Figure 4.3 Effect of surface chemistry on water transport in graphene-based laminar membranes. From (a) to (c): water transport in a GO channel with ordered surface chemistry (a); (b) HRTEM image of the sp^2 regions on GO channel wall (scale bar, 2 nm) and; ice-like water structure sandwiched in this channel (c). From (d) to (f): water transport in a GO channel with random surface chemistry (d); (e) HTREM image of sp^2–sp^3 regions on GO channel wall and disorder water structure in this channel (f). Source: (b) and (e) Zettl et al. [35]/with permission of John Wiley & Sons, Inc. (c) and (f) Zheng et al. [39]/with permission of American Chemical Society. From (g) to (i): (g) water transport in GO channels with symmetrical (slow-deposited, top) and asymmetrical (fast-deposited, bottom); and their water permeation performance comparison in (h). Source: Xu et al. [38]/with permission of American Chemical Society; (i) a typical parabolic-shape water velocity profile obtained in simulated GO channels mimicking real-world situations. Source: Huang et al. [43]/with permission of Springer Nature.

interrupted when moving forward. Therefore, the ice array will glide undisturbedly within nanochannels before coming to inter-edge gaps, where it melts and then reforms in the next layer to realize directional rapid water movement [33].

Nonetheless, either the "slip-flow" theory or "ice-involved" flowing mode is established on two theoretical prerequisites, namely (i) a spatial sp^2 region should locate in the middle of GO sheets, while sp^3 clusters mainly align at the edge, and (ii) the sp^2 regions of neighboring GO sheets should be perfectly overlapped. Such assumptions for an ordered chemical structure are, however, incompatible with the direct HRTEM imaging of GO nanosheets from Zettl et al., who revealed that sp^2 and sp^3 regions are randomly distributed on GO surface (Figure 4.3d,e) [35]. This disordered surface chemistry should readily discourage the slip flow from two aspects. Primarily, the emergence of sp^3 patches in sp^2 region breaks down the contiguous network that is necessary for frictionless slipping. More essentially, since hydrogen bonds are easily formed between the hydrogen atoms of water molecules and electronegative atoms in the vicinity (less than 3 Å away from O, H, and F), close contact of water with oxidized sp^3 regions can induce a strong interaction to drag down its transport [36].

Therefore, to accurately mimic the water passage through sp^2–sp^3 regions, Wei et al. simulated the flow through 1–2 nm wide sp^2–sp^2 and sp^2–sp^3 channels for comparison [37]. In contrast to pure sp^2 channels, where flow rate could be enhanced by more than 1600 times, the sp^2–sp^3 hybridized channel only permitted a limited eightfold improvement. As expected, oxygenated groups in the hybridized channels are able to capture the adjacent water molecules through hydrogen bonds and cause a "side-pinning effect," impeding significant velocity enhancement. In line with this simulation finding, Yu and coworkers further experimentally verified that water transport can be hindered if the pristine regions of top and bottom GO sheets are mismatched [38]. By simply tuning the GO suspension depositing speed, they obtained interlayer channels with symmetrical (slow-deposited) and asymmetrical (fast-deposited) surface chemistry (Figure 4.3g). In the fast-deposited channels, the sp^2 region of top layers is more likely to reach a face-to-face overlap with the sp^3 regions of bottom nanosheets. The staggered stacking of the fast-deposited membranes generates extra hydrogen bonds, which gives rise to four-time water flux reduction compared with the membranes prepared by slow deposition (Figure 4.3h). Later simulation by Mi and coworkers further proved that the alignment of ice crystal near GO surface could be immediately disrupted by those functional groups to lose its solid structure, which defies the "gliding-as-ice" theory as well (Figure 4.3f) [39].

With more physicochemical characterizations for GO surface chemistry made available and detailed discussion on surface chemistry-mediated water transport provided, another flowing model based on "relative water movement" has been raised [40–42]. In this model, the sp^3 regions on GO channel walls are able to lock a considerable number of nearby water molecules through hydrogen bonds. The locked water further stops the in-plane movement of other unlocked water molecules near sp^2 regions, thus forming a first hydration layer whose mobility is severely undermined at GO–water interface [41]. However, the bonding between water and functional groups is short-ranged and only works on the already-immobilized first layers. That is to say, water layers further away from the GO surface are less affected and permitted to migrate with low resistance. This relative motion of one active liquid water layer over another immobilized one was experimentally corroborated by Talyzin et al. by a facile freezing operation [42]. In the procedure of freezing a wet GO membrane, water trapped inside would spontaneously come out from the membrane due to volume change. While water near the surface was found firmly attached to GO surface, additional water layers at the center of the channel could escape at much higher velocity. Assisted by the driving force from a capillary-like pressure, the relatively free flow of these water layers was deemed to potentially speed up overall water transport. Such flowing mode is also consistent with the water velocity profiles in a series of previous studies considering the real-world distribution of oxygen-containing functional groups in GO channels. In these profiles, the velocity of water exponentially increases as it drifts away from GO–water interface, showing a parabolic-shape distribution (Figure 4.3i) [43].

4.2.3 External-Environment-Mediated Transport

While the transport of water in graphene-based membranes is primarily controlled by membrane geometry and surface chemistry, these intrinsic features are subject to environments where the membranes work. To apply these membranes for water separation purposes, thus, will unavoidably be influenced by a range of extrinsic factors, such as solution chemistry, pressure, and electricity [44, 45]. These environmental stimuli have been found to modulate water permeation directly or indirectly by changing the above physiochemical properties.

4.2.3.1 Solution Chemistry Effects

As the major media where water separation is achieved, solution with complex chemistry (pH, ionic strength [IS]) imposes sound controlling effects on the transporting characteristics of graphene-based membranes. The most notable effect is water-induced swelling in which membrane channels enlarge when experiencing a shift from dry to wet state (Figure 4.4a) [46]. This phenomenon occurs once water is soaked in GO channels in aqueous environments. Not only will the incorporated water pull away two adjacent nanosheets, but also it will cause the dissociation of carboxyl groups ($-COOH \rightarrow -COO^- + H^+$) to impart negative charges on GO nanosheets. As-charged GO sheets continue repelling each other and further enlarge the channel until these repulsive forces (electrostatic repulsion and hydration force) are balanced by inter-sheet van der Waals attraction. Depending on the content of functional groups on GO building blocks, fully hydrated membranes can have swollen interlayer channels whose size range from 1.3 to 6 nm, giving rise to increased water permeance by varying degrees [39, 47].

Considering the close relation among membrane surface charge, pH, and ionic strength (IS), it is envisaged that an alteration of pH and/or IS will affect water permeation indirectly via *d*-spacing control. By acidifying solution for lower pH or adding excessive electrolyte for higher IS will lead to a drop of effective membrane charges because of carboxyl protonation ($-COO^- + H^+ \rightarrow -COOH$) or ionic screening effect, respectively [48]. Electrical double layers (EDLs) generated by these charged nanosheets will thereby attenuate to have a smaller thickness (Debye length, λ_D). As a dynamic strength measure of electrostatic interactions between GO layers, a decrease in λ_D will imply the concurrent *d*-spacing reduction (Figure 4.4b). Consequently, to concentrated NaCl solution from 1 to 10 mM will be accompanied by a steep flux fall from over 70 to below 10 l m^{-2} h^{-1} bar^{-1} in the work by Huang et al. (Figure 4.4c) [44].

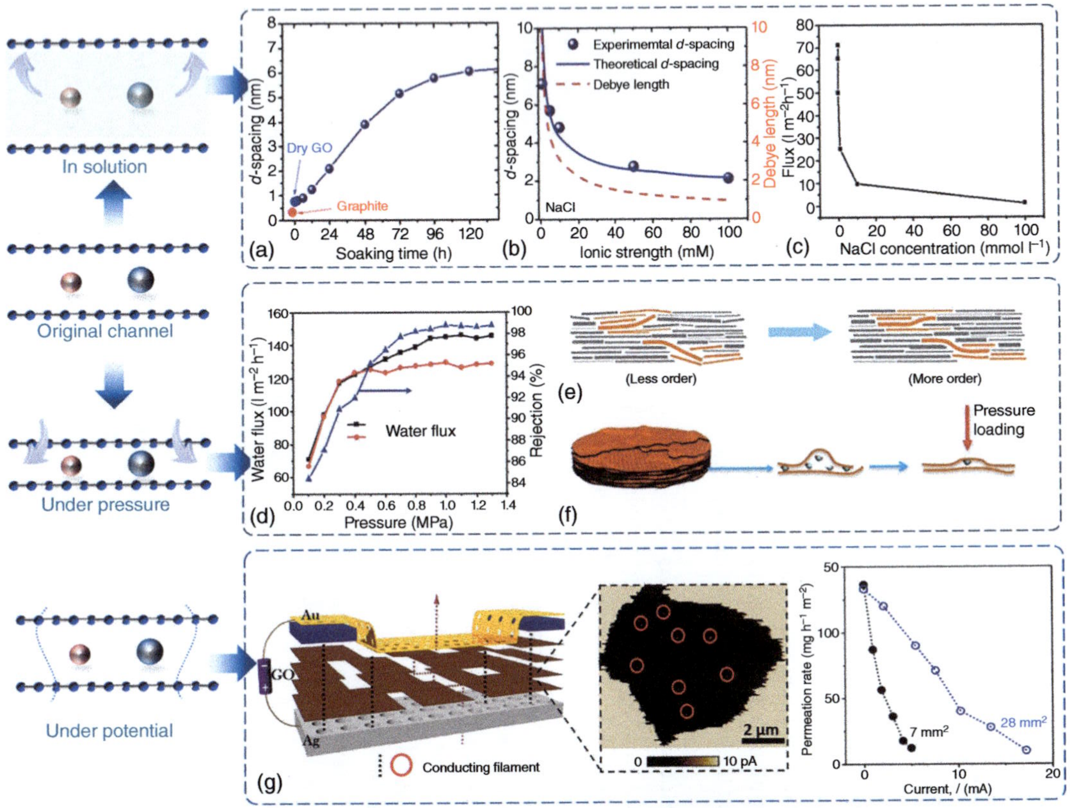

Figure 4.4 Effects of solution chemistry and pressure on the structure and water transporting features of GO membranes. (a) Swelling of GO membranes in water. (b) Relation of ionic strength with Debye length and *d*-spacing. Source: Modified from Wang et al. [47]. (c) Relation of ionic strength (NaCl solution concentration) with water flux. (d) Relation of water flux with applied pressure. This is explained by (e) microscale structural changes and (f) nanoscale structural changes. Source: (c), (d), and (f) Huang et al. [44]. Reproduced with permission of Royal Society of Chemistry. (e) Chong et al. [51]. Reproduced under the terms of the Creative Commons Attribution 4.0 International License. (g) Formation of conductive filaments throughout GO membranes and (h) potential-mediated water permeation. Source: Zhou et al. [45]. Reproduced with permission of Springer Nature.

4.2.3.2 Applied Pressure Effects

External pressure, as one of the most common driving forces for filtrations, is ubiquitous in water separation processes, including reverse osmosis and nanofiltration [49]. Owing to the relatively low out-of-plane bending stiffness of GO nanosheets (<1.4 eV), membranes constructed by these building blocks are compressible once mechanical forces are applied from vertical direction [50]. Geometrical deformations may occur in micro- and/or nanoscale in response to varying pressure to modulate trans-membrane water permeation. A study led by Peng discovered that water flux (recorded at stable state) climbed linearly with applied pressure rising from 0 to 0.3 MPa, but the increasing rate slowed down when the pressure once further elevated to 1.3 MPa (Figure 4.4d) [44] The linear pressure–flux relation in the low-pressure range can be ascribed to the moderate rigidity of GO sheets so that pressure is fully converted to drive water flow. The non-linear relation in the high-pressure range, on the other hand, is considered to involve membrane compactions in both micro and nanoscale. Microscale compaction happens because of improved nanosheet stacking. It is believed that GO membranes assembled by imperfect methods, mostly vacuum-assisted filtration, display a loosely stacked configuration [51]. In addition to regular interlayer channels, this configuration contains plenty of empty spaces and cavities as extra water pathways. Under pressure, however, these non-ideal paths will be eliminated as nanosheets will rearrange to be more densely and orderly packed, minimizing the water leakage (Figure 4.4e). Nanoscale changes can also take place when sufficient force is exerted. Besides collapsing those untenable nanowrinkles, heavy load pushes neighboring sheets to get closer despite their mutual repulsions, and the narrowed channels will then impose considerably higher energy barrier to water transport (Figure 4.4f) [44].

4.2.3.3 Applied Potential Effects

Electrical conductivity is another intrinsic property of GO nanosheets and laminar membranes assembled by them, and this property allows their potential use as electrodes. Therefore, externally applied potential or current can conduct through to induce additional electric fields in the membrane to regulate water diffusion in 2D channels. Such electrical control over water transport has been achieved in Zhou's latest work by mediating water ionization degree in graphene-based membranes. Through a controllable breakdown method, nanoscale filaments are obtainable in GO membranes at the membrane thickness direction (Figure 4.4g) [45]. Even with a minor potential applied, these conductive filaments can generate electric fields as high as over 10^7 V m^{-1} and dissociate surrounding water molecules into H_3O^+ and OH^-. Through strong hydrogen bond with more water around them, these as-produced ions then form large hydrated clusters to stuff the channel and hinder the free passage of water. In response to escalating current density from 0 to 20 mA, in-channel ions will instantly aggregate and reduce water permeance from ~150 to 10 mg m^{-2} h^{-1}(Figure 4.4h).

4.2.4 Guest-Material-Mediated Transport

With the water-selective nano or subnanoscale channels, pristine graphene-based membranes have indeed found themselves applicable in a few water separation processes. Nevertheless, the increasing separation demands of water from a wider range of solutes make it necessary to further modify membranes for optimal water-solute selectivity in specific settings. The most common way for modification is to introduce guest materials into host GO membranes. Because of the large surface area and rich active sites, GO can incorporate 0D (quantum dots, molecules, and ions) [52], 1D (nanotubes and nanothreads) [53], and 2D (nanosheets) [54] via crosslinking, bending, and even *in situ* growth. The reconstituted multicomponent layered structure not only benefits from improved stability but also has its channel size adjusted accordingly for desired separation outcomes.

4.2.4.1 Stabilizing Effects

As can be implied by Section 4.2.3.1, stability proves a critical parameter for graphene-based membranes working in aqueous media as it resists unwanted swelling (caused by water) and compaction (caused by strong crossflow or hydraulic force) to ensure reliable and predictable separation performances. One of the most straightforward stabilizing strategies is to introduce "linking forces" between host nanosheets by inserting or growing diminutive species (Figure 4.5a) [55]. Owing to the versatile chemical nature of graphene-based membranes, namely their sp^2–sp^3 mixed surface, multiple crosslinking routes have been developed, including (i) chemical bonding, particularly between carboxyl and hydroxyl groups of GO with amino- and carboxyl-containing crosslinkers through esterification, dehydration condensation, and nucleophilic addition reactions [56].; (ii) electrostatic attractions between negatively charged GO nanosheets and positive materials [57]; (iii) intermolecular forces that take the advantage of conjugated π-electron cloud (sp^2) and oxygenated functional groups (sp^3) for π–π interaction and hydrogen bonding, respectively [53]. A good example of crosslinker-stabilized GO composite membranes is an "all-carbon" architecture established by Chen and coworkers by spacing partly reduced GO nanosheets with carbon nanotubes (CNTs) [53]. These carbon-based spacers, once assimilated into both top and bottom nanosheets through π–π stacking, can not only firmly hold the membrane but also well maintain the interlayer channels due to their high rigidity. Therefore, such CNT–rGO hybrid membranes show superior resistance toward swelling and turbulent flow in pressure-driven separations. The other strategy is to blend host GO and other guest 2D nanosheets. This design aims to avoid the face-to-face contact of GO nanosheets, so their interlayer repulsive forces can be minimized to cause swelling and destabilization [58]. Considering material compatibility, graphene (Gr), rGO, and other carbonaceous nanosheets are ideal candidate guests to be introduced. Compared to single-component GO membranes, the hybrid membranes will have a large fraction of GO–GO channels replaced by GO–Gr and GO–rGO channels. The presence of less functionalized channels amid GO network will help suppress interlayer repulsion yet strength π–π attraction for enhanced stability (Figure 4.5b) [54].

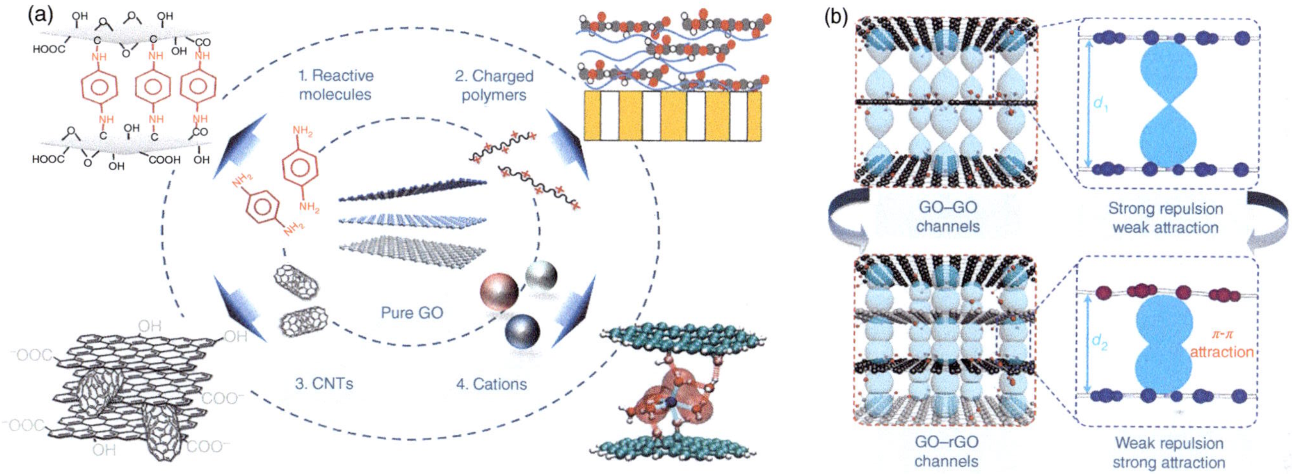

Figure 4.5 Effects of guest materials in GO membranes stabilization. (a) Crosslinking GO membranes via reactive molecules, charged polymers, CNTs, and cations. (b) Blended membranes composed of GO nanosheets with rGO nanosheets. Source: Kang et al. [6]. Reproduced with permission of Wiley-VCH.

4.2.4.2 Size-Controlling Effects

Maintaining the channel size of graphene-based membranes at a stable and appropriate level is the prerequisite for effective separations. To meet specific separating requirements, the intercalation of guest materials with varied dimensions as size modifiers is necessary. Following this concept, early works first endeavored to crosslink GO membranes using monomers with adjustable *d*-spacing, such as diamine (NH_2–R–NH_2) and dicarboxylic (HOOC–R–COOH) molecules [59]. As the two terminal groups of –NH_2 and –COOH covalently bridge GO nanosheets via –NH–CO– and —COO— bond, respectively, the molecular length of R chain in the middle can be designed for either enlarged or narrowed channels. While this tactic offers a wide size range to manipulate mass transport, it mainly works in nanoscale. To exercise accurate *d*-spacing control within subnanoscale, where critical water-solute separations, like desalination, happen, Chen et al. developed a cation-tailoring method [60]. By treating GO membranes with a series of solutions (containing K^+, Na^+, Mg^{2+}, Ca^{2+}, and Al^{3+}), cations could settle between nanosheets through the interaction with both sp^2 and sp^3 regions. This method enables *d*-spacing modulation by the precision of 1 Å and the membrane crosslinked by smaller ions (Na+) will block any other ions with larger hydrated size (Mg^{2+}, Ca^{2+}, and Al^{3+}) to make highly precise water-ion separations possible. It is also worth mentioning that these guest materials, besides being effective size modifiers, also play another role as membrane stabilizers.

4.2.4.3 Surface-Chemistry-Modifying Effects

Accompanied by the grafting of guest materials is the introduction of their chemical properties into the composite membranes. The grafting modifies channel surface chemistry in terms of hydrophilicity and functional groups and thus alters the way water is transported. In an attempt to modify wettability, Lai managed to interlink GO laminates with diamine monomers, including aliphatic ethylenediamine (EDA), butylenediamine (BDA), and aromatic *p*-phenylenediamine (PPD) [59]. While PPD-linked channels possessed the largest channel size (10.4 Å compared to 8.7 Å) among the three, they unexpectedly yielded the lowest water uptake, which is inconsistent with size-control transporting mechanism. This inconsistency is believed to stem from the chemical difference of these monomers. Compared to the aliphatic EDA and BDA, PPD contains aromatic rings and inclines to anchor at nanochannel entrance, and thereby the hydrophobic surface prohibits water from streaming in. In opposite, the introduction of guest materials with high hydrophilicity will possibly contribute to enhanced water flux. This has been verified by Yang and Grey who interlinked GO membrane with sulfosuccinic acid (SSA) molecules (containing two carboxyl groups and one sulfonic group) [61]. While the –COOH acts as the crosslinking handles, the pendant hydrophilic –SO_3H shows a good affinity with water, and such affinity becomes even stronger once the sulfonic groups are deprotonated in water (–SO_3H → –SO_3^- and H^+) to take in more water. Besides the conventional wettability-oriented modification route, the surface chemistry can also be modified by introducing and programming various functionalities to achieve faster water transport. This concept inspires an emerging trend to incorporate artificial aquaporin (aquaporin (AQR), ultrafast water channels in biological cells) into graphene-based membranes [62]. It involves the synthesis of a peptide with a few well-sequenced amino groups, and the choice of these amino groups, as well as their special sequence, will lead to water recognition capability. By programming two precursor amino acids arginine (R) and phenylalanine (F), Lee et al. managed to replicate the core water-affinitive function of aquaporin in their prepared peptides. Once materialized and tethered into GO channels, these peptides can serve as water-selective filters to facilitate water passage by a great degree without sacrificing much solute-rejecting ability. Considering the limited accuracy to program artificial aquaporins, a more recent study directly tethered AQR-carrying red blood cells on GO nanosheets to build up composite membranes. With the naturally programmed aquaporin introduced, the membranes showed a very high NaCl rejection of 99.1% while still achieving a decent water flux around 8 $l\,m^{-2}\,h^{-1}\,bar^{-1}$ [63].

4.2.4.4 Smart Gating Effects

Smart gating is a process in which the channel properties of membranes are altered by external signals (e.g. pH, temperature, and light) so that trans-membrane mass transport can be turned up (or on) and down (or off) [64–67]. Its significance has been gradually acknowledged with the increasing need for water separation in more complex settings. One of such needs is simultaneous water collection and solute recovery from multicomponent sources (e.g. those containing monovalent ions, divalent ions, and dyes). Due to the physicochemical difference of solutes, previous treatment schemes rely on several successive filtration operations on separate membranes, which is time-, labor-, and material-consuming (Figure 4.6a). With the assistance of smart gating, continuous and nonintrusive control on channel size or surface chemistry is feasible to realize various separating purposes on the same membrane [66].

Up till now, the construction of smart gating GO membranes is via hosting stimuli-responsive guest materials. As a proof of concept, Zhao and Liu established a temperature-gated GO membrane by combining GO nanosheets with temperature-sensitive polymerized *N*-isopropylacrylamide (PNIPAM) [66]. When covalently anchored inside GO channels, these guests can change their molecular configuration according to temperature to adjust channel size (Figure 4.6b,c). When the membrane is treated with temperature below the polymer lower critical solution temperature (LCST), its chain will display a swollen coil structure as a result of hydrogen bonds between PNIPAM and water, leading to larger *d*-spacing. By contrast, if the temperature goes beyond LCST, the PNIPAM–water hydrogen bond will be replaced by PNIPAM–PNIPAM intra/intermolecular hydrogen bonds to cause shrunken channels. Correspondingly, water permeance can be gated in the range from 25 to 50 °C, showing a reverse shift between 12.4 and 1.8 $l\,m^{-2}\,h^{-1}\,bar^{-1}$ (Figure 4.6d,e). Notably, along with the channel shrinkage in rising temperatures, a series of solutes of different sizes can be removed stepwise.

Apart from temperature, pH is another common yet important signal for gated transport [68, 69]. It has been proved that the introduction of oppositely charged materials can induce a pH-channel chemistry dependence in nanochannels to

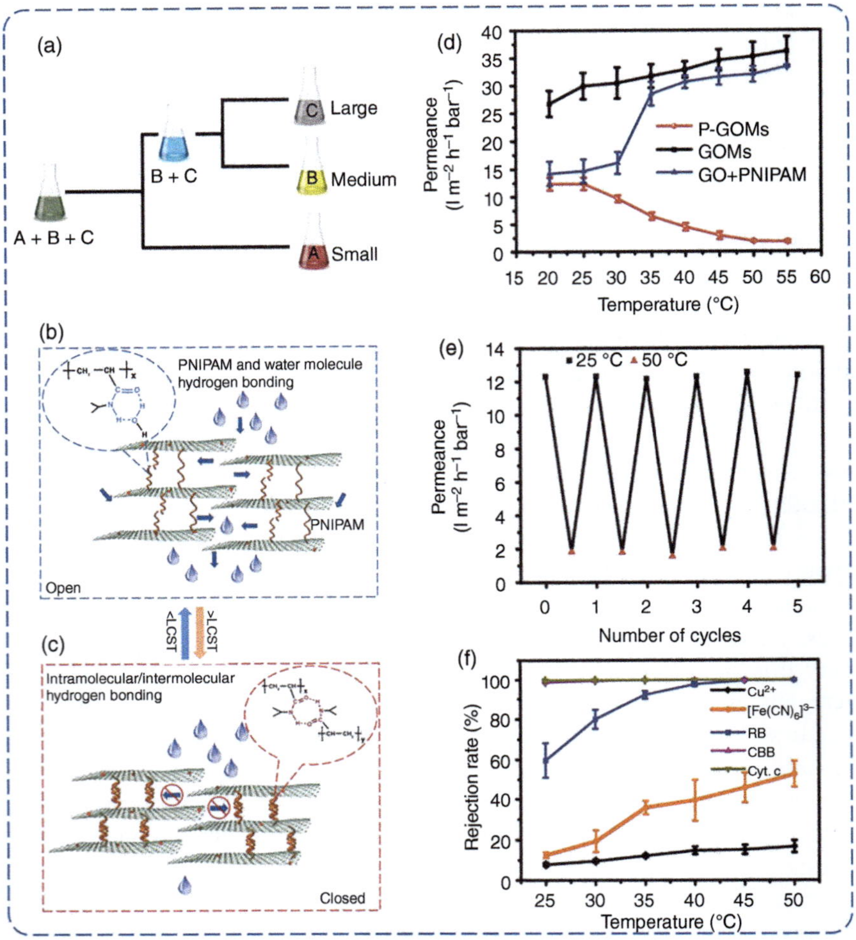

Figure 4.6 Temperature-gated mass transport through GO–PNIPAM membranes. (a) schematic illustration of step-wise water separation in complex water source by smart gating. (b) and (c) Channel size change due to the molecule configuration swift of PNIPAM chain in response to temperature. (d) and (e) Water permeance change in GO–PNIPAM membranes following temperature rise and fall. (f) Incremental solute removal by GO–PNIPAM membranes in elevating temperature. Source: Modified from Hong et al. [64]. Reproduced under the terms of the Creative Commons Attribution 4.0 International License.

regulate water transport. A recent study by Andreeva and Novoselov developed a composite membrane built up by sandwiching positively charged polyethyleneimine (PEI, bearing amino groups) among negatively charged GO nanosheets [65]. As a weak polycation, PEI changes its degree of ionization according to local pH and renders the membrane more permeable at lower pH (~2) and less permeable at higher range (>4), realizing a pH-gated water transport (Figure 4.7a,b). This gating effect primarily stems from the varied channel wettability caused by the different solution environments. In a more acidic media, the protonation of amino groups ($-NH_2 + H^+ = -NH_3^+$) occurs and helps the polymer stretch along with the channel

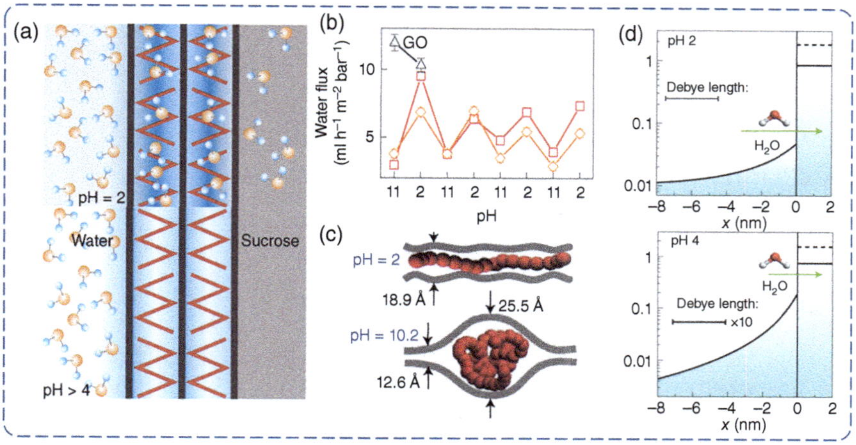

Figure 4.7 pH-gated mass transport through GO–PEI membranes. (a) Water transport in GO–PEI membranes in pH = 2 (top, fast) and pH = 4 (bottom, slow) and thus (b) pH-gated water flux in response to solution pH change. (c) PEI chain configuration inside GO channels at varied pH states. (d) The discrepancy of inner and outer positive charges creates osmotic pressure for extra water intake into GO–PEI membranes (pH = 2 or 4). Two GO nanosheets are set at $x = 0$ and $x = 2$ to mimic a positive channel, while the region at $x < 0$ is the diffusion layer of the EDL induced by GO outer surface. Source: Lee et al. [63]. Reproduced with permission of Springer Nature.

direction, providing more hydrophilic hopping sites for continuous transport (Figure 4.7c). Once the solution goes more neutral, however, the deprotonation of $-NH_3^+$ will cause the coiling and increased hydrophobicity of PEI chains, and thus hinder water intake to the channel. Interestingly, the charge distribution at different pH values also explains the controllable water permeation. Unlike pure GO membranes with only negative charges, GO–PEI membranes are designed to contain both negative and positive ones. While the outer GO surface still carries more negative charges ($-COOH = -COO^- + H^+$), the inner channels are predominantly positive charged ($-NH_2 + H^+ = -NH_3^+$). Considering that the EDL induced by outer GO surface will be mainly composed of counter ions (positively charged), the inner–outer positive charge density discrepancy will form an extra osmotic pressure to draw water into the membrane (Figure 4.7d). Such discrepancy reaches its maximum at pH = 2 but reduces at higher pH because of less positive inner channel ($-NH_3^+ + OH^- = -NH_2 + H_2O$) and more negative outer surface ($-COOH + OH^- = -COO^- + H_2O$), resulting in smaller osmotic pressure and less water intake.

To conclude, the transport of water in graphene-based membranes is a multifactor-controlled process due to the unique layered membrane structure. Primarily, its permeation is closely relevant to membrane geometry and surface chemistry as they decide the uptake, traveling distance and speed through the membranes. These intrinsic membrane features are further subjected to extrinsic interferences, including working environmental stimuli and sometimes the introduction of guest materials. By carefully manipulating these external transport-controlling factors, mass transport will become more tunable in graphene-based membranes to enable various water separation applications with distinct requirements, as will be discussed in Section 4.3.

4.3 Graphene-based Membrane Water Separation Applications

4.3.1 Nanofiltration

Nanofiltration is a low-pressure-driven membrane operation aiming to remove species whose molecular size is roughly around 1 nm. It allows the separation of water from multivalent ions (e.g. Ca^{2+}, Mg^{2+}, and Cu^{2+}) and soluble organics to realize water softening, heavy metal removal, and purification [70]. When fully hydrated in aqueous working environments, GO membranes normally possess a *d*-spacing of at least 1.2 nm, which gives a desired effective channel size of around 0.85 nm by subtracting the thickness of a GO nanosheet (~0.35 nm). The appropriate channel size explains the popularity in recent years to apply graphene-based membranes in nanofiltration field (as listed in Table 4.1, also for reverse osmosis and forward osmosis). Although a direct comparison between the separation performance of these membranes is unreliable due to drastic differences in terms of start material property, feed solution composition, and testing conditions, a progressive analysis is still practical and necessary to reflect the incremental development of graphene-based membranes.

Much discouraged by swelled interlayer channels and poorly controlled thickness (Figure 4.8a), initial graphene-based membranes only exhibited moderate water flux of 30 $l\,m^{-2}\,h^{-1}\,bar^{-1}$ and poor rejection of less than 80% even just against medium-sized dyes (Evans Blue 1.2×3.1 nm) [44]. Since the ultimate outcomes of filtration are expected to be high permselectivity (both high water permeability and solution rejection), following membranes were mainly designed through two strategies [3]. A number of studies implemented crosslinking and reduction strategies to reduce *d*-spacing to build up membranes with better solute-rejecting capability (increased selectivity, Figure 4.8b) [57, 72, 73, 82]. One of such endeavors was to mutually stack GO nanosheets with 1,3,5-benzenetricarbonyl trichloride (trimesoyl chloride (TMC)) monomers [49]. These diminutive molecules hydrolyzed to yield three carboxyl groups and held host GO membranes via —COO— bonds. With the *d*-spacing reduced from ~1.2 to 1 nm, the membrane showed a decent dye rejection (Rhodamine-WT, 95%) and relatively low divalent rejection of 46% while maintaining water flux at around 20 $l\,m^{-2}\,h^{-1}\,bar^{-1}$. To further improve ion-rejecting ability, Li and Zhang treated GO membranes in mild heating (80 °C) to yield even more reduced channels of only 0.95 nm [71]. In comparison to membrane treatment in higher temperature, a mild heating atmosphere well preserved the layered structure, resulting in good Na_2SO_4 rejection of 91% and water flux of around 4 $l\,m^{-2}\,h^{-1}\,bar^{-1}$.

The other strategy, however, attempted to construct more permeable (increased permeability) graphene-based membranes by incorporating additional water pathways, including large-sized nanowrinkles and nanopores (Figure 4.8c) [18, 28, 43].

Table 4.1 Water separation performance of a few representative graphene-based membranes.

Hybrid membranes	Stability mechanism	Geometrical factors	Membrane process and test condition	Membrane performance	Reference
Pure GO	N.A	N.A	NF (1 bar) 15 μM EB	F: 30 l/MHbar R(EB): 80% pH = 10	[44]
GO/TMC	–COO– between –COOH from TMC and –OH from GO	$d \sim 1$ nm $h \sim 50$ nm	NF (~3.4 bar), 10 mM Na_2SO_4 7.5 mg l^{-1} RWT	F: 18 l/MHbar R(Na_2SO_4): 46% R(RWT): 95%	[49]
rGO (mildly reduced)	Increased π–π stacking due to more sp^2 regions	$d \sim 0.95$ nm $h \sim 20$ nm	NF (~4 bar) 1 g l^{-1} Na_2SO_4	F: 4 l/MHbar R(Na_2SO_4): 91%	[71]
rGO (air-reduced)	Increased π–π stacking due to more sp^2 regions	$d \sim 1.05$ nm $h \sim 25$ nm	NF (~6 bar) 1 g l^{-1} NaCl	F: 1.1 l/MHbar R(NaCl): 83%	[72]
rGO/K^+ (partly reduced)	Increased π–π stacking due to more sp^2 regions and K^+-GO complexion	$d \sim 0.63$ nm $h \sim 25$ nm	NF (~6 bar) 1 g l^{-1} NaCl	F: 0.6 l/MHbar R(NaCl): 91%	[73]
GO with nanowrinkles	N.A.	$d \sim 2.5$ nm $h \sim 1850$ nm	NF (~1 bar) 1 mM l^{-1} RB	F: 279 l/MHbar R(RB): 87%	[43]
Porous rGO (lightly reduced)	Increased π–π stacking due to more sp^2 regions	$d \sim 1$ nm $h \sim 22$ nm	NF (~5 bar) 20 mM Na_2SO_4 20 mM MB	F: 21.8 l/MHbar R(Na_2SO_4): 60% R(MB): 99.2%	[28]
rGO/MWCNT	π–π stacking due to more sp^2 regions	$h \sim 64$ nm	NF (~5 bar) 10 mM Na_2SO_4 20 mM DY	F: 11.3 l/MHbar R(Na_2SO_4): 81% R(MB): 99%	[74]
Porous rGO (highly reduced)	Increased π–π stacking due to more sp^2 regions	$d \sim 0.66$ nm $h \sim 15$ nm	NF (~6 bar) 20 mM Na_2SO_4	F: 40 l/MHbar R(Na_2SO_4): 90%	[75]
GO/ZIF-8	Crystal growth-induced link	$d \sim 0.93$ nm $h \sim 50$ nm	NF (~1 bar) 0.1 g l^{-1} MB	F: 50 l/MHbar R(MB): 100%	[76]
Shear-aligned GO	Increased π–π stacking due to order structure	$d \sim 1.2$ nm $h \sim 200$ nm	NF (~0.5 bar) 0.01 g l^{-1} MR	F: 40 l/MHbar R(MR): 100%	[77]
Symmetrical GO	Increased π–π stacking due to order structure	$d \sim 0.97$ nm $h \sim 4.7$ nm	NF (~bar) 50 mM $MgSO_4$	F: 20 l/MHbar R($MgSO_4$): 60%	[38]

(continued)

Table 4.1 (Continued)

Hybrid membranes	Stability mechanism	Geometrical factors	Membrane process and test condition	Membrane performance	Reference
GO/Phenolic nanomesh	π–π stacking, –COO– between GO and mesh	$d \sim 0.64$ nm $h \sim 260$ nm	NF (~1 bar) 500 mg Na_2SO_4	F: 165.6 l/MHbar R(Na_2SO_4): 97%	[78]
GO/C_3N_4	Decreased electrostatic repulsion between GO	$d \sim 0.42$ nm $h \sim 2200$ nm	NF (~5 bar) 20 mM Na_2SO_4	F: 33.5 l/MHbar R(Na_2SO_4): 92.5%	[58]
GO/PAH	Electrostatic attraction between negative GO and positive PAH	N.A.	RO (~20.7 bar) 6.7 mM $MgCl_2$ 6.7 mM Na_2SO_4	F: 2.5 l/MHbar R($MgCl_2$): 90% R(Na_2SO_4): 70%	[79]
GO/Graphene/Ca^{2+}	π–π stacking force and GO–Ca^{2+} complexion	$d \sim 1.1$ nm $h \sim 30$ nm	RO (~50 bar) 2 g l^{-1} NaCl	F: 0.35 l/MHbar R(NaCl): 85%	[80]
rGO/Tannic Acid	π–π stacking force between rGO and aromatic tannic acid	$d \sim 0.95$ nm $h \sim 100$ nm	RO (~10 bar) 0.5 g l^{-1} NaCl	F: 4 l/MHbar R(NaCl): 90%	[81]
GO/Red blood cells (aquaporin)	N.A.	$d \sim 2$ nm $h \sim 700$ nm	RO (~8 bar) 1 g l^{-1} NaCl	F: ~8 l/MHbar R(NaCl): 99.1%	[63]
GO/K^+	GO–K^+ complexion hydrogen bond network	$d \sim 1.1$ nm $h \sim 750$ nm	FO Draw: 0.25 mM NaCl + KCl Feed: 0.25 mM KCl	F: 0.1 l/MH R(NaCl): 99%	[60]
GO/Graphene	π–π stacking of GO and Graphene	$d \sim 1.0$ nm $h \sim 5$ μm	FO Draw: 3 M Sucrose Feed: 0.1 mM NaCl	F: 0.5 l/MH R(NaCl): 99%	[11]
rGO (fully reduced)	Increased π–π stacking due to more sp^2 regions	$d \sim 0.34$ nm $h \sim 100$ nm	FO Draw: 1 M NaCl Feed: water	F: 25 l/MH R(NaCl): 1.2 g/MH	[13]
rGO (fully reduced, hexane treated)	Increased π–π stacking due to more sp^2 regions	$d \sim 0.34$ nm $h \sim 100$ nm	FO Draw: 1 M NaCl Feed: water	F: 8 l/MH R(NaCl): 1.2 g/MH	[17]
rGO (fully reduced, ethanol treated)	Increased π–π stacking due to more sp^2 regions	$d \sim 0.34$ nm $h \sim 100$ nm	FO Draw: 1 M NaCl Feed: water	F: 50 l/MH R(NaCl): 1.2 g/MH	[17]

Notes: NF: nanofiltration; RO: reverse osmosis; FO: forward osmosis; TMC: 1,3,5-benzenetricarbonyl trichloride; MWCNTs: multi-walled carbon nanotubes; ZIF-8: zeolitic imidazolate framework-8; PAH: poly (allylamine hydrochloride); EB: Evans Blue; RWT: Rhodamine-WT; RB: Rhodamine B; MB: Methylene Blue; DY: Direct Yellow; MR: Methylene Red; F: water flux; R: rejection rate; l/MHbar: $l\,m^{-2}\,h^{-1}\,bar^{-1}$; l/MH: $l\,m^{-2}\,h^{-1}$; g/MH: $g\,m^{-2}\,h^{-1}$. d: interlayer spacing; h: membrane thickness.

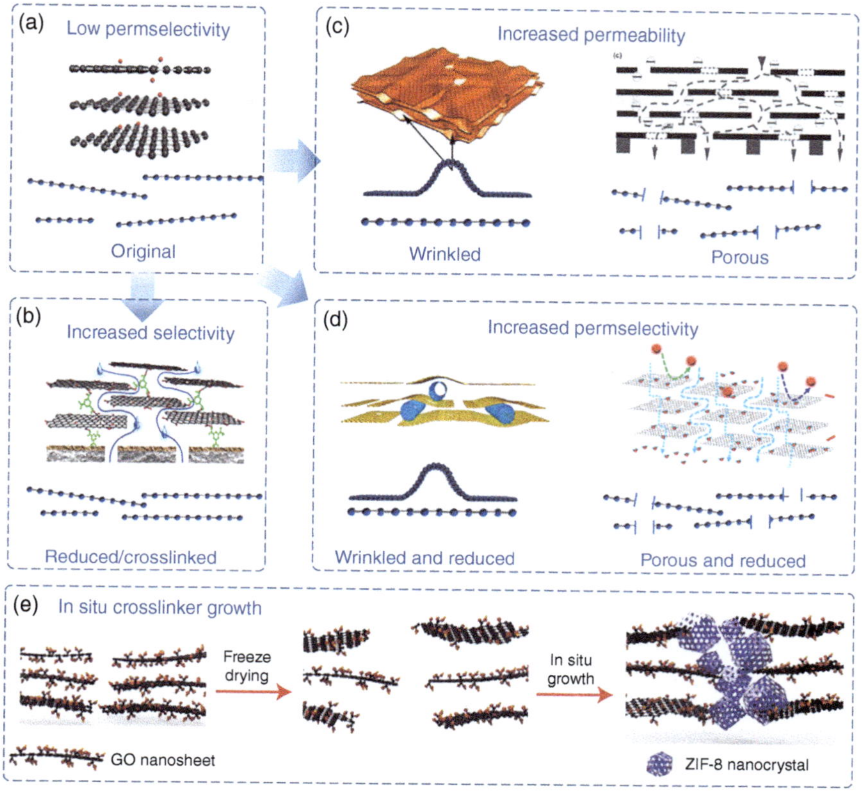

Figure 4.8 Design of development of graphene-based membranes for nanofiltration. (a) Original GO membranes with low permselectivity. (b) Membranes with increased selectivity by reduction and crosslinking. Source: (a) Hu et al. [49]/with permission of American Chemical Society. (c) Membrane with increased permeability by introducing nanowrinkles and nanopores. Source: (b) Adapted from Huang et al. [43]. (c) Adapted from Han et al. [28]. (d) Membranes with increased permselectivity by combining the strategies in (b) and (c). Source: (d) Adapted from Meng et al. [82]. Adapted from Li et al. [71]. (e) Membranes with increased permselectivity by a less-interruptive *in situ* growth method. Source: (e) Zhang et al. [76]/with permission of Springer Nature.

An exemplified work by Huang et al. managed to introduce plenty of nanowrinkles into GO membranes through self-sacrifice-templating method [43]. This involved the insertion of 1D sacrificial $Cu(OH)_2$ nanostrands with a diameter of 2.5 nm between GO nanosheets during membrane preparation and their dissolution in a post-etching operation. These as-prepared curved nanowrinkles tremendously enhanced water throughput to ~280 $m^{-2}\ h^{-1}\ bar^{-1}$ and Rhodamine B (RB) rejection to near 90%. Similar phenomena were also observable when in-plane pores, normally with a diameter of 2–3 nm, were massively produced on GO nanosheets and used as shortcuts, which improves water flux at the cost of solute rejection of a certain degree [28].

Noticing this obvious “trade-off” effect between membrane permeability and selectivity, further studies started to merge the above two routes, hoping to establish graphene-based membranes with more balanced separation performances and increased permselectivity. To this end, two rational designs have been raised by combining ion-rejecting reduced interlayer channels with water-permeable nanowrinkles or nanopores (Figure 4.8d) [74, 75]. To test the effectiveness of these designs, Han and Gao established a multi-walled carbon nanotube (MWCNT)-GO graphene-based membrane by intercalating multi-walled carbon nanotubes into partially reduced GO nanosheets [74]. 2D channels in the as-prepared membrane featured both MWCNT-supported enlarged regions and reduced narrow regions. On one hand, the enlarged and frictionless CNT regions opened channels for facilitated water transport and helped the membrane reach a flux of $11.3\,l\,m^{-2}\,h^{-1}\,bar^{-1}$. On the other hand, the narrowed parts, while still being water-permeable, could deny the entry of solutes into the channel to achieve a very good rejection toward not only dyes (>99% for Direct Yellow) but also divalent salts (84% for Na_2SO_4). Meanwhile, the combination of narrowed channel with open pores also worked, as reflected in a design by Li et al. [75] In this design, graphene-based membranes were stacked by GO nanosheets with in-plane pores (pGO, pore diameter ~3 nm) and treated by gentle thermal reduction. While the carefully controlled interlayer channels (*d*-spacing of 0.69 nm) were responsible for solute blocking, those pore structures in different layers, when connected with others, could form straightforward pathways throughout the membrane as fast tracks for water passage. Therefore, a satisfactory water flux of $\sim 40\,l\,m^{-2}\,h^{-1}\,bar^{-1}$ and Na_2SO_4 rejection of around 90% could be achieved in this membrane, being superior to a typical rGO membrane with much lower flux of only $1.5\,l\,m^{-2}\,h^{-1}\,bar^{-1}$ yet close *d*-spacing (0.65 nm) and salt rejection (93%). Considering the difficulty for precise control on pore formation, more recent studies improved the "channel-pore" design by replacing pGO nanosheets with materials possessing uniform intrinsic pores. One such work by Lan and Liu mixed GO with phenolic nanomeshes. Owing to the abundant number of aromatic rings and hydroxyl groups, these meshes could bond tightly with GO via π–π attraction and covalent bonding to narrow interlayer spacing down to 0.64 nm. The proper mesh size and the hydrophilic nature greatly favored water permeation, leading to very high water permeance up to $165.6\,l\,m^{-2}\,h^{-1}\,bar^{-1}$ while maintaining Na_2SO_4 rejection of 97% [78].

Regardless of the chosen strategies, the previous design and fabrication of high-performance graphene-based membranes for nanofiltration purposes somehow relies on the introduction of guest materials, like in many of the above-mentioned studies. While it proves simple and convenient, the operation to include external materials into host membranes unavoidably causes disruption to the layered structure and potentially undermines membrane separation ability. To deal with this problem, a novel *in situ* crosslinker growth route has been raised by An and Zhang to grow zeolitic imidazolate framework-8 (ZIF-8) crystals throughout GO membranes (Figure 4.8e) [76]. The preparation was achieved by first immersing frozen GO membranes in mixed MeOH precursor solution so that the precursors would stay at inter-edge gaps rather than streaming into interlayer channels. Once these precursors coordinated with edge-clustered carboxyl groups on GO nanosheets, crystal nucleation was initiated and the crystal seeds would continue growing until the gap space was stuffed with ZIF-8 crystals. In addition to providing sufficient mechanical strength to the membrane, the positioning of ZIF-8 around nanosheet edge also sealed any possible microscale defects for unwanted leakage. Taking advantages of the non-interruptive and defect-sealing crosslinkers, the ZIF-8-GO hybridized membranes demonstrated stable and excellent nanofiltration performance even under crossflow-operating mode, achieving a long-term flux of $50\,l\,m^{-2}\,h^{-1}\,bar^{-1}$ and complete block toward Methylene Blue.

4.3.2 Reverse Osmosis

In theory, reverse osmosis shares the same working principle with nanofiltration but operates for a more demanding goal of isolating almost all the solutes (e.g. monovalent ions, such as K^+ and Na^+) from water. The high separation standard requires any promising membrane candidates possess channels below 1 nm (preferably <0.7 nm to remove K^+). To utilize graphene-based membranes for reverse osmosis thus always needs the narrowing of their effective channel size [79]. To this end, a tightly crosslinked graphene-based membrane based on multiple intermolecular interactions was devised by Morelos-Gomez et al. [80]. Different from the majority of crosslinked membranes enabled by only single attractive force, this composite membrane consists of GO, few-layered graphene (FG) and Ca^{2+}, and the increased GO–FG π–π stacking force and GO–Ca^{2+} coupling firmly hold neighboring layers together to control the d-spacing below 1.1 nm (effective size below 0.76 nm). Operated under 50 Bar, this membrane can achieve NaCl rejection up to 85% and water flux of $0.5\,l\,m^{-2}\,h^{-1}\,bar^{-1}$. Such low water output is rather predictable once it is related to highly occupied channel interior after Ca^{2+} crosslinking. To bypass this trouble, Guan et al. endeavored to decrease effective channel size by a "one side" decoration method with the assistance of tannic acid. As a polyphenol, tannic acid possesses multiple aromatic rings and phenol groups [81]. While the aromatic rings enable its π–π combination with one side of GO channel wall, the phenol groups will either reduce nearby oxygenated groups or stay pendant, resulting in a reduced and stuffed channel with appropriate effective size and hydrophilic surface. Since the tannic acid is considered only residing at one side of the channel, more space is left for water transport than in the case of highly crosslinked channels. Therefore, a decent water flux of $\sim 4\,l\,m^{-2}\,h^{-1}\,bar^{-1}$ is possible for this membrane while its NaCl blocking remains as strong as 90%.

4.3.3 Forward Osmosis

Unlike the last two water separation applications, forward osmosis works free of hydraulic pressure and drives water flow by osmotic pressure. By placing a well-designed membrane between solutions of different concentrations, water will be drawn from the low (feed side, usually water source) to the high-concentration side (draw solution), and then recovered to realize water separation [83]. Despite the major difference in working mechanisms, membranes qualified for FO, by and large, follow the same rules as for NF and RO, albeit with smaller thickness to minimize internal concentration polarization (ICP). A successful implementation of graphene-based membranes for FO has been illustrated by Fang and coworkers by cationic controlling [60]. With K^+ inserted and settled in GO membranes through a hydrogen-bonding network, the membrane, using 0.25 M NaCl as draw, could attract water by 0.1 l m^{-2} h^{-1} while permitting almost no Na^+ to reversely pass, leading to a nominal salt rejection of over 99%. Similarly, Abraham and coworkers dispersed graphene (Gr) using GO as the surfactant and assembled them into a GO–Gr laminar membrane. By loading graphene nanosheets from 0 to 18 wt% (with respect to GO weight) in the blended membrane, its interlayer distance reduced from 14 to 10.2 Å. Because of the narrow and fixed channels, the as-obtained membrane has high NaCl rejection (over 90%) but again relatively low flux (~0.5 l m^{-2} h^{-1}) [11].

It is worth mentioning that despite the good ion-blocking ability shown by the above membranes, they do suffer from low flux. One fundamental reason can be

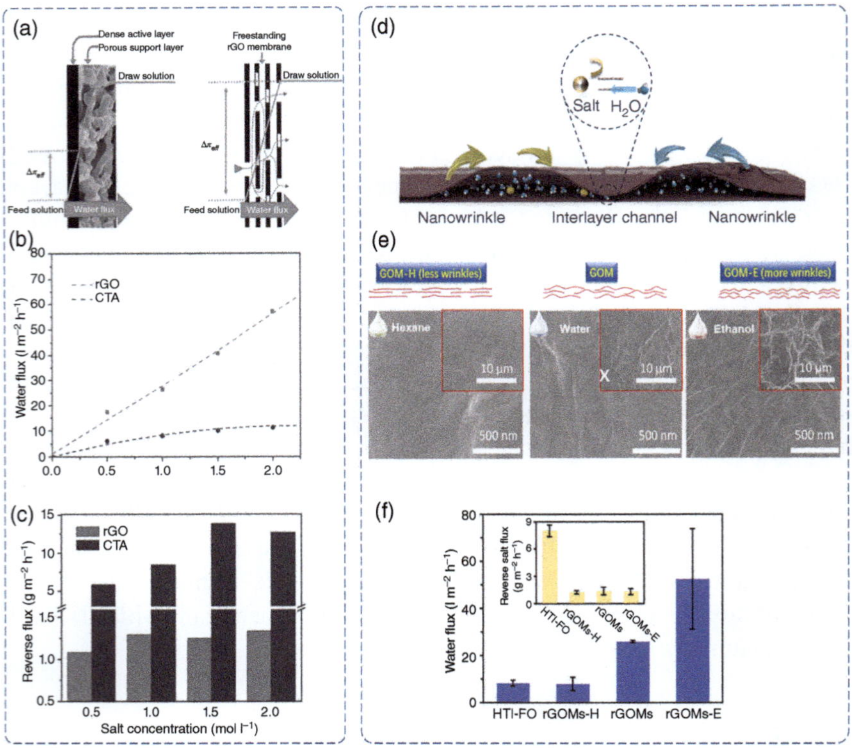

Figure 4.9 Design of development of graphene-based membranes for forward osmosis. (a) Comparison of osmotic pressure gradient profiles with thick (left) and thin (right) membranes. (b) Water flux and (c) reverse salt flux through freestanding rGO membranes, referenced by a commercial CTA membrane. Source: Liu et al. [13]/with permission from John Wiley & Sons, Inc. (d) Schematics of mass transport in nanowrinkle-involved transporting networks in rGO membrane. (e) Varied wrinkle densities on graphene-based membranes mediated by hexane (left), water (middle), and (ethanol), and (f) their corresponding separation performance in FO process. Source: Kang et al. [17]/with permission of John Wiley & Sons, Inc.

the excessive thickness that prevents the formation of effective osmotic pressure gradient across two sides (Figure 4.9a). Therefore, Zhang and coworkers developed ultrathin and fully reduced rGO membranes for FO process [13]. Being freestanding without substrate, this 100 nm-thick membrane could yield a very high water throughput of ~25 l m^{-2} h^{-1} and low reverse salt flux of 1.2 g m^{-2} h^{-1} with 1 M NaCl solution in the draw side (Figure 4.9b,c). The result is somewhat surprising because the nanochannels of fully reduced rGO are long considered impermeable [57]. Their following investigation then attributed the abnormal water yield to the existence of intrinsic nanowrinkles [17]. Instead of being composed of only typical interlayer channels, the transporting network in graphene-based membranes comprises nanowrinkles bridged by interlayer channels (Figure 4.9d). While the nanowrinkles act as fast water tracks, those narrow channels block most of salt passage yet allow water passage at some irregular stacking regions. As a proof of concept, they further managed to yield varied wrinkle density by a "solvent-controlling" drying method where hexane and ethanol were added on freshly prepared GO membranes, resulting in either less or more wrinkles (Figure 4.9e). With more nanowrinkles created in ethanol-treated membranes, the water flux in FO can correspondingly increase to 50 l m^{-2} h^{-1}, while the reverse salt flux remains almost unchanged (Figure 4.9f).

4.4 Conclusions and Perspectives

To conclude, water separation remains to be a critical process to purify and collect water from various sources for multidisciplinary usage. Membranes, particularly 2D membranes composed of graphene-based materials, are a promising candidate for possessing nano and subnanoscale channels in their unique layered structure. These channels, with their adjustable size and versatile chemistry, are able to control mass transport (water in this section) in terms of species uptake at membrane entrance and species flowing inside the membrane. External interferences, here defined as working environments and guest materials, can indirectly regulate mass transport by modifying the physicochemical properties of membrane channels. These modifications, in turn, enable a large variety of tools and methods to devise graphene-based membranes toward desired transporting characteristics for nanofiltration, reverse osmosis, forward osmosis, and more application in the future.

Despite these inspiring discoveries from previous studies, to practically use graphene-based membranes for any of these mentioned or potential water-related separations does await a few critical breakthroughs at different levels: (i) From starting material itself. While the mechanical or chemical synthesis of graphene-based materials has become so prevalent and convenient in lab scale, to massively produce quality single-layered graphene/GO/rGO nanosheets with controllable features (e.g. lateral size, in-plane porosity, and functionalization degree) has always been a multidisciplinary obstacle. (ii) From membrane assembly. More importantly, preparing membranes free of structural defects (pinholes, curling, and macrovoids) is currently unachievable. To ensure the even distribution of nanosheets on substrate or over already-deposited nanosheets is a prerequisite to yield expected separation outcome as a minor irregular stacking during membrane preparation has been found severely compromising membrane selectivity [84]. While the nanosheets in vacuum-assisted filtration method often exhibit Brownian-like motion, directional forces, such as those generated by shear alignment, electrical or magnetic fields, may offer a way leading to ordered nanosheet piling up [77, 85]. (iii) From membrane modifications and post-treatments, it has become clear that the use of pure and pristine graphene-based membranes in all water-related separations is rather limited due to unmatched size, surface chemistry, and unwanted swelling effect. Appropriate modifications or post-treatments on the membrane, particularly by guest materials, prove necessary in most cases. To select proper materials, to ensure efficient host–guest binding, and to avoid membrane structure disruption during the binding may present further problems. With these hurdles transcended, graphene-based membranes are more likely to find themselves used in current water as well as emerging separation-related applications.

References

1 Alvarez, P.J.J., Chan, C.K., Elimelech, M. et al. (2018). Emerging opportunities for nanotechnology to enhance water security. *Nat. Nanotechnol.* 13: 634–641.

2 Mauter, M.S. et al. (2018). The role of nanotechnology in tackling global water challenges. *Nat. Sustain.* 1: 166–175.

3 Park, H.B., Kamcev, J., Robeson, L.M. et al. (2017). Maximizing the right stuff: The trade-off between membrane permeability and selectivity. *Science (80-.)* 356: 1138–1148.

4 Liu, G., Jin, W., and Xu, N. (2016). Two-dimensional-material membranes: a new family of high-performance separation membranes. *Angew. Chem. Int. Ed.* 55: 13384–13397.

5 Liu, G., Jin, W., and Xu, N. (2015). Graphene-based membranes. *Chem. Soc. Rev.* 44: 5016–5030.

6 Kang, Y., Xia, Y., Wang, H., and Zhang, X. (2019). 2D Laminar membranes for selective water and ion transport. *Adv. Funct. Mater.* 1902014: 1902014.

7 Koltonow, B.A.R. and Huang, J. (2016). Two-dimensional nanofluidics. *Sceince* 351: 1395–1397.

8 Muscatello, J., Jaeger, F., Matar, O.K., and Müller, E.A. (2016). Optimizing water transport through graphene-based membranes: insights from nonequilibrium molecular dynamics. *ACS Appl. Mater. Interfaces* 8: 12330–12336.

9 Hu, S. et al. (2019). Complete steric exclusion of ions and proton transport through confined monolayer water. *Science (80-.)* 363: 145–148.

10 Nair, R.R., Wu, H.A., Jayaram, P.N. et al. (2012). Unimpeded permeation of water through helium-leak-tight graphene-based membranes. *Science (80-.)* 335: 442–444.

11 Abraham, J. et al. (2017). Tunable sieving of ions using graphene oxide membranes. *Nat. Nanotechnol.* 12: 546–550.

12 Su, Y. et al. (2014). Impermeable barrier films and protective coatings based on reduced graphene oxide. *Nat. Commun.* 5: 1–5.

13 Liu, H., Wang, H., and Zhang, X. (2015). Facile fabrication of freestanding ultra-thin reduced graphene oxide membranes for water purification. *Adv. Mater.* 27: 249–254.

14 Amadei, C.A., Montessori, A., Kadow, J.P. et al. (2017). Role of oxygen functionalities in graphene oxide architectural laminate subnanometer spacing and water transport. *Environ. Sci. Technol.* 51: 4280–4288.

15 Shen, X., Lin, X., Yousefi, N. et al. (2014). Wrinkling in graphene sheets and graphene oxide papers. *Carbon, NY* 66: 84–92.

16 Wei, Y., Wang, B., Wu, J. et al. (2013). Bending rigidity and Gaussian bending stiffness of single-layered graphene. *Nano Lett.* 13: 26–30.

17 Kang, Y. et al. (2020). The role of nanowrinkles in mass transport across graphene-based membranes. *Adv. Funct. Mater.* 30: 1–7.

18 Li, D. et al. (2011). Controllable corrugation of chemically converted graphene sheets in water and potential application for nanofiltration. *Chem. Commun.* 47: 5810.

19 Woo, M.W. et al. (2019). The effects of edge functional groups on water transport in graphene oxide membranes. *ACS Appl. Mater. Interfaces* 11: 8483–8491.

20 Chen, B., Jiang, H., Liu, X., and Hu, X. (2017). Molecular insight into water desalination across multilayer graphene oxide membranes. *ACS Appl. Mater. Interfaces* 9: 22826–22836.

21 Cohen-Tanugi, D., Lin, L.C., and Grossman, J.C. (2016). Multilayer nanoporous graphene membranes for water desalination. *Nano Lett.* 16: 1027–1033.

22 Cho, Y.H. et al. (2017). Water and ion sorption, diffusion, and transport in graphene oxide membranes revisited. *J. Membr. Sci.* 544: 425–435.

23 Chen, J. et al. (2015). Size fractionation of graphene oxide sheets via filtration through track-etched membranes. *Adv. Mater.* 27: 3654–3660.

24 Geng, H. et al. (2017). Size fractionation of graphene oxide nanosheets via controlled directional freezing. *J. Am. Chem. Soc.* 139: 12517–12523.

25 Yang, Q. et al. (2017). Ultrathin graphene-based membrane with precise molecular sieving and ultrafast solvent permeation. *Nat. Mater.* 16: 1198–1202.

26 Li, H. et al. (2013). Ultrathin, molecular-sieving graphene oxide membranes for selective hydrogen separation. *Science (80-.)* 342: 95–98.

27 Saraswat, V. et al. (2018). Invariance of water permeance through size-differentiated graphene oxide laminates. *ACS Nano* 12: 7855–7865.

28 Han, Y., Xu, Z., and Gao, C. (2013). Ultrathin graphene nanofiltration membrane for water purification. *Adv. Funct. Mater.* 23: 3693–3700.

29 Yang, Q.-H., Huang, J., Yeh, C.-N. et al. (2015). On the origin of the stability of graphene oxide membranes in water. *Nat. Chem.* 7: 166–170.

30 Sun, P. et al. (2015). Ultrafast liquid water transport through graphene-based nanochannels measured by isotope labelling. *Chem. Commun.* 51: 3251–3254.

31 Deng, J. et al. (2017). Mechanism of water transport in graphene oxide laminates. *Chem. Sci.* 343: 740–742.

32 Radha, B. et al. (2016). Molecular transport through capillaries made with atomic-scale precision. *Nature* 538: 222–225.

33 Boukhvalov, D.W., Katsnelson, M.I., and Son, Y.W. (2013). Origin of anomalous water permeation through graphene oxide membrane. *Nano Lett.* 13: 3930–3935.

34 Algara-Siller, G. et al. (2015). Square ice in graphene nanocapillaries. *Nature* 519: 443–445.

35 Zettl, A. et al. (2010). Determination of the local chemical structure of graphene oxide and reduced graphene oxide. *Adv. Mater.* 22: 4467–4472.

36 Willcox, J.A.L. and Kim, H.J. (2017). Molecular dynamics study of water flow across multiple layers of pristine, oxidized, and mixed regions of graphene oxide: effect of graphene oxide layer-to-layer distance. *J. Phys. Chem. C* 121: 23659–23668.

37 Wei, N., Peng, X., and Xu, Z. (2014). Understanding water permeation in graphene oxide membranes. *ACS Appl. Mater. Interfaces* 6: 5877–5883.

38 Xu, W.L. et al. (2017). Self-assembly: a facile way of forming ultrathin, high-performance graphene oxide membranes for water purification. *Nano Lett.* 17: 2928–2933.

39 Zheng, S., Tu, Q., Urban, J.J. et al. (2017). Swelling of graphene oxide membranes in aqueous solution: characterization of interlayer spacing and insight into water transport mechanisms. *ACS Nano* 11: 6440–6450.

40 Ban, S. et al. (2016). Insight into the nanoscale mechanism of rapid H_2O transport within a graphene oxide membrane: impact of oxygen functional group clustering. *ACS Appl. Mater. Interfaces* 8: 321–332.

41 Chen, B., Jiang, H., Liu, X., and Hu, X. (2017). Observation and analysis of water transport through graphene oxide interlamination. *J. Phys. Chem. C* 121: 1321–1328.

42 Talyzin, A.V., Hausmaninger, T., You, S., and Szabó, T. (2014). The structure of graphene oxide membranes in liquid water, ethanol and water-ethanol mixtures. *Nanoscale* 6: 272–281.

43 Huang, H. et al. (2013). Ultrafast viscous water flow through nanostrand-channelled graphene oxide membranes. *Nat. Commun.* 4: 2979.

44 Huang, H. et al. (2013). Salt concentration, pH and pressure controlled separation of small molecules through lamellar graphene oxide membranes. *Chem. Commun.* 49: 5963–5965.

45 Zhou, K.G. et al. (2018). Electrically controlled water permeation through graphene oxide membranes. *Nature* 559: 236–240.

46 Talyzin, A.V., Sharifi, T., Yu, J. et al. (2015). Structure of graphene oxide membranes in solvents and solutions. *Nanoscale* 7: 15374–15384.

47 Wang, Z. et al. (2017). Understanding the aqueous stability and filtration capability of MoS_2 membranes. *Nano Lett.* 17: 7289–7298.

48 Saraswat, V. et al. (2017). Extrinsic cation selectivity of 2D membranes. *ACS Nano* 11: 1340–1346.

49 Hu, M. and Mi, B. (2013). Enabling graphene oxide nanosheets as water separation membranes. *Environ. Sci. Technol.* 47: 3715–3723.

50 Borysiuk, V.N., Mochalin, V.N., and Gogotsi, Y. (2018). Bending rigidity of two-dimensional titanium carbide (MXene) nanoribbons: a molecular dynamics study. *Comput. Mater. Sci.* 143: 418–424.

51 Chong, J.Y., Wang, B., Mattevi, C., and Li, K. (2018). Dynamic microstructure of graphene oxide membranes and the permeation flux. *J. Membr. Sci.* 549: 385–392.

52 Thebo, K.H. et al. (2018). Highly stable graphene-oxide-based membranes with superior permeability. *Nat. Commun.* 9: 1–8.

53 Goh, K. et al. (2015). All-carbon nanoarchitectures as high-performance separation membranes with superior stability. *Adv. Funct. Mater.* 25: 7348–7359.

54 Xi, Y.H. et al. (2016). Graphene oxide membranes with strong stability in aqueous solutions and controllable lamellar spacing. *ACS Appl. Mater. Interfaces* 8: 15557–15566.

55 Mi, B. (2014). Graphene Oxide Based Membranes for Ionic and Molecular Sieving. *Science (80-.)* 343: 740–743.

56 Yang, J. et al. (2018). Self-assembly of thiourea-crosslinked graphene oxide framework membranes toward separation of small molecules. *Adv. Mater.* 30: 1–8.

57 Hu, M. and Mi, B. (2014). Layer-by-layer assembly of graphene oxide membranes via electrostatic interaction. *J. Membr. Sci.* 469: 80–87.

58 Wu, Y. et al. (2021). 2D heterostructured nanofluidic channels for enhanced desalination performance of graphene oxide membranes. *ACS Nano* 15: 7586–7595.

59 Lai, J.-Y. et al. (2014). Cross-linking with diamine monomers to prepare composite graphene oxide-framework membranes with varying *d*-spacing. *Chem. Mater.* 26: 2983–2990.

60 Chen, L. et al. (2017). Ion sieving in graphene oxide membranes via cationic control of interlayer spacing. *Nature* 550: 1–4.

61 Yang, G. et al. (2019). Functionalizing graphene oxide framework membranes with sulfonic acid groups for superior aqueous mixture separation. *J. Mater. Chem. A* 7: 19682–19690.

62 Hwang, Y.Y. et al. (2018). Facilitated water transport through graphene oxide membranes functionalized with aquaporin-mimicking peptides. *Adv. Mater.* 30: 1705944.

63 Lee, C.S., Kim, I., Jang, J.W. et al. (2021). Aquaporin-incorporated graphene-oxide membrane for pressurized desalination with superior integrity enabled by molecular recognition. *Adv. Sci.* 8: 1–10.

64 Hong, S. et al. (2020). Photothermoelectric response of $Ti_3C_2T_x$MXene confined ion channels. *ACS Nano* 14: 9042–9049.

65 Andreeva, D.V. et al. (2021). Two-dimensional adaptive membranes with programmable water and ionic channels. *Nat. Nanotechnol.* 16: 174–180.

66 Liu, J. et al. (2017). Bioinspired graphene membrane with temperature tunable channels for water gating and molecular separation. *Nat. Commun.* 8: 1–8.

67 Lv, R. et al. (2018). Aqueous stable Ti_3C_2 MXene membrane with fast and photoswitchable nanofluidic transport. *ACS Nano* 12: 12464–12471.

68 Zhang, L., Ghaffar, A., Zhu, X., and Chen, B. (2019). Stable graphene-based membrane with pH-responsive gates for advanced molecular separation. *Environ. Sci. Technol.* 53: 10398–10407.

69 Zhang, Z. et al. (2021). Bioinspired graphene oxide membranes with pH-responsive nanochannels for high-performance nanofiltration. *ACS Nano* 15: 13178–13187.

70 Zhang, Y., Zhang, S., and Chung, T.S. (2015). Nanometric graphene oxide framework membranes with enhanced heavy metal removal via nanofiltration. *Environ. Sci. Technol.* 49: 10235–10242.

71 Li, Y. et al. (2020). Mild annealing reduced graphene oxide membrane for nanofiltration. *J. Membr. Sci.* 601.

72 Yuan, S. et al. (2020). Minimizing non-selective nanowrinkles of reduced graphene oxide laminar membranes for enhanced NaCl rejection. *Environ. Sci. Technol. Lett.* 7: 273–279.

73 Yuan, S., Li, Y., Xia, Y. et al. (2021). Stable cation-controlled reduced graphene oxide membranes for improved NaCl rejection. *J. Membr. Sci.* 621: 118995.

74 Han, Y., Jiang, Y., and Gao, C. (2015). High-flux graphene oxide nanofiltration membrane intercalated by carbon nanotubes. *ACS Appl. Mater. Interfaces* 7: 8147–8155.

75 Li, Y. et al. (2019). Thermally reduced nanoporous graphene oxide membrane for desalination. *Environ. Sci. Technol.* 53: 8314–8323.

76 Zhang, W.H. et al. (2021). Graphene oxide membranes with stable porous structure for ultrafast water transport. *Nat. Nanotechnol.* 16: 337–343.

77 Akbari, A. et al. (2016). Large-area graphene-based nanofiltration membranes by shear alignment of discotic nematic liquid crystals of graphene oxide. *Nat. Commun.* 7: 10891.

78 Lan, Q. et al. (2021). Chemically laminating graphene oxide nanosheets with phenolic nanomeshes for robust membranes with fast desalination. *Nano Lett.* 21: 8236–8243.

79 Oh, Y. et al. (2017). Understanding the pH-responsive behavior of graphene oxide membrane in removing ions and organic micropollulants. *J. Membr. Sci.* 541: 235–243.

80 Morelos-Gomez, A. et al. (2017). Effective NaCl and dye rejection of hybrid graphene oxide/graphene layered membranes. *Nat. Nanotechnol.* 12: 1083–1088.

81 Guan, K., Jia, Y., Lin, Y. et al. (2021). Chemically converted graphene nanosheets for the construction of ion-exclusion nanochannel membranes. *Nano Lett.* 21: 3495–3502.

82 Meng, N. et al. (2018). A low-pressure GO nanofiltration membrane crosslinked via ethylenediamine. *J. Membr. Sci.* 548: 363–371.

83 Jin, L., Wang, Z., Zheng, S., and Mi, B. (2018). Polyamide-crosslinked graphene oxide membrane for forward osmosis. *J. Membr. Sci.* 545: 11–18.

84 Ritt, C.L., Werber, J.R., Deshmukh, A., and Elimelech, M. (2019). Monte Carlo simulations of framework defects in layered two-dimensional nanomaterial desalination membranes: implications for permeability and selectivity. *Environ. Sci. Technol.* 53: 6214–6224.

85 Lin, F. et al. (2017). Orientation control of graphene flakes by magnetic field: broad device applications of macroscopically aligned graphene. *Adv. Mater.* 29: 1–6.

5

Graphene-Based Membranes for Ions Separation

Guoke Zhao[1,2] *and Hongwei Zhu*[1]

[1] *Tsinghua University, School of Materials Science and Engineering, 1 Tsinghua Yuan, Haidian District, Beijing, 100084, P. R. China*
[2] *SINOPEC Beijing Research Institute of Chemical Industry, 14 Beisanhuan Road (East), Chaoyang District, Beijing, 100013, P. R. China*

5.1 Introduction

Global water scarcity has become an increasingly serious issue in the twenty-first century. Desalination of seawater and brackish water to produce clean water is a feasible strategy to solve the water shortage problem. Compared with thermal-driven desalination technologies, membrane-based desalination process is more efficient with less energy consumption. Additionally, they are environment-friendly and can be operated in a continuous mode [1]. Currently, the reverse osmosis (RO) process is the most extensively industrialized membrane desalination technology worldwide. Under an applied trans-membrane pressure, water molecules permeate through the RO membrane while salt ions are rejected, generating desalted water. Membrane is the critical component in the RO process. Developed in the late 1970s, the polyamide thin-film composite (PA TFC) membrane is now the most commonly used RO membrane in desalination plants, owing to the ease of fabrication, low cost, and flexibility. Over the past four decades, great efforts have been made and the PA TFC RO membranes have witnessed many developments in terms of reducing the working pressure, improving membrane water flux and selectivity, as well as relieving membrane fouling. However, the membrane water flux is still far from desirable and they are vulnerable to harsh chemical and thermal working environments [2]. Novel materials need to be explored for fabricating efficient, durable, and cost-effective desalination membranes. Over the past decades, advances in nanomaterials have inspired researchers to design novel membranes based on the unique transport characteristics at nanoscale [3].

Ideally, a separation membrane should be as thin as possible to minimize the transport resistance and maximize its permeability. Controlled pore dimensions and functionalities are desired for better selectivity. The mechanical strength should be strong enough so that it can maintain structural integrity under working conditions. Chemical inertness should be desirable so that the membrane can work

Two-Dimensional-Materials-Based Membranes: Preparation, Characterization, and Applications, First Edition. Edited by Gongping Liu and Wanqin Jin.
© 2022 WILEY-VCH GmbH. Published 2022 by WILEY-VCH GmbH.

in harsh environments with good stability. Graphene, with atomic thickness, has been proved to possess excellent mechanical strength and good chemical stability, making it a promising candidate for fabricating energy-efficient membranes [4, 5]. However, intrinsic graphene was found to be impermeable to all gas molecules, including helium. It provides the world's thinnest separation barrier that is only one atom thick. The π orbitals of graphene form a dense electron cloud that blocks the gap in its aromatic rings, making the intrinsic graphene not suitable for separation applications. To enable graphene as a separation membrane, it should be perforated with fine nanopores so that water molecules are permeable while ions and unwanted substances are rejected. Theoretical simulations and experimental investigations have been carried out to shed light on the transport behaviors and influence factors of ions through nanoporous graphene. Water molecules permeate through nanoporous graphene with 2–3 orders magnitude higher than conventional RO membranes. Nanoporous graphene, with well-defined subnanopores, is promising for desalination. Another alternative method is to assemble graphene sheets into a laminated structure. The interlayer channels act as the molecular sieving and transport pathways of ions and small molecules. Generally, these lamellar membranes are fabricated with graphene oxide (GO), a derivative of graphene. The outstanding mass transport properties, including ultrafast water permeability, ion selectivity as well as the cost-effective fabrication of GO sheets, and easy membrane assembly techniques, make GO-based membranes a hotspot in the field of separation over the past decade. In this chapter, these three kinds of graphene-based membranes for ions separation are reviewed. Recent progresses about graphene-based membranes for desalination, including fabrication, structural design, and transport mechanisms, are discussed. In addition, photons and external field-induced ions transport behaviors in GO-based membranes are also summarized.

5.2 Single-Layer Graphene

Intrinsic graphene is impermeable to all gases and atoms but exhibits high permeability to thermal protons. When subnanometer pores are deliberately introduced in a controlled way, nanoporous graphene is also used as a separation membrane, whose ion transport properties are related to the pore size, pore chemistry, and external stimuli, such as pressure and electric field. Pore-drilling methods, such as oxygen plasma etching and ion bombardment, have been used to validate the potential of single-layer graphene for desalination experimentally. However, the practical realization of using single-layer graphene for scalable desalination applications is highly challenging owing to the technical difficulties, including the large-area and defect-free fabrication of graphene, the difficulty of the nanopore-drilling techniques, and the mechanical strength problems of the drilled graphene film.

5.2.1 Theoretical Calculations

The impermeability of intrinsic graphene to gas molecules was confirmed by Bunch et al. in 2008 [6]. They suspended mechanically exfoliated graphene sheets over predefined wells of SiO_2 support. The leaking rate of the sealed gas was found to be independent of the membrane thickness, indicating that the gas molecules transport through the graphene/substrate interface. Later, theoretical calculations proved that intrinsic graphene is impermeable to all gases and atoms under ambient conditions [7]. However, Geim and coworkers reported graphene's high permeability to thermal protons [8]. The measured proton current was linearly dependent on the applied bias voltage. The transport mechanisms were not elucidated in their work. Subsequently, Acthyl et al. studied the aqueous proton transport behaviors across single-layer graphene in the absence of external electrical field at room temperature and proposed that the transport pathways were –OH-terminated atomic defects [9]. Density functional theory (DFT) calculation results indicated that the activation barrier for aqueous protons to pass through intrinsic, defect-free graphene was quite high, over 3.8 eV. When the atomic defects were terminated with –OH, hydrogen bonding would be formed, facilitating aqueous proton transfer across graphene at room temperature. Similar results were reported by Walker et al. [10]. They studied the diffusion-driven aqueous proton transport behaviors across graphene sheets prepared via chemical vapor deposition (CVD) method. When the defects were sealed with atomic layer deposition (ALD), the ionic current was found to be decreased, while membrane's selectivity of H^+/Cl^- was increased. These reports demonstrated that the defects were critical for graphene's separation and inspired more researchers to carry out investigations on the transport properties of graphene sheets with artificially introduced nanopores.

To enable graphene as a separation membrane, subnanometer pores should be introduced in a controlled way so that water molecules are able to pass through while ions and unwanted solutes are rejected. Generally, the transport behaviors of water molecules at nanoscale differ from what they are like in the bulk. Considering this, Zhu et al. investigated the microscopic transport behaviors of water molecules through nanoporous graphene with various pore sizes using molecular dynamic (MD) simulations [11]. The water flux exhibited a linear relationship with the pore area when the pore diameter was larger than 1.5 nm. However, a nonlinear relationship was identified when the pores were smaller than 1.5 nm. With the pore size approaching the size of water molecules, the size effect and the water-membrane interaction led to the transition of water flow pattern from continuum flow to discrete flow, which further resulted in the above difference.

The functionalization of nanopores in single-layer graphene leads to different ion selectivity. F- and N-terminated nanopores favored the passage of cations, while H-terminated nanopores facilitated the passage of anions [12]. The results were ascribed to the difference in the charge properties induced by the termination atoms. F and N functionalization made the nanopores negatively charged while the hydrogenated pores were slightly positively charged. The nanopores also exhibited size selectivity. Ions with weaker coupling to water molecules in their hydration shells passed through faster because part of their hydration shells could be removed by the pore rims. The importance of Donnan effect in nanoporous graphene was also confirmed by Zhao et al. [13]. The selective transport behavior of ions through charged nanoporous graphene was investigated using MD calculations. The negative charges on the pore rims remarkably enhanced the passage of K^+ while impeded the passage of Cl^-. Interestingly, it was demonstrated that the selective ability of charged nanoporous graphene could be preserved even if the pore sizes were much larger than the hydrated diameters of ions and size sieving failed to work under this condition.

In addition to atomic termination, functional groups were theoretically introduced to see the resulting effects. Using MD simulations, Cohen-Tanugi and Grossman found that the desalination performance of nanoporous graphene was related to the pore size, pore chemistry, and applied pressure [14]. The hydrophilic hydroxyl groups on the pore edges would double the water flux. Meanwhile, membrane selectivity was compromised owing to the formation of hydrogen bonds between hydroxyl groups and the water molecules in the hydration shell of the ions. The water flux of the optimized nanoporous graphene was several orders higher than that of the commercial RO membranes, indicating that single-layer nanoporous graphene is a promising candidate for developing high-performance desalination membranes, as shown in Figure 5.1a–c. Nitrogen atom is considered as an excellent doping element because of its similar atomic size to carbon. Up to now, various methods have been tried to fabricate N-doped nanoporous graphene with controlled configurations. N-doped graphene, with different functionalities and doping levels, was theoretically investigated as RO membranes [16]. The pyridinic nanoporous graphene exhibited promising desalination performance. The curvature of nanoporous graphene is another intrinsic parameter affecting its ion transport behaviors. Ebrahimi investigated the effects of pore size, membrane curvature, and pressure on the desalination performance of Si-terminated nanoporous graphene using MD simulations [17]. The permeability of salt ions and water molecules were both enhanced with increased radius of membrane curvature.

Overall, the ion transport properties of nanoporous graphene are related to the pore size, pore chemistry, and external stimuli, such as pressure and electric field. By finely tuning these parameters, nanoporous graphene with desired ion permeability and selectivity is expected to be obtained. Inspired by protein channels, He et al. designed nanopores in graphene sheets that could discriminate between Na^+ and K^+ by modifying the pore size and the functionalities of the pore rims, as shown in Figure 5.1d [15]. When the pore rims were decorated with carbonyl groups, the selectivity of K^+ over Na^+ was identified owing to the stronger coordination. However, Na^+ was preferentially transported when the pore rims were decorated with carboxyl groups. The binding between Na^+ and carboxyl groups was stronger, leading to faster passage. When a high voltage was applied to nanoporous graphene with carboxyl groups, the selectivity of the membrane was switched from favoring Na^+ to K^+. In this case, the binding of Na^+ was destabilized by the applied electric field, which

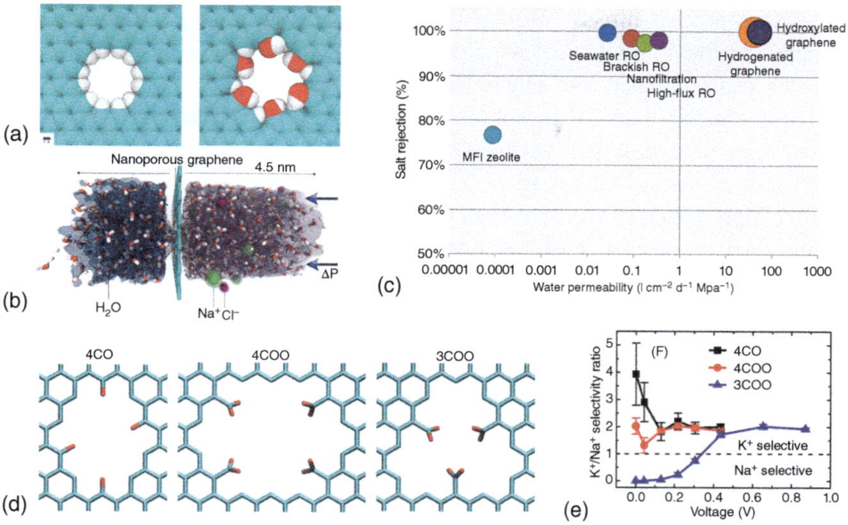

Figure 5.1 (a) Graphene nanopores terminated with hydrogen atoms and –OH groups. (b) Side view of the simulation model. (c) Comparison of the performance of functionalized nanoporous graphene and other current existing membranes, indicating the potential of nanoporous graphene for high-performance desalination membranes. Source: Cohen-Tanugi and Grossman [14]/with permission of American Chemical Society. (d) The simulation models of functionalized nanoporous graphene modified with four carbonyl, four carboxylate, and three carboxylate groups, respectively. (e) K^+/Na^+ selectivity ratio of the functionalized nanoporous graphene under different transmembrane voltages. Source: He et al. [15]/with permission of American Chemical Society.

slowed its passage. The K^+/Na^+ selectivity ratio of the functionalized nanoporous graphene under different transmembrane voltages is displayed in Figure 5.1e.

Generally, the working pressure of a commercial polymeric RO membrane is lower than 10 MPa. However, the employed pressure in the above-discussed theoretical models was as high as 100–200 MPa to improve the calculation efficiency. In a work carried out by Cohen-Tanugi and Grossman, they answered the question that whether the excellent permeability of nanoporous graphene under high pressures can be preserved under common RO pressures using MD simulations [18]. The permeability was found to remain constant over a wide range even down to very low pressures, indicating the promising application of nanoporous graphene for practical desalination. Although many theoretical calculations confirmed the promising application of single-layer nanoporous graphene as high-performance desalination membranes, their fabrication remains a big challenge. Compared with single-layer graphene, multi-layered graphene films are much easier to get. This fact inspired researchers to figure out whether the desalination performance of single-layer nanoporous graphene can be maintained if the material is no longer one atom thick. Later, the same group carried out MD simulations to investigate how the interplay between the nanopores on different layers affect the water and salt permeability through multi-layered graphene membranes [19]. Variables, including pore size, pore offset, and layer separation distance, were found to affect the separation process directly. When the layer separation is larger than 0.8 nm, membrane's permeability decreases linearly with increasing layer number. Meanwhile, salt rejection is increased due to higher diffusion resistance. When the layer separation is smaller than 0.8 nm, pore size and pore offset become more crucial in determining the overall performance of the multi-layered graphene membrane. Their work demonstrated the potential of multi-layered nanoporous graphene for fabricating high-performance desalination membranes. They can retain most of the exceptional properties of single-layer graphene and offer more cost-effective availability.

5.2.2 Experimental Validations

The experimental validation of the promising desalination performance of single-layer nanoporous graphene was carried out by Surwade et al. Oxygen plasma etching was used to introduce nanometer-sized pores on suspended single-layer graphene synthesized by CVD, as illustrated in Figure 5.2a,b [20]. The pore sizes were in the range of 0.5–1 nm and most of the pore rims were terminated with silicon atoms because of the silicon nitride support employed in the oxygen plasma etching process. Using pressure as the driving force, excellent water permeability was achieved with nearly 100% salt rejection. Experimentally, their work confirmed the effectiveness of single-layer nanoporous graphene for desalination applications.

In addition to oxygen plasma etching, ion bombardment combined with oxidative etching was also employed to introduce subnanometer pores in single-layer graphene [22]. Ion bombardment was used to introduce reactive and isolated defects that were then enlarged in the subsequent etching process to get nanopores with diameters ~0.4 nm. Nanopores drilled with this method were negatively charged so that they exhibited cation-selective transport properties. Larger pores were generated with increasing etching time. The results indicated that nanoporous graphene may also be used to separate dye molecules based on steric effects. Ideally, the deliberately introduced subnanometer pores should be the only transport pathways so that desired selectivity can be achieved. However, nonselective defects are often inevitable during the growth and transfer process of CVD graphene. These defective sites act as leakage pathways that severely decrease the membrane selectivity. Therefore, reducing and sealing the intrinsic defects are important before the pore-drilling process. Considering this, the same group proposed a method to seal the nonselective defects and then enable the sealed single-layer graphene as an active separation membrane [21]. As shown in Figure 5.2c–e, ALD of hafnia was employed to seal the nanometer-sized intrinsic defects. The ALD reaction started at point defects, grain boundaries, and surface wrinkles, owing to the higher surface energy. After the sealing process, they introduced subnanometer pores in the same way as described before. As the nonselective defects can be sealed before desirable-sized nanopores are introduced, their work offers a strategy to prepare single-layer nanoporous graphene membrane with a relatively large area. Although experiments have confirmed single-layer nanoporous graphene as a promising candidate for desalination, the reliable control of the pore dimension and the

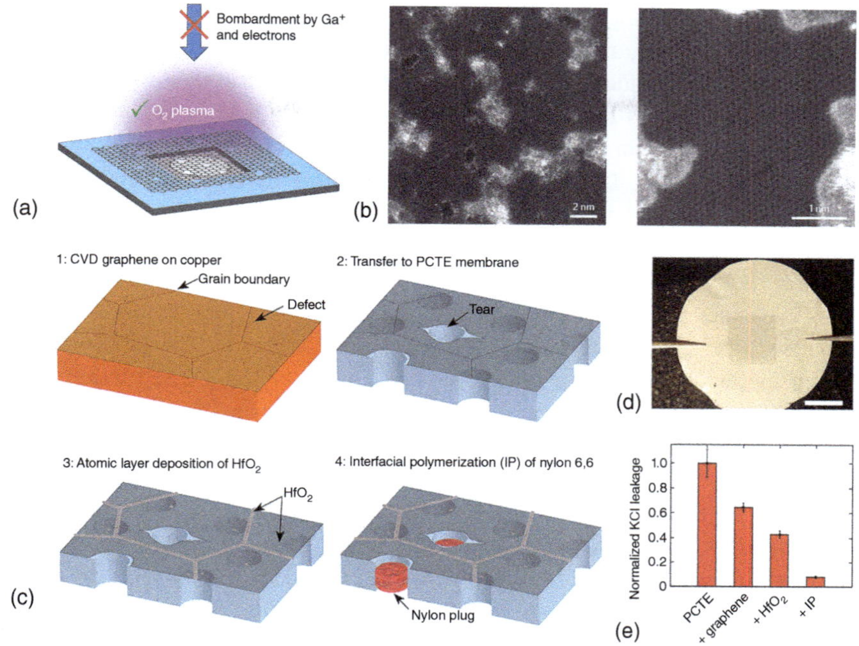

Figure 5.2 (a) Schematic illustration of the single-layer graphene suspended on a 5 μm sized hole. (b) Aberration-corrected STEM characterization of the single-layer graphene with nanopores introduced with oxygen plasma. The pore size is ~1 nm. Source: Surwade et al. [20]/with permission of Springer Nature. (c) Illustration of the graphene fabrication and defect sealing procedure. Grain boundaries and pinholes are formed during the CVD process. Tears and wrinkles are introduced during the transfer process. The nanoscale defective sites are sealed with ALD of hafnia. Larger defects are then sealed via interfacial polymerization. (d) Digital image of the sealed large-area graphene membrane. The scale bar is 0.5 cm. (e) The measured KCl leakage through the control membranes in each successive step. The final leakage rate of the sealed graphene membrane is ~8%. Source: O'Hern et al. [21]/with permission of American Chemical Society.

scalable fabrication are still challenging. Compared with CVD graphene, GO offers much easier availability. It is highly desirable that single-layer GO sheets can act as the precursor material to prepare single-layer nanoporous graphene membranes. Recently, Raffone et al. proposed a new strategy to prepare nanoporous graphene membranes from single-layer GO sheets based on theoretical calculations [23]. The randomly distributed functional groups were found to form clusters when GO sheets were annealed under mild conditions. The clusterization process led to the formation of nanopores with controlled sizes.

Generally, it is accepted that nanopores, with narrow and uniform size distribution, are necessary for fabricating high-performance desalination membranes, with no exception for nanoporous graphene. However, Rollings et al. found that the selectivity of K^+ over Cl^- in an electrodialysis process could be retained even when the size of the pores in graphene was as large as 20 nm [24]. Nanopores with negatively charged rims were introduced in graphene using electrical pulse method. The selectivity ratio of K^+/Cl^- was found to be above 100 at pH 8 even when the pores were slightly larger than 20 nm. The deprotonated functional groups in nanoporous graphene (both intrinsic and introduced ones during the pore-drilling process) made the membrane negatively charged at neutral pH values. Electrostatic interactions between the membrane surface and cations led to the increased local concentration of cations near the membrane surface and thus the cation-selective current. Different from other works that highlight the careful control of the nanopore dimensions, this work suggests that strict pore-drilling limitations are not necessary. Current pore-drilling technologies are mature enough to fabricate nanoporous graphene membranes with pores ranging from 1 to 20 nm that are able to act as efficient cation-exchange electrodialysis membranes.

Using high-energy bombardment to introduce nanopores can be taken as the top-down method. The control of pore dimensions is of limited resolution and precision. Considering this, a bottom-up method to grow nanoporous graphene from small molecular building blocks is more desirable. Bieri et al. synthesized a 2D polyphenylene networks by surface-promoted aryl–aryl coupling of molecular building blocks with designed nanostructure [25]. The as-obtained periodic polymer network exhibited single-atom wide pores. By tailoring the molecular structure of the building blocks, honeycomb-like periodic networks with different pore dimensions can be fabricated. The bottom-up method offers a route to control the nanopores with atomic precision.

Both theoretical and experimental investigations have confirmed the potential of single-layer graphene for desalination. However, there still exist various technological limitations. First, it relies on the fabrication of large-area and high-quality graphene film. The nonselective defects introduced during the growth and transfer of CVD graphene should be carefully avoided. Second, regular pores with high density and narrow size distribution are still hard to control using current pore-drilling techniques. Third, stress concentration is likely to be generated during the pore-drilling process, leading to a severe decrease in the mechanical strength of the membrane.

5.3 Graphene Oxide Membranes

Lamellar GO membranes exhibit ultrafast water permeability, indicating the promising for desalination applications. In addition, the difference in the interaction strength between metal cations and GO membranes leads to selective ion transport properties. The permeation characteristics of GO membranes are highly dependent on the interlayer nanostructure so that the intrinsic properties and the stacking pattern of GO sheets should be well controlled to facilitate the water/ion transport. The interlayer spacing of GO membranes can be confined through cross-linking, reduction, and vertical pressure application to improve the desalination performance via molecular sieving enhancement. What's more, the increased Donnan effect caused by surface charge modification is beneficial for the improvement of membrane ion rejection. The ion transport through GO membranes modulated by external fields is also discussed.

5.3.1 Structure of Graphene Oxide and Graphene Oxide Membranes

GO can be taken as a highly oxidized form of graphene. They can be prepared by liquid-phase exfoliation of graphite using strong oxidation agents. The controversy about the exact chemical structure of GO sheets has existed for decades. No unambiguous model has been reached to describe the type and location of the functional groups. Lerf et al. proposed that the hydroxyl and epoxy groups are mainly decorated on the basal plane of GO sheets, while the carboxylic groups are located at the edges and also as carbonyl defects within the sheets [26, 27]. The rich functional groups on the basal plane and at the edges make them negatively charged in solution state owing to the ionization of carboxyl groups, which effectively prohibits them from aggregation. In addition, they may act as active sites for various postmodification reactions. The intrinsic hydrophilicity also contributes to their good dispersivity in water without the help of any stabilizing agents. The atomic structure and properties of GO sheets are highly dependent on their synthesis process. Compared with graphene, GO is a more attractive choice for water treatment applications owing to its better hydrophilicity.

The GO dispersion is readily processed into lamellar GO membranes through filtration or evaporation. Various characterizations have confirmed the layer-by-layer structure of GO membranes. The hierarchical structure of GO membranes makes the system rather complex. The transport pathways in GO membranes are composed of the 2D nanochannels between adjacent GO layers, the intrinsic in-plane defects, and the inter-edge slits. This indicates that the size, chemical functionalities, as well as stacking uniformity of GO sheets, directly affect the separation performance of the assembled GO membranes. In addition, GO sheets consist of both hydrophobic and hydrophilic regions in their structure. Their interactions with bound water and intercalated water molecules, as well as the hydration layer of ions, make the transport behaviors of these components in confined GO membranes much different from those in the bulk. What's more, the sizes of GO sheets are often not uniform and the stacking pattern of GO membranes deviates from the ideal 2D lamellar model. The surface ripples and wrinkles may lead to disordered stacking, which further results in short transport pathways within GO membranes. Water and solutes tend to pass through these defective sites with smaller transmembrane resistance, leading to decreased membrane selectivity. The complexity of the system makes the investigations on the correlation between the membrane microstructure and its transport characteristics rather difficult.

5.3.2 Ultrafast Water Permeability

The ultrafast water vapor transport across sub-micrometer-thick GO membranes was reported by Nair et al. [28]. The freestanding GO membranes were found to be impermeable to various organic solvents and standard gases, including helium. However, they exhibited unimpeded permeation to water vapor. The authors proposed that the 2D nanocapillaries formed between closely packed GO sheets provided low-friction flow of water vapor. Subsequently, the transmembrane properties of liquid water through GO membranes were investigated by Sun et al. using an isotope labeling technique [29]. The diffusion coefficient of liquid water through GO membranes was 4–5 orders magnitude higher than that of the bulk diffusion. This observation was in agreement with the case of water vapor. They proposed that the low-friction tortuous sp^2 nanocapillaries in the GO membranes were the major transport pathways.

The underlying mechanism was studied by using first-principles calculations. The results indicated that hexagonal ice bilayer was formed between the stacked GO layers [30]. The melting transition of ice at the flake edges leads to the anomalous water transport behavior in GO membranes, as shown in Figure 5.3a,b. This may provide a reasonable explanation for the ultrafast water permeation through GO membranes. Later, theoretical calculations and experimental observations further confirmed the existence of ordered water structures between nanoconfined graphene sheets [31]. As shown in Figure 5.3c, the formation of square ice in graphene nanocapillaries was observed with high resolution transmission electron microscope (HRTEM) characterization. The unique ice phase has interesting kinetic and dynamic transport behaviors compared with traditional molecular liquids.

There are two premises for the above transporting mechanisms. For one, the single graphene layer flake is impermeable for any liquids or gases. For the other, the pristine graphene channels are the major water transport pathways while the oxygen-containing functional groups act as nanospacers to help the nanocapillary formation. However, the existence of a continuous percolation network has not been validated by any experimental observations. The wrinkled morphology, surface functional groups, as well as their interactions with water molecules in the confined 2D nanocapillaries, make the system rather complicated. Wei et al. demonstrated that the ultralow friction at water/graphene interface was prohibited by the side-pinning effect caused by the oxidized regions in GO sheets [32]. They attributed the ultrafast water permeation to the porous and defective microstructures (holes, spaces between GO flakes, wide channels formed

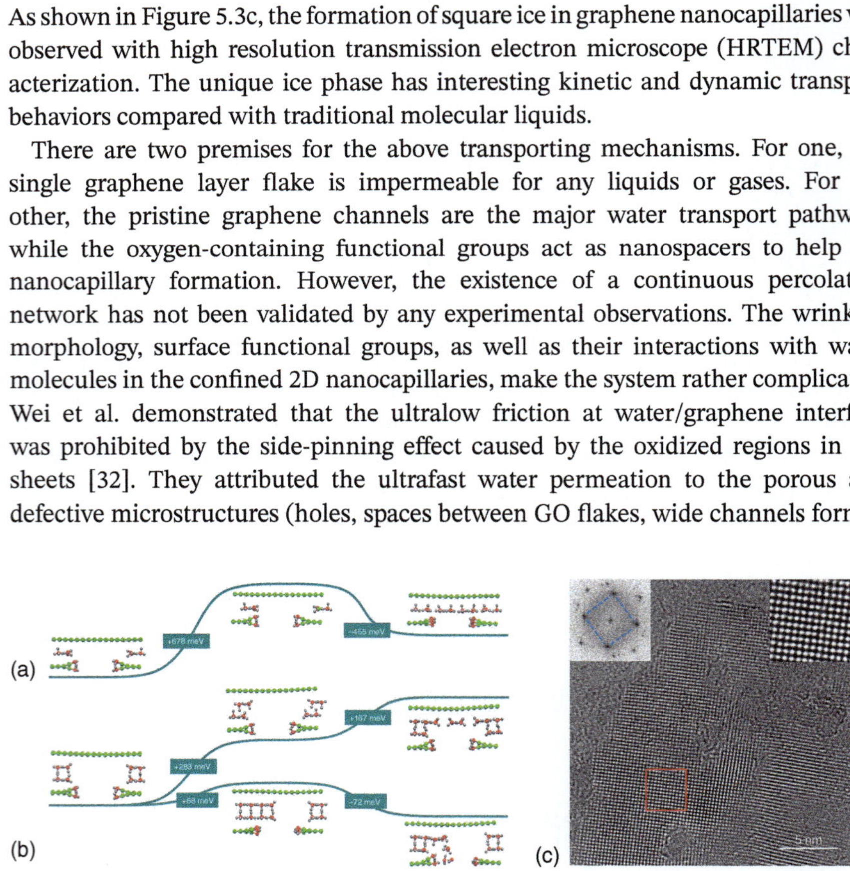

Figure 5.3 Energy costs and optimized atomic structures of migrations of (a) ice monolayer and (b) ice bilayer along zigzag direction over hydroxyl passivated graphene pore edges, respectively. Source: Boukhvalov et al. [30]/with permission of American Chemical Society. (c) HRTEM confirmation of the square ice in graphene capillary. The top left insert is the Fourier transform of the entire image. The top right insert shows a magnified image of the area outlined in red. Source: Algara-Siller et al. [31]/with permission of Springer Nature.

by surface wrinkles, and functional groups) of GO membranes. Later, the same group proposed that the noncontinuum collective diffusion of water may explain the fast molecular transport, especially under low humidity [33].

Up to now, there is still no consensus on the water transport pathways within lamellar GO membranes. Saraswat et al. proposed that the major pathways were the pinholes on GO sheets and the voids formed during membrane fabrication [34]. They prepared GO sheets with similar chemical properties but different lateral sizes (2 orders of magnitude difference). Interestingly, the water permeance through GO membranes assembled with these GO sheets was nearly the same. The difference in water permeance was expected to be ~100 if water molecules need to travel along with the entire lateral dimension of GO sheets. The experimental observation can be repeated with simulations only with the assumption that pinholes are present in GO sheets and they allow for the transmembrane flow. Stacking voids are inevitable during the membrane-forming process because of the wrinkles and empty spaces. These defective sites offer much shorter water transport pathways with lower resistance, leading to fast permeability. Overall, further efforts are needed to elucidate the water transport mechanisms within GO membranes. The most direct and convincing method is to *in situ* characterize the water transmembrane process.

5.3.3 Ion Selectivity

The ultrafast water permeability of GO membranes discussed above indicates that they are promising for desalination applications because higher water flux means higher efficiency. In addition, membrane selectivity is another important indicator of membrane performance. The selective ion penetration properties of GO membranes were systematically investigated by Sun et al. [35]. They prepared freestanding GO membranes by drop-casting method and measured the permeation of various ions under concentration gradient-driven conditions. Sodium ions were found to permeate through GO membranes quickly. However, heavy metal ions, such as Cu^{2+}, were nearly blocked. They attributed the phenomenon to the different interaction strength between these ions and GO membranes, as illustrated in Figure 5.4. Transition metal ions tend to interact with sp^3 regions with stronger coordination interactions while main group metal ions lacking *d* electrons prefer to interact with sp^2 regions via weaker cation–π interactions. Later, they demonstrated the selectivity of GO membranes between alkali and alkaline earth cations [36]. In this case, the selectivity arises from the balance between cation–π interactions and the dehydration effects, which is similar to the transporting mechanism of ions across cellular membranes. Increasing the lateral size of GO sheets and the thickness of GO membranes, the ion transmembrane permeation decreased obviously, indicating that the ions were more effectively captured by the increased amount of active sites. GO membranes exhibited intrinsic high water/ion selectivity under diffusion-driven separation processes [37]. However, the desalination performance was found to be poor under pressure-driven conditions. Water-ion interactions would be weakened when external pressure is applied, leading to decreased selectivity. To increase the

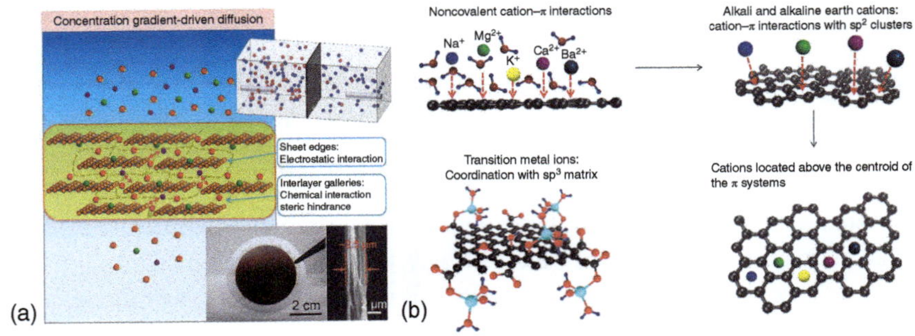

Figure 5.4 (a) Illustration of the desalination mechanisms of GO membranes under concentration gradient-driven conditions. The bottom right corner shows the digital image and the cross-sectional SEM image of the GO membranes. (b) Different interactions between metal cations and GO sheets. Alkali and alkaline earth cations interact with the sp^2 regions through cation–π interactions. Transition metal ions tend to coordinate with the sp^3 matrix. Source: Sun et al. [36]/with permission of American Chemical Society.

desalination performance of GO membranes under pressure-driven conditions, various efforts have been made, which are discussed in Sections 5.3.4–5.3.7.

In addition to the intrinsic ion selectivity, ion-recognizable copolymers have been deliberately introduced into GO membranes to facilitate the transport of certain ions [38]. The poly(*N*-isopropylacrylamide-*co*-benzo-15-crown-5-acrylamide) (PNB) copolymers were introduced into partially reduced GO membranes. Owing to the complexation interactions between the polymer units and K^+ ions near the lower critical solution temperature of the polymer, the passage of K^+ ions was greatly enhanced, resulting in the increased K^+/Na^+ selectivity. By introducing other facilitators into GO membranes, composite membranes with specific selectivity are expected to be obtained.

5.3.4 Microstructure Optimization for Desalination

Generally, the influence factors for ions transport across GO membranes include the intrinsic properties of GO sheets (lateral size, oxidation extent, and type of functional groups) and ions (charge and hydration), the microstructure of the assembled GO membranes (membrane thickness and stacking uniformity), their interactions and the external field (pressure, photons, electrical, and magnetic field). The physicochemical properties of GO sheets directly affect the ion transport characteristics of the assembled GO membranes [39]. Larger lateral sizes and more functional groups indicate longer pathways and more active sites. In addition, the interlayer nanostructure of GO membranes also plays an important role in determining the permeation characteristics. Functional groups tend to aggregate on GO sheets, forming island-like distribution. As a result, the stacking manner in which adjacent GO layers are assembled varies under different membrane fabrication conditions. Xu et al. found that GO sheets preferred to self-assemble uniformly with functional groups facing each other while sp^2 regions forming low friction channels under low deposition rate, as shown in Figure 5.5 [40]. When the deposition process was short, the nanostructure was randomly packed, leading to simultaneously

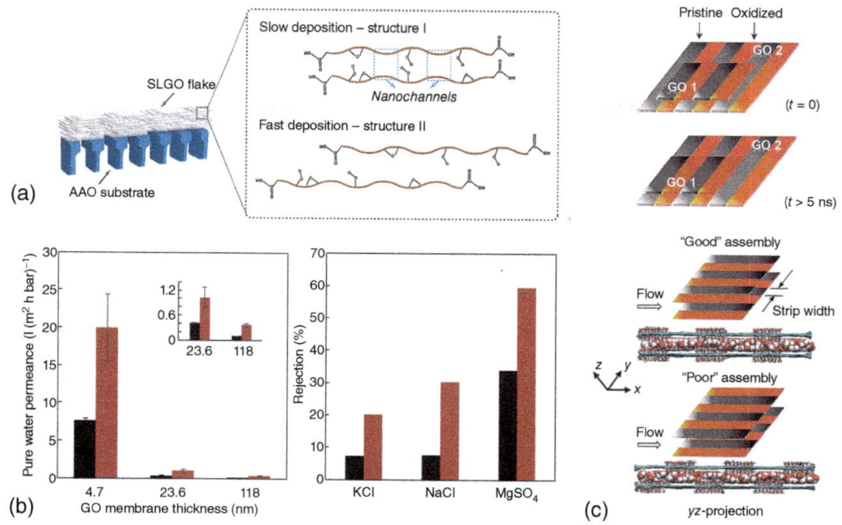

Figure 5.5 (a) Illustration of the proposed nanostructures of the GO membranes prepared by slow and fast deposition rates. In structure I, the oxygen-containing functional groups on adjacent GO sheets self-assemble into an ordered interlayer structure. In structure II, the functional groups are arranged in a random manner. (b) Water permeation and salt rejection of the GO membranes prepared by slow (red column) and fast (black column) deposition rates. (c) Sketches of the system configurations used for studying the water transport properties in theoretical simulations. Source: Xu et al. [40]/with permission of American Chemical Society.

decreased water flux and salt rejection. By simply controlling the deposition rate of the GO sheets, 2.5–4 times faster water permeability and 1.8–4 times higher salt rejection were achieved.

Doping was also proved to be an effective strategy to tune the nanostructure of GO membranes. The atomic doping content and bonding configuration contributed to the tunable interlayer spacing of the obtained GO membranes. Nitrogen doping with varying doping contents and bonding configurations were introduced with hydrothermal method. It was found that the pyridinic-N was the most influential configuration owing to the highest polarization and adsorption energy [41]. This work offers an alternative strategy to improve the ion separation performance of GO-based membranes.

Recently, Li et al. carried out MD simulations to reveal the influence factors of GO membranes' ion transport properties and proposed reasonable design principles for high-performance GO desalination membranes [42]. The complex interactions (electrostatic interactions, hydrogen bonding, cation–π interactions, and vdW force) between water/ion and the functional groups on GO sheets impede their transmembrane transport. The formation of continuous non-oxidized sp^2 regions facilitates water/ion transport. To improve the desalination performance of GO membranes, the interlayer space should be controlled to obtain desired water permeability and ion rejection. In addition, the GO sheets are expected to pack in the way that their non-oxidized regions face each other.

5.3.5 Interlayer Spacing Control for Desalination

5.3.5.1 Cross-Linking

The size-dependent molecular sieving of GO membranes was demonstrated by Joshi et al. [43]. They prepared micrometer thick GO membranes by vacuum filtration and tested their separation performance for various ions and small molecules under concentration gradient-driven conditions. When immersed in water, the hydrated GO membranes could act as precise and ultrafast molecular sieves, allowing the passage of solutes with hydrated diameter smaller than 0.9 nm. In addition, the diffusion rates were thousands of times faster than the expected value. They attributed the phenomenon to the high capillary pressure in the nanoconfined space between adjacent GO layers. Their work demonstrates that the interlayer spacing control of GO membranes is crucial to achieve size-dependent selectivity. To use GO membranes for desalination, physical or chemical confinement should be employed to narrow the interlayer spacing, down to 0.7 nm, so that water molecules can pass through, while ions are rejected.

GO membranes tend to adsorb water and swell in aqueous environment owing to the intrinsic hydrophilicity of GO sheets. The swelling behavior of GO membranes was investigated using quartz crystal microbalance with dissipation and ellipsometry [44]. The interlayer spacing was accurately quantified. The initial *d*-spacing of dry GO membranes was 0.76 nm, which increased to 6–7 nm when they were immersed in pure water, as shown in Figure 5.6a,b. When salts were present, the equilibrium *d*-spacing was decreased to ~2 nm because of the charge-screening effect. As the functional groups on GO sheets impede the fast water transport in GO membranes, the authors proposed a route to optimize the desalination performance of GO membranes. They should be firstly cross-linked with rigid spacers to obtain ideal *d*-spacing and then reduced to create ultrafast water channels.

With the presence of salts in the surrounding environment of GO membranes, the interlayer spacing differs from that in pure water. The interactions between GO membranes and salt solutions lead to changes in their interlayer spacing, which further result in tunable ion sieving effect [46]. When immersed in pure water, the interlayer spacing of GO membranes was about 4.8 nm. However, it increased to 6.1 nm in Na_2SO_4 while decreased to 2.8 nm in $MgSO_4$. The interactions between divalent Mg^{2+} ions and carboxylate groups in GO sheets are stronger than those between Na^+ and carboxylate groups. What's more, the hydrated diameter of Mg^{2+} is smaller than Na^+. As a result, Mg^{2+} ions joined the adjacent GO layers and formed a more compact membrane structure with smaller interlayer spacing. In addition to the interlayer spacing, cations also affect the charge properties of GO membranes by adsorbing on the membrane surface. The higher tendency of Mg^{2+} ions to adsorb on the negatively charged GO sheets leads to stronger charge neutralization compared with Na^+ ions, resulting in lower salt rejection. The interactions between cations and GO sheets and their influences on the membrane performance cannot be ignored because the presence of cations is inevitable when GO membranes are used for desalination.

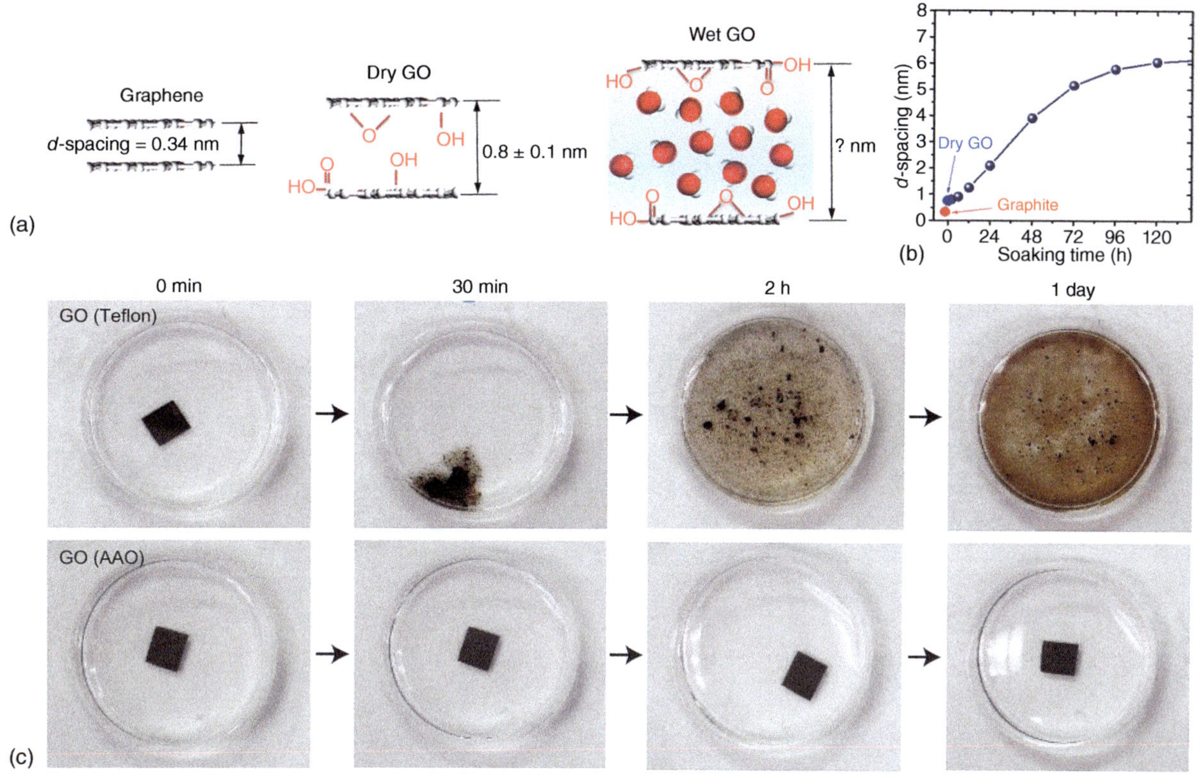

Figure 5.6 (a) Illustration of the *d*-spacing of graphene, dry GO, and fully hydrated GO membranes. (b) Calculated *d*-spacing of a wet GO membrane with increasing socking time. Source: Zheng et al. [44]/with permission of American Chemical Society. (c) Digital images showing that GO membranes deposited on Teflon support tend to disintegrate in water. GO membranes deposited on AAO support remain intact when immersed in water, exhibiting much better structural integrity. Source: Yeh et al. [45]/with permission of Springer Nature.

Generally, different types of salts coexist in a membrane separation system. The mutual interactions between these salts and the membranes differ from those in simple systems with only two components. What's more, the pH of the practical feed solution varies. Considering this, Pei et al. investigated the permeation properties of GO membranes under concentration gradient-driven conditions with single and mixed solutions of acid and Cu^{2+} salts as the feed solutions [47]. The mutual influences of H^+ and Cu^{2+} on each other when they were transported through GO membranes were analyzed. The type of anion in both acids (H_2SO_4 vs. HCl) and copper salts (Cu_2O_4 vs. $CuCl_2$) was found to affect the permeation of cations obviously. The permeation rate of H^+ was 16 times larger than that of Cu^{2+} with the same anion (H_2SO_4 vs. $CuSO_4$ and HCl vs. $CuCl_2$), which was ascribed to the much smaller hydrated radius of H^+ ions as well as the weaker interactions with GO sheets. Further experiments demonstrated that the permeation rate of H^+ decreased by 1.6 times and that of Cu^{2+} decreased by an order of magnitude when they coexisted. They proposed that the shrinkage of the interlayer spacing of GO membranes and the competitive adsorption of H^+ against Cu^{2+} ions contributed to the inhibitory phenomenon.

In addition to swelling, the structural integrity of GO membranes in water is another problem. It is expected that the negatively charged GO sheets pull each other in water, resulting in the tendency of GO membranes to disintegrate, as shown in Figure 5.6c. However, the structural integrity of GO membranes can be well maintained when immersed in water [45]. Yeh et al. ascribed this phenomenon to the cross-linking effects caused by the unintentionally introduced metal ions, such as Al^{3+} and Mn^{2+}, during the preparation of GO sheets or the fabrication of GO membranes. Their work suggests that metal cations may act as cross-linkers to relieve the swelling of GO membranes in aqueous environment. Subsequently, Chen et al. reported the interlayer spacing control of GO membranes with metal cations in angstrom precision [48]. GO membranes cross-linked by one type of cation could effectively reject other ions with larger hydrated diameters, as shown in Figure 5.7a,b. The freestanding GO membranes were prepared via drop-casting method and then immersed in various salt solutions. The interlayer spacing of KCl treated GO membranes was the smallest, 1.14 nm. All the other ions with larger hydrated volumes were selectively rejected. DFT calculations revealed that cations were preferentially adsorbed at the sites where functional groups and aromatic rings coexist. The coordination interactions between cations and the functional groups and the cation–π interactions between the cations and the delocalized π electrons contributed to the cross-linking effect.

Small molecules, such as diamine monomers were also tried to introduce chemical bonding between GO layers. Different diamine monomers, including ethylenediamine, butylenediamine, and *p*-phenylenediamine, were used to cross-link GO membranes for stability improvement and interlayer spacing control, as shown in Figure 5.7c,d [49]. The cross-linked GO membranes exhibited good stability with varying interlayer spacing from 0.87 to 1.04 nm. To get more uniform packing structure, GO sheets were first grafted with oxalic acid molecules to introduce more active sites and then cross-linked by ethylenediamine. The reduced

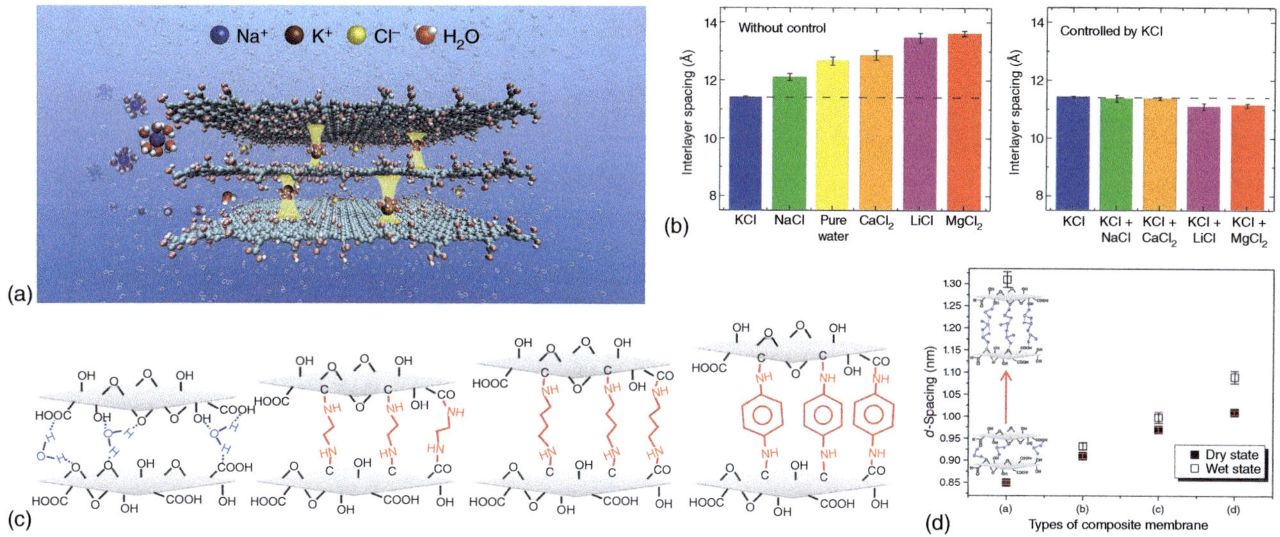

Figure 5.7 (a) Illustration of the K^+ cross-linked GO membrane. Other cations with larger hydrated diameters are rejected while water molecules can pass through. (b) Interlayer spacing of GO membranes immersed in water and various salt solutions, respectively (left). Interlayer spacing of GO membranes immersed in KCl solution followed by immersion in various salt solutions (right). Source: Chen et al. [48]/with permission of Springer Nature. (c) Illustration of the GO membranes cross-linked by different diamine monomers. (d) Changes in d-spacing of the GO membranes cross-linked by different diamine monomers when they are fully immersed in water. Source: Hung et al. [49]/with permission of American Chemical Society.

wrinkling and restricted swelling led to the significantly improved separation performance of the assembled GO membranes.

Recently, $Mg(OH)_2$ and GO quantum dots (GO QDs) have been employed separately to introduce cationic or covalent cross-linking to GO membranes [50]. In the case of $Mg(OH)_2$, the electrostatic interactions between Mg^{2+} and negatively charged functional groups interlinked the adjacent GO layers. As for GO QDs, they stitched the GO sheets by esterification. In addition, the small-sized GO QDs created more subnanochannels and formed an interlocked structure in the membranes. The inter-linked GO-based membranes showed excellent stability in aqueous environment and good ion separation efficiency.

5.3.5.2 Reduction

The interlayer spacing can also be narrowed by chemical or thermal reduction of GO sheets to remove the functional groups. When the interlayer spacing is greatly decreased, the reduced graphene oxide (rGO) membranes can even exhibit barrier properties, which means that they are impermeable to all gases and liquids [51]. Kim et al. investigated the changes in water/ion permeation during the transition from GO membranes to impermeable rGO membranes [52]. They fabricated GO membranes with different amounts of oxygen functional groups by controlling the thermal reduction conditions. It was found that rGO membranes with interlayer spacing smaller than 0.96 nm were not permeable to water. Initial GO membranes exhibited similar permeability to most ions. However, rGO membranes with the interlayer spacing of 0.96 nm demonstrated size and charge-dependent ion permeability. Al^{3+} ions were blocked owing to the high free-energy barrier. Anions with larger hydrated volumes exhibited an obvious cutoff at a certain interlayer spacing.

Yang et al. investigated the ion separation performance of GO membranes exposed to hydroiodic acid steam for different time durations [53]. The interlayer spacing of GO membranes was found to be dependent on the reduction degree. With the increase in the exposure time, the interlayer spacing was decreased and the distribution of the nanochannels was narrowed. As a result, the ion rejection rate was obviously increased, but water permeability was decreased simultaneously. GO sheets tend to form more corrugation if the reduction conditions are aggressive. It was reported that highly flat GO sheets could be obtained under mild reduction conditions so that the 2D nanochannels in the assembled rGO membranes were uniformly arranged [54]. The interlayer spacing was precisely controlled at ~0.8 nm to separate mono/multivalent cations. The nanowrinkles in rGO membranes also caught the attention of Yuan et al. [55]. They thermally reduced GO membranes under different atmospheres and revealed the underlying role of nanowrinkles for ion separation. When GO membranes were reduced in vacuum or HI vapor, nonselective transport pathways for water and ions were formed by the nanowrinkles. However, the 2D nanochannels became narrower and smoother when GO membranes were thermally treated in air, leading to increased salt rejection and water permeance. Their work suggests that the mild and slow reduction process is critical for the ion separation of rGO membranes.

5.3.5.3 External Pressure

Generally, a polymeric membrane would experience a compaction process under pressure before the water flux becomes steady. Similar phenomenon was also observed in the case of GO membranes. Under hydraulic pressure, the water flux decreased while the salt rejection increased significantly. GO membranes were compacted and the interlayer spacing became narrower under pressure. The formation of surface wrinkle from initial wrinkle under external pressure is shown in Figure 5.8a. The results indicated that applying an external pressure may be an alternative strategy for interlayer spacing control [56]. Physical confinement was employed by Abraham et al. to prepare GO membranes with interlayer spacing ranging from 0.64 to 0.98 nm [57]. GO membranes were stored in different humidity to get various interlayer spacings. After that, they were encapsulated in epoxy, which provided physical confinement to restrict membrane swelling, as illustrated in Figure 5.8b. The permeation of ions was found to decrease exponentially with

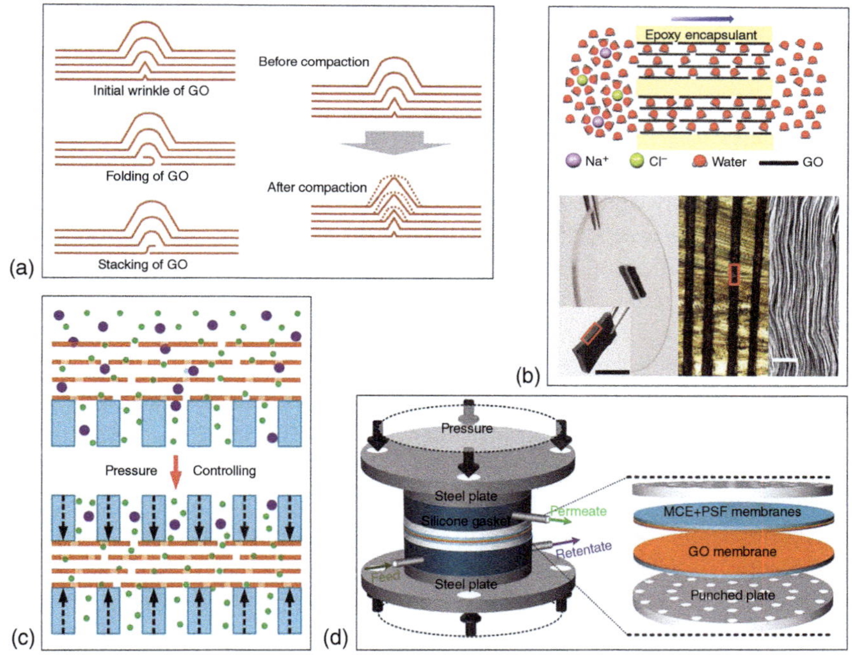

Figure 5.8 (a) Illustration of the formation of surface wrinkle from initial wrinkle. Source: Wei et al. [56]/with permission of Elsevier. (b) Illustration of the ion/water permeation through the planar 2D nanocapillaries between graphene layers. Digital image of a GO membrane physically confined by epoxy (bottom left). The scale bar is 5 mm. Optical image of the cross-sectional area. The GO membrane is about 100 μm thick (bottom middle). SEM image of the cross-sectional area. The scale bar is 1 μm (bottom right). Source: Abraham et al. [57]/with permission of Springer Nature. (c) Illustration of the desalination process by swelled and compressed GO membrane, respectively. (d) Illustration of the permeation cell employed for desalination evaluation. Source: Li et al. [58]/with permission of American Chemical Society.

decreasing interlayer spacing. Meanwhile, water transport was much less affected by the *d*-spacing.

The concept of external pressure regulation was introduced by Li et al. to control the interlayer spacing of GO membranes [58]. They applied an external force vertical to the lamellar GO sheets to restrain membrane swelling, as shown in Figure 5.8c,d. The compressed GO membranes exhibited high rejection to various salts, including KCl (94%), NaCl (97%), and $CaCl_2$ (98%), with desired water permeance (~25 LMH bar^{-1}). The interlayer spacing at 6 MPa was characterized to be about 0.95 nm. Using this method, GO membranes can even work under cross-flow conditions.

5.3.6 Charge Modification for Desalination

Steric hindrance and Donnan effect are the two main rejection mechanisms of GO membranes. In addition to decreasing the interlayer spacing of GO membranes, control of their charge properties has been proved to be another strategy for improving the desalination performance. The highly charged membrane surface would repel co-ions while impeding the passage of counter-ions so that the overall charge of the solutions remains balanced. Zhang et al. tuned the surface charge of GO membranes by immersing them into selected polyelectrolytes with different functional groups and charge properties, as shown in Figure 5.9 [59]. The polyelectrolytes were poly diallyl dimethyl ammonium (PDDA), polyethylene imine (PEI), polyallylamine hydrochloride (PAH), polystyrene sulfonate (PSS), polyacrylic acid (PAA), and sodium alginate (SA). The as-obtained membranes preserved the initial lamellar structure with an ultrathin polyelectrolyte-coating layer on the surface. By controlling the intensity and amount of the functional groups (amine/sulfonic/carboxyl/hydroxyl groups), the surface charge properties of the coated membranes could be well tuned. With the membrane charge changing from highly positive to highly negative, the permeability of $MgCl_2$ experienced a continuous increase while that of Na_2SO_4 witnessed a gradual decrease. The positively charged surface tends to repel AB2-type salts, while negatively charged GO membranes tend to reject A2B-type salts. The ionizable functional groups of the polyelectrolyte layer could tailor the electrostatic interactions between the membrane surfaces and the charged ions, leading to controlled ion transport. As the coating layer was very thin, it would not affect the fast water transport of the initial GO membranes. As a result, the charge-modified GO membranes demonstrated excellent ion rejection and high water permeance simultaneously.

Highly charged nanochannels were also created by Song et al. [60] with the help of polyelectrolytes. The ion transport behavior of the modified GO membranes with positive and negative charges, respectively, was then studied. Four model salts with different Z^+/Z^-, including $MgCl_2$, $MgSO_4$, NaCl, and Na_2SO_4, were selected. The results indicated that salt rejection was strongly dependent on the charge density of the membranes and the valence and size of the hydrated ions. Positively charged GO membranes showed the highest rejection to $MgCl_2$ (88.1%) while negatively charged GO membranes exhibited the highest rejection to Na_2SO_4 (90.3%). They ascribed the

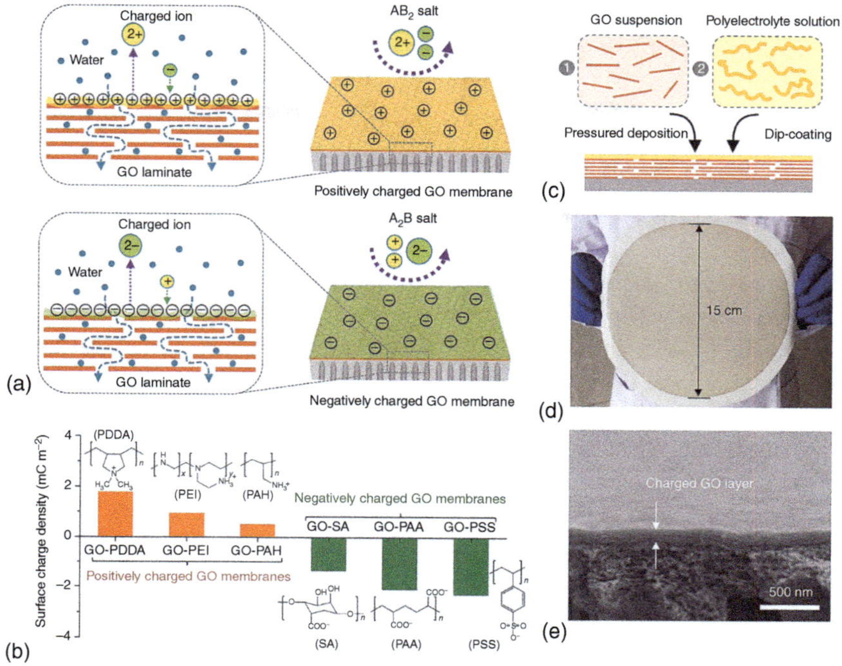

Figure 5.9 (a) Illustration of the surface charged GO membranes by coating polyelectrolytes with different charge properties. Positively charged GO membranes tend to reject AB2-type salts effectively while negatively charged GO membranes can exclude A2B-type salts. (b) Surface charge densities of the modified GO membranes. The molecular structures of the employed polyelectrolytes are also shown inside. (c) Illustration of the preparation process of the surface-charged GO membranes. The surface coating is prepared through dip-coating method. (d) Digital image of a large-area surface-modified GO membrane. (e) Cross-sectional SEM image of a surface charged GO membrane on PAN substrate. Source: Zhang et al. [59]/with permission of Springer Nature.

phenomenon to the electrostatic repulsion between the membrane surfaces and the co-ions. Manipulation of the charge properties of GO membranes may be used to accurately separate solutes that differ in charge and mass.

Modifying the surface charge of GO membranes, with polyelectrolytes, enables the membranes to reject co-ions effectively but their rejection to counter-ions is still low. If the coating polyelectrolyte is positive/negative, the modified membranes will repel positive/negative ions. It is more desirable that one modification treatment endows GO membranes with high rejection to different salts. In other words, the surface charge of GO membranes is expected to be reversible rather than fixed. Recently, charged-gated GO membranes modified by a CO_2 responsive polymer (poly(*N*,*N*-dimethylaminoethyl methacrylate) (PDMAEMA)) have been reported [61]. Upon bubbling CO_2 or Ar into the solution, the modified membranes exhibited positively charged or negatively charged surfaces, resulting in high rejection to $MgCl_2$ and Na_2SO_4, respectively. When CO_2 was bubbled in the solution, the protonation of tertiary amine groups on PDMAEMA endowed the GO membranes with positive charges, leading to increased rejection to divalent cation-containing salts. Using Ar to remove CO_2, the GO membranes recovered to the initial negatively charged state, exhibiting desirable rejection to divalent anion-containing salts.

5.3.7 External Field Modulated Ion Transport

Ion transport in GO membranes is not only influenced by the membrane itself but also affected by external field. The modulation of ion transport through a separation membrane by external field may be achieved in two ways—by changing the ion transport process or by altering the structure of the membranes. Generally, ion transport characteristics would be affected by applying an external electrical field across the membranes. The ion transport in GO membranes under concentration gradient-driven conditions was found to be dependent on the external electric and magnetic field [62]. A forward electric field facilitated the penetration while a reverse electric field decreased the penetration. However, the penetration was enhanced regardless of the direction of the applied magnetic field. They affected the ion transport in GO membranes in different ways. The electric field dominated the ion migration in solutions while the magnetic field altered the nanostructure of GO membranes. This work demonstrated that applying an external electric field may be used to enhance the ion selectivity of GO membranes.

The confined 2D nanochannel network in GO membranes coupling with the external electric field may lead to some interesting ion transport behaviors. An interfacial electrical double layer (EDL) without bulk solution can be formed between GO layers. In addition, the adjacent GO layers can act as the channel wall and the gating electrode simultaneously. This allows the *in situ* modulation of the EDL with an external electric field. By applying a low potential, an evident change in the ion population and distribution within the EDL was observed [63]. The influences of EDL induced by a gate potential on the ion transport characteristics of GO membranes under concentration gradient-driven conditions were investigated by Cheng et al. The diffusion rates were reversibly modulated and enhanced by 4–7 times by the applied potential (only 0.5 V). The confined nanochannels, the concentration gradient, and the external potential bias formed a unique environment, in which short-range ion–ion correlations were markedly increased. Their work indicated that the ion transport in nanoconfined channels could be effectively modulated by a small external voltage. With the improvement in the atomically precise nanochannels and nanopores, as well as the field-effect control of ion transport, more smart and functional devices may be developed.

Recently, heterogeneous GO membranes have been fabricated, and the ion transport behaviors driven by an applied voltage were investigated [64]. The GO (negatively charged) and polyetherimide-grafted GO (positively charged) sheets were vacuum filtrated on a porous substrate sequentially to form heterogeneously charged GO membranes. Instead of the typical ohmic behavior of an intrinsic GO membrane, a unique nonlinear I–V curve was observed, indicating evident ion current rectification. Further studies revealed that the rectification performance was dependent on the electrolyte concentration. A rectification factor of 108 was achieved by modifying the charge properties of the GO membranes by tuning the pH of the electrolyte. A concentration-polarized mechanism was employed to explain the current rectification in the heterogeneous GO membranes. When a downward potential bias is applied, the K^+ and Cl^- ions tend to migrate to the junction region. Since the ions are more active under high voltage, the ion concentration in the junction does not increase linearly with the applied voltage. K^+ and Cl^- ions tend to stay in the bulk solution when an upward voltage is applied, leading to ion depletion in the junction region. The charge accumulation and depletion in the junction region result in the rectification behavior.

5.3.8 Ion Transport Through Planar GO Laminates

In the above discussions, the applied concentration gradient, electrical or magnetic field, and pressure are vertical to the aligned GO sheets. Ions transport follows the applied gradient. Inspired by lipid membranes that pump ions through against their concentration gradients, Yang et al. reported a net cationic flow through planar GO membranes by using asymmetric light illumination, as shown in Figure 5.10a–c [65]. In their system, the ions move uphill against their concentration gradients with rates orders of magnitude faster than that in simple diffusion. A photoelectric response in GO membranes was observed without an applied voltage. A net ionic current was detected when the GO strip was irradiated locally. The direction of the ionic current was from the non-illuminated side to the illuminated side. To eliminate thermal effects, GO membranes were heated from one direction and no net ion transport was observed. Further experiments confirmed that the antigradient ion transport pumped by external photons could be achieved with various mono and divalent ions. Under illumination, photo-induced electrons and holes tend to migrate to the non-irradiated direction driven by their concentration gradients. Since holes possess a faster transport rate than electrons in GO membranes, a potential difference is, therefore, established, with the low electric potential in the illuminated area. Their work demonstrated an alternative strategy to control the ion transport behaviors in 2D planar nanochannels in a noninvasive way.

Very recently, the same group has developed an inverted T-shape water extraction system using planar heterogeneous GO membranes (PHGOM), in which ~100% salt rejection and water flux as high as 1529 LMH bar^{-1} were simultaneously achieved [66]. The negatively charged GO sheets and positively charged polyethyleneimine-grafted GO sheets were put in a filtration vessel and separated by a plastic clapboard. The solutions were then vacuum filtrated to fabricate the heterogeneous GO membranes in a dual-flow process, illustrated in Figure 5.10d. The obtained lamellar GO membranes were composed of three segments, negatively charged, near neutral, and positively charged, from left to right (Figure 5.10e). The oppositely charged lamellar GO heterojunction exhibited a bi-unipolar transport behavior, rejecting both cations and anions. Ions were depleted in the neutral zone so that deionized water could be extracted. Considering this, a PHGOM-based desalination device was fabricated. With an applied bias, both Na^+ and Cl^- ions were dragged out of the membrane. Pure water was then extracted from the central

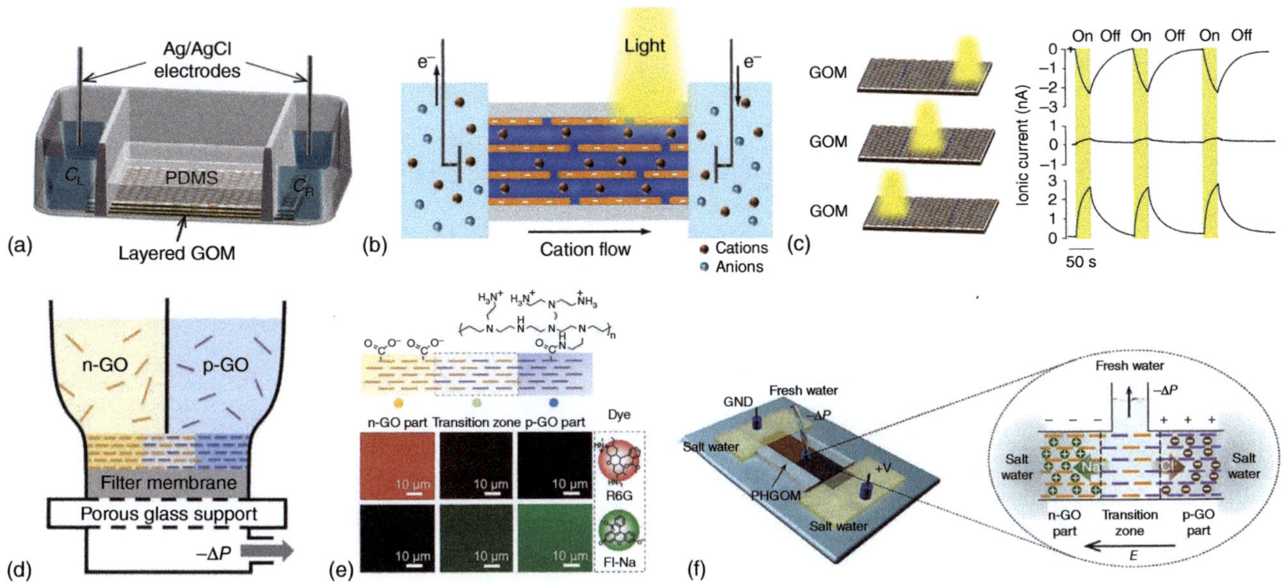

Figure 5.10 (a) Illustration of the photo-induced ion transport device. (b) Illustration of the cation flow generated in GO membrane induced by asymmetric light illumination. (c) The measured photocurrent when the light illumination was applied on the right, middle, and left side, respectively; Source: Yang et al. [65]/Springer Nature/Licensed under CC BY 4.0. (d) Illustration of the dual-flow filtration setup used to prepare the heterogeneous GO membrane. (e) Fluorescence images of the heterogeneous GO membrane, confirming the opposite charges of the separate parts. (f) Illustration of the mechanisms for deionized water extraction from the transition zone of the heterogeneous GO membrane. Source: Wen et al. [66]/with permission of John Wiley and Sons.

transition zone, as shown in Figure 5.10f. In a general membrane desalination process, size sieving and Donnan effect are the two dominant salt rejection mechanisms. However, in this unique system, ions are removed from water based on the electric field-induced ion concentration depletion effect. There is no need to decrease the interlayer spacing of GO membranes for the improvement of salt rejection but often with the unavoidable compromise in membrane permeance. In addition to the excellent desalination performance, the system also exhibited low energy consumption and good working stability.

5.4 Summary and Perspective

Graphene-based nanomaterials provide us a unique platform to investigate the transport characteristic of fluids and ions under nanoscale confinement. Their transport behaviors are much different from what they are in the bulk. Based on the unique mass transfer characteristics, graphene-based membranes are attractive candidates to revolutionize current polymer-based desalination technology. Both theoretical calculations and experimental investigations have confirmed the great promise of graphene-based membranes for desalination. For the practical application of graphene-based membranes, some important issues should be addressed. The major limitation is the low-cost availability of high-quality graphene and graphene-based membranes in large quantities.

Nanopores introduced to single-layer graphene are expected to be uniformly distributed with a narrow size distribution so that a high rejection rate can be obtained. Nonselective defects introduced during the synthesis of graphene, such as the grain boundaries and other intrinsic defects, as well as those introduced during the processing of graphene, should be avoided as much as possible. Solutes tend to pass through these defective sites owing to the lower transmembrane resistance, which is detrimental to the selectivity of the membranes. The experimental investigations on nanoporous graphene in the literature were all carried out using small-sized graphene so that the abovementioned defects can be eliminated. This further reflects the technical difficulties to fabricate large area and defect-free single-layer graphene for scalable desalination applications. At present, CVD is considered as the most promising method for the preparation of large-area single-layer graphene with relatively low cost. However, the formation of intrinsic defects is inevitable and thus the area of defect-free graphene rarely exceeds 1 mm. The CVD graphene is usually polycrystalline phase with point defects and grain boundaries. In addition, nonselective large defects and pores are introduced during the transfer process of CVD graphene. What's more, the current nanopore-drilling techniques are not mature enough to finely tune the physicochemical properties of the pores. Although various theoretical studies have confirmed the potential of single-layer porous graphene as high-performance desalination membranes, the practical realization is still highly challenging.

Lamellar GO membranes provide an alternative approach for the practical application of graphene-based membranes. Compared with single-layer graphene, GO sheets are much easier to fabricate in a large quantity via low-cost exfoliation method. The rich functional groups on the basal planes and edges of GO sheets make them well dispersed in water. Various liquid phase film-forming methods, such as vacuum filtration, drop coating, and spin coating can be employed to fabricate lamellar GO membranes. The ultrafast water permeability and selective ion transport properties of GO membranes make them highly promising in the field of separation. The transport channels in lamellar GO membranes are the 2D interlayer capillaries between adjacent GO layers and the in-plane structural defects in GO sheets. Many investigations have proved the vital role of the interlayer galleries in dominating the separation performance of GO membranes. To enable GO membranes for desalination, efforts should be devoted to decreasing the interlayer spacing and inhibiting the swelling of GO membranes. Following this idea, various methods, including chemical reduction, cross-linking with ions or small molecules, and applying external pressure, have been tried to improve the salt rejection of GO membranes in aqueous environment. In addition to steric hindrance, Donnan effect is another important rejection mechanism for GO membranes. Modifying the charge properties of GO sheets has been demonstrated as an effective strategy to fabricate high-performance GO-based desalination membranes. Recently, investigations on ion transport through GO membranes modulated by external electric or magnetic fields have been reported. The 2D-confined nanochannels coupling with the external field lead to some interesting ion transport behaviors. Studies about ion transport through finely designed GO-based nanostructures, such as heterogeneously charged GO membranes and planar GO membranes may pave the way for novel functional devices in the near future.

Currently, the limitations of the practical applications of GO-based membranes mainly include the following aspects—(i) the swelling and structural stability of GO membranes in aqueous environment. The stable pore structure and structural integrity are the prerequisites for their practical applications. However, the intrinsic structure of GO sheets makes the assembled membranes tend to swell and disintegrate in water. Generally, dead-end filtration systems are used to test the separation performance of GO membranes at lab scale. The cross-flow systems that are commonly used in industry have higher requirements on membrane structural stability. Mechanical constraints and chemical bonding that can be applied on a large scale should be considered; (ii) scale processing of large-area GO membranes with controlled nanochannels. Instead of flat sheet membranes, membrane modules are more preferred for industrial applications; (iii) deep understanding of the transport mechanisms through the hierarchical pathways—2D interlayer nanochannels, intrinsic in-plane defects, and slit-like pores. Based on this, the properties of GO sheets, as well as the nanostructure of the lamellar GO membranes, can be designed and controlled for better separation performance; (iv) precise control of the interlayer spacing down to angstrom scale. At present, it is easy to increase the interlayer spacing of GO membranes with the intercalation of nanospacers. However, narrowing the interlayer spacing with precise control so that ions can be efficiently rejected is still challenging. (v) membrane fouling. Charged contaminants tend to be adsorbed by the functional groups on GO sheets, leading to decreased membrane permeability with time. Recently, ion transport characteristics of GO-based membranes under external stimuli, including photons, potential bias, and magnetic field, have been investigated. Heterogeneously charged and planar structured GO membranes have been fabricated, which exhibited novel mass transport properties. The molecular transport process through 2D nanochannels coupling with other processes, such as sensing and energy harvesting, may generate novel smart devices with unique functionalities. The rapid growth in the number of publications regarding graphene-based membranes indicates their great potential to be explored. More relevant efforts are on the way.

References

1 Elimelech, M. and Phillip, W.A. (2011). The future of seawater desalination: energy technology and the environment. *Science* 333: 712–717.

2 Park, H.B., Kamcev, J., Robeson, L.M. et al. (2017). Maximizing the right stuff: the trade-off between membrane permeability and selectivity. *Science* 356 (6343): 1137.

3 Hinds, B.J., Chopra, N., Rantell, T. et al. (2004). Aligned multiwalled carbon nanotube membranes. *Science* 303 (5654): 62–65.

4 Liu, G.P., Jin, W.Q., and Xu, N.P. (2015). Graphene-based membranes. *Chem. Soc. Rev.* 44 (15): 5016–5030.

5 Perreault, F., Fonseca de Faria, A., and Elimelech, M. (2015). Environmental applications of graphene-based nanomaterials. *Chem. Soc. Rev.* 44 (16): 5861–5896.

6 Bunch, J.S., Verbridge, S.S., Alden, J.S. et al. (2008). Impermeable atomic membranes from graphene sheets. *Nano Lett.* 8 (8): 2458–2462.

7 Berry, V. (2013). Impermeability of graphene and its applications. *Carbon* 62: 1–10.

8 Hu, S., Lozada-Hidalgo, M., Wang, F.C. et al. (2014). Proton transport through one-atom-thick crystals. *Nature* 516 (7530): 227–230.

9 Achtyl, J.L., Unocic, R.R., Xu, L. et al. (2015). Aqueous proton transfer across single-layer graphene. *Nat. Commun.* 6: 6539.

10 Walker, M.I., Braeuninger-Weimer, P., Weatherup, R.S. et al. (2015). Measuring the proton selectivity of graphene membranes. *Appl. Phys. Lett.* 107 (21): 213104.

11 Zhu, C.Q., Li, H., and Meng, S. (2014). Transport behavior of water molecules through two-dimensional nanopores. *J. Chem. Phys.* 141: 18C528.

12 Sint, K., Wang, B., and Kral, P. (2008). Selective ion passage through functionalized graphene nanopores. *J. Am. Chem. Soc.* 130: 16448–16449.

13 Zhao, S.J., Xue, J.M., and Kang, W. (2013). Ion selection of charge-modified large nanopores in a graphene sheet. *J. Chem. Phys.* 139: 114702.

14 Cohen-Tanugi, D. and Grossman, J.C. (2012). Water desalination across nanoporous graphene. *Nano Lett.* 12 (7): 3602–3608.

15 He, Z., Zhou, J., Lu, X. et al. (2013). Bioinspired graphene nanopores with voltage-tunable ion selectivity for Na^+ and K^+. *ACS Nano* 7 (11): 10148–10157.

16 Chen, Q. and Yang, X. (2015). Pyridinic nitrogen doped nanoporous graphene as desalination membrane: molecular simulation study. *J. Membr. Sci.* 496: 108–117.

17 Ebrahimi, S. (2016). Influence of curvature on water desalination through the graphene membrane with Si-passivated nanopore. *Comput. Mater. Sci.* 124: 160–165.

18 Cohen-Tanugi, D. and Grossman, J.C. (2014). Water permeability of nanoporous graphene at realistic pressures for reverse osmosis desalination. *J. Chem. Phys.* 141: 074704.

19 Cohen-Tanugi, D., Lin, L.C., and Grossman, J.C. (2016). Multilayer nanoporous graphene membranes for water desalination. *Nano Lett.* 16 (2): 1027–1033.

20 Surwade, S.P., Smirnov, S.N., Vlassiouk, I.V. et al. (2015). Water desalination using nanoporous single-layer graphene. *Nat. Nanotechnol.* 10 (5): 459–464.

21 O'Hern, S.C., Jang, D., Bose, S. et al. (2015). Nanofiltration across defect-sealed nanoporous monolayer graphene. *Nano Lett.* 15 (5): 3254–3260.

22 O'Hern, S.C., Boutilier, M.S., Idrobo, J.C. et al. (2014). Selective ionic transport through tunable subnanometer pores in single-layer graphene membranes. *Nano Lett.* 14 (3): 1234–1241.

23 Raffone, F., Savazzi, F., and Cicero, G. (2019). Controlled pore generation in single-layer graphene oxide for membrane desalination. *J. Phys. Chem. Lett.* 10 (23): 7492–7497.

24 Rollings, R.C., Kuan, A.T., and Golovchenko, J.A. (2016). Ion selectivity of graphene nanopores. *Nat. Commun.* 7: 11408.

25 Bieri, M., Treier, M., Cai, J. et al. (2009). Porous graphenes: two-dimensional polymer synthesis with atomic precision. *Chem. Commun.* 45: 6919–6921.

26 Lerf, A., He, H., Forster, M. et al. (1998). Structure of graphite oxide revisited. *J. Phys. Chem. B.* 102: 4477–4482.

27 Erickson, K., Erni, R., Lee, Z. et al. (2010). Determination of the local chemical structure of graphene oxide and reduced graphene oxide. *Adv. Mater.* 22 (40): 4467–4472.

28 Nair, R.R., Wu, H.A., Jayaram, P.N. et al. (2012). Unimpeded permeation of water through helium-leak-tight graphene-based membranes. *Science* 335 (6067): 442–444.

29 Sun, P.Z., Liu, H., Wang, K.L. et al. (2015). Ultrafast liquid water transport through graphene-based nanochannels measured by isotope labelling. *Chem. Commun.* 51 (15): 3251–3254.

30 Boukhvalov, D.W., Katsnelson, M.I., and Son, Y.W. (2013). Origin of anomalous water permeation through graphene oxide membrane. *Nano Lett.* 13 (8): 3930–3935.

31 Algara-Siller, G., Lehtinen, O., Wang, F.C. et al. (2015). Square ice in graphene nanocapillaries. *Nature* 519 (7544): 443–445.

32 Wei, N., Peng, X.S., and Xu, Z.P. (2014). Understanding water permeation in graphene oxide membranes. *ACS Appl. Mater. Interfaces* 6 (8): 5877–5883.

33 Jiao, S. and Xu, Z. (2017). Non-continuum intercalated water diffusion explains fast permeation through graphene oxide membranes. *ACS Nano* 11 (11): 11152–11161.

34 Saraswat, V., Jacobberger, R.M., Ostrander, J.S. et al. (2018). Invariance of water permeance through size-differentiated graphene oxide laminates. *ACS Nano* 12 (8): 7855–7865.

35 Sun, P.Z., Zhu, M., Wang, K.L. et al. (2013). Selective ion penetration of graphene oxide membranes. *ACS Nano* 7 (1): 428–437.

36 Sun, P.Z., Zheng, F., Zhu, M. et al. (2014). Selective trans-membrane transport of alkali and alkaline earth cations through graphene oxide membranes based on cation-π interactions. *ACS Nano* 8 (1): 850–859.

37 Sun, P.Z., Ma, R.Z., Deng, H. et al. (2016). Intrinsic high water/ion selectivity of graphene oxide lamellar membranes in concentration gradient-driven diffusion. *Chem. Sci.* 7 (12): 6988–6994.

38 Hu, J.Q., Liu, Z., Deng, K. et al. (2019). A novel membrane with ion-recognizable copolymers in graphene-based nanochannels for facilitated transport of potassium ions. *J. Membr. Sci.* 591: 117345.

39 Paneri, A. and Moghaddam, S. (2015). Impact of synthesis conditions on physicochemical and transport characteristics of graphene oxide laminates. *Carbon* 86: 245–255.

40 Xu, W.L., Fang, C., Zhou, F. et al. (2017). A facile way of forming ultrathin high-performance graphene oxide membranes for water purification. *Nano Lett.* 17 (5): 2928–2933.

41 Song, J.H., Yu, H.W., Ham, M.H. et al. (2018). Tunable ion sieving of graphene membranes through the control of nitrogen-bonding configuration. *Nano Lett.* 18 (9): 5506–5513.

42 Li, W., Zhang, L., Zhang, X. et al. (2020). Atomic insight into water and ion transport in 2D interlayer nanochannels of graphene oxide membranes: implication for desalination. *J. Membr. Sci.* 596: 117744.

43 Joshi, R.K., Carbone, P., Wang, F.C. et al. (2014). Precise and ultrafast molecular sieving through graphene oxide membranes. *Science* 343: 752–754.

44 Zheng, S., Tu, Q., Urban, J.J. et al. (2017). Swelling of graphene oxide membranes in aqueous solution: characterization of interlayer spacing and insight into water transport mechanisms. *ACS Nano* 11 (6): 6440–6450.

45 Yeh, C.N., Raidongia, K., Shao, J. et al. (2015). On the origin of the stability of graphene oxide membranes in water. *Nat. Chem.* 7 (2): 166–170.

46 Baskoro, F., Wong, C.B., Kumar, S.R. et al. (2018). Graphene oxide-cation interaction: inter-layer spacing and zeta potential changes in response to various salt solutions. *J. Membr. Sci.* 554: 253–263.

47 Pei, J.X., Huang, L., Jiang, H.F. et al. (2019). Inhibitory effect of hydrogen ion on the copper ions separation from acid solution across graphene oxide membranes. *Sep. Purif. Technol.* 210: 651–658.

48 Chen, L., Shi, G.S., Shen, J. et al. (2017). Ion sieving in graphene oxide membranes via cationic control of interlayer spacing. *Nature* 550 (7676): 380–383.

49 Hung, W.S., Tsou, C.H., De Guzman, M. et al. (2014). Cross-linking with diamine monomers to prepare composite graphene oxide-framework membranes with varying *d*-spacing. *Chem. Mater.* 26 (9): 2983–2990.

50 Padmavathy, N., Behera, S.S., Pathan, S. et al. (2019). Interlocked graphene oxide provides narrow channels for effective water desalination through forward osmosis. *ACS Appl. Mater. Interfaces* 11 (7): 7566–7575.

51 Su, Y., Kravets, V.G., Wong, S.L. et al. (2014). Impermeable barrier films and protective coatings based on reduced graphene oxide. *Nat. Commun.* 5: 4843.

52 Kim, J., Lee, S.E., Seo, S. et al. (2019). Near-complete blocking of multivalent anions in graphene oxide membranes with tunable interlayer spacing from 3.7 to 8.0 angstrom. *J. Membr. Sci.* 592: 117394.

53 Yang, E., Ham, M.H., Park, H.B. et al. (2018). Tunable semi-permeability of graphene-based membranes by adjusting reduction degree of laminar graphene oxide layer. *J. Membr. Sci.* 547: 73–79.

54 Xi, Y.H., Liu, Z., Li, J. et al. (2018). Graphene-based membranes with uniform 2D nanochannels for precise sieving of mono-/multi-valent metal ions. *J. Membr. Sci.* 550: 208–218.

55 Yuan, S., Li, Y., Xia, Y. et al. (2020). Non-selective nanowrinkles of reduced graphene oxide laminar membranes for enhanced nacl rejection. *Environ. Sci. Technol. Lett.* 7 (4): 273–279.

56 Wei, Y., Zhang, Y.S., Gao, X.L. et al. (2016). Declining flux and narrowing nanochannels under wrinkles of compacted graphene oxide nanofiltration membranes. *Carbon* 108: 568–575.

57 Abraham, J., Vasu, K.S., Williams, C.D. et al. (2017). Tunable sieving of ions using graphene oxide membranes. *Nat. Nanotechnol.* 12 (6): 546–550.

58 Li, W.B., Wu, W.F., and Li, Z.J. (2018). Controlling interlayer spacing of graphene oxide membranes by external pressure regulation. *ACS Nano* 12 (9): 9309–9317.

59 Zhang, M.C., Guan, K.C., Ji, Y.F. et al. (2019). Controllable ion transport by surface-charged graphene oxide membrane. *Nat. Commun.* 10 (1): 1253.

60 Song, X.X., Zambare, R.S., Qi, S. et al. (2017). Charge-gated ion transport through polyelectrolyte intercalated amine reduced graphene oxide membranes. *ACS Appl. Mater. Interfaces* 9 (47): 41482–41495.

61 Yin, C.C., Dong, L.L., Wang, Z.G. et al. (2019). CO_2-responsive graphene oxide nanofiltration membranes for switchable rejection to cations and anions. *J. Membr. Sci.* 592: 117374.

62 Sun, P.Z., Zheng, F., Wang, K.L. et al. (2014). Electro- and magneto-modulated ion transport through graphene oxide membranes. *Sci. Rep.* 4: 6798.

63 Cheng, C., Jiang, G.P., Simon, G.P. et al. (2018). Low-voltage electrostatic modulation of ion diffusion through layered graphene-based nanoporous membranes. *Nat. Nanotechnol.* 13 (8): 685–690.

64 Fei, W.W., Xue, M.M., Qiu, H. et al. (2019). Heterogeneous graphene oxide membrane for rectified ion transport. *Nanoscale* 11 (3): 1313–1318.

65 Yang, J.L., Hu, X.Y., Kong, X. et al. (2019). Photo-induced ultrafast active ion transport through graphene oxide membranes. *Nat. Commun.* 10 (1): 1171.

66 Wen, Q., Jia, P., Cao, L.X. et al. (2020). Electric-field-induced ionic sieving at planar graphene oxide heterojunctions for miniaturized water desalination. *Adv. Mater.* 32 (16): 1903954.

6

Graphene-Based Membranes for Pervaporation

Jing Zhao and Wanqin Jin

Nanjing Tech University, College of Chemical Engineering, State Key Laboratory of Materials-Oriented Chemical Engineering, 30 Puzhu South Road, Nanjing 211816, P. R. China

6.1 Introduction

Pervaporation (PV), as typical membrane technology for liquid mixture separation, has been developed as an important alternative to conventional distillation and adsorption technologies. In PV process, the liquid feed contacts with the surface of a semipermeable membrane, and the permeate in vapor form can be collected on the downstream side of the membrane [1–3]. The components in feed solution show different solubilities and diffusivities in membrane, thus achieving efficient separation. Apart from the advantages of small footprint, flexibility, and easy operation, pervaporation is particularly advantageous in low-operating temperature, low-energy input, environmental-friendliness, and independence on thermodynamic vapor–liquid equilibrium. Therefore, PV is more suitable for the separation of azeotropic mixtures, thermally sensitive components, and the removal of dilute solvents [2, 3].

Nowadays, the eager demands for high-performance membrane separation motivate the development of advanced membrane materials. Generally, an ideal membrane should integrate the following features—[4, 5] (i) the thin thickness to confer low transport resistance (high flux); (ii) the well-defined transport channels with matching physical structure and chemical microenvironment to realize precise separation (high separation factor); (iii) the robust structure to prevent membrane rupture in harsh separation conditions and long-term operation (high stability). Two-dimensional (2D) materials with atomic thickness, exemplified by graphene and its derivative graphene oxide (GO), have emerged as promising nanobuilding blocks for next-generation high-performance separation membranes due to their capability to construct ultrathin selective layers with controllably ordered nanochannels, featuring extraordinary permeation properties [5–7]. Commonly, the graphene-based membrane can be categorized into two main types—porous single-layer graphene membrane and laminar GO membrane. The study about porous single-layer graphene membrane is still in its infant stage and mainly focuses

Two-Dimensional-Materials-Based Membranes: Preparation, Characterization, and Applications, First Edition. Edited by Gongping Liu and Wanqin Jin.
© 2022 WILEY-VCH GmbH. Published 2022 by WILEY-VCH GmbH.

on gas separation and ion/dye rejection, while the application in pervaporation has not been reported yet. By contrast, laminar GO membranes with interlayer galleries have demonstrated great potential in pervaporation owing to the facile fabrication and high separation efficiency. Moreover, GO contains abundant oxygen-containing functional groups, such as hydroxyl, epoxy, and carboxyl, which confer rich modification pathways for material optimization and structural regulation.

Pervaporation technology is commonly applied in dehydration of solvents, removal/recovery of low-concentration volatile organic compounds (VOCs) from water, and separation of organic liquid mixtures. To date, the research of laminar GO membrane for pervaporation primarily focuses on solvent dehydration, due to the intrinsic hydrophilicity of GO nanosheet and its excellent dispersity in polar solvents. Compared with organic solvent molecules, water molecule with higher polarity is more prone to be adsorbed on the hydrophilic membrane surface, and the smaller size endows water molecule with lower diffusion resistance in GO channels, thus achieving the water-preferential permeation. Besides the traditional applications, pervaporation technology shows great potential in desalination of high-salinity water in recent years, which is driven by the chemical potential difference, thus avoiding the high operation pressure in reverse osmosis process. GO membrane has attracted more and more attention in pervaporation desalination due to its excellent performance.

In 2012, Geim and coworkers first reported the peculiar unimpeded permeation of water molecules through the two-dimensional channels between orderly stacked GO nanosheets [8]. Since then, GO membranes have received worldwide research enthusiasm to explore their potential in water-selective permeation. However, the assembly of GO nanosheets relies on weak hydrogen bonds and π–π interactions. During the operation in aqueous environment, the abundant hydrophilic hydroxyl, epoxy, and carboxyl groups on GO nanosheets could form strong hydration in the presence of water molecules, thus impairing the interlayer bonding, leading to the swelling and even disintegration of GO membrane [9]. The resulting GO membrane usually shows a low separation factor and poor long-term stability, especially for the subnanometer-scale separations in pervaporation. To achieve high-performance separation, it is demanded to incorporate strong interlayer bonding to improve membrane stability, precisely tune the channel size to intensify size discrimination effect, and regulate the chemical environment in GO channels to facilitate water permeation. This chapter provides a comprehensive overview of the progress of graphene-based membranes (laminar GO membranes) for pervaporation, with emphasis on the structure–performance relationship and the structure manipulation strategies.

6.2 Mass-Transport Mechanism

6.2.1 Mass Transport in Pervaporation Process

In pervaporation process, the feed side is commonly kept at atmospheric pressure, while the permeate side maintains a low partial pressure of permeating molecules using vacuum or sweep gas to generate a partial pressure difference (chemical potential difference) as the driving force of molecular permeation across membrane [2]. It has been widely recognized that the mass transport in pervaporation membrane conforms to the solution-diffusion mechanism. According to this mechanism, the permeation of molecules through membrane consists of three steps—(i) sorption of permeating molecules from the feed liquid onto the upstream surface of membrane; (ii) diffusion of permeating molecules through the membrane; (iii) desorption of permeating molecules to vapor phase on the downstream of the membrane. Among them, the first two steps are dominant for the mass-transport process, while the impact of the last step is negligible. Separation is achieved between different permeants due to the difference in the amount of molecules that dissolve in the membrane and the rate at which the molecules diffuse through the membrane. The sorption is a thermodynamic process, and the diffusion is a kinetic process, which not only depends on the property and structure of membrane materials but also closely relates to the interactions among permeating molecules and the interactions between membrane and permeating molecules.

6.2.2 Mass Transport in GO Membrane

The special transport behavior of water molecules in GO membrane has gained extensive investigations since its first report. For PV dehydration applications, the key is to study the selective permeation behavior of water molecules in the presence of solvent molecules. With water/alcohol mixture as a representative, the impacts of the physical structure and chemical environment of GO channels on the molecular transport have been well studied mainly through molecular dynamics simulations, and the separation mechanism is tentatively proposed. These studies are expected to provide guidelines for the design and construction of high-performance GO membranes for pervaporation dehydration.

Yang and coworkers studied the permeation of ethanol–water mixtures through single-layer GO with the effective diameter of the in-plane pores varying in the range of 5.5–9.0 Å [10]. The GO nanosheets functionalized with hydroxyl and epoxy groups display comparable affinity toward water and ethanol, and the complete rejection for ethanol can be achieved via size sieving when the pore size is 5.5 Å. Our group investigated the impacts of pore size and O/C ratio [11]. Higher O/C ratio improves the affinity between water and membrane, then facilitates the preferential adsorption of water on membrane and enhances water/ethanol selectivity. With a high oxidization degree and a suitable pore size, infinite water/ethanol selectivity can be obtained. Besides the intrinsic oxygen-containing groups on GO, Yang and coworkers tuned the functional groups on the periphery of GO pores and investigated the influences on water/ethanol permeation [12]. Compared with the GO pore functionalized with COOH (non-ionized carboxyl), the GO–COO^- (ionized carboxyl) pore shows highly selective water transport even with the pore size (~10 Å) larger than ethanol molecule. This phenomenon could be ascribed to the stronger water affinity of COO^- group, which promotes the preferential adsorption of water and consequently blocks the permeation of ethanol molecules, thus improving the water–ethanol separation efficiency. Similar results in experiments have also been demonstrated [13]. The ionic sulfonate groups in GO channels confer higher water flux and water/butanol separation factor than their nonionic counterpart. Qiao and coworkers tuned the proportion of ionized COOH group at the edge of GO pores to investigate the impacts of ionization degree [14]. As the pore becomes more negatively charged, the occupancy of water molecules inside the pore gradually increases with high water mole fraction sites being localized near the functional groups (Figure 6.1a), while the number of ethanol molecules

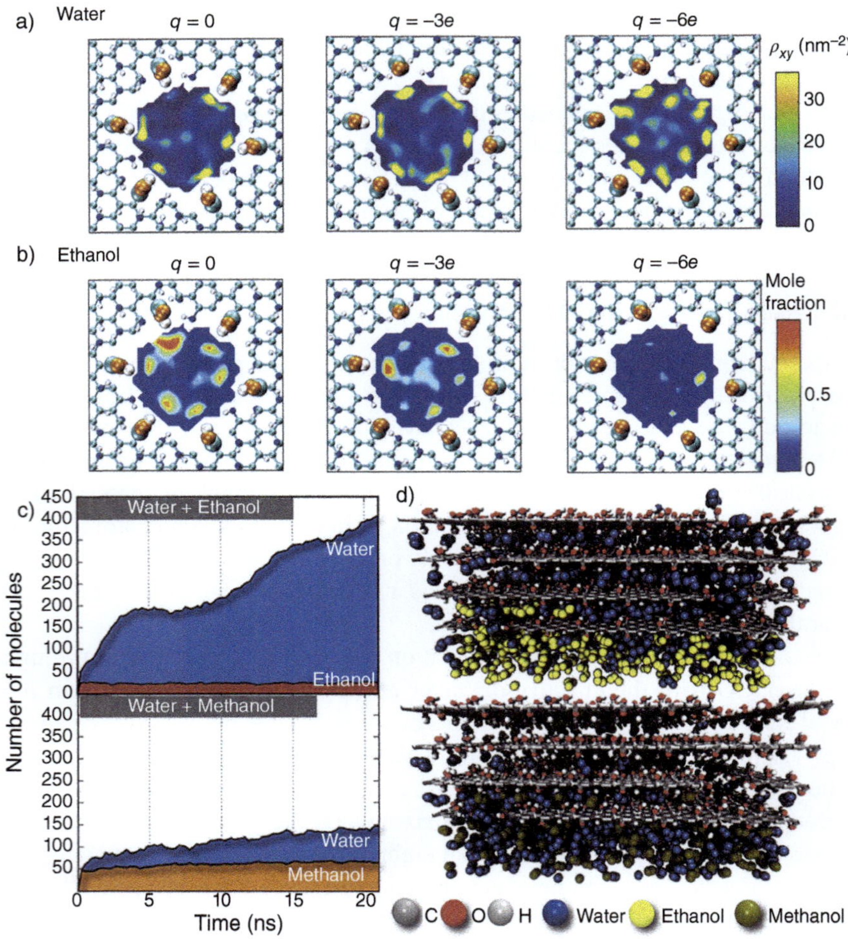

Figure 6.1 (a) The *xy*-plane number density map of the water molecules inside 1.12 nm-wide pores with different net charges. (b) The *xy*-plane mole fraction map of ethanol molecules. The carboxyl groups (neutral and ionized) along the pore's edge are highlighted as large spheres (hydrogen, carboxyl oxygen, oxygen, and carbon atoms are in white, orange, blue, and cyan color, respectively). Source: [14]/with permission of Elsevier. (c) The number of water, ethanol, and methanol molecules permeating the GO membrane with interlayer distance of 7 Å and O/C ratio of 35%. (d) MD snapshots showing the membrane being occupied mainly by water molecules (blue beads), while alcohols are rejected by the membrane. Source: Damasceno Borgers et al. [15]/with permission of Elsevier.

inside the pore remarkably decreases (Figure 6.1b). Therefore, with the increase of the ionization degree, the water flux is enhanced and the ethanol flux is reduced. Due to the large pore size (1.12 nm) in the simulation, the neutral pore ($q = 0$) shows a low water/ethanol selectivity of about 1.5, which is increased to 6.5 after ionization ($q = -6e$), close to that achieved in neutral pore with a diameter of 0.68 nm. This result reveals that improving the hydrophilicity of GO pores through ionization could extend the range of pore size for efficient separation. The reported experimental data further evidence this conclusion—the GO membrane with channel size larger than both water and solvent molecules could achieve high-performance separation due to the existence of hydrophilic groups [13, 16].

All the above studies employed porous single-layer GO as a simplified model, which actually possesses different channel properties with the practical laminated structure, such as channel dimensionality. Galvao and coworkers probed the water/ethanol and water/methanol separation through multilayer GO membranes with different interlayer distances and oxidation degrees (Figure 6.1c,d) [15]. For water–ethanol system, the water percentage in GO membrane was up to 95%, while the value was about 67% for water–methanol case. They found that besides the size exclusion and membrane affinity, the 2D H-bond network formed by the monolayer water molecules in GO channels also contributes to the fast water permeation and effective separation. Yang and coworkers simulated the competitive insertion/adsorption behavior of water and ethanol molecules into the nanochannels of layered GO structures [17]. An increase in the oxidization level of GO can improve the water intercalation. They proposed that the high resistance for ethanol inside GO nanochannels may be the critical factor leading to low ethanol permeability.

Although there are large discrepancies between simulation results and experimental data, the above theoretical studies indicate the critical roles of channel size and chemical structure (affinity) in the water-selective permeation in GO membrane [10, 12, 14, 18]. It can be concluded that the pervaporation separation through GO membrane still conforms to the solution-diffusion mechanism, while the size exclusion effect also makes some contribution if the pore size is in the suitable range. Due to the small size difference between water and organic solvent molecules, it is difficult to achieve such precise regulation of channel size in practical experiments, especially in aqueous environment. Comparatively, the tuning of chemical structure is much easier to be realized than the size regulation in Å scale. Through functionalization of GO channels, the chemical environment for molecular permeation can be optimized, thus improving the preferential adsorption and diffusion of water molecules. In this case, a high-efficiency separation can be achieved with a wider channel size, which is in favor of obtaining the higher permeate flux.

In terms of pervaporation desalination, Hou and coworkers proposed a four-stage mechanism based on the experimental results to describe the water transport process in GO membrane (Figure 6.2a)—[18] surface adsorption of liquid water, water diffusion under concentration gradient, phase change from liquid to vapor, and water vapor transport through GO laminates. As to the hydrated salt ions, the penetration of anions can be restrained by electrostatic repulsion and size exclusion effects, while the cations gradually diffuse into the GO laminates. The cations

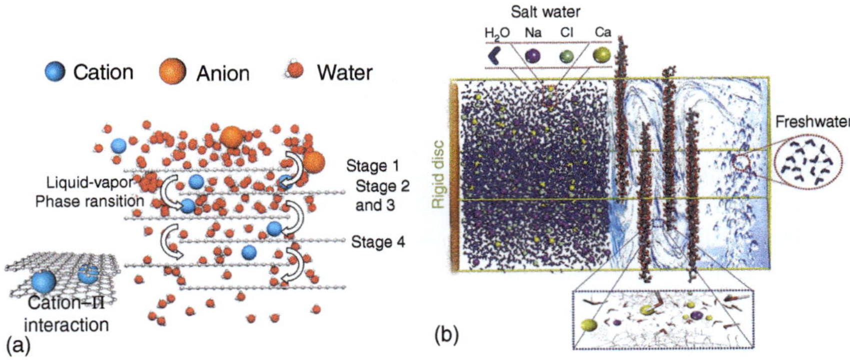

Figure 6.2 (a) Schematic diagram of the brine desalination process in GO membrane. Source: Li et al. [18]/with permission of American Chemical Society. (b) MD simulation model of freestanding GO membrane consisting of a heated fixed wall on the left and four stacked GO nanosheets. Source: Cha-Umpong et al. [19]/with permission of Elsevier.

inside GO channels play the role of interlayer cross-linking after water evaporation, thus stabilizing the GO structure and inhibiting the further transport of cations to achieve almost complete rejection for salts. Razmjou and coworkers investigated the transport of water molecules and hydrated cations through the GO membrane in a temperature-assisted simulation system similar to the pervaporation desalination process (Figure 6.2b) [19]. It is found that the cross-linked cations inside GO channels create an accessible passage for water molecules while blocking other cations. In addition, high temperature can assist with the binding between the hydrated cations and the oxygen functional groups on GO nanosheets by reducing the hydration shell of cations. Therefore, the water permeation increases with temperature, while the transport of cations remains negligible. The simulation results provide a guideline for modifying GO membrane with cations and operating at high temperature to optimize desalination performance.

6.3 Progresses in GO Membranes for Pervaporation

The excellent water transport property in GO membrane has attracted wide attention and considerable research since its first report [8]. Afterward, pristine GO membranes were prepared via various assembly techniques to explore their potentials in pervaporation dehydration and desalination processes [16, 20, 21]. Chu and coworkers first deposited GO laminates on a thin-film nanofibrous composite (TFNC) support via spin coating or vacuum filtration method [22]. The 93-nm-thick GO membrane shows a permeation flux of 2.2 kg m^{-2} h^{-1} and a water/ethanol separation factor of 308 (70 °C, water content in feed of 20 wt%). Our group achieved the successful assembly of GO nanosheets on ceramic hollow fiber support via the vacuum suction method [16], which showed a permeation flux of 1.7 kg m^{-2} h^{-1} and a separation factor of 743 for water/dimethyl carbonate separation (25 °C, water content in feed of 2.6 wt%). The separation performance of pristine GO membranes

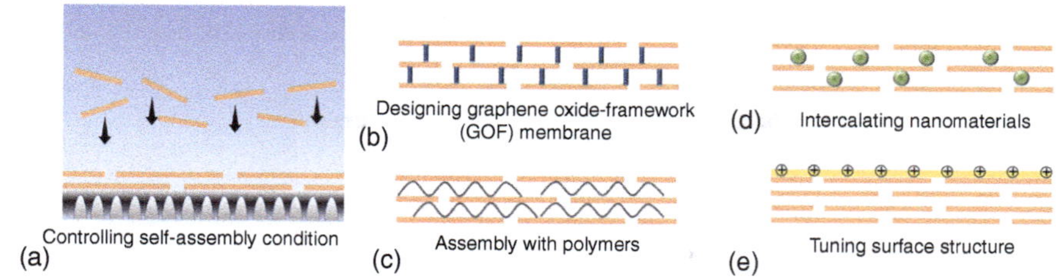

Figure 6.3 Summary of the structural manipulation strategies of GO membranes for pervaporation.

is far from what is expected, which can be essentially ascribed to the following reasons: (i) The packing of GO nanosheets is highly relevant to the assembly method, driving force, nanosheet structure, and composition [21, 23, 24]. Disordered packing increases the minimal thickness needed for defect-free GO layers to confer high resistance for molecular transport, and deteriorates the size exclusion effect of GO channels that result in low separation accuracy. (ii) The weak interlamellar interactions and the resultant swollen channel spacing in aqueous environment lead to a low separation factor and poor long-term stability [9]. (iii) Although there are abundant hydrophilic groups on GO nanosheets, the water-preferential adsorption property is not good enough [25]. In recent years, remarkable progresses have been achieved in the structural manipulation and performance optimization of GO membrane, as shown in Figure 6.3.

6.3.1 Controlling Self-Assembly Condition

Due to the ultrathin thickness and high aspect ratio of two-dimensional GO nanosheets, they are prone to form laminar membranes through self-assembly. The type and strength of the driving force for self-assembly exert crucial impacts on the packing behavior of nanosheets and the microstructure of the as-prepared GO membranes [26, 27]. Hung and coworkers fabricated GO laminar membranes via pressure-, vacuum-, and evaporation-assisted self-assembly techniques (PASA, VASA, and EASA) (Figure 6.4a) [21], which provide different driving forces for the self-assembly of GO nanosheets, thus influencing the microstructure of GO membranes and the separation performance for water/1-butanol mixture. Compared with the constant-pressure filtration in PASA, VASA is a variable-pressure filtration process and the pressure decreases with GO deposition. Therefore, the GO membrane fabricated by PASA shows a denser GO layer with a smoother surface and a shorter interlayer spacing (Figure 6.4b,c). Different from the downward driving force in PASA and VASA, the GO assembly in EASA is driven by a weak upward force arising from liquid evaporation, as a result, the GO-assembled layers are highly random with some loops (Figure 6.4d), leading to the highest surface roughness and interlayer distance. In addition, with random assembly, more hydrophobic parts on GO nanosheets are exposed, leading to the transition of membrane surface characteristics from hydrophilic to hydrophobic. As a result, the GO membrane with PASA process exhibits the highest pervaporation separation

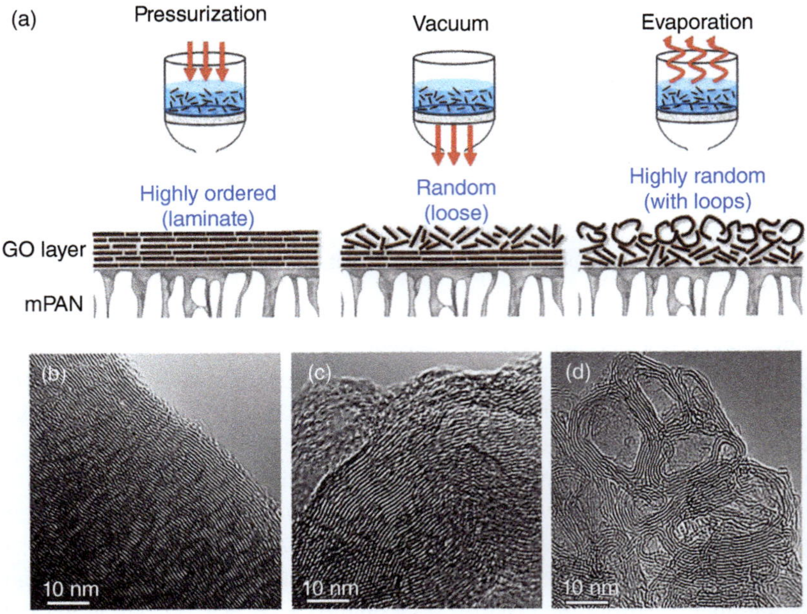

Figure 6.4 (a) Schematic diagrams of composite GO membranes fabricated through PASA, VASA, and EASA. Cross-sectional TEM images of GO assembly layers in (b) GO_{PASA}, (c) GO_{VASA}, and (d) GO_{EASA}. Source: Tsou et al. [21]/with permission of Elsevier.

performance (concentration of water in permeate of 99.6 wt% and permeation flux of 2.54 $kg\,m^{-2}\,h^{-1}$) due to its ordered, dense structure and favorable hydrophilicity. The value of the operation pressure in pressure filtration determines the packing density and the regularity of the GO nanosheets [28]. Higher pressure is expected to increase the packing density, while it may also lead to the formation of defects due to the inadequate time for orderly assembly at too high deposition rate. Therefore, there should be an optimal value for the operation pressure that could realize the balance between packing density and regularity so that to obtain the highest separation performance. Similarly, the driving force in VASA could also be adjusted via changing the vacuum degree, so that to tune the microstructure of GO laminates [23]. With slow deposition rate, a more thermodynamically favored interlayer structure can be formed with the oxidized regions on neighboring GO nanosheets assembling, and the non-oxidized areas constructing fast water transport channels. By contrast, the relatively random assembly of GO nanosheets at fast deposition rate leads to mismatches between oxidized and non-oxidized regions on neighboring GO nanosheets, which are unfavorable for fast water permeation.

6.3.2 Designing Graphene Oxide-Framework (GOF) Membrane

Membranes with well-designed porous structures in subnanometer scale, such as zeolite membrane, metal–organic-frameworks (MOFs) membrane, and carbon molecular sieves (CMS) membrane, hold great potentials in precise molecular separation via size exclusion effect. However, there still exist some critical problems for the practical applications of these membranes, such as brittleness and microdefects. Comparatively, the GO membrane with two-dimensional mass-transport channels shows better film-forming and mechanical properties, yet the poor size-sieving ability due to the channel swelling in aqueous environment is the key issue [9]. In this light, intercalating cross-linkers via strong interactions (such as covalent bond) to construct graphene oxide-framework (GOF) membrane provide an effective strategy, which could improve the structural stability of GO channels through strong chemical bonding and meanwhile precisely tuning the interlayer *d*-spacing to improve size exclusion effect via employing cross-linkers with different physical structures (such as molecular size and spatial configuration) [29].

Hung et al. first designed GOF membranes with diamine molecules, including ethylenediamine (EDA), butylenediamine (BDA), and *p*-phenylenediamine (PPD), through pressure-assisted self-assembly technique [29]. The diamine molecules are chemically bonded to both GO nanosheets and cellulose acetate (CA) support via forming C—N covalent bonds, thus impeding the stretching of interlayer *d*-spacing during pervaporation and achieving excellent membrane stability. The disparity of interlayer *d*-spacing between the dry and wet states is reduced to 0.02 nm for GOF membranes, while it is 0.46 nm for unmodified GO membranes. Due to the different molecular sizes of diamine cross-linkers, the interlayer *d*-spacing in wet state can be tuned in the range of 0.93–1.09 nm. The GO–EDA/CA membrane possessing the smallest interlayer *d*-spacing confers an excellent water/ethanol separation performance with the permeation flux of 2.30 kg m^{-2} h^{-1} and the water concentration in permeate high to 99.8 wt% at 80 °C. Chung and coworkers developed a series of aldehyde-functionalized GOF membranes with improved separation performance for ethanol dehydration [30]. Positron annihilation spectroscopy (PAS) was employed to qualitatively characterize the evolution of membrane microstructure, which revealed the denser structure of GOF membranes compared with the pristine GO membrane. Moreover, the operation stability of GOF membranes, at a relatively high temperature (60 °C), was evaluated for the first time. The membrane performance keeps stable within a long period of 160–200 hours.

In most of the studies, the interlayer *d*-spacing in dry state is expanded compared with pristine GO after incorporating cross-linkers. To obtain GOF membranes with fixed narrow 2D channels, Sun and Zeng et al. employed thiourea (TU) as the cross-linker, which possesses small size (three-atom molecular skeleton), multiple reactive terminal groups (dual-NH_2 and –C=S), and mild reducing ability (Figure 6.5) [31]. With a facile hydrothermal synthesis, stable TU-bridged GO laminates can be formed through the dehydration condensation reactions between –NH_2/–C=S groups on TU and –COOH groups on GO, as well as the nucleophilic additions between –NH_2 groups on TU and C–O–C groups on GO. Due to the small size of TU molecules and the deoxygenation reactions between GO and TU, the interlayer *d*-spacing of the resultant TU–GOF membrane decreases from 0.83 nm of pristine GO to 0.74 nm, with the channel size

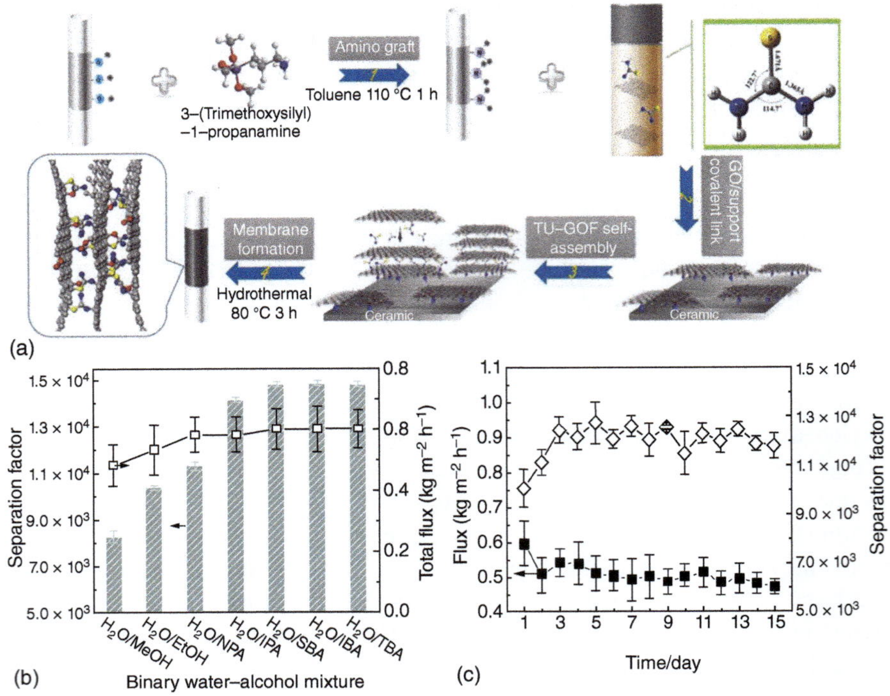

Figure 6.5 (a) Schematic of the self-assembly of TU–GOF membrane. (b) Pervaporative separation of water–alcohol systems with TU–GOF membrane (50 °C, 10% H_2O in feed). (c) Stability of TU–GOF membrane during dehydration of ethanol (50 °C, 10% H_2O in feed). Source: Yang et al. [31]/with permission of John Wiley and Sons.

of 0.34 nm, thus conferring excellent sieving properties for alcohols and ions. For water/alcohol separation with C2–C4 alcohols, the water concentration in permeate is higher than 99.9 wt%, corresponding to the water/alcohol separation factor higher than 10 000. Even for the water/methanol mixture with much higher separation difficulty, the separation factor reaches 8260, far superior to that of the reported membranes, demonstrating the extraordinary exclusion effect of TU–GOF membrane for the primary alcohols. To further improve membrane stability, the ceramic support is functionalized with amino groups in advance, so that to form covalent bond with GO layers. The covalent links between GO–TU and GO–support endow the membrane with desirable operation stability. The representative cross-linkers in GO membranes for pervaporation are listed in Figure 6.6 with the reactive groups highlighted. The interlayer cross-linking is commonly achieved via the covalent reactions between amino-epoxy/carboxyl [29, 32, 33], sulfhydryl–carboxyl [31], acyl chloride–hydroxyl/carboxyl [34, 35], aldehyde–hydroxyl [30], isocyanate–hydroxyl/carboxyl [36], carboxyl–hydroxyl [37], boronic acid–hydroxyl [38], and so on.

Besides covalently bonded cross-linkers, metal ions also exhibit promising results in stabilizing membrane structure due to the formation of multiple interactions with GO nanosheets [39], while their subnanometer sizes endow the possibility

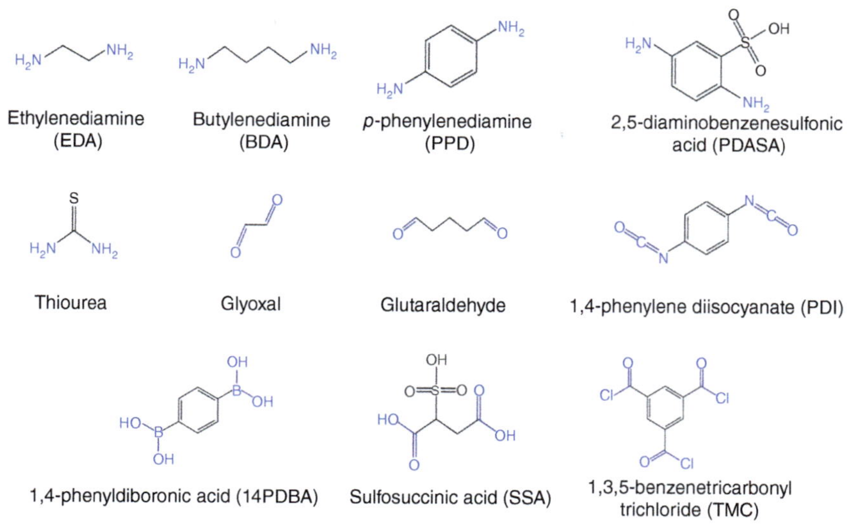

Figure 6.6 The representative cross-linkers in GO membranes for pervaporation.

in tuning interlayer spacing [40]. It has been evidenced that metal ions can form coordination interaction with carboxyl groups, thus bridging neighboring GO nanosheets laterally since most of the carboxyl groups on GO nanosheets are at the edges (Figure 6.7) [41]. In addition, the metals ions as Lewis acids could promote the ring-opening reaction of the epoxy groups on GO to form hydroxyl groups. Metal ions coordinate with the carbonyl and hydroxyl groups on the basal plane of GO nanosheets, and form cation–π interactions with the aromatic rings on GO nanosheets, then intercalate into the interlayer channels in GO membrane. Due to the multiple interactions between GO and metal ions, the direct doping of metal

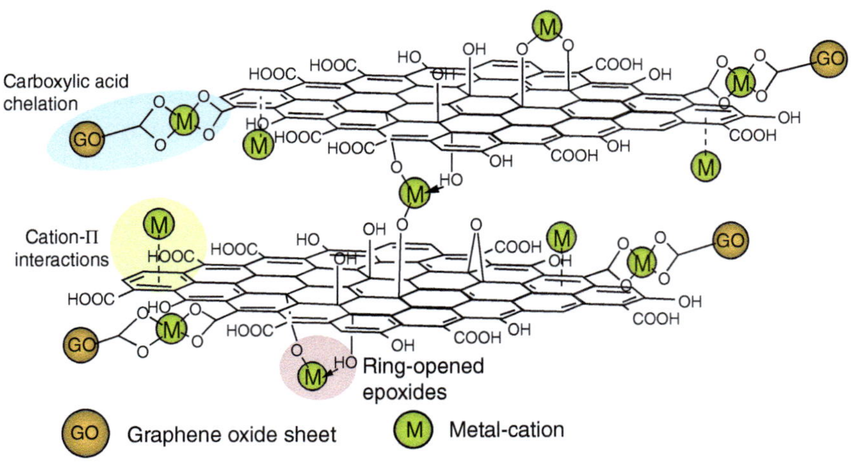

Figure 6.7 Proposed model of the bond locations between metal ions and GO nanosheets. Source: Ching et al. [41]/with permission of Elsevier.

ions into GO suspension will result in flocculation and even precipitates, thus making it impossible to fabricate ultrathin, defect-free GO membranes via direct filtration of the mixture. Our group employed a substrate-incorporation method to fabricate GO–metal ion membranes [42]. Hydrolyzed polyacrylonitrile (PAN) was utilized as the substrate and loaded with metal ions in advance. During the pressure-assisted self-assembly process, the metal ions diffuse upwards into the suspension via concentration gradient, while the GO nanosheets move downwards to stack on the substrate under pressure. Herein, the metal ions not only play a role in stabilizing the GO laminates in aqueous environment but also contribute to the strengthened interfacial bonding between GO and substrate. Zhao and Chen et al. [41] reported an in-depth study on the impacts that different cations exert on the water permeation behaviors in GO membranes. Monovalent (K^+ and Na^+), divalent (Cu^{2+} and Mn^{2+}), and trivalent (Al^{3+}, Fe^{3+}, and La^{3+}) cations were incorporated into GO membranes via post-immersion method. For cations with the same valence state, the smaller hydration radius and larger electronegativity are favorable to the formation of stronger bonding with GO and narrower interlayer spacing. Amongst the different valence states, water permeation is dominated by the cation's ability to form and retain a hydration shell. The cation with a large charge density would prolong the water duration inside the outer hydration shell, and then slow the water permeation rate.

The intercalation of cross-linkers remarkably improves the separation accuracy and stability of GO membranes, yet it is implemented with a penalty of decreased permeation flux [29]. On one hand, the channel size during pervaporation test is narrowed, and part of the channel is occupied by the cross-linkers, thus leading to higher diffusion resistance. On the other hand, many hydrophilic groups on GO are consumed through reacting with cross-linkers, thus impairing the interactions between water and GO membrane and inhibiting the water preferential adsorption. From the above analysis, it can be concluded that single regulation of the physical structure of GO channels cannot overcome the trade-off effect between permeation flux and separation factor. To address this issue, our group proposed the strategy of synergistically manipulating the physical structure and chemical environment in GO channels [13]. The non-ionized hydrophilic groups on pristine GO interact with water molecules through hydrogen bonds. Comparatively, the ionic groups can form electrostatic interactions with water, which have higher bond strength and lower directionality [43], thus adsorbing more water molecules, conferring higher water affinity, and constructing a more favorable chemical environment for water transport. Therefore, the multifunctional sodium 1,4-phenylenediamine-2-sulfonate (PDASA(Na^+)) was incorporated into GO channels, wherein the amine groups covalently bonded with GO through nucleophilic addition reaction to achieve interlayer cross-linking and ensure robust membrane structure, and the ionized sulfonate groups acted as water transport promoters to facilitate water-selective sorption and expedite water-preferential diffusion (Figure 6.8). Compared to the control membrane with PPD as cross-linker, the PDASA(Na^+)/GO membrane achieves the simultaneously improved permeation flux from 1.85 to 2.88 $kg\,m^{-2}\,h^{-1}$

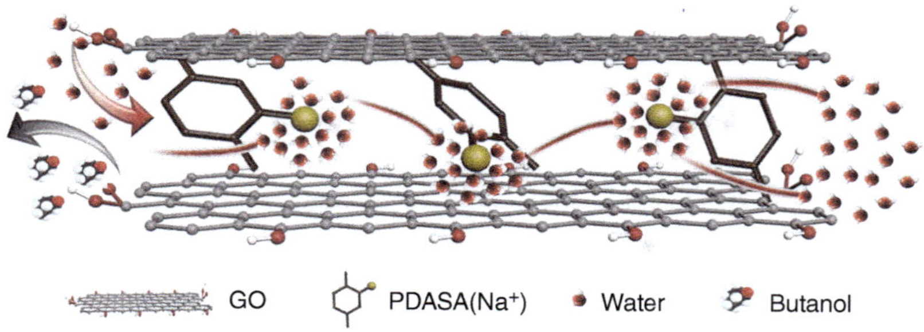

Figure 6.8 Schematic diagram of the molecular transport in GO membrane with optimized physical structure and chemical environment. Source: Liang et al. [13]/with permission of John Wiley and Sons.

and separation factor from 164 to 1366, successfully overcoming the trade-off effect and demonstrating the critical role of sulfonate group for water permeation. The study of sulfosuccinic acid (SSA) covalently linked GO membrane shows a similar improvement of water/alcohol separation performance [37]. The remarkably enhanced membrane hydrophilicity from the sulfonic groups and the narrowed interlayer channels confers the membrane with ultrahigh water permeation flux with outstanding separation accuracy toward solvent dehydration. These works demonstrate the effective strategy of functionalizing GOF membrane with additional hydrophilic groups (especially ionic hydrophilic groups) to obtain high-performance membranes for water-selective pervaporation.

6.3.3 Assembly with Polymers

The brick-and-mortar structure in nature (such as nacre) offers an ingenious inspiration for improving the structural stability of GO membranes [44, 45]. The well-stacked alternating layers of 2D inorganic nanomaterial and organic compound, as well as the robust interfacial bonding, lead to exceptional mechanical properties [45, 46]. It can be envisaged that highly stable GO membranes can be achieved via assembling GO with suitable polymers. In addition, the existence of polymer could promote the orderly assembly of GO nanosheets via providing molecular interactions as the driving force, and meanwhile, reduce the generation of microdefects in GO laminates [26]. Due to the various polar groups and the negatively charged carboxylate groups on GO nanosheets, there often exist multiple interactions between GO and polymer, including covalent bond, hydrogen bond, and electrostatic attraction [47–49]. Although the interlayer *d*-spacing of GO membrane may not be precisely tuned as achieved by small molecules or metal ions, the polymer confers a better chemical environment for water transport. The operation methods of GO assembly with polymer mainly include filtration-assisted assembly of GO–polymer mixture and layer-by-layer (LbL) self-assembly. Comparatively, the direct mixing and filtration method is more facile to operate, whereas flocculation and even precipitation may occur if the binding between GO and polymer is strong and fast, for instance, the electrostatic attraction between GO and cationic polyelectrolyte (PEL) [50]. In this case, LbL self-assembly is more suitable to obtain uniform and defect-free GO membrane [47, 49, 51].

Polyethyleneimine (PEI) is one of the most widely used PEL for membrane separation, which possesses abundant amino and ammonium groups to act as covalent reaction, hydrogen bond, and electrostatic interaction sites for assembly and confer high hydrophilicity [52]. Feng and coworkers prepared the GO/PEI multilayer membranes on a chlorine-treated thin-film composite polyamide substrate through LbL self-assembly and applied them in the dehydration of ethanol, isopropanol, and ethylene glycol, as well as the desalination of high-salinity water [47, 51, 52]. The membranes show a stable performance over a period of 210 hours. On this basis, our group proposed a zwitterionic modification strategy to further optimize the chemical environment in GO channels (Figure 6.9) [53]. After employing zwitterionic graphene oxide (ZGO) to replace GO and assembly with PEI, the permeation flux and water/ethanol separation factor increase by 47.4% and 2.63-fold, respectively. The abundant zwitterions (charged quaternary ammonium and sulfonate groups) on GO nanosheet and the increased PEI molecules after zwitterionic functionalization construct stable water channels with high-density ionic hydrophilic groups, which inhibit the strong coupling between water and ethanol, adsorb water molecules through hydrogen bond or ionic solvation, and repel ethanol molecules due to the existence of hydration layer. As a result, the ultrathin ZGO/PEI membrane acquires an exceptionally high-separation performance with a permeation flux of 3.23 $kg\,m^{-2}\,h^{-1}$ and water/ethanol separation factor of 2248 (70 °C, water content in feed of 10 wt%).

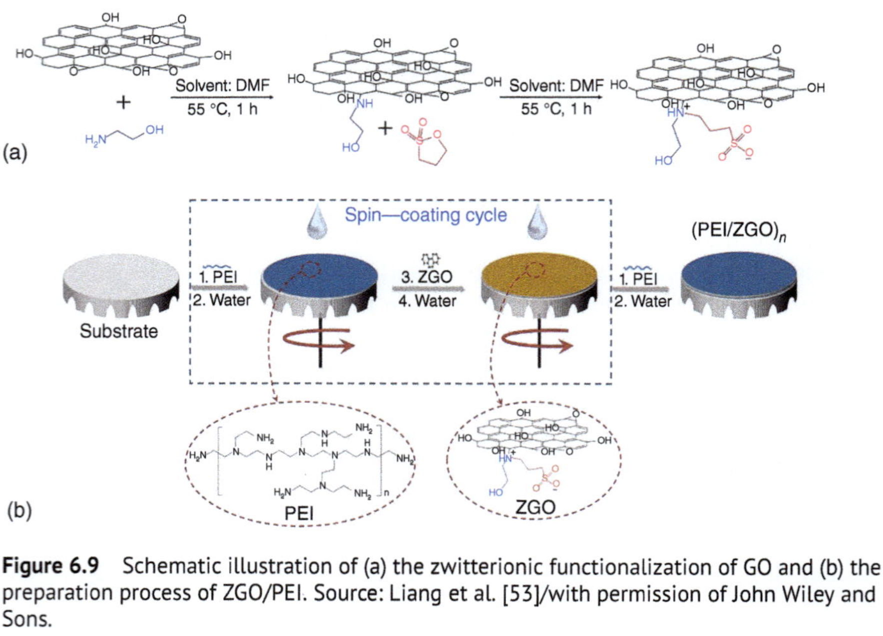

Figure 6.9 Schematic illustration of (a) the zwitterionic functionalization of GO and (b) the preparation process of ZGO/PEI. Source: Liang et al. [53]/with permission of John Wiley and Sons.

Chitosan (CS) is a widely utilized membrane material for pervaporation dehydration due to its high water affinity [34, 54]. Upon mixing of CS and GO, the occurrence of ionic complexation between the negatively charged carboxylate groups on GO and the positively charged protonated amines on CS results in severe aggregation. To address this issue, Hung et al. performed hydrothermal reduction to remove most carboxylate groups on GO and then successfully inserted the hydrophilic CS molecular chains into the reduced graphene oxide (rGO) laminates (Figure 6.10) [50]. Although rGO forms black precipitates in water, the mixture of the hydrophobic rGO with CS exhibits a favorable dispersion, which can be ascribed to the hydrophilic protective layer at rGO surface arising from the weak electrostatic force between CS and rGO. Compared with the random structure in GO/CS membrane, the rGO/CS membrane shows a more orderly stacked structure with a narrowed interlayer spacing, thus achieving an excellent molecular-sieving performance with the water/methanol separation factor high to ~4500. More importantly, the uniform

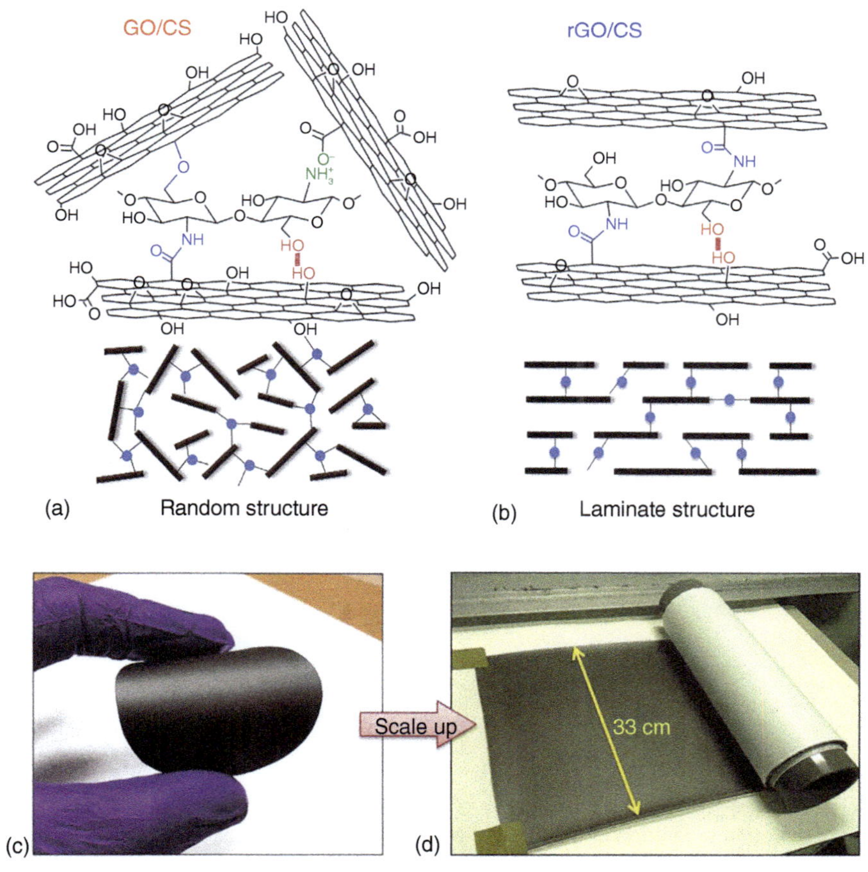

Figure 6.10 Proposed chemical interactions in GO/CS (a) and rGO/CS (b). Photographs of rGO/CS composite membranes fabricated by batch (c) and in large scale (d). Source: Hung et al. [50]/with permission of Elsevier.

rGO/CS mixture suspension without aggregates ensures simple membrane fabrication with the conventional solution casting technique. Through this method, the continuous membrane preparation on large scale can be realized with a membrane width of 33 cm. Liu et al. [54] also employed CS as a modifier to tune the interlayer structure of GO membrane via LbL self-assembly. The strong electrostatic attraction between CS and GO enables the formation of a well-stacked and defect-free laminar separation layer and improves the swelling resistance of GO membrane so that to obtain outstanding separation performance for dehydration of C2–C4 alcohols. The as-prepared GO/CS membrane exhibits permeation fluxes of 2.35, 2.98, and 4.69 $kg\,m^{-2}\,h^{-1}$ for water/ethanol, water/isopropanol, and water/*n*-butanol mixtures, respectively, as well as separation factors for the same mixtures of 3390, 5790, and 4680, respectively (70 °C, water content in feed of 10 wt%). The above results demonstrate the great potential of PEL modification in obtaining high-performance GO membranes for pervaporation.

Imitating the composition of the natural nacre with highly ordered brick-and-mortar structure, stable GO-based membranes can be obtained with minute amount of poly(vinyl alcohol) (PVA) as the interlayer cross-linker (Figure 6.11) [55]. Due to the strong interfacial interactions arising from the abundant covalent and hydrogen bonds between PVA and GO, the expanding of interlayer *d*-spacing from dry to wet

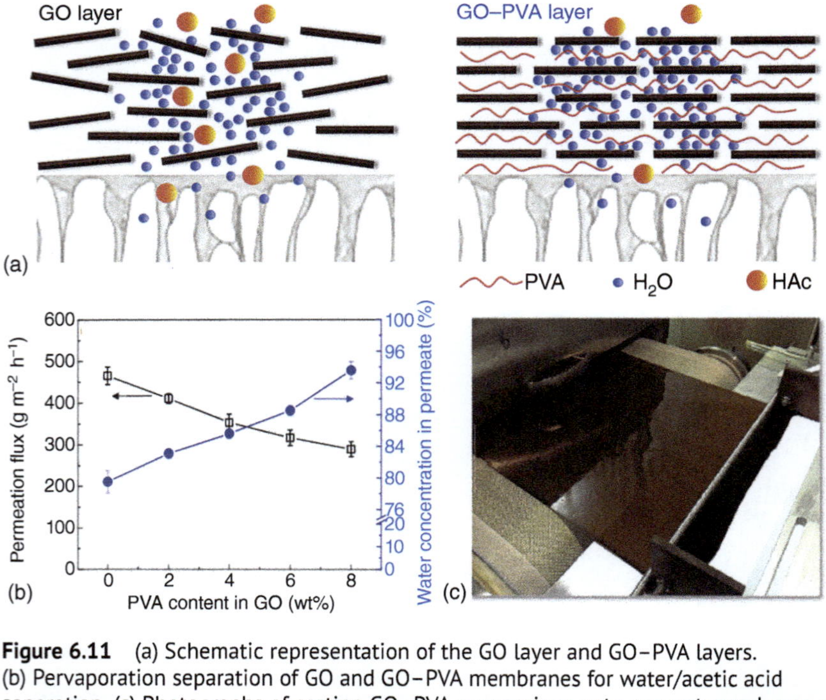

Figure 6.11 (a) Schematic representation of the GO layer and GO–PVA layers. (b) Pervaporation separation of GO and GO–PVA membranes for water/acetic acid separation. (c) Photographs of casting GO–PVA suspensions onto support membrane on large scale. Source: Lecaros et al. [55]/with permission of Elsevier.

state can be lowered from 2.59 to 0.38 Å, and a more ordered lamellar structure can be obtained. For the pervaporation separation of water/acetic acid mixture, which is rather challenging due to the strong coupling effect, the GO–PVA membrane shows outstanding performance with the water concentration in permeate increasing from 80 wt% of pristine GO membrane to 94 wt% (40 °C, water content in feed of 10 wt%). In addition, the incorporation of PVA could increase the viscosity of GO solution, thus enabling the simple membrane preparation via solution-casting method with a low GO concentration on large-scale.

Hung et al. proposed a novel design of smart electric-responsive membrane through assembly of poly(vinylidene fluoride) (PVDF) and graphene (Figure 6.12) [56]. The electrostatic interaction between the polar C—F bond and the mobile π electron cloud of graphene induces the highly β phase of PVDF, and the resultant piezoelectric property. With applied voltage, the free volume of electroactive PVDF will increase, thus decreasing the size of the molecular transport channel at PVDF–graphene interface, achieving high rejection for alcohols. Herein, the size-sieving effect from the PVDF–graphene interfacial channels dominates the water/alcohol separation instead of the conventional solution-diffusion mechanism. The membrane exhibits an obvious electrical response between the on/off external

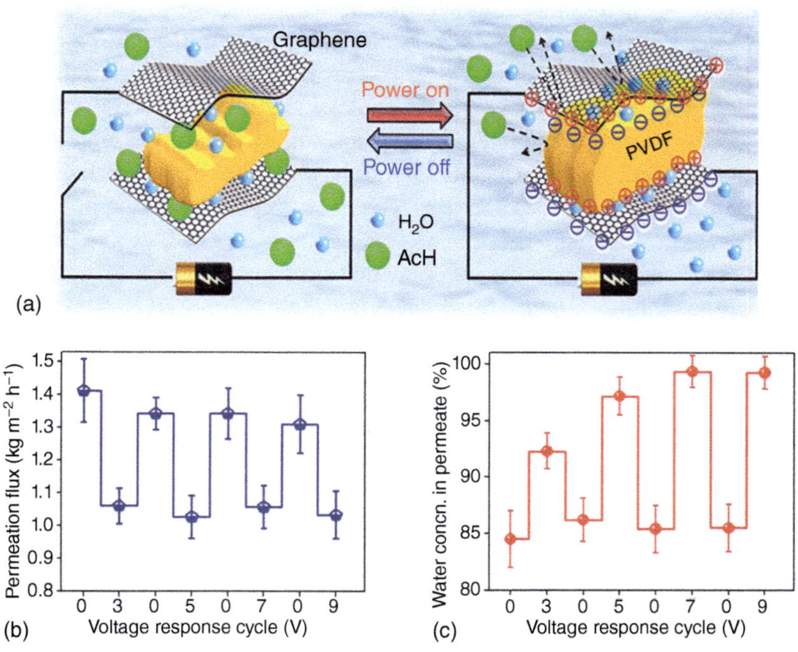

Figure 6.12 (a) Cartoons for demonstration of the mechanism for the pervaporation performance of tuning PVDF/graphene via electric fields. Performance of the PVDF/graphene membrane in the dehydration of MeOH under repeated on/off external voltages: (b) permeation flux, (c) water concentration in permeate. Source: Hung et al. [56]/with permission of American Chemical Society.

voltages. When the external voltage is on, high selectivity is obtained with the decreased permeation flux. For the challenging water/methanol separation, the membrane achieves exceptionally high performance with the permeation flux of 1.35 $kg\,m^{-2}\,h^{-1}$ and the separation factor of ~1500 (30 °C, water content in feed of 10 wt%).

6.3.4 Intercalating Nanomaterials

Most of the above structural manipulation strategies have achieved remarkable improvements in separation factor, while the enhancement in permeation flux is more challenging. In principle, higher permeation flux can be obtained via enlarging the interlayer distance and reducing the transport resistance with nanomaterial intercalation. Due to the flexibility of GO nanosheets, the interlayer *d*-spacing is not equivalent to the size of the intercalated materials. It is possible to maintain the subnanometer separation precision of GO membrane when increasing permeation flux with a nanoscale material, yet the size, structure, and chemical composition of the nanomaterial must be carefully tuned to ensure the ordered stacking of GO nanosheets and avoid the generation of microdefects [57–59].

The super-hydrophilic MOF UiO-66(Zr)-$(COOH)_2$ was intercalated into GO laminates via pressure-assisted self-assembly technique. The sufficiently small particle size (20–30 nm) avoids the formation of nonselective defects in GO layers. More importantly, the super-hydrophilicity of MOF improves the preferential sorption of water molecules on membrane surface and then blocks ethyl acetate (EA) molecules, while the porous MOF provides preferential diffusion pathways for water molecules. As a result, the MOF-intercalated GO membrane shows a 159% higher permeation flux (from 0.94 to 2.42 $kg\,m^{-2}\,h^{-1}$) and a 244% higher water/EA separation factor (from 2833 to 9751) compared with pristine GO membrane (30 °C, water content in feed of 2 wt%). In recent work, the spherical polyelectrolyte brush (SPB), with a highly hydrophilic and flexible PEL brush layer outside the polystyrene (PS) core, has been introduced into the interlayer channels of GO membrane (Figure 6.13) [57]. Besides analyzing the improved water sorption due to the hydrophilicity of intercalated materials as demonstrated in other reported works, this work reveals the critical role of polymer chain flexibility in water-selective transport for the first time. By comparing the water transport properties in the GO laminates with stretched SPB or agglomerated SPB via experiments and molecular simulations, it has been evidenced that SPB with stretched chains could enlarge the water transport channels, and the flexible hydrophilic chains show higher water transport capacity, thus achieving fast water permeation. With optimized hydrophilicity and flexibility, the SPB intercalation achieves concurrently remarkable enhancement in permeation flux and water/butanol separation factor, with the permeation flux reaching 5.23 $kg\,m^{-2}\,h^{-1}$ and the separation factor high to 8000 (70 °C, water content in feed of 10 wt%), which lies in the most selective separation membranes reported to date. To achieve more uniform nanomaterial intercalation, Wang and coworkers proposed an *in situ* site-directed synthesis strategy to incorporate Prussian blue (PB) nanoparticles into rGO lamellar membrane [59]. Small PB nanoparticles with size

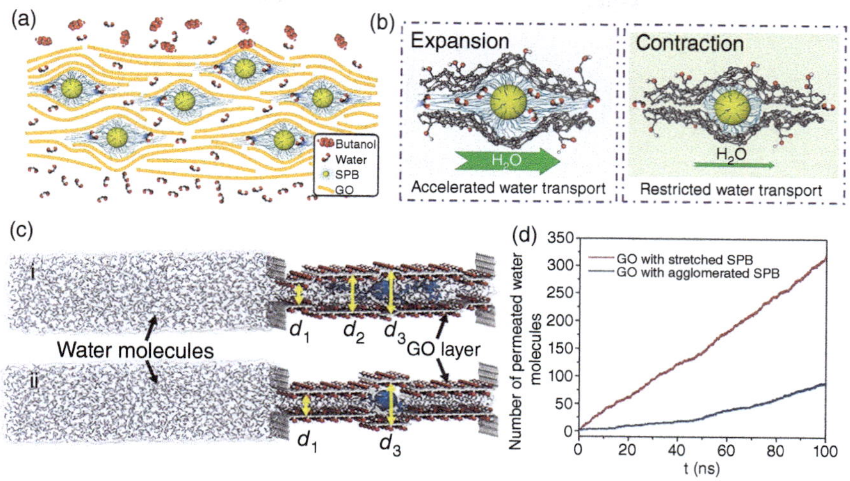

Figure 6.13 (a) Molecular transport in the PAA–SPB/GO membrane. (b) Influence of the SPB state on water transport. (c) Schematic of the modeled structure of the GO composite membrane system, where (i) and (ii) correspond to GO membranes with stretched or agglomerated brush chains, respectively. (d) The number of permeated water molecules of the simulated water transport process for the different GO composite membranes over 100 ns. Source: Dai et al. [57]/with permission of John Wiley and Sons.

distribution centering at 6–10 nm were formed on the functionalized region of GO nanosheets. The resultant rGO–PB membrane shows 25 times higher water flux than the unmodified rGO membrane in pervaporation desalination process.

6.3.5 Tuning Surface Structure

Since the molecular transport in GO membrane for pervaporation is mainly determined by the solution-diffusion mechanism, the selective adsorption of molecules from feed mixture onto membrane surface is a critical factor influencing the subsequent diffusion process. The ultrafast water diffusion in GO interlayer channels would be compromised if the adsorbed water molecules on GO membrane surface are insufficient [25]. With this regard, our group proposed a bioinspired surface hydrophilization strategy. An ultrathin (<10 nm) water-capturing polymer layer (CS) was deposited on the surface of GO membrane, which showed much stronger water-sorption ability than the pristine GO membrane due to its higher hydrophilicity and polymeric feature (Figure 6.14a,b) [25]. The enhanced water-sorption ability from the polymeric layer brings ultrafast water transport channels in GO laminates into full play, thus achieving the highly selective water permeation with an outstanding water flux of over $10\,\mathrm{kg\,m^{-2}\,h^{-1}}$ and water/butanol separation factor of about 1500 for 90 wt% *n*-butanol/water mixtures at 70 °C, exceeding the performance upper-bound of state-of-the-art membranes for butanol dehydration. The function of this elaborately designed membrane structure was further demonstrated through comparison with hybrid and sandwiched structured membranes of CS and GO. The following study with sodium alginate (SA) as surface layer displays

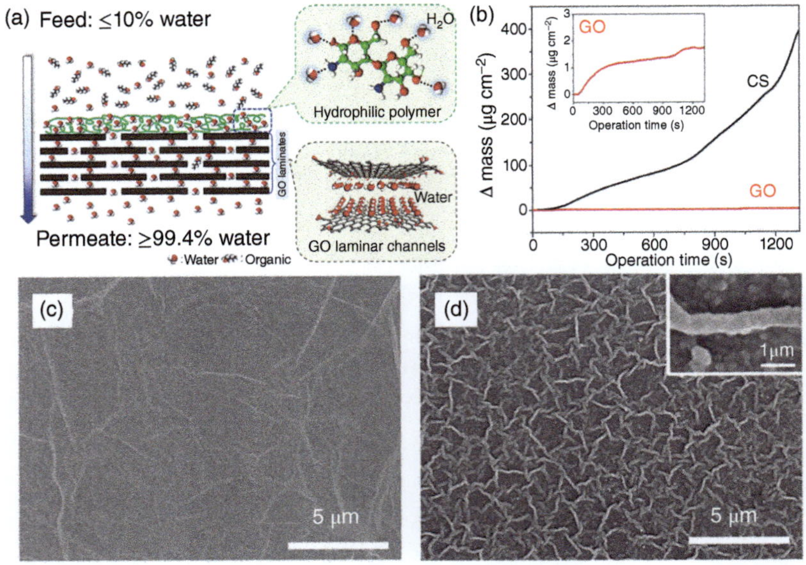

Figure 6.14 (a) Schematic of the water-organic separation process using the synergistic effect of a hydrophilic polymer and GO laminates. (b) Water-sorption ability of CS and GO measured by quartz crystal microbalance (QCM). The inset is an enlarged figure for the water-sorption ability of GO. Source: Huang et al. [25]/with permission of John Wiley and Sons. Surface morphologies of (c) GO and (d) PA–GO membranes. Source: Zhao et al. [60]/with permission of American Chemical Society.

similar improvement in water/ethanol separation performance [61, 62]. When the SA deposition amount is 0.2 $g\,m^{-2}$, the water content in permeate remarkably increases from 82.37 to 99.88 wt% [62]. Zhang and coworkers deposited an ultrathin PEL layer onto the surface of sulfonated graphene oxide (SGO) membrane and investigated the impacts of the different functional groups in PEL layer [63]. After incorporating sulfonate groups into the PEL layer containing carboxylate groups, the permeation flux is increased from 2.96 to 3.67 $kg\,m^{-2}\,h^{-1}$ with the water content in permeate remaining at 99.9 wt%. This result can be ascribed to the higher hydration capacity of sulfonate groups than carboxylate groups, which enrich more water molecules onto the membrane surface and promote water transport. Besides optimization of the surface chemical property, Zhang and coworkers constructed a thin and stripe-structure polyamide (PA) layer on top of GO membrane through interfacial polymerization (Figure 6.14c,d) [60]. On one hand, the PA thin layer contributes to protecting the GO layer from swelling. On the other hand, the enhanced hydrophilicity and rough surface morphology provide more active sites for adsorption and permeation of water molecules. The water flux could reach up to 26.7 $kg\,m^{-2}\,h^{-1}$ with the NaCl rejection of 99.99% under 70 °C.

Chemically modifying GO nanosheets before assembly could realize the simultaneous manipulation of the membrane surface properties, as well as the physical structure and chemical environment inside interlayer channels [64, 65]. Lee and coworkers grafted zwitterionic copolymer, poly(glycidyl methacrylate-sulfobetaine methacrylate) (poly(GMA-SBMA)), onto GO nanosheets due to its high water affinity from the positively and negatively charged moieties on polymer chains [64]. Compared with pristine GO membrane, the zwitterionic GO membrane shows much higher hydrophilicity with the water contact angle decreasing from 64° to 30°. The synergistic effect of the zwitterionic surface and the zwitterionic channel results in a remarkable enhancement in permeation flux from 0.67 to 1.18 $kg\,m^{-2}\,h^{-1}$ with the water concentration in permeate increasing from 98.4 to 99.4 wt% for water/isopropanol separation (25 °C, water content in feed of 30 wt%). In this work, the zwitterionic copolymer was synthesized in advance and then grafted onto GO nanosheets via the covalent reaction between the epoxy ring in poly(GMA-SBMA) and the hydroxyl group on GO. Due to the limited amount of hydroxyl groups on GO and the steric hindrance for reaction, the grafting amount was not high enough with the optimal S content of 2.99%. As an alternative, the free-radical polymerization pathway could achieve the *in situ* grafting of SBMA monomers on the carbon–carbon double bonds of GO, thus achieving much higher loading amount of zwitterionic units with the S content of 7.66% [66]. Pan and coworkers tuned the microstructure of GO nanosheets and the interlayer channel size via grafting polydopamine (PDA) on GO [65]. Through controlling reaction time, a series of PDA–GO nanosheets with different PDA loadings can be obtained, which were subsequently assembled into membranes via vacuum filtration. Dopamine molecules and PDA oligomers deposit on GO nanosheets via covalently reacting with the functional groups on GO and physical stacking. The as-formed "raised points" structure increases the surface roughness and enlarges the interlayer channels. With prolonged reaction, the gaps between the "raised points" are gradually filled with PDA, leading to decreased roughness, narrowed interlayer spacing, and improved size exclusion effect. Compared with pristine GO membrane, the PDA–GO membrane achieves a 39% enhancement in permeation flux with the water concentration in permeate up to 99.7 wt% for water/isopropanol separation. Nevertheless, the membrane shows poor separation efficiency for water/ethanol mixture.

6.4 Summary and Perspective

In this chapter, we summarized the progress of graphene-based membranes for pervaporation processes with a focus on the structure–performance relationship and the structure manipulation strategies. We first discussed the molecular transport mechanism through the confined 2D nanochannels in GO membranes, based on the simulation studies. Following this, we analyzed the progress of GO membranes from the viewpoints of five main strategies used for improving the performance of GO membranes, that is, controlling the assembly conditions, fabricating GO framework, assembly with polymers, intercalating nanomaterials, and tuning the surface structure. Since the first report in 2013, around 50 studies have been reported regarding GO pervaporation membranes and impressive progresses have been made. Overall, the GO membranes demonstrate ultrahigh separation performance with high permeation flux and separation factor/rejection as well as excellent long-term stability, and many of them break the performance record from the literature. The prospect of scaling up GO-based pervaporation membranes is attractive. Nevertheless, many challenges remain unresolved.

The production expense of high-quality GO aqueous dispersion, which is used for fabricating GO membrane, is high. This will undoubtedly increase the cost of GO membranes. To reduce the cost, fabricating GO membrane with smaller thickness or improving the synthesis route toward low-cost GO dispersion is desired.

The molecular transport mechanisms in GO or functional GO membranes require more studies, which can give insights into the design of high-performance GO membranes for pervaporation. The GO nanosheets contain randomly distributed oxidized areas and non-oxidized areas. This means the GO nanochannels are disordered in both chemical structure and physical structure, making it difficult to characterize the channel structure and investigate nanofluidic phenomenon in GO membranes. In the future, high-resolution (atom-scale) characterization tools and molecular simulation are suggested to be applied to solve this issue. Emphasis on water state and water transport in the GO nanochannels is highly recommended.

The 2D morphology with large lateral size and thin thickness of GO nanosheets imparts GO with excellent thin-film-formation capability. Ultrathin GO membranes with thickness smaller than 100 nm have been demonstrated in many studies by various methods, including vacuum-assisted assembly, solution-casting, or spray-coating. All these fabrication methods are scalable. However, almost all the reported GO membranes for pervaporation are fabricated on centimeter-scale. Whether these methods can enable meter-scale large-area GO membrane with reproducible high-performance remains unanswered.

The permeation flux of GO membranes can be further increased. Generally, the lateral size of GO nanosheets is μm-scale, which benefits the fabrication of thin membranes. However, the large GO nanosheets considerably extend the molecular diffusion path, thus decreasing the molecular permeance. Although the nanopores generated by the strong oxidation conditions during the synthesis of GO can increase the molecular permeance, the density of transport channel in the through-plane direction seems not high enough. To overcome this issue, strategies that can be adopted include etching the GO to introduce more nanopores and hybridizing GO with other porous nanosheets made of zeolite, MOFs, or covalent–organic frameworks (COFs).

Pure GO membranes are not stable in polar liquid media owing to the swelling of GO membranes. To solve this problem, introducing modifiers as cross-linker, such as small molecules, polymers, and nanoparticles, into the GO interlayer gallery is indispensable to maintain the structural stability in liquid. From the viewpoint of practical application, the structure of GO membrane should be as simple as possible to simplify the fabrication of the membranes. In this regard,

GO membrane intercalated with small molecules or polymers has higher prospect of practical application. Inspired by the structural properties of nacre, addition of a small amount of polymer can greatly improve the mechanical stability of the GO membranes. Moreover, polymer can reduce the defects in GO membranes. These striking features would be beneficial for the scaling up of GO/polymer membranes.

The long-term stability of GO membranes in the time span of 10 000 hours in the working conditions needs to be measured to evaluate the potential of practical application of GO membranes. Currently, most of the long-term stability tests for GO pervaporation membranes are performed with a time span shorter than 200 hours using high-purity liquid as the feedstock. Whether the GO membranes are stable for longer time in real conditions remains unanswered. GO nanosheets have a risk of getting reduced in electron-rich environment, which can result in remarkable decrease of the molecular permeance through the GO membranes.

References

1 Song, Y., Pan, F., Li, Y. et al. (2019). Mass transport mechanisms within pervaporation membranes. *Front. Chem. Sci. Eng.* 13 (3): 458–474.

2 Liu, G. and Jin, W. (2021). Pervaporation membrane materials: recent trends and perspectives. *J. Membr. Sci.* 636: 119557.

3 Ong, Y.K., Shi, G.M., Le, N.L. et al. (2016). Recent membrane development for pervaporation processes. *Prog. Polym. Sci.* 57: 1–31.

4 Shen, J., Liu, G., Han, Y. et al. (2021). Artificial channels for confined mass transport at the sub-nanometre scale. *Nat. Rev. Mater.* 6 (4): 294–312.

5 Cheng, Y., Pu, Y., and Zhao, D. (2020). Two-dimensional membranes: new paradigms for high-performance separation. *Membranes* 15 (15): 2241–2270.

6 Liu, G., Jin, W., and Xu, N. (2015). Graphene-based membranes. *Chem. Soc. Rev.* 44 (15): 5016–5030.

7 Liu, G., Jin, W., and Xu, N. (2016). Two-dimensional-material membranes: a new family of high-performance separation membranes. *Angew. Chem. Int. Ed.* 55 (43): 13384–13397.

8 Nair, R.R., Wu, H.A., Jayaram, P.N. et al. (2012). Unimpeded permeation of water through helium-leak-tight graphene-based membranes. *Science* 335 (6067): 442–444.

9 Zheng, S., Tu, Q., Urban, J.J. et al. (2017). Swelling of graphene oxide membranes in aqueous solution: characterization of interlayer spacing and insight into water transport mechanisms. *ACS Nano* 11 (6): 6440–6450.

10 Yu, T., Bai, L., Xu, Z. et al. (2018). Molecular simulation of permeation behaviour of ethanol/water molecules with single-layer graphene oxide membranes. *Mol. Simul.* 44 (10): 840–849.

11 Liu, Q., Wu, Y., Wang, X. et al. (2019). Molecular dynamics simulation of water-ethanol separation through monolayer graphene oxide membranes: significant role of O/C ratio and pore size. *Sep. Purif. Technol.* 224: 219–226.

12 Yu, T., Xu, Z., Liu, S. et al. (2018). Enhanced hydrophilicity and water-permeating of functionalized graphene-oxide nanopores: molecular dynamics simulations. *J. Membr. Sci.* 550: 510–517.

13 Liang, F., Liu, Q., Zhao, J. et al. (2020). Ultrafast water-selective permeation through graphene oxide membrane with water transport promoters. *AlChE J.* 66 (2): 11.

14 Fang, C., Wu, H., Lee, S.-Y. et al. (2018). The ionized graphene oxide membranes for water-ethanol separation. *Carbon* 136: 262–269.

15 Damasceno Borges, D., Woellner, C.F., Autreto, P.A.S. et al. (2018). Insights on the mechanism of water-alcohol separation in multilayer graphene oxide membranes: Entropic versus enthalpic factors. *Carbon* 127: 280–286.

16 Huang, K., Liu, G., Lou, Y. et al. (2014). A graphene oxide membrane with highly selective molecular separation of aqueous organic solution. *Angew. Chem. Int. Ed.* 53 (27): 6929–6932.

17 Zhan, W., Xu, Z., and Yang, X. (2020). Molecular interlayer intercalation of ethanol-water mixture towards GO laminated membranes. *Sep. Purif. Technol.* 233: 116029.

18 Li, L., Hou, J., and Chen, V. (2019). Pinning down the water transport mechanism in graphene oxide pervaporation desalination membranes. *Ind. Eng. Chem. Res.* 58 (10): 4231–4239.

19 Cha-Umpong, W., Hosseini, E., Razmjou, A. et al. (2020). New molecular understanding of hydrated ion trapping mechanism during thermally-driven desalination by pervaporation using GO membrane. *J. Membr. Sci.* 598: 117687.

20 Liang, B., Zhan, W., Qi, G. et al. (2015). High performance graphene oxide/polyacrylonitrile composite pervaporation membranes for desalination applications. *J. Mater. Chem. A* 3 (9): 5140–5147.

21 Tsou, C.-H., An, Q.-F., Lo, S.-C. et al. (2015). Effect of microstructure of graphene oxide fabricated through different self-assembly techniques on 1-butanol dehydration. *J. Membr. Sci.* 477: 93–100.

22 Yeh, T.-M., Wang, Z., Mahajan, D. et al. (2013). High flux ethanol dehydration using nanofibrous membranes containing graphene oxide barrier layers. *J. Mater. Chem. A* 1 (41): 12998–13003.

23 Xu, W.L., Fang, C., Zhou, F. et al. (2017). Self-assembly: a facile way of forming ultrathin, high-performance graphene oxide membranes for water purification. *Nano Lett.* 17 (5): 2928–2933.

24 Kim, H.W., Yoon, H.W., Yoon, S.M. et al. (2013). Selective gas transport through few-layered graphene and graphene oxide membranes. *Science* 342 (6154): 91–95.

25 Huang, K., Liu, G., Shen, J. et al. (2015). High-efficiency water-transport channels using the synergistic effect of a hydrophilic polymer and graphene oxide laminates. *Adv. Funct. Mater.* 25 (36): 5809–5815.

26 Shen, J., Liu, G., Huang, K. et al. (2016). Subnanometer two-dimensional graphene oxide channels for ultrafast gas sieving. *ACS Nano* 10 (3): 3398–3409.

27 Shen, J., Liu, G., Huang, K. et al. (2015). Inside back cover: membranes with fast and selective gas-transport channels of laminar graphene oxide for efficient CO_2 capture. *Angew. Chem. Int. Ed.* 54 (2): 697.

28 Tang, Y.P., Paul, D.R., and Chung, T.S. (2014). Free-standing graphene oxide thin films assembled by a pressurized ultrafiltration method for dehydration of ethanol. *J. Membr. Sci.* 458: 199–208.

29 Hung, W.-S., Tsou, C.-H., De Guzman, M. et al. (2014). Cross-linking with diamine monomers to prepare composite graphene oxide-framework membranes with varying *d*-spacing. *Chem. Mater.* 26 (9): 2983–2990.

30 Hua, D., Rai, R.K., Zhang, Y. et al. (2017). Aldehyde functionalized graphene oxide frameworks as robust membrane materials for pervaporative alcohol dehydration. *Chem. Eng. Sci.* 161: 341–349.

31 Yang, J., Gong, D., Li, G. et al. (2018). Self-assembly of thiourea-crosslinked graphene oxide framework membranes toward separation of small molecules. *Adv. Mater.* 30 (16): e1705775.

32 Qian, Y., Zhang, X., Liu, C. et al. (2019). Tuning interlayer spacing of graphene oxide membranes with enhanced desalination performance. *Desalination* 460: 56–63.

33 Qian, Y., Zhou, C., and Huang, A. (2018). Cross-linking modification with diamine monomers to enhance desalination performance of graphene oxide membranes. *Carbon* 136: 28–37.

34 Zhang, X., Zhang, M.X., Ding, H. et al. (2019). Double-crosslinked GO interlayer framework as a pervaporation hybrid membrane with high performance. *ACS Omega* 4 (12): 15043–15050.

35 Qin, Z., Wang, L., Zhang, W. et al. (2017). A novel surface cross-linked GO-based membrane with superior separation performance. *RSC Adv.* 7 (85): 54213–54221.

36 Feng, B., Xu, K., and Huang, A. (2016). Covalent synthesis of three-dimensional graphene oxide framework (GOF) membrane for seawater desalination. *Desalination* 394: 123–130.

37 Yang, G., Xie, Z., Cran, M. et al. (2019). Functionalizing graphene oxide framework membranes with sulfonic acid groups for superior aqueous mixture separation. *J. Mater. Chem. A* 7 (34): 19682–19690.

38 Li, G., Shi, L., Zeng, G. et al. (2015). Sharp molecular-sieving of alcohol-water mixtures over phenyldiboronic acid pillared graphene oxide framework (GOF) hybrid membrane. *Chem. Commun. (Camb)* 51 (34): 7345–7348.

39 Park, S., Lee, K.S., Bozoklu, G. et al. (2008). Graphene oxide papers modified by divalent ions - enhancing mechanical properties via chemical cross-linking. *ACS Nano* 2 (3): 572–578.

40 Chen, L., Shi, G., Shen, J. et al. (2017). Ion sieving in graphene oxide membranes via cationic control of interlayer spacing. *Nature* 550 (7676): 380–383.

41 Ching, K., Lian, B., Leslie, G. et al. (2020). Metal-cation-modified graphene oxide membranes for water permeation. *Carbon* 170: 646–657.

42 Guan, K., Liu, Q., Zhou, G. et al. (2019). Cation-diffusion controlled formation of thin graphene oxide composite membranes for efficient ethanol dehydration. *Sci. China Mater.* 62 (7): 925–935.

43 Wu, J., Lin, W., Wang, Z. et al. (2012). Investigation of the hydration of non-fouling material poly(sulfobetaine methacrylate) by low-field nuclear magnetic resonance. *Langmuir* 28 (19): 7436–7441.

44 He, G., Xu, M., Zhao, J. et al. (2017). Bioinspired ultrastrong solid electrolytes with fast proton conduction along 2D channels. *Adv. Mater.* 29 (28): 1605898.

45 Podsiadlo, P., Kaushik, A.K., Arruda, E.M. et al. (2007). Ultrastrong and stiff layered polymer nanocomposites. *Science* 318 (5847): 80–83.

46 Li, Y.Q., Yu, T., Yang, T.Y. et al. (2012). Bio-inspired nacre-like composite films based on graphene with superior mechanical, electrical, and biocompatible properties. *Adv. Mater.* 24 (25): 3426–3431.

47 Halakoo, E. and Feng, X. (2020). Layer-by-layer assembled membranes from graphene oxide and polyethyleneimine for ethanol and isopropanol dehydration. *Chem. Eng. Sci.* 216: 115488.

48 Sun, J., Qian, X., Wang, Z. et al. (2020). Tailoring the microstructure of poly(vinyl alcohol)-intercalated graphene oxide membranes for enhanced desalination performance of high-salinity water by pervaporation. *J. Membr. Sci.* 599: 117838.

49 Zhao, J., Zhu, Y., Pan, F. et al. (2015). Fabricating graphene oxide-based ultrathin hybrid membrane for pervaporation dehydration via layer-by-layer self-assembly driven by multiple interactions. *J. Membr. Sci.* 487: 162–172.

50 Hung, W.-S., Chang, S.-M., Lecaros, R.L.G. et al. (2017). Fabrication of hydrothermally reduced graphene oxide/chitosan composite membranes with a lamellar structure on methanol dehydration. *Carbon* 117: 112–119.

51 Halakoo, E. and Feng, X. (2020). Layer-by-layer assembly of polyethyleneimine/graphene oxide membranes for desalination of high-salinity water via pervaporation. *Sep. Purif. Technol.* 234: 116077.

52 Halakoo, E. and Feng, X. (2020). Self-assembled membranes from polyethylenimine and graphene oxide for pervaporation dehydration of ethylene glycol. *J. Membr. Sci.* 616: 118583.

53 Liang, F., Zheng, J., He, M. et al. (2021). Exclusive and fast water channels in zwitterionic graphene oxide membrane for efficient water–ethanol separation. *AlChE J.* 67 (7): e17215.

54 Liu, S., Zhou, G., Guan, K. et al. (2021). Dehydration of C2–C4 alcohol/water mixtures via electrostatically enhanced graphene oxide laminar membranes. *AlChE J.* 67 (6): aic17170.

55 Lecaros, R.L.G., Mendoza, G.E.J., Hung, W.-S. et al. (2017). Tunable interlayer spacing of composite graphene oxide-framework membrane for acetic acid dehydration. *Carbon* 123: 660–667.

56 Hung, W.-S., Ho, S.-Y., Chiao, Y.-H. et al. (2020). Electrical tunable PVDF/graphene membrane for controlled molecule separation. *Chem. Mater.* 32 (13): 5750–5758.

57 Dai, L., Xu, F., Huang, K. et al. (2021). Ultrafast water transport in two-dimensional channels enabled by spherical polyelectrolyte brushes with controllable flexibility. *Angew. Chem. Int. Ed.* 60 (36): 19933–19941.

58 Guan, K.C., Liu, Q., Ji, Y.F. et al. (2018). Precisely controlling nanochannels of graphene oxide membranes through lignin-based cation decoration for dehydration of biofuels. *ChemSusChem* 11 (14): 2315–2320.

59 Ma, H., Chen, X., Mohammed, S. et al. (2020). A thermally reduced graphene oxide membrane interlayered with an *in situ* synthesized nanospacer for water desalination. *J. Mater. Chem. A* 8 (48): 25951–25958.

60 Zhao, X., Tong, Z., Liu, X. et al. (2020). Facile preparation of polyamide–graphene oxide composite membranes for upgrading pervaporation desalination performances of hypersaline solutions. *Ind. Eng. Chem. Res.* 59 (26): 12232–12238.

61 Zhao, D., Zhao, J., Ji, Y. et al. (2018). Facilitated water-selective permeation via PEGylation of graphene oxide membrane. *J. Membr. Sci.* 567: 311–320.

62 Hao, W., Tong, Z., Liu, X. et al. (2020). Optimizing nanostrands-inserted graphene oxide membrane with polyelectrolyte protective layer for enhanced ethanol pervaporation dehydration. *Sep. Purif. Technol.* 251: 117322.

63 Tong, Z., Liu, X., and Zhang, B. (2020). Sulfonated graphene oxide based membranes with enhanced water transport capacity for isopropanol pervaporation dehydration. *J. Membr. Sci.* 612: 118446.

64 Ang, M.B.M.Y., Gallardo, M.R., Dizon, G.V.C. et al. (2019). Graphene oxide functionalized with zwitterionic copolymers as selective layers in hybrid membranes with high pervaporation performance. *J. Membr. Sci.* 587: 117188.

65 Zhang, W.Z., Jia, J., Qiu, Y.Y. et al. (2018). Polydopamine-grafted graphene oxide composite membranes with adjustable nanochannels and separation performance. *Adv. Mater. Interfaces* 5 (8): 1701386.

66 Zhao, J., Zhu, Y., He, G. et al. (2016). Incorporating zwitterionic graphene oxides into sodium alginate membrane for efficient water/alcohol separation. *ACS Appl. Mater. Interfaces* 8 (3): 2097–2103.

7

Two-Dimensional-Materials Membranes for Gas Separations

Yunpan Ying, Hadi Rouhani, Dechao Wang, and Dan Zhao

National University of Singapore, Department of Chemical and Biomolecular Engineering, 4 Engineering Drive 4, Singapore 117585, Singapore

7.1 Introduction

As important industrial processes, gas separations through membrane technologies are highly dependent on high throughput (permeance or flux), separation capability (selectivity or separation factor), and stability (pressure, chemical, thermal, etc.) [1]. As the former (high throughput) directly impacts the process footprint, its realization is of high importance. The most straightforward way of increasing throughput is to decrease the membrane thickness [2]. However, the fabrication of ultrathin membranes through traditional chemical growth, polymer solution casting, or some other methods remains a daunting challenge.

Since the discovery of graphene by Novoselov, Geim, and coworkers in 2004 [3], the concept of "two-dimensional (2D) materials" has attracted great interest. In recent years, expanding scopes of low-dimensional materials, including zeolites [4–7], graphene-based materials [8–11], graphitic carbon nitride (g-C_3N_4) [12, 13], metal–organic frameworks (MOFs), [14–17] covalent–organic frameworks (COFs), [18–20] MXenes, [21–24] transition-metal dichalcogenides (TMDs) [25–28], and other materials [29–31], have been discovered, investigated, and developed for various applications (Figure 7.1).

2D materials have aroused great attention as membrane materials because of their ability to be made with atomic thicknesses. This attribute makes them ideal building blocks for ultrathin membrane fabrication. The introduction of size-selective nanochannels and/or nanopores into 2D materials provides a route toward simultaneous excellent selectivity and remarkable separation permeance, promising new opportunities for molecular (e.g. gas and liquid) separations. In this chapter, we review the more recent developments with a specific focus on membrane materials, structural design, membrane preparation methods, distinctive membrane transport characteristics, and industrially appealing gas-separation processes. In addition, the challenges and opportunities of 2D-materials membranes for industrial applications are briefly discussed.

Two-Dimensional-Materials-Based Membranes: Preparation, Characterization, and Applications, First Edition. Edited by Gongping Liu and Wanqin Jin.
© 2022 WILEY-VCH GmbH. Published 2022 by WILEY-VCH GmbH.

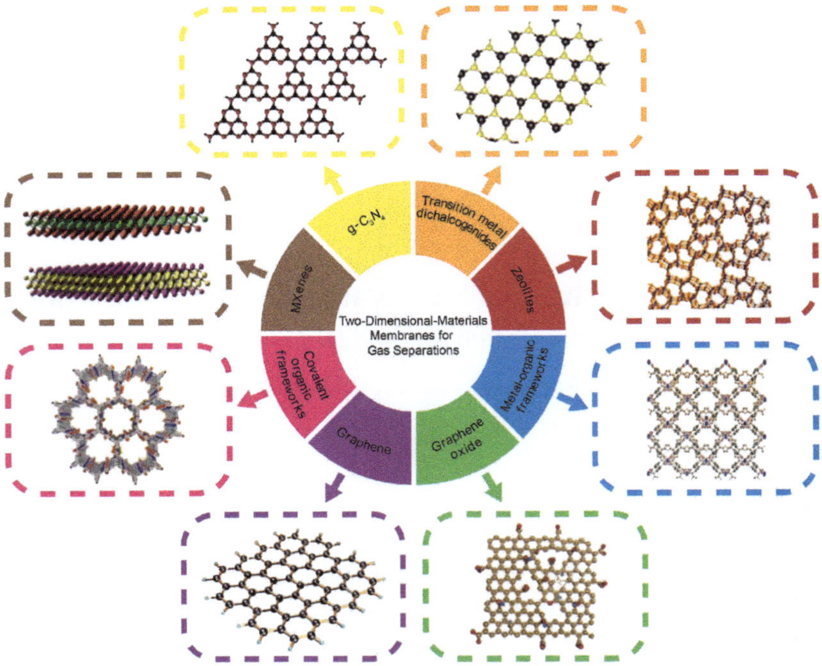

Figure 7.1 Different types of promising 2D-materials membranes for gas separations. Source: Yunpan Ying, Hadi Rouhani, Dechao Wang, and Dan Zhao.

7.2 2D-Materials Membranes

Since the structure determines the separation performance, the differences of the pore/slit structure, pore size, layer spacing distance, and physiochemical properties from various 2D materials heavily affect the overall separation performance for special gas-separation applications. Hence, the selection of 2D materials is critical. Advanced porous materials, such as MOFs and COFs, in particular, have emerged as 2D structures with tailored functionalities, pore size, and microenvironment that can be used as membrane building blocks to meet the various separation process requirements. In this section, we focus on the structures, functionalities, transport pathways, separation mechanisms, and breakthrough performance of various 2D-materials membranes, and highlight their unique advantages for special separation systems.

7.2.1 Zeolites

Zeolites are a family of porous, crystalline materials constructed from silica-based precursors [32–34]. They possess a uniform pore size ranging from 0.25 nm to a few nanometers, regular pore shape, and high stability. The molecular-sieving performance for gas mixture separation can be realized by suitably choice of the in-plane pore size and pore shape (Figure 7.2a) [34, 35]. Typically, zeolites are synthesized

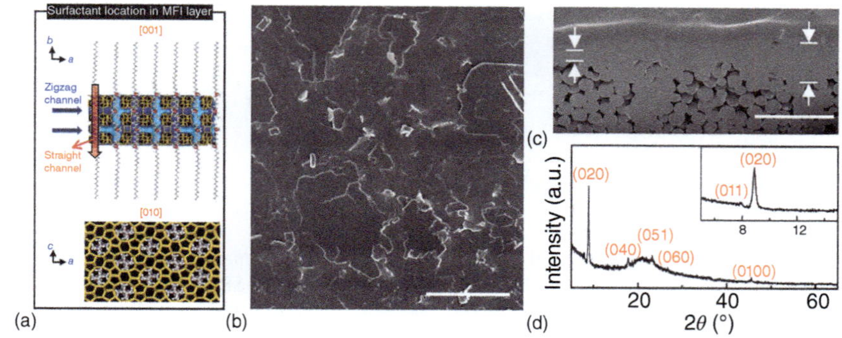

Figure 7.2 (a) Proposed structure for a single MFI nanosheet. Source: Choi et al. [35]/with permission of Springer Nature. (b) Top view, and (c) cross-sectional SEM images of an MFI membrane synthesized by MFI nanosheets and subsequent gel-free intergrowth. Scale bars show 2 μm. (d) XRD pattern of the MFI membrane in (b) and (c) (a.u., arbitrary units). Source: Jeon et al. [4]/with permission of Springer Nature.

by hydrothermal method under the assistance of a structure-directing agent. 2D zeolite nanosheets with a thickness equivalent to the structure's unit-cell dimension can be used to prepare molecular-sieving membranes with excellent vapor and gas-separation performance. Due to the in-plane porosity in 2D zeolite nanosheets, the gas transport can avoid the tortuous pathways in the layer-layer spaces and show high permeability. The successful preparation of 2D zeolite membranes mainly depends on (i) the preparation of stable zeolite nanosheet suspension that contains high-aspect-ratio nanosheets with homogeneous thickness and lateral dimension and (ii) the efficient methods for depositing the suspended zeolite nanosheets onto porous supports and controlling the nanosheet orientation in the resultant membranes.

Tsapatsis' group has reported several examples of 2D zeolite membranes for vapor and gas separations [4]. In a recent breakthrough, they have developed a bottom-up synthesis method to produce zeolite nanosheets with high-aspect ratio, improved yield, and lower cost. The mordenite framework inverted (MFI) nanosheets with only 2.5-unit-cell thickness (about 5 nm) were synthesized using a single rotational intergrowth that triggered the growth of nanocrystals. Subsequently, these nanosheets were vacuum-filtrated onto porous sintered silica fiber supports to prepare thin and defect-free coatings. High-flux and ultra-selective MFI membranes can be prepared by the intergrowth of coated nanosheets. Compared to the MFI membranes that were constructed from the nanosheets prepared by a top-down method, these membranes displayed better separation performance (Figure 7.2b,c,d).

In another study, Agrawal et al. [36] used the top-down method to exfoliate the sodalite precursor RUB-15 zeolite, possessing hydrogen-sieving six-membered rings (6-MRs) of SiO_4 tetrahedra, into ultrathin crystalline nanosheets with a thickness of 0.8 nm and lateral size of 205 ± 62 nm. However, the thin membranes (as filtrated) prepared by filtration of the exfoliated nanosheets suspension did not display a good separation performance (H_2/N_2 selectivity is about 20). This can be attributed to the

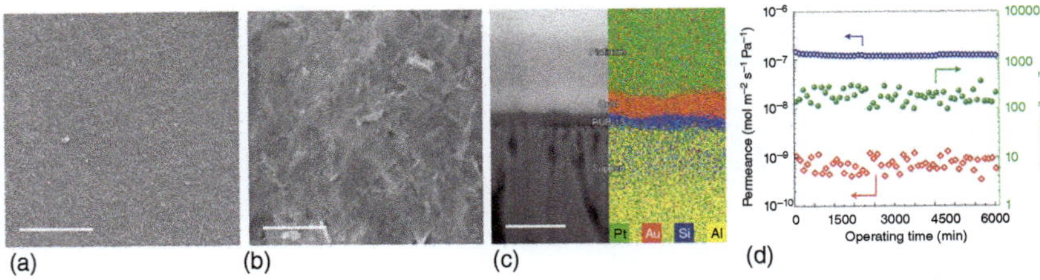

Figure 7.3 (a) Low-magnification and (b) high-magnification SEM images of the as-filtered nanosheet films. Scale bars: (a) 5 μm; (b) 500 nm. (c) Cross-sectional SEM image of the film (left) and elemental mapping overlaid on top of the SEM image (right). The underlying alumina support is represented by the letter Al. Scale bar, 500 nm. (d) A 100-hours stability test with an equimolar H_2/CO_2 mixture feed in the presence of 4% water vapor at 200 °C. Source: Dakhchoune et al. [36]/with permission of Springer Nature.

possibility of gas transport through the relatively large intersheet gaps instead of the 6-MR apertures. The large intersheet gaps were subsequently sealed by the condensation of the terminal silanol groups on RUB-15 nanosheets. As a result, the resultant 2D RUB-15 membrane (after calcined) reaches a high H_2/CO_2 selectivity up to 100, making it a viable candidate for precombustion carbon capture (Figure 7.3).

7.2.2 Graphene-Based Materials

Graphene is single-layer graphite, in which sp^2 carbon atoms are located in an ordered honeycomb lattice. Due to the narrowing effect of graphene π-orbitals, in-plane pores in graphene have a size of only 0.64 Å, which is significantly smaller than the kinetic diameters of most gas molecules, even water (2.65 Å) and helium (2.6 Å) [37–39]. As such, defect-free graphene is impermeable to gases and cannot be directly used for separations [40, 41]. On the other hand, the introduction of porosity in graphene has allowed its implementation in gas separation. Graphene-based materials were categorized into two groups, nanoporous graphene membranes and graphene oxide (GO) membranes that are included in lamellar membranes. These graphene derivatives allow other unique characteristics of graphene, such as the atomic thickness, barrier property, extraordinary mechanical strength (about TPa), and chemical inertness, to be introduced into membrane separation.

7.2.2.1 Nanoporous Graphene

Nanoporous graphene is composed of some nanopores onto the atomically thin graphene, which can illustrate high separation performance due to high pore density, molecular-level pore size, minimum resistance due to ultrathin atomic thickness, and narrow pore size distribution (Figure 7.4a) [42, 45, 46]. The gas transport mechanism through a graphene nanosheet membrane is mainly molecular sieving. Some studies showed the mechanism of nanoporous graphene from both points of molecular dynamics simulation and experiment.

From the simulation viewpoint, density functional theory (DFT) calculations, conducted by Jiang et al., were used to study the permeability and selectivity of graphene nanosheets using pores in the subnanometer size region. They could

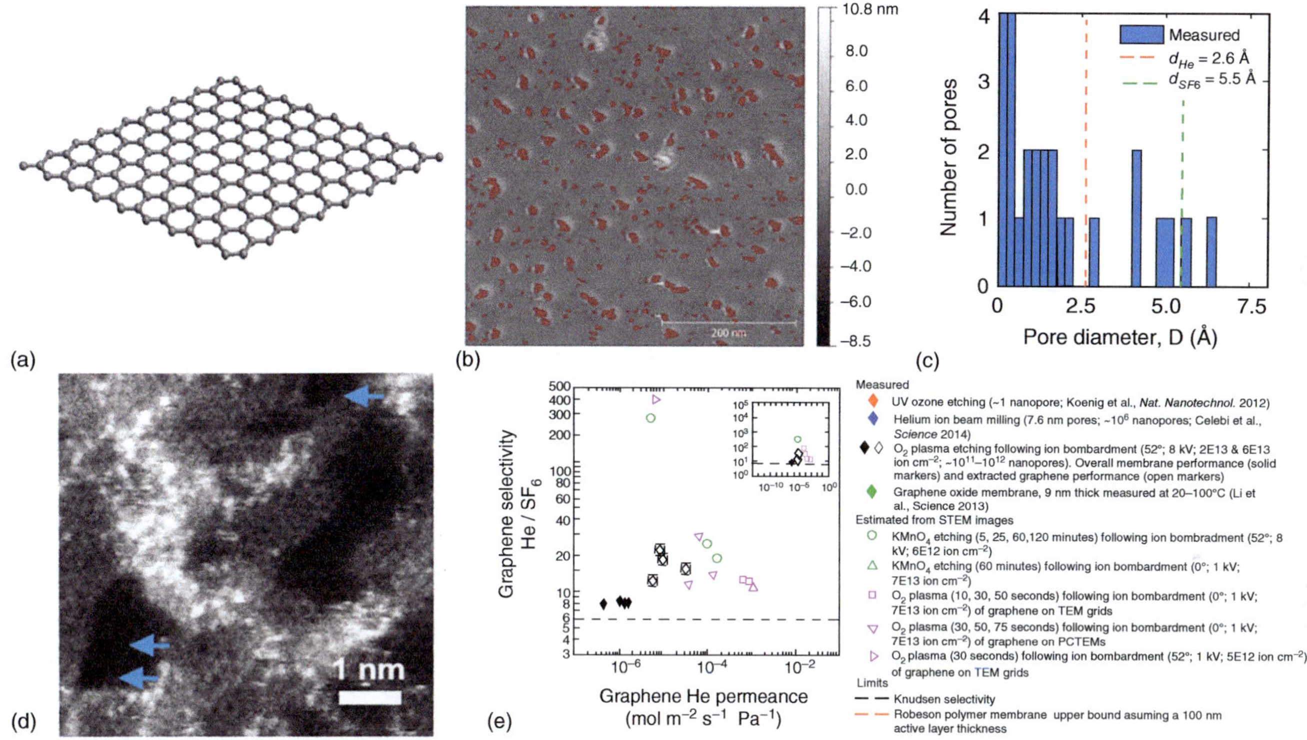

Figure 7.4 (a) Single-layer graphene structure. Source: Schrier [42]/with permission of American Chemical Society. (b) AFM image of the graphene membrane prepared by ultraviolet etching (red areas). Source: [43]. Koenig S P et al., (2012)/with permission of Springer Nature. (c) Pore size distribution estimated from STEM images of graphene on a TEM grid bombarded with gallium ions at 5×10^{12} ions cm^{-2}, 1 kV accelerating voltage, 52° incidence, and plasma-etched for 30 seconds. (d) STEM image of pores created in graphene by ion bombardment followed by oxygen plasma etching. (e) Overall membrane performance and extracted nanoporous graphene performance, in comparison to results from the literature. Source: Boutilier et al. [44]/with permission of American Chemical Society.

reach H_2/CH_4 selectivity of 10^8 or 10^{24} for pores that are functionalized by nitrogen atoms (pore dimension is 3.0 Å × 3.8 Å) or passive pores (pore width is 2.5 Å) [47]. In another work, Jiang and coworkers tried to examine the nanographene sheet pores for various kinds of gas separations [48]. Molecular dynamic simulation illustrated that gas flux followed the trend of kinetic diameters of the guests ($H_2 > CO_2 > N_2 > Ar > CH_4$). The authors claimed size-selective effect of the membrane due to the very high CO_2/CH_4 ideal selectivity attained (up to 6.6×10^7). Due to the difficulties in generating ultrasmall pores (<4 Å) with sizes approaching the kinetic diameters of gases of interest, Dai et al. developed a new approach to carefully adjust pores by covering nanoporous thin membranes with an ionic liquid monolayer [49]. They demonstrated that while the graphene nanosheets with a pore size of 6 Å do not show any CO_2/CH_4 or CO_2/N_2 selectivity, the pore size of the membrane may be decreased by the carbon dioxide-philic anion from the ionic liquid, leading to selectivities of 42 and 5.2 for CO_2/CH_4 and CO_2/N_2, respectively.

Taking these interesting results from simulations into account, enormous research efforts were made to create nanoporous thin membranes for gas separations. For this purpose, numerous studies have endeavored to generate pores in graphene nanosheets by various methods, such as ultraviolet (UV)-induced oxidative etching [43, 50, 51], focused electron beam, [52–54] oxygen plasma etching, [55, 56] focused ion beam irradiation (FIB) [57, 58], and micro-wave combustion, [59] with the view of maintaining its high mechanical and chemical stabilities.

Bunch et al. [45] recorded the pore evolution of bilayer graphene by UV light radiation ($\lambda_1 = 185$ nm, $\lambda_2 = 254$ nm) (Figure 7.4b). This process led to the development of holes with subnanometer size (3.5 Å, 150 steps). The pore size is consistent with experimentally measured gas leakage rates across species with different kinetic diameters. Mahurin and coworkers [55] fabricated single-layer graphene via chemical vapor deposition (CVD) method on a copper foil catalyst and suspended it into a microchip of silicon nitride device subsequently. Exposure of the suspended graphene to the oxygen plasma for just 1.5 seconds led to the creation of pores between 0.5 and 1 nm. The duration required for the plasma-etching approach was significantly shorter than when utilizing UV as an etchant. In another example, Karnik et al. produced subnanometer pores in graphene at a density of ~1011 cm^{-2} by ion bombardment which was followed by oxygen plasma etching (Figure 7.4c,d) [44]. The as-prepared membranes demonstrated outstanding H_2/CO_2 and H_2/CH_4 separation performances that provided evidence of molecular sieving across centimeter-scale nanoporous graphene (Figure 7.4e).

Although the single-layer and bilayer graphene were successfully fabricated, the presence of extrinsic defects (created when graphene was transferred to the substrate) and intrinsic defects (created during physical exfoliation or CVD processes) was a major barrier for the scale-up of such membrane architectures. Recently, Agrawal's group has acquired a suspended and crack-free graphene film by a nanoporous-carbon-assisted transferring method that had a relatively large area (1 mm^2) [60]. These graphene films demonstrated high H_2/CH_4 selectivity (25) and H_2 permeance about 4.1×10^{-7} mol m^{-2} s^{-1} Pa^{-1}. These were derived from inherent defects formed during the CVD process. Furthermore, the H_2 permeance and H_2/CH_4 selectivity were boosted up to 300% and 150%, respectively, by ozone functionalization-based etching and pore-modification chemistry.

7.2.2.2 Graphene Oxide

GO is generated from graphene/graphite with different groups, such as hydroxyl, epoxy, and carboxyl [61, 62]. GO is normally produced by oxidizing graphite using the Hummers method [63], followed by exfoliation into sheets via mechanical or thermal treatments. On the edge and basal planes of GO, there are different oxygen-containing hydrophilic functional groups, such as carbonyl, carboxyl, hydroxyl, and epoxy groups, which serve as targets for cross-linking agents [64, 65]. rGO is made from GO by eliminating the oxygen-containing groups via thermal or chemical reduction (Figure 7.5a) [69]. It is easier to manipulate GO and rGO in comparison to graphene due to the improved processability of these materials in most solvents. Vacuum filtration [70, 71], layer-by-layer (LbL) deposition [72, 73], spin-coating [67, 74], or drop-casting [75] have been used to construct GO/rGO nanosheets into lamellar membranes. Unlike nanoporous atomically thin membranes, which have intrinsic nanopores that provide rapid molecular or ion transport pathways, GO- and rGO-based laminate membranes have two distinct penetration paths—(i) in-plane structural defects and slit-like nanopores produced by edge-to-edge nanosheet interactions and (ii) plane-to-plane interlayer nanochannels [40].

GO- and rGO-based membranes provide significant gas-separation potential. Many GO-based membranes separate guest molecules by accumulated nanochannels between adjoining nanosheets because the in-plane structure is largely nonporous. The interaction between oxygen and transport guests groups enhanced the selective separation procedure [76]. Zhao et al. produced an ultrathin GO membrane using spin-coating procedure [67]. These GO membranes showed a remarkable molecular-sieving separation mechanism with an H_2/CO_2 selectivity of 240 and H_2 permeance up to 3.4×10^{-7} mol m^{-2} s^{-1} Pa^{-1} (Figure 7.5b,c,d,e). Variable temperature tests and single gas permeation tests indicate possible gas diffusion through GO interlayer galleries and dominating molecular-sieving separation mechanism.

Several studies have been sought to tune nanochannel and surface properties of GO to improve the separation performance. For instance, Jin et al. employed properly constructed external forces to accurately adjust the nanochannel of graphene membranes to ~0.4 nm [62]. The external forces directed the uniform in-plane and plane-to-plane stacking of the GO nanosheets during the vacuum spin coating procedure. These membranes had an optimum selectivity of 260 and H_2 permeance of 3.9×10^{-7} mol m^{-2} s^{-1} Pa^{-1} in H_2/C_3H_8 separation experiments. The permeance was increased by ~30% compared to the membrane prepared by Yu and coworkers [77] The authors also found the adjoining layer spacing of the GO nanosheets to correlate with the gas-separation efficiency. Later, Yu and coworkers modified the surface characteristics of GO nanosheets using piperazine [68]. Vacuum-assisted coating

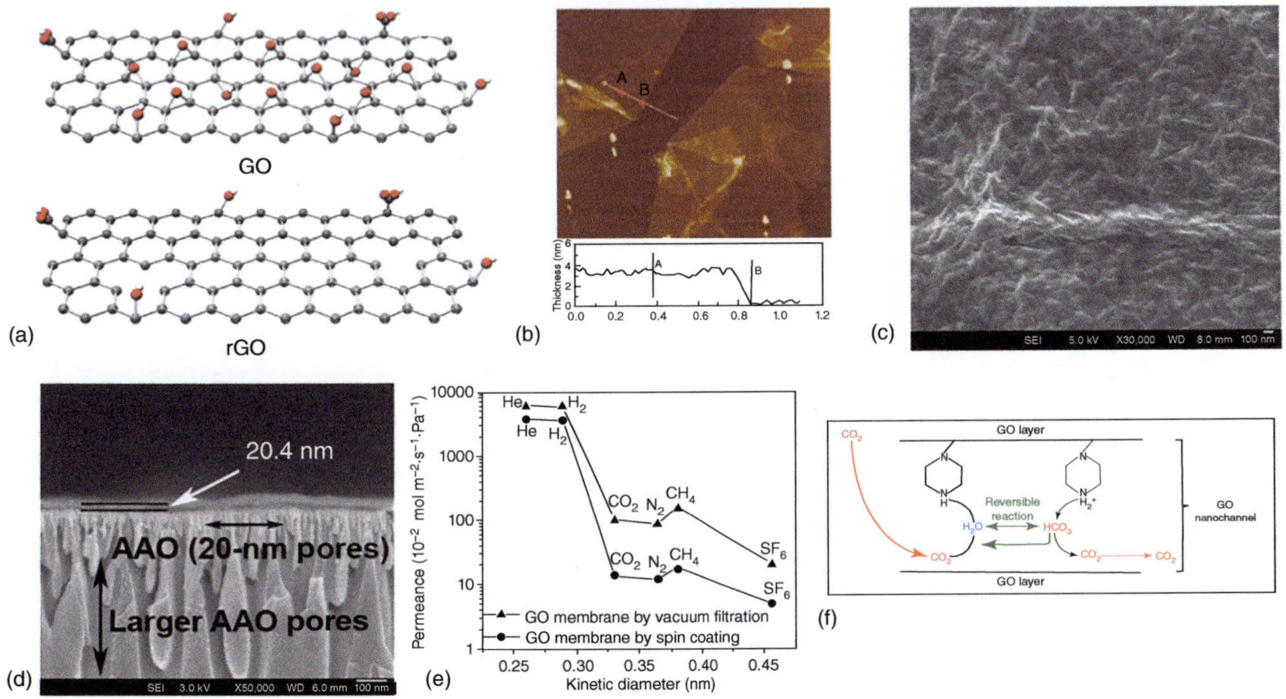

Figure 7.5 (a) Single-layer GO and rGO structures. Source: Toda et al. [66]/with permission of Elsevier B.V. (b) AFM images of exfoliated GO nanosheets with corresponding height profiles. (c) SEM image of GO membrane prepared by spin coating. (d) Cross-sectional SEM image of a GO membrane prepared by spin coating. (e) Permeance of single gases through GO membranes prepared by vacuum filtration and spin coating. Source: Chi et al. [67]/with permission of American Chemical Society. (f) Structure diagram and CO_2 facilitated transport method of the ultrathin GO-based hollow-fiber membranes. Source: Modified from Zhou et al. [68].

was used to change modified GO nanosheets on hollow fibers. The CO_2 permeance was enhanced to be 3.4×10^{-7} mol m^{-2} s^{-1} Pa^{-1} with a CO_2/N_2 selectivity of 680 in the wet CO_2/N_2 separation. The secondary amine groups in piperazine facilitated CO_2 transportation (Figure 7.5f).

7.2.3 MOFs

MOFs are crystalline porous materials constructed by coordination bonds between metal ions or clusters and organic linkers [78]. Due to their high porosity, tunable pore size and structure, and diverse chemical functionalities, MOFs have demonstrated significant potential in several applications, including gas storage and separation, drug delivery, chemical sensing, and catalysis [79–81]. Layered structures were commonly observed in MOFs [82, 83], suggesting the possibility of manufacturing ultrathin 2D MOF membranes using MOF nanosheets to accomplish quick and highly selective separation via the uniform porosity in MOFs.

Yang et al. used a hybrid ball milling and ultrasonication approach to exfoliate layered MOF crystals termed poly[Zn_2(benzimidazole)$_4$] into 1-nm-thick sheets [83]. The pore size can be somewhat greater than other 2D materials due to the flexibility of the MOF structure, which has an orifice size of c. 0.21 nm as determined from its model structure. Ultrathin 2D MOF membranes were made by hot-drop-coating procedure and demonstrated strong H_2/CO_2 molecular sieving with approximately 2500 GPU of H_2 permeance and selectivity of 260. More impressively, these 2D MOF membranes synthesized at higher temperatures showed an anomalous rise in H_2 permeability along with H_2/CO_2 selectivity, eliminating the trade-off phenomenon as seen in traditional polymeric membranes. In a follow-up study, Yang et al. delaminated another layered MOF, [Zn_2(benzimidazole)$_3$(OH)(H_2O)]$_n$ (Figure 7.6a,b), and used the resultant nanosheets to prepare sub-10 nm-thick 2D MOF membranes (Figure 7.6c,d) [84]. The layered MOF has a pore dimension of 0.29 nm between H_2 (0.289 nm) and CO_2 (0.33 nm). The as-constructed MOF membrane shows H_2 permeance of about 2400 GPU and an H_2/CO_2 selectivity of 166, verifying the strong molecular sieving impact in nanosheet membranes (Figure 7.6e,f).

Due to the brittleness of MOF nanosheets, there is a high probability of nanosheet damage under conventional exfoliation techniques, such as ultrasonication. To address this issue, Zhao et al. used a moderate freeze–thaw technique to laminate layered MAMS-1 (mesh-adjustable molecular sieve-1) MOF crystals into nanosheets [85]. The MAMS-1 material showed two possible gas transport routes with an orifice size of 0.29 nm along with the (001) plane and orifice size of 0.5 nm along with the (100) plane. The authors hypothesized these routes to undertake roles of gas discrimination and rapid gas transport respectively, which combined to give excellent separation performance. The shear force derived from the volumetric alteration in solvents through freeze–thaw cycles can assist to disrupt the poor *van der Waals* connections in layers of MAMS-1 and cause MAMS-1 to exfoliate into ultrathin nanosheets without damaging the in-plane MOF structure. The 2D MAMS-1 membranes were developed using a hot-drop casting process for

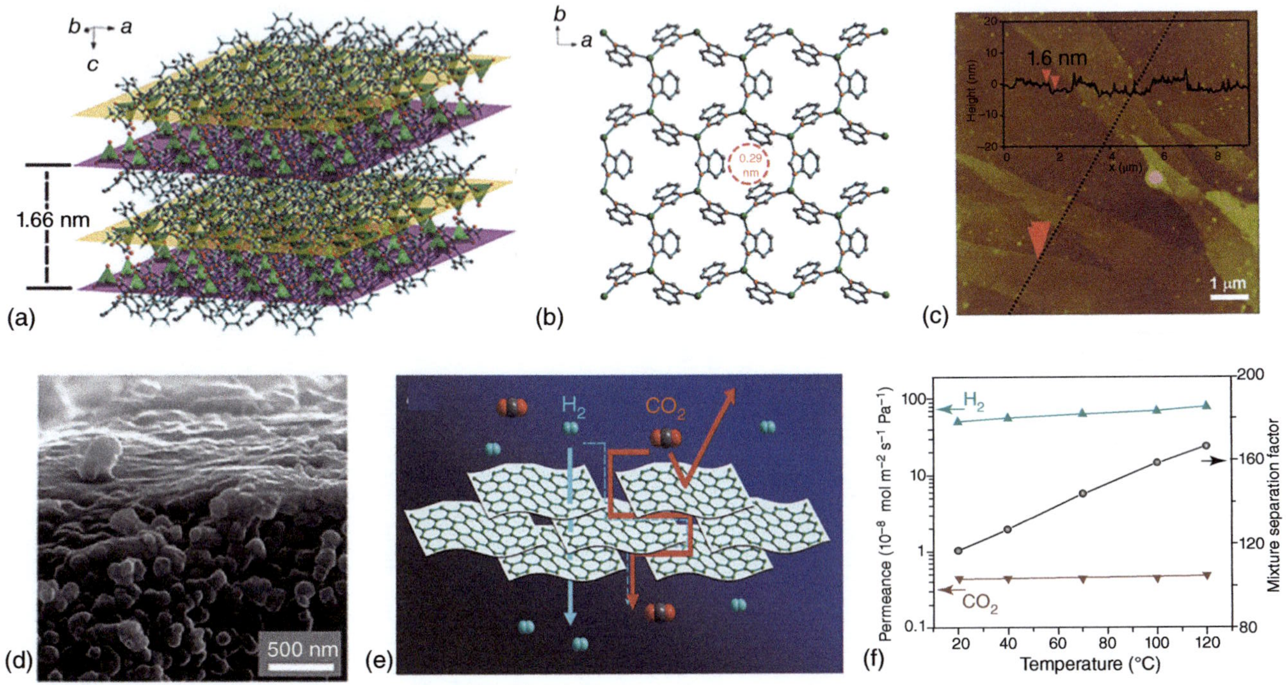

Figure 7.6 (a) Four-layered stacking diagram of $Zn_2(Bim)_3$ precursors along the c-axis. H white, O red, Zn green, N orange, C gray. The layers with benzimidazole ligands along c-axis are depicted in purple, the Zn coordination polyhedra are displayed in green, and the others in yellow. (b) Single-layered nanosheet with the benzo-rings upwards, highlighting the triply-linked coordination of Zn nodes with benzimidazole ligands. N orange, Zn green, C gray. (c) AFM image and height profile of $Zn_2(Bim)_3$ nanosheets. (d) Cross-section view of the $Zn_2(Bim)_3$ nanosheet membranes prepared at 200 °C. (e) Representing hypothesis of gas separation through porous $Zn_2(Bim)_3$ nanosheets. (f) Effect of various temperatures on H_2/CO_2 permeance and separation factor of a $Zn_2(Bim)_3$ nanosheet membrane. Source: Peng et al. [84]/with permission of Wiley-VCH GmbH.

these nanosheets. By altering the content of nanosheet suspension utilized in membrane manufacturing, membrane thickness could be readily controlled from 4 to 40 nm. The 40-nm-thick 2D MAMS-1 membrane has H_2 permeance of 715 GPU and H_2/CO_2 selectivity of 245 and was stable up to 10 000 minutes of operation. Moreover, the as-prepared membranes display inverted thermally switchable molecular-sieving processes due to the flexibility of tert-butyl groups in MAMS-1, opening the way for the construction of smart membranes with stimuli-responsive features for specific usage.

7.2.4 COFs

COFs are porous crystalline polymers consisting of organic linkers connected by strong covalent connections [86] (Figure 7.7a) and have huge applications in gas storage and separation [88–90], energy storage [91, 92], and heterogeneous catalysis [93–95]. Due to their huge surface area, well-defined pores, and excellent stability, COFs are highly attractive as potential membrane materials. However, very few reports are available on supported 2D COF gas-separation membranes since the pores of COFs (>1 nm) are considerably bigger than the kinetic diameters of gases, such as H_2 (2.9 Å), CO_2 (3.3 Å), N_2 (3.6 Å), and CH_4 (3.8 Å). This precludes any prospect of molecular-sieving separation. While many COFs have very large pores, the effective COF pore size in 2D COF membranes may be reduced by stacking offsets. For example, in the bulk phase, COFs with staggered stacking mode or alternating layers have smaller pore sizes compared to COFs with eclipsed stacking with no layer offset. If stacking phenomena may be effectively controlled during COF design or the crystallization process, "tight" membranes with pore sizes similar to gas molecules may be accessed.

Zhao et al. [18] implemented oppositely charged ultrathin 2D ionic covalent–organic nanosheets (iCONs) to reduce the pore size of 2D COF membrane. The TpEBr@TpPa-SO_3Na hybrid membranes were fabricated through LbL assembly of two kinds of iCONs with opposite charges and different pore sizes (Figure 7.7b). The resulting membranes have optimal stacking patterns, decreased aperture size, and compact dense structure without losing thickness control due to the staggered packing of iCONs with strong electrostatic interactions, making them appropriate for molecular-sieving gas separation. At 423 K, TpEBr@TpPa-SO_3Na membrane has H_2 permeance of 2566 GPUs and an H_2/CO_2 separation factor of 22.6, exceeding the Robeson upper bound and demonstrating long-term hydrothermal stability (Figure 7.7c).

In another study, Caro and coworkers created COF–COF bilayer composite membranes via a temperature-swing solvothermal method by controlling the development of imine-based COF-LZU1 and azine-based ACOF-1 layers, sequentially (Figure 7.7d,e) [87]. Due to the creation of tiny, interlaced pores (0.3–0.5 nm), such bilayer composite membranes demonstrated substantially greater separation selectivity for H_2 over CO_2, N_2, and CH_4, giving a novel approach to modify the pore size of COFs. They reported that the H_2 permeance can reach 2.45×10^{-7} mol m^{-2} s^{-1} Pa^{-1}, and the ideal separation selectivities of H_2 from

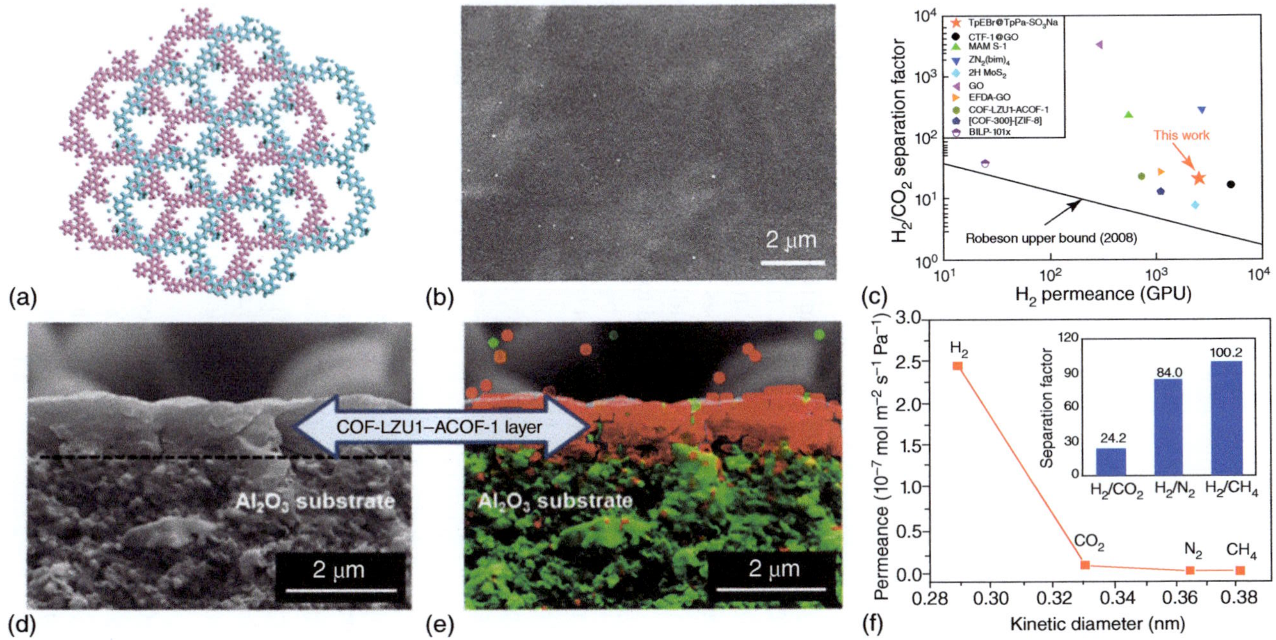

Figure 7.7 (a) Scheme of the iCON membrane with staggered packing pattern. (b) SEM image of TpEBr@TpPa-SO_3Na iCON membrane. (c) Comparison of H_2/CO_2 separation performance of iCONs with other membranes (assuming a membrane thickness of 100 nm). Source: Ying et al. [18]/with permission of American Chemical Society. (d) Cross-sectional view of the COF-LZU1–ACOF-1 bilayer membrane. (e) EDXS mapping of the cross-sectional view of COF-LZU1–ACOF-1 bilayer membrane. (f) Single-gas permeance of the COF-LZU1–ACOF-1 bilayer membrane as a function of the kinetic diameter of permeating molecules at 298 K, 1 bar. Source: Fan et al. [87]/with permission of American Chemical Society.

CO_2, N_2, and CH_4 are 26.7, 88.7, and 105.0, respectively, which are much higher than the corresponding Knudsen constants (Figure 7.7f). By utilizing such mentioned methods, the researchers do not only enhance gas-separation performance but also manipulate the pore sizes of COF membranes.

7.2.5 g-C_3N_4

g-C_3N_4 is a graphite-like layered material. It is made up of triazine units joined by tertiary amino groups in each layer (Figure 7.8a) [98]. This new emerging material has high pore density, suitable pore size, and thus minimal mass-transfer resistance. Because the *van der Waals* interaction between layers of g-C_3N_4 is weak, bulk g-C_3N_4 powder can be easily exfoliated into ultrathin nanosheets. g-C_3N_4 exhibits narrow interlayer spacing (3.2 Å) and orderly in-plane triangular pores (3.11 Å) as a potential 2D material, which may aid the size-sieving process in the resulting membranes [99].

Agrawal and coworkers [96] demonstrated a reproducible and safe ambient pressure synthesis method that produces several grams of stacked poly(triazine imide) (PTI, a type of crystalline g-C_3N_4) platelets in a single batch with a 70% yield over the precursor. Solvent exfoliation under dry conditions resulted in single-layer PTI nanosheets that exhibited the characteristic triangular nanopores of g-C_3N_4 (Figure 7.8b). It was confirmed that PTI nanopores can filter He and H_2 from larger molecules, according to gas permeation experiments. As a result, PTI nanosheet–based membranes produced by the scalable filter coating approach have the ability of high-temperature H_2 sieving, with H_2 permeance of 1500 GPUs, as well as H_2/CO_2, H_2/N_2, and H_2/CH_4 selectivities of 10, 50, and 60 at 250 °C, respectively (Figure 7.8c,d).

Wang et al. used a simple GO-modification technique to effectively produce g-C_3N_4 nanosheet membranes with outstanding hydrogen permselectivity (Figure 7.8e) [97]. The nonselective defects in g-C_3N_4 membranes were efficiently remedied by covalent contacts between the terminal –COOH in GO and $-NH_2$/–NH in g-C_3N_4, contributing to high hydrogen selectivity (Figure 7.8f). The g-C_3N_4–GO_{30} membrane illustrated ideal selectivity of H_2 to CO_2, N_2, C_3H_6, and C_3H_8 of 39.2, 19.3, 221, and 306, respectively, suggesting excellent molecular-sieving performance (Figure 7.8g). A small quantity of GO was introduced to correct nonselective defects in the g-C_3N_4 nanosheets. The g-C_3N_4 endowed the g-C_3N_4–GO membranes with molecular-sieving properties. Furthermore, the H_2 permeance of the g-C_3N_4–GO_{30} membrane was 2.16×10^{-7} mol m^{-2} s^{-1} Pa^{-1}, indicating that g-C_3N_4 is a promising material for gas separation.

7.2.6 MXenes

MXenes are 2D-layered transition-metal carbonitrides and carbides made by selectively etching the A element from the MAX phases [100, 101]. $M_{n+1}AX_n$ ($n = 1, 2, 3$) is the precursor for MXenes, where M represents an early transition metal, A represents an element from group IIIA/IVA, and X represents C and/or N (Figure 7.9a). More than 60 distinct pure MAX phases have been found to date, with $Ti_3C_2T_x$

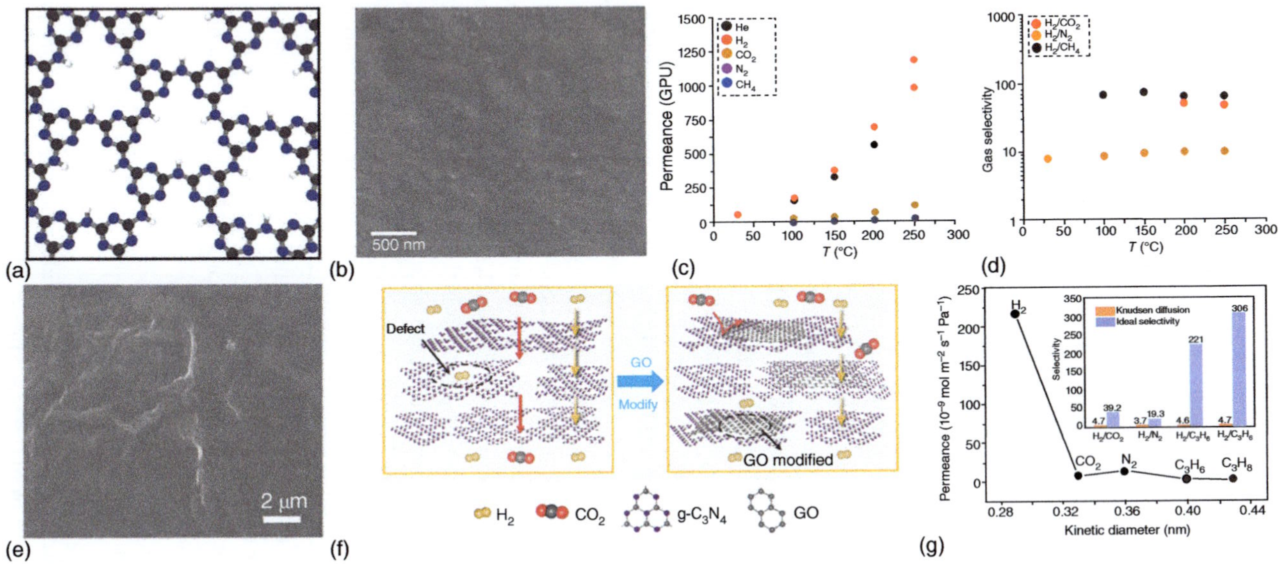

Figure 7.8 (a) Structure of the PTI layer (g-C_3N_4). (b) SEM image of the PTI nanosheet film. (c) Permeance and (d) selectivity of PTI nanosheet–based membrane. Source: Villalobos et al. [96]/with permission of American Association for the Advancement of Science. (e) Surface SEM image of the g-C_3N_4–GO_{30} membrane. (f) Schematic of gas molecular transport via g-C_3N_4 nanosheet membranes modified with GO and its process. (g) Single gas permeation of the g-C_3N_4–GO_{30} membrane at room temperature. Source: Zhou et al. [97]/with permission of Elsevier B.V.

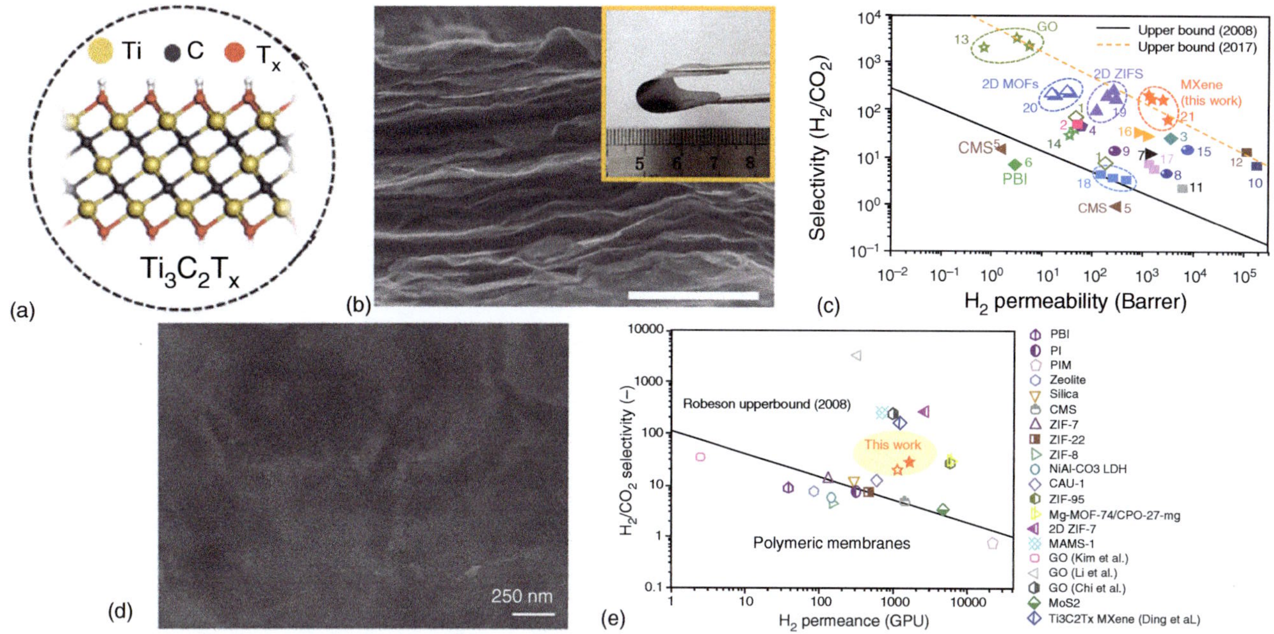

Figure 7.9 (a) Structures of the MXene $Ti_3C_2T_x$ [102]/with permission of John Wiley and Sons. (b) Cross-sectional SEM image of an MXene membrane. Source: Ding et al. [21]/with permission from Elsevier B. V. (c) Comparison of H_2/CO_2 separation performance of the MXene membrane with state-of-the-art gas separation membranes. The black line shows the Robeson 2008 upper bound of polymeric membranes and the orange dashed line demonstrates 2017 upper bound of the reported membranes. Source: Ding et al. [21]/Springer Nature/CC BY-4.0. (d) Surface SEM image of an MXene nanofilm. (e) Comparison of H_2/CO_2 separation performance of MXene nanofilms with state-of-the-art gas separation membranes. The open red star symbol illustrates the mixed gas permeation (50 : 50 H_2/CO_2) at 1 bar and 25 °C, while the solid red star symbol shows the single gas permeation at 1.5 bar and 25 °C. Source: Shen et al. [102]/with permission of John Wiley and Sons.

being the most frequently researched MXene, in which T represents the termination groups (O, F, and/or OH) and X is the number of termination groups [103]. MXene nanosheets can be processed into 2D membranes for size-selective separations in a way similar to those of GO nanosheets.

Gas separations based on the molecular-sieving effect might potentially benefit from 2D MXene membranes that have subnanometer channels. For effective gas separation, Wang et al. produced 2D MXene membranes by filtrating single-layer $Ti_3C_2T_x$ nanosheets onto porous substrates (Figure 7.9b) [21]. The free interlayer spacing (approximately 0.35 nm) in these intact and dense 2D MXene membranes can cause intense molecular sieving between small gas molecules (e.g. He and H_2) and big gas molecules (e.g. N_2, O_2, and CH_4). Due to the strong interactions between CO_2 and the terminal groups in MXene layers, 2D MXene membranes have high CO_2-philicity, which significantly decreases the rate of CO_2 diffusion in the membranes. The resulting 2D MXene membranes have outstanding H_2/CO_2 selectivity (>160) and exceptional H_2 permeability (>2200 Barrer), making them a promising candidate for effective H_2 purification (Figure 7.9c).

Furthermore, Jin and coworkers created 2D MXene membranes (20-nm thick) for gas separation with excellent molecular-sieving performance (e.g. H_2/C_3H_8 selectivity of c. 110) (Figure 7.9d) [102]. They speculated that the molecular-sieving effect of ultrathin 2D MXene membranes was due to the in-plane slit-like nanopores rather than interlayer nanochannels. The membrane selectivity was maintained when the interlayer free distance was reduced from 0.52 to 0.38 nm after vacuum heating treatment. Moreover, these 2D MXene membranes were functionalized with polyethylenimine (PEI) and borate molecules, significantly improving CO_2 solubility inside the membranes. The synthesized membranes have a CO_2/CH_4 selectivity of 15.3 and high CO_2 permeance of 350 GPU (Figure 7.9e).

7.2.7 Other 2D Materials

Some other 2D materials have also been investigated as suitable membrane materials for diverse separation applications, inspired by the amazing success in the above-mentioned 2D membranes. Boron nitride (BN) [104–106], TMDs [26, 107], and layered double hydroxides [108, 109] are examples of such 2D materials. Despite the well-defined interlayer nanochannels, these nanosheets may be readily formed into 2D membranes, and most of them demonstrate good gas transport properties.

BN is a representative 2D crystal with strong covalent sp^2 bonds in the plane that has high thermal conductivity, mechanical strength, and chemical stability (Figure 7.10a). BN nanosheets, also known as white graphene, are made up of a few layers of atomically thin hexagonal BN planes and have a structure similar to graphene [29]. Due to its intrinsic features, BN is a new and promising option for creating separation membranes that have superior overall performance. Chen et al. [104] fabricated a BN membrane with a unique nanoconfinement effect for effective ethylene/ethane separation. The BN membrane is endowed with numerous percolating nanochannels due to the horizontal and inclined self-assembly of 2D BN nanosheets. Reactive ionic liquids (RILs) were used to decorate nanochannels to

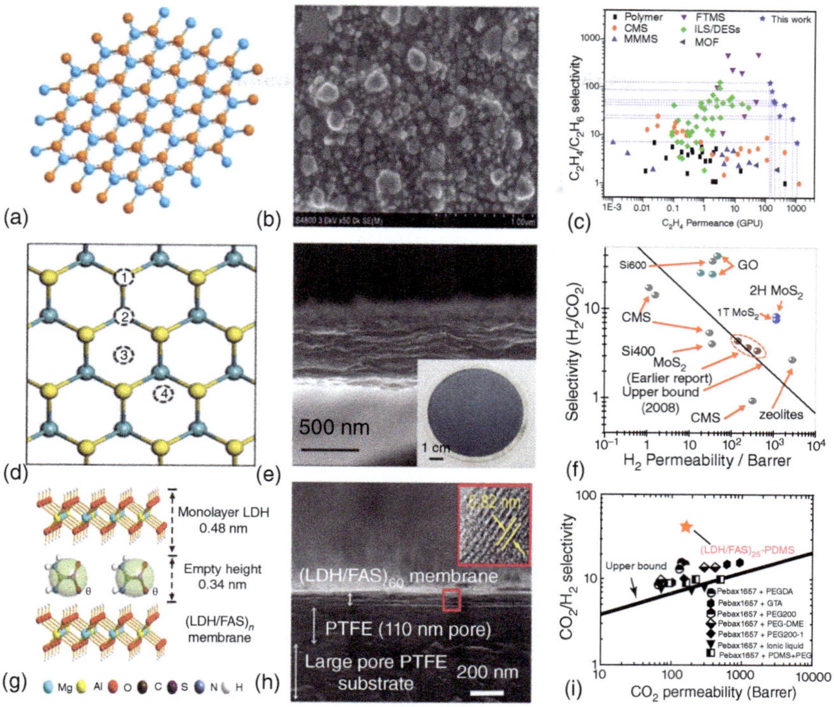

Figure 7.10 (a) Scheme of the h-BN structure. Source: Lu et al. [29]/with permission of Springer Nature. (b) Surface SEM images of the RIL-BN/Nylon membrane. (c) Comparison of C_2H_4/C_2H_6 separation performance with the state-of-the-art membranes, including polymer membranes, MMMs, CMS, FTMs, MOFs, IL, or DES-based membranes. Source: Dou et al. [104]/with permission of Wiley-VCH GmbH. (d) The structure of $WSe_2/MoSe_2$ nanosheet. Source: Yu et al. [110]/with permission of Royal Society of Chemistry. (e) Cross-sectional FESEM image of the MoS_2 membrane. (f) Comparison of H_2/CO_2 gas separation performance of the MoS_2 membrane (1T) and MoS_2 membrane heated at 160 °C (2H) with other membranes reported in the literature. Source: Achari et al. [25]/with permission from Royal Society of Chemistry. (g) Schematic diagram of the LDH/FAS. (h) Cross-sectional SEM image of $(LDH/FAS)_{60}$ membrane. (i) Comparison of CO_2/H_2 separation performance of $(LDH/FAS)_{25}$-PDMS membrane with states-of-art literatures. Source: Xu et al. [111]/with permission of Springer Nature.

adjust their sizes and create the nanoconfinement effect (Figure 7.10b). As a result, the membranes achieved remarkable separation performance with superhigh C_2H_4 permeance of 138 GPU and a C_2H_4/C_2H_6 selectivity of 128 as well as substantially enhanced long-term stability for 180 hours (Figure 7.10c).

TMDs are a kind of layered material with MX_2 formula, where X stands for chalcogen (Se, S, or Te), and M stands for transition-metal atoms, including Mo, Hf, Nb, W, Re, and so on (Figure 7.10d). The interlayer bonds are maintained by *van der Waals* forces, whereas the intralayer bonding has a mixed covalent–ionic nature [25]. In terms of the structural compatibility and *van der Waals* interaction, 2D TMDs have the ability to create nanosheet layers, which are suitable for membrane fabrication. MoS_2, in particular, has defect sites that can interact with CO_2 molecules, making it an excellent material for CO_2 selective gas separation [110]. In this regard, Eswaramoorthy et al. [25] investigated gas permeation performance and efficiency of H_2/CO_2 separation via submicrometer thick MoS_2 membranes. The membranes were fabricated by vacuum filtration of MoS_2 nanosheet solution (Figure 7.10e) and they demonstrated 1175 Barrer for H_2 permeability and 8.29 for ideal H_2/CO_2 selectivity (Figure 7.10f).

Layered double hydroxides (LDHs) are ideal candidates for high-performance membranes because they have homogeneous interlayer galleries and numerous hydroxyl groups covalently linked with 2D host layers (Figure 7.10g). Due to their well-defined interlayer galleries, LDHs show excellent potential for separation processes, achieving precise molecular sieving. O'Hare et al. [111] reported (LDH/FAS)n-PDMS hybrid membranes that were fabricated by self-assembly of unilamellar LDH nanosheets and formamidine sulfinic acid (FAS), followed by spray coating with a PDMS layer (Figure 7.10h). The membranes contained 2D subnanometre channels, demonstrating CO_2 permeation of 7748 GPU and CO_2 selectivities over H_2, N_2, and CH_4 as high as 43, 86, and 62, respectively (Figure 7.10i).

7.3 Preparation of 2D Nanosheets

To realize high-performance separation on 2D-materials membranes, high-quality 2D nanosheets are highly appreciated as building blocks. These nanosheets must be carefully assembled to construct the channels governing the transport behavior in the final membrane configuration. The size, defect, thickness, and surface function of the nanosheets are all important factors for membrane preparation. Nanosheets with large lateral size and ultrathin thickness are beneficial to decrease the amount of intraplanar, less selective gaps, and increase the internanosheet interactions [83, 112]. The 2D nanosheet fabrication approach may typically be classed into bottom-up and top-down techniques (Figure 7.11). In the bottom-up technique, 2D nanosheets are directly synthesized from basic building units, while 2D nanosheets are exfoliated from the bulk in the top-down technique.

7.3.1 Top-Down Method

Top-down strategy can be a useful technique when the interlayer interactions, such as hydrogen bonding, *van der Waals* forces, and π–π stacking, are weak. Therefore, the exfoliation will be feasible in this case because most of the layered materials are built from the stacking along with the vertical direction. Intercalation and mechanical exfoliations are the most frequent category of top-down methods [113].

In the intercalation exfoliation method, molecules and ions are inserted between the layers to reduce layer interactions or heighten the surface charge to be better dispersed in solvents. Li^+, Na^+, and K^+ are the cations that are usually intercalated into TMDs and these cations help to exfoliate nanosheets (MoS_2 nanosheets are usually achieved by this method). For instance, Eswaramoorthy et al. [25] exfoliated bulk

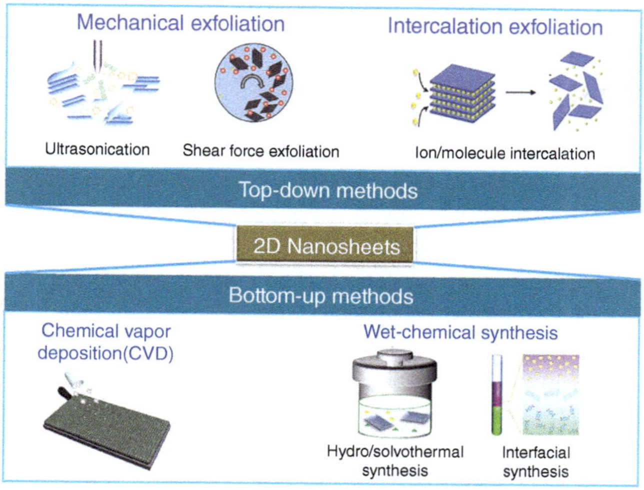

Figure 7.11 Different synthesis types of 2D nanosheets. Source: Wang et al. [113]/with permission of Royal Society of Chemistry.

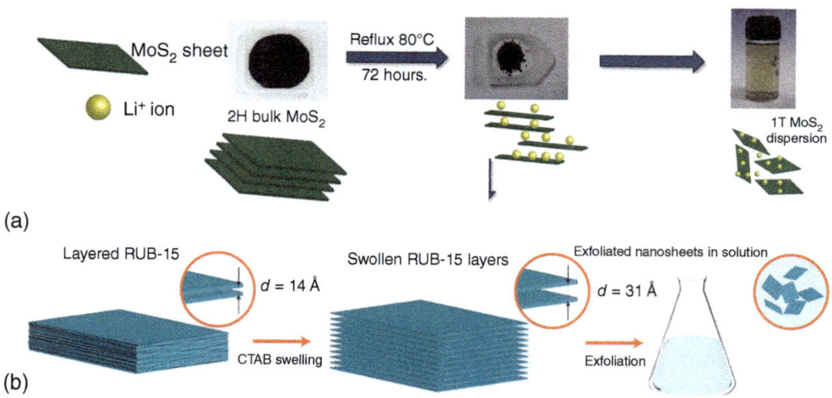

Figure 7.12 (a) Scheme of the preparation of lithium intercalated MoS_2 dispersion. Source: Achari et al. [25]/with permission of Royal Society of Chemistry. (b) Scheme of the exfoliation of bulk RUB-15 to achieve nanosheets. Source: Dakhchoune et al. [36]/with permission of Springer Nature.

MoS_2 successfully via intercalating lithium between the bulk layers (Figure 7.12a). This circumstance was done by dissolving bulk MoS_2 in dry hexane, followed by adding *n*-butyl lithium under an inert atmosphere. Refluxing the mixture at 80 °C for 72 hours resulted in monolayer MoS_2 nanosheets.

Agrawal et al. [36] exfoliated RUB-15, a type of zeolite, successfully by intercalating cetyltrimethylammonium bromide (CTAB) to increase the interlayer spacing of bulk RUB-15 from 1.4 to 3.1 nm, with a corresponding expansion of the gallery spacing from 0.6 to 2.3 nm. The swollen layers were then exfoliated by using low molecular weight polystyrene and a polymer melt compounding method (Figure 7.12b). While the exfoliation technique of intercalation yields stable nanosheet dispersions in a wide scope of solvents, this method is time-consuming (up to three days). Also, in order to permit ion transport into the interlayers, a smaller-sized flake (<1 μm) is usually preferred [114]. Meanwhile, intercalation exfoliation may introduce some impurities to the exfoliated nanosheets. This can be a concern, especially when dealing with oxidative chemicals or explosive intercalators. If subsequent washing steps are incorporated to remove the residual chemical substances, the ultimate performance of membranes may be negatively influenced.

Mechanical exfoliation is a simple treatment approach that uses mechanical force to overcome the weak interlayer forces of 2D materials. The "Scotch-tape" strategy is the original technique that is used in the isolation and characterization of 2D samples from their bulk crystalline samples [3]. Although this method is effective for small-scale preparation, the scalability of this approach is relatively poor. Therefore, shear force and ultrasonication have been explored as more user-friendly and environment-friendly options with higher throughput than the "Scotch-tape" method [114]. Typically, after utilizing the above methods, the colloidally stable nanosheets contain a limited number of stacked layers (less than 10), which may be further separated as one-layer 2D nanosheets by successive centrifugation [115].

A careful solvent selection is also crucial for proper exfoliation, along with other factors, such as processing length, rotation rate, and sonication power. The main parameters are polarity and solvent molecular size. Yang's group [84] reported one of the first successful examples of MOF exfoliation by using a combination of low-energy wet ball milling and ultrasonication (Figure 7.13a,b). Their method yielded samples with approximately 1-nm thickness, 1.5-μm lateral size. Most importantly, the MOF samples retained crystallinity after exfoliation. The *n*-propanol and methanol mixture played a crucial role as the solvent in successful exfoliation. To stabilize the exfoliated nanosheets, *n*-propanol was employed through hydrophobic alkanes to connect to the surface of MOF nanosheets, whereas methanol, the smaller molecules, could diffuse to interlayers and aid the exfoliation. In other exfoliation processes, this co-solvent technique should also be effective.

Freeze–thaw and gas exfoliation are other strategies of mechanical exfoliation that solve the problem of nanosheet damage and fragmentation [67, 85, 116]. Ogino et al. [116] synthesized dispersed GO nanosheets in water via repetitive simple freeze–thaw cycles (Figure 7.13c). These cycles involve rapidly freezing a graphite oxide-containing aqueous solution and then thawing the resulting solid. After six freeze–thaw cycles, graphite oxide was successfully exfoliated with a yield of around 80%. The freeze–thaw approach produces GO nanosheets that are less fragmented throughout the exfoliation process with a lateral size of at least 3-fold larger than GO produced by sonication. Zhao et al. [67, 85] successfully exfoliated 2D MOF nanosheets (MAMS-1) and GO nanosheets via freeze–thaw strategy to produce suitable 2D membranes for gas separation. Zhu et al. [117] reported an effective approach for the scalable production of few-layered h-BN nanosheets (BNNS) with a novel gas exfoliation of bulk h-BN in liquid N_2 (Figure 7.13d). The foundation of this technique was a combination of cryogenic L-N_2 gasification and

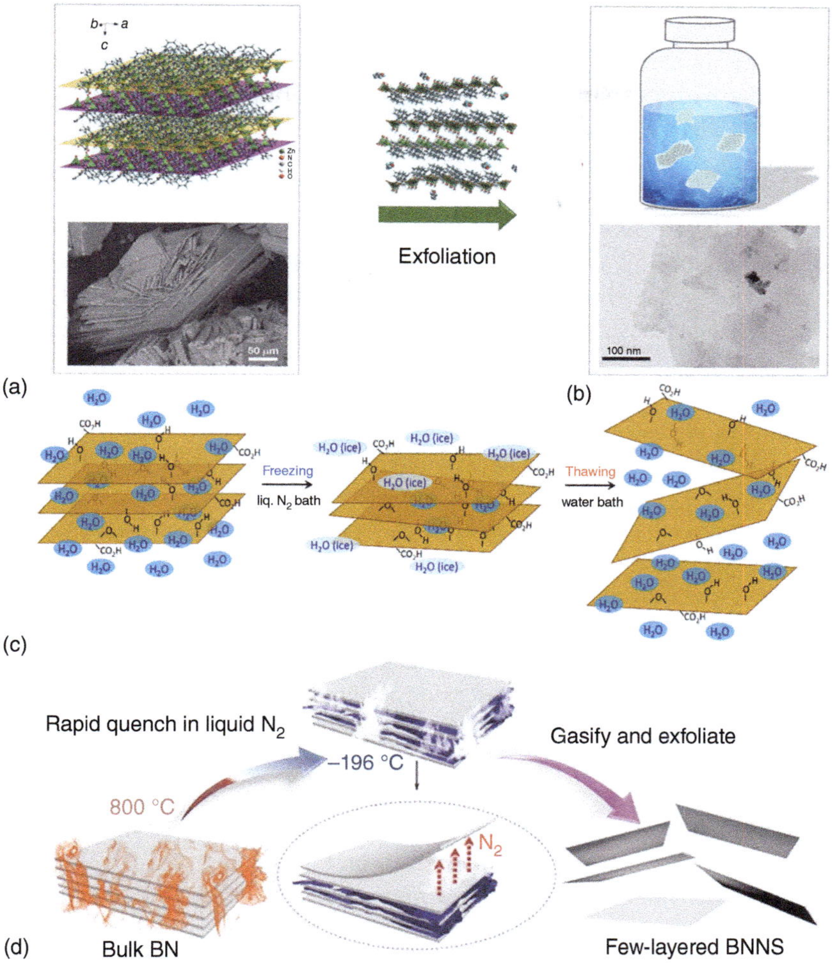

Figure 7.13 (a) Top: four-layered stacking diagram of $Zn_2(Bim)_3$ precursors along *c*-axis; bottom: SEM image of the layered precursor particles. (b) Top: $Zn_2(Bim)_3$ nanosheets dispersed in equivoluminal methanol/*n*-propanol mixture; bottom: TEM image of $Zn_2(Bim)_3$ nanosheets. Source: Peng et al. [84]/with permission of John Wiley and Sons. (c) Exfoliation of GO via a rapid freeze–thaw method. Source: Ogino et al. [116]/with permission from American Chemical Society. (d) Gas exfoliation of h-BN triggered by thermal expansion. Source: Zhu et al. [117]/with permission of John Wiley and Sons.

high-temperature-triggered bulk h-BN expansion to exfoliate the h-BN. After 10 cycles, the generated BNNS predominantly consisted of less than five atomic layers, with a high mass yield of 16–20%.

Due to its easy operation and universality, top-down techniques are commonly utilized through low yield. The resulting solution-processed nanosheets can make it easier to blend, cast, and function for target membrane usage. Nevertheless, some top-down techniques may have limited throughput, damage nanosheets, and add impurities to the exfoliated nanosheets.

7.3.2 Bottom-Up Method

The top-down approach applies to already isolated or crystallized bulk-layered materials. Alternatively, nanosheets may be assembled directly from molecular precursors using bottom-up methods. In principle, practically all 2D materials may be isolated using this approach. The bottom-up techniques include wet-chemical synthesis and CVD growth. h-BN, graphene, and TMDs are fabricated by CVD growth epitaxially on substrates, such as SiO_2 and SiC. The CVD process makes it possible to produce highly crystalline 2D nanosheets with controlled thickness and scalable size that can be employed as membrane building blocks. On the other hand, for 2D nanosheets made from linkers and nodes (e.g. COFs, zeolites, and MOFs), the prevailing approach is wet-chemical synthesis. The process of synthesis is comparable to the bulk materials, and a successful fabrication depends on the selective growth of materials in a single direction. Hydro/solvothermal synthesis and interfacial synthesis are the most well-known strategies in the bottom-up technique.

Solvo/hydrothermal syntheses are commonly performed under high temperatures and pressures to produce a diverse scope of nanosheets, including 2D metal chalcogenides, metal oxides, MOFs, etc. Structure-directing agents or surfactants are typically used to aid the production of nanosheets. The product thickness, dimensions, and crystallinity can be regulated using the synthesis parameters, such as the concentration of the predecessor, reaction time/temperature, and type of solvent and surfactant. Recently, the development of 2D nanosheets has been effectively controlled by the introduction of additional agents. MFI is a type of zeolite that is widely implemented as membrane material, but they are sensitive to defects. Therefore, the synthesis of these materials by bottom-up method can eliminate the probability of nonselective defects on the structure. Tsapatsis' group [4] utilized bis-1,5(tripropyl ammonium)pentamethylene diiodide (dC5) to control the growth of MFI along with crystallographic *b* axis of the material (Figure 7.14a). Micrometer-size MFI nanosheets were obtained with approximately 5-nm thickness. The MFI nanosheets showed abundant straight orthogonal micropores on the sheet surface, which acted as selective channels for gas permeation.

Interfacial synthesis is another bottom-up strategy that is appropriate for both thin membranes and nanosheets [120]. In this process, the interaction between various monomers is limited to the interfacial mass transfer zone, which ensures a low thickness. The interfacial synthesis approach has been demonstrated on liquid/solid, liquid/liquid as well as liquid/air interfaces. As the diffusion rates between the monomers and the interface regulate the structural integrity of the prepared nanosheets, the intermediary layer should be judiciously chosen [121]. Gascon and coworkers [118] isolated Cu(BDC) nanosheets instead of the more common bulk 3D structure by buffering the reactant diffusion using multiple liquid layers (three liquid layers, mixtures of co-solvent (*N*, *N*-dimethyl formamide [DMF], and CH_3CN) at appropriate ratios (Figure 7.14b). The bottom layer and top layer contained the organic ligand (1,4-benzene-dicarboxylic acid [BDC]) and metal precursor ($Cu(NO_3)_2$), respectively. The slow diffusion of the reactants

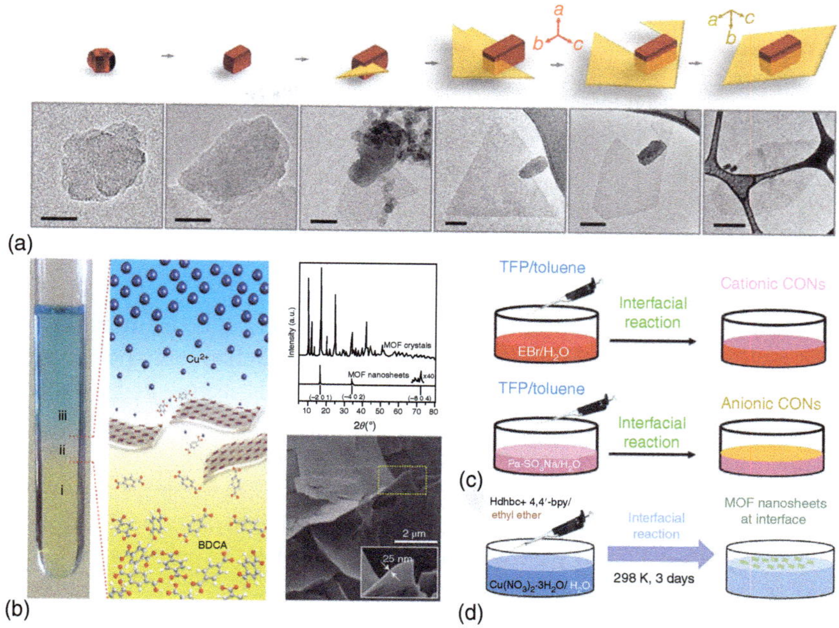

Figure 7.14 Synthesizing 2D nanosheets by a bottom-up strategy. (a) Modulated synthesis for plate-like MFI creation. The top photos and the bottom TEM images demonstrate the growth process of nanosheets. Source: Jeon et al. [4]/with permission of Springer Nature. (b) synthesis of CuBDC nanosheets by interfacial reaction. The left section shows various liquid layers spatial arrangement. The right section illustrates SEM and XRD of the created MOF sheets. Source: Rodenas et al. [118]/with permission of Springer Nature. (c) Scheme of the TpEBr and TpPa-SO_3Na nanosheets fabrication. Source: Ying et al. [18]/with permission of American Chemical Society. (d) Scheme of the few-layer Cu(dhbc)$_2$(bpy)·H_2O MONs preparation at the air/water interface. Source: Ying et al. [119]/with permission of Wiley-VCH GmbH.

resulted in the formation of thin MOF layers (5–25-nm thicknesses and 0.5–4-mm lateral dimensions) in the middle layer. This general synthesis methodology may be modified to manufacture various types of MOF nanosheets by changing the relevant components.

Zhao et al. [18] successfully synthesized two types of iCONs (TpEBr and TpPa-SO_3Na) by bottom-up strategy. In this regard, an aqueous solution containing ethidium bromide and trifluoroacetic acid was poured into a polytetrafluoroethylene dish as the subphase, and toluene solution with 1,3,5-triformylphloroglucinol was subsequently spread on the subphase dropwise to cover the surface. The reaction system was left undisturbed for 48 hours at room temperature and large-area thin nanosheets (TpEBr) appeared on the water surface. In another study, Zhao and coworkers [119] also reported the preparation of MOF nanosheets (Cu(dhbc)$_2$(bpy) H_2O) at air/water interface by pouring Cu(NO_3)$_2$·3H_2O aqueous solution into a polytetrafluoroethylene dish as a bottom phase and then, adding diethyl ether solutions (containing 4,4′-bpy and Hdhbc) consecutively on the aqueous bottom phase. The MOF nanosheets were generated successfully after three days.

Using the bottom-up approach, it is possible to prepare large-area nanosheets with controllable thickness, lesser flaws without strong force-generating fragmentation, or morphological damage. However, it is exceedingly challenging to find a suitable reaction system (including the surfactant, structure-directing agent, solvent and the ratio of co-solvent, etc.) to produce the nanosheets when using the bottom-up method.

7.4 Preparation of 2D-Materials Membranes

High-quality 2D nanosheets are the important building blocks of 2D membranes and a careful approach for membrane preparation to produce the channels. The selection of a proper preparation method for membrane production is an essential point and affects membrane performance. Up till now, membrane preparation strategies are divided into two main groups, top-down and bottom-up techniques. In the top-down method, nanosheets are firstly prepared and subsequently deposited on a suitable substrate. In the bottom-up preparation method, on the other hand, nanosheet membranes can be prepared directly from the seeds (nanoparticles [NPs] or nanosheets) or seed-free methods. In this section, we elaborately focus on 2D membrane preparation strategies.

7.4.1 Top-Down Method

2D nanosheets have the unique feature of being solution processable. Their suspensions may be readily transformed into lamellar structures by various procedures [106, 122]. It is possible to prepare lamellar membranes parallel to the support with the nanosheets so that the mass transport resistance can be decreased either by developing suitable in-plane pores or by reducing the membrane thickness [74]. Interlocked and parallel layers with reduced edge-to-edge nonselective gaps should be maintained in a perfect lamellar membrane structure. Filtration, coating, and LbL can be employed to prepare the membranes in a top-down manner.

7.4.1.1 Filtration-Assisted Assembly

Filtration-assisted assembly method is one of the most commonly reported procedures for the preparation of lamellar membranes. This method can be classified into two techniques based on the driving force, namely, vacuum filtration and pressure filtration. In vacuum filtration, a downward vertical force is implemented to assist 2D nanosheets to be interlocked on the substrate, while an upward vertical force has the role of interlocking nanosheets on the substrate during pressure filtration. By changing the deposition quantities in both techniques, the thickness of the membrane can be efficiently adjusted from the nanometer to the submicrometer.

Vacuum-assisted self-assembly (VASA) is the most common filtration method suitable for all kinds of nanosheets (Figure 7.15a). For example, Jiang and coworkers [123] prepared a GO–MOF composite 2D membrane by vacuum filtration technique. First, they synthesized GO flakes and CuBDC nanosheets separately and dispersed them together to achieve a concentrated suspension. Subsequently, CuBDC–GO membranes were made by stacking mixed nanosheet dispersions

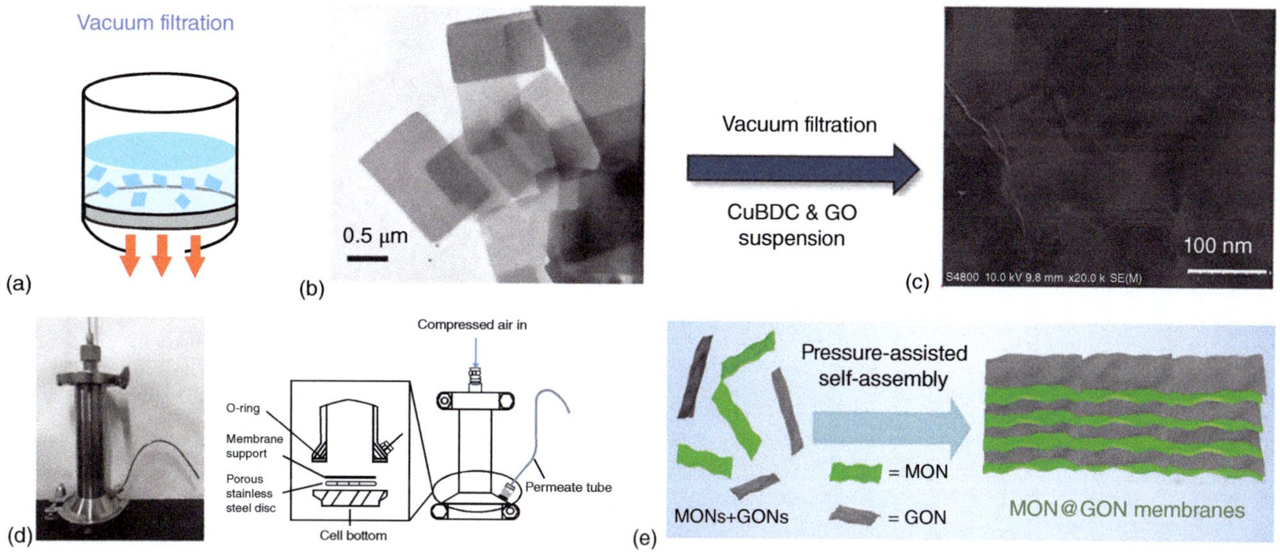

Figure 7.15 (a) Scheme of vacuum filtration method. Source: Wang et al. [113]/with permission of Royal Society of Chemistry. (b) TEM image of CuBDC nanosheets. (c) Top-view SEM image of CuBDC-GO membrane. Source: [124]. Yang F et al., (2018)/with permission of Royal Society of Chemistry. (d) PASA setup for the membrane preparation and internal configuration. (e) Scheme of MON@GON membranes prepared by compositing MONs with GONs through PASA method. Source: Ying et al. [119]/withpermission of Wiley-VCH GmbH.

onto Al_2O_3 support or nylon-6 for 1.5 hours via a vacuum filtration system using 0.05 MPa of driving force (Figure 7.15b,c). Yu et al. [77] and Jiang et al. [124] also implemented the vacuum filtration strategy to interlock GO and GO–PEGDA suspension on the anodic aluminum oxide (AAO) and polyethersulfone (PES) substrates, respectively. These membranes were evaluated for gas separations.

Pressure-assisted self-assembly (PASA) is also reported in the literature as an effective top-down method for membrane fabrication. In this method, the driving force (pressure) can be conveniently tuned to control the membrane structure and compactness. For instance, Zhao et al. [119] fabricated MON@GON membranes by the pressure filtration method. For this purpose, certain volumes of metal-organic nanosheet (MON) suspension and graphene oxide nanosheet (GON) suspension were mixed in isopropanol and then shaken for 5 min to reach a homogeneous suspension. Finally, the suspension was filtrated out by operating the PASA filtration process (Figure 7.15d) at a constant pressure difference of 1 bar on AAO or α-Al_2O_3 as substrate (Figure 7.15e). Hung's group [125] utilized PASA method to stack GO flakes on the modified polyacrylonitrile (PAN) substrate, creating GO_{PASA}/mPAN membranes. PASA was operated at a pressure difference (ΔP) of 760 Torr.

7.4.1.2 Coating

Different coating processes, including spray coating, drop coating, spin coating, casting, etc., have been used for the assembly of 2D nanosheets into lamellar membranes. A successful uniform coating depends on the substrate smoothness, applied evaporation process, and surface tension of the coating solution.

Among the mentioned methods, spin coating (Figure 7.16a) might use shear and centrifugal forces to regulate nanosheet construction, generating a well-interlocked, organized lamellar structure [74]. The substrate rotation speed and coating time determine the performance and morphology of the membrane. As coating strategies can be simply operated and adjusted, they have been used extensively in 2D membrane preparation. For instance, Pan et al. used spin coating to assemble 2D MOF

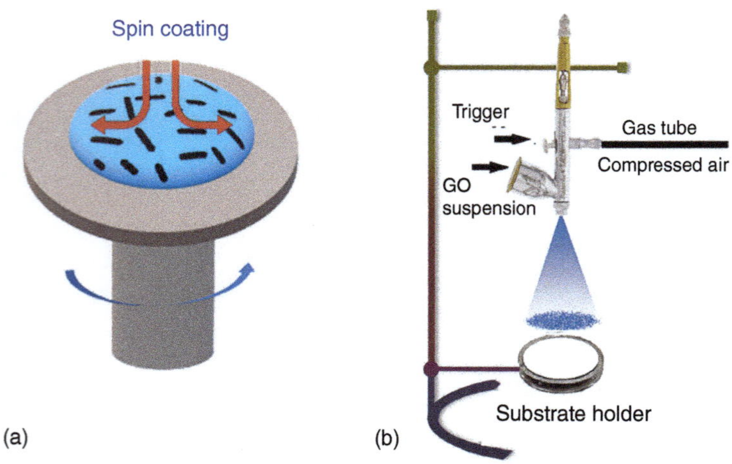

Figure 7.16 (a) Scheme of the spin-coating technique. Source: Wang et al. [113]/with permission of Royal Society of Chemistry. (b) Scheme of GO membrane synthesized by spray coating method. Source: Ibrahim et al. [126]/with permission of Elsevier B.V.

nanosheets atop PAN substrate assisted by sodium alginate [127]. Lin and coworkers [126] utilized spray-coating technology for the deposition of GO solution on the hydrophilic polyester track-etched (PETE) substrates preparation (Figure 7.16b). However, the GO membranes created by spray coating displayed more evident and denser extrinsic wrinkles than VASA-prepared GO membranes. This suggests that although spray coating may be advantageous for large-scale production of 2D membranes, it may cause inhomogeneity in the resulting membranes, lowering their performance. The solution-casting method, which is frequently utilized in polymer membrane fabrication, is another coating process, which holds scale-up potential and can produce highly homogeneous membranes. This approach is now also used for the production of highly ordered lamellar membranes [122].

7.4.1.3 Layer-by-Layer Assembly

The LbL assembly method is mostly used to deposit different types of materials on the substrate. This technique depends mostly on interactions among the neighboring layers, including electrostatic interactions, covalent interactions, or hydrogen bonding [128]. The LbL technique can regulate the thickness of the selected layer accurately with different numbers of deposition cycles and is useful to insert stabilizing forces between the layers [40, 128]. The resulting membranes can, therefore, stay stable in organic or aqueous environments. However, the application of this technology needs material interactions and therefore takes a long time to prepare.

Zhao and coworkers [18] used an LbL assembly method, Langmuir–Schaefer (LS) transfer, to prepare thin iCON membranes on α-Al_2O_3 substrate (Figure 7.17a). The support was initially contacted face-down at the interface with TpEBr nanosheets, then pulled out until it was entirely separated from the surface. The initial iCON layer on the support was brushed with compressed air to remove surface water and then heated at 60 °C for 10 minutes under a dynamic vacuum. The anionic TpPa-SO_3Na iCONs were subsequently transferred via the same technique as for the first TpEBr layer (Figure 7.17b). As a result, they were able to transfer multiple layers of different iCONs onto the support by using this approach.

7.4.2 Bottom-Up Method

The bottom-up method is a direct synthesis technique that circumvents some weaknesses of the top-down method, such as structural deterioration, inefficient intercalation, morphological damage, interleaved pore arrays, and randomly

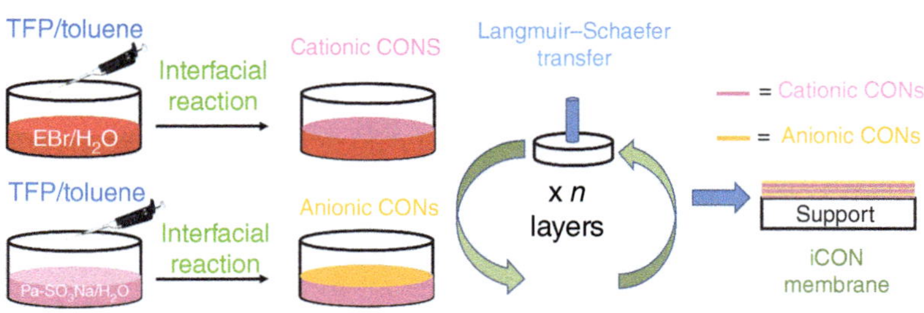

Figure 7.17 Scheme of the Layer-by-Layer assembly of TpEBr@TpPa-SO_3Na iCON membranes via a Langmuir–Schaefer transfer manner. Source: Ying et al. [18]/with permission of American Chemical Society.

orientation of the obtained nanosheet membranes. The bottom-up approach is straightforward to prepare and scale-up and may result in high-aspect-ratio membranes with higher yield and lower cost. Therefore, researchers pay much attention to this strategy nowadays [129].

Zhang and coworkers [130] fabricated a $Zn_2(bIm)_4$ ZIF nanosheet tubular membrane based on GO-guided self-conversion of ZnO nanoparticles (NPs) (Figure 7.18a). They created the novel tubular membrane in three steps. Firstly, they prepared a sol containing zinc to coat the inner wall of the porous tube by using a dip-coating technique. They then dried the coating at 100 °C for one hour to obtain a ZnO layer. Secondly, a dual-layer ZnO NPs@GO was assembled on the porous tube by coating the ZnO NP layer with GO solution and drying at 100 °C for four hours. Finally, the ZnO NPs@GO tube was placed in the synthesis solution, containing benzimidazole (bIm), NH_4OH, MeOH, and toluene to form ZIF nanosheets at 100 °C for the desired reaction time. The prepared 2D MOF membranes in tubular form via the bottom-up method displayed good H_2/CO_2 separation performance.

The same group [131] successfully synthesized 2D ZIF membranes on porous hollow-fiber substrate by bottom-up method (Figure 7.18b). They dip coated a solution of $Zn(OAc)_2$ and monoethanolamine in ethylene glycol monomethyl ether on α-Al_2O_3 tubular substrate. After drying the dip-coated tubular substrate at 100 °C for one hour, it was then calcined at 400°C in air for two hours. Finally, the ZnO NPs-coated porous Al_2O_3 tube was located in synthesis solution comprising

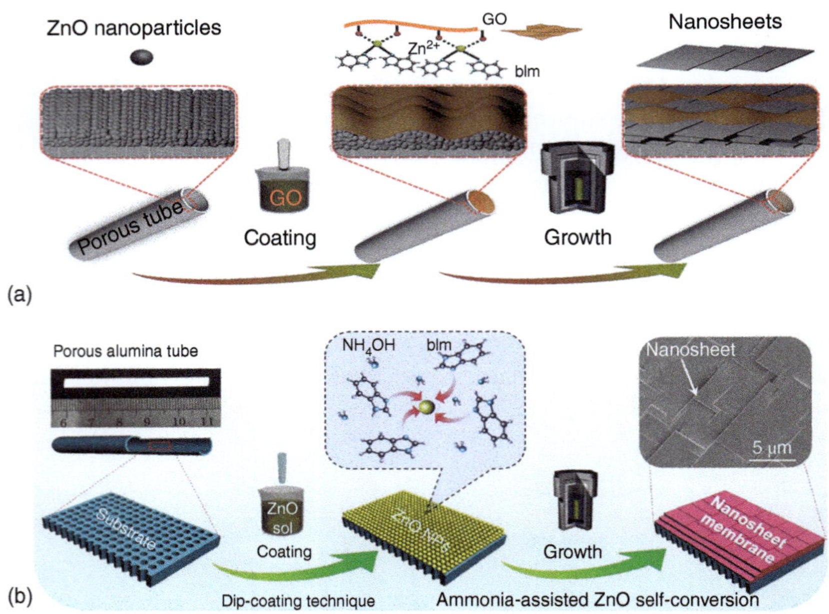

Figure 7.18 (a) Scheme of the highly oriented $Zn_2(bIm)_4$ nanosheet membranes prepared via the ZnO self-conversion growth in a GO confined space. Source: Li et al. [130]/Royal Society of Chemsitry/CC BY-3.0. (b) Scheme of 2D ZIF nanosheet membranes prepared by bottom-up method. Source: Li et al. [131]/with permission of Springer Nature.

bIm, NH_4OH, methanol, and toluene with specific molar composition to produce nucleation sites for the growth of a uniform nanosheet membrane at 100 °C. The resulting ZIF 2D nanosheet tubular membrane has high H_2/CO_2 separation performance and only a 50-nm thickness.

7.5 Gas Separations

7.5.1 H_2/CO_2, H_2/N_2, and H_2/CH_4 Separations

H_2 is a sustainable and high-density energy carrier that respects the environment. Thanks to the clean energy release mechanism with water as the only product, H_2 receives tremendous attention as a green energy carrier. Nowadays, more than 85% of H_2 are manufactured by steam gas reforming, while the remainder is based on biomass and coal conversion [132]. However, the gas impurities in the mixture are a crucial issue for H_2 generation. There are some gases, such as CO_2, N_2, and CH_4, in H_2 generation processes that should be eliminated to collect high-purity H_2. Accordingly, several industrial processes have been used to purify H_2 from these gases. For example, syngas purification, recovering and reusing hydrogen from the synthetic ammonia process, and steam methane reforming are currently utilized to remove H_2 from CO_2, N_2, and CH_4, respectively [132, 133]. However, the mentioned industrial methods are not able to purify H_2 from these gases. Furthermore, for environmental protection and energy development, the effective separation of H_2 from those mixed gases (CO_2, N_2, and CH_4) is critically essential. A comprehensive review of gas-separation system (H_2/CO_2, H_2/N_2, and H_2/CH_4) by 2D-materials membranes is shown in Tables 7.1, 7.2, and 7.3, respectively.

7.5.2 CO_2/N_2 and CO_2/CH_4 Separations

The huge CO_2 emissions from human activities, in recent decades, have raised major concerns since it is one of the main sources of global warming as a greenhouse gas. The atmospheric CO_2 level has grown rapidly from 280 ppm to more than 400 ppm since pre-industrial times to 2019 [10]. Carbon capture and storage attempts to minimize CO_2 emissions by selectively capturing and storing it in appropriate locations from combustion-generated gas blends [141]. The separation of CO_2/N_2 and CO_2/CH_4 is of major relevance for carbon capture and natural gas purification, respectively. Natural gas has been considered one of the most important energy sources due to its large supply and clean combustion process. The world demand for natural gas is predicted to be 32 billion cubic feet in 2040, with the majority used in petrochemicals and power stations [142]. Raw natural gas is, however, a complex combination of hydrocarbons and unwanted contaminants, including H_2S, CO_2, N_2, and H_2O. The level of CO_2 in natural gas, depending on the source, can be varied from 0.06 to 42.66 mol% and should be lowered to less than 2 mol%, to fulfill the pipeline specifications. Removing CO_2 from flue gas and natural gas through membrane technology has seen quick progress. A comprehensive review of gas-separation performance on these systems (CO_2/N_2 and CO_2/CH_4) by 2D-materials membranes is shown in Tables 7.4 and 7.5, respectively.

Table 7.1 H_2/CO_2 separation performance of 2D nanosheet-based membranes.

2D nanosheet	Nanosheet preparation method	Support	Membrane preparation method	Thickness (nm)	H_2 permeance (GPU)	Separation factor/ideal selectivity*	Test condition (trans-membrane pressure and temperature)	Refs.
TMD (MoS_2)	Chemical exfoliation	α-Al_2O_3	Vacuum filtration	60 35 17	2448 7045 27 433	4.4* 3.7* 3.4*	$\Delta P = 0.2$ bar $T = 35\,°C$	[134]
MXene ($Ti_3C_2T_x$)	HF etching + sonication	AAO	Vacuum filtration	N.A	1105	167	$\Delta P = 0$ bar $T = 25\,°C$	[21]
MXene ($Ti_3C_2T_x$)	HCl and LiF etching + sonication	α-Al_2O_3	Vacuum filtration	20	1195 1582	20 27*	$\Delta P = 0.5$ bar $T = 25\,°C$	[102]
MOF ($Zn_2(bim)_4$)	Wet-ball milling + sonication	α-Al_2O_3	Hot-drop coating	N.A	2687	291	$\Delta P = 0$ bar $T = 25\,°C$	[83]
MOF (MAMS-1)	Freeze–thaw exfoliation	α-Al_2O_3	Hot-drop coating	12	5762–7642	30–40	$\Delta P = 0$ bar $T = 25\,°C$	[85]
MOF ($Zn_2(bim)_4$)	Wet-ball milling + sonication	α-Al_2O_3	Hot-drop coating	10	2390	166	$\Delta P = 0$ bar $T = 120\,°C$	[84]
GO	Sonication	α-Al_2O_3	Vacuum filtration	9	~300	3400	$\Delta P = 0$ bar $T = 20\,°C$	[77]
iCONs (TpEBr + TpPa-SO_3Na)	Interfacial reaction	α-Al_2O_3	Layer-by-Layer assembly	41	2566	22.6	$\Delta P = 0.2$ bar $T = 150\,°C$	[18]
COF (CTF-1) + GO	Wet-ball milling + ultrasonication	Isopore cellulose acetate	Vacuum filtration	100	5075	17.4	$\Delta P = 3$ bar $T = 25\,°C$	[135]
COF (CTF-BTD) + GO	Wet-ball milling + ultrasonication	AAO	Vacuum filtration	140–260	655.6	43.1*	$\Delta P = 1$ bar $T = 150\,°C$	[136]

GO	Sonication	α-Al_2O_3	External force driven assembly	N.A	~0.0011	~33	$\Delta P = 0.2$ bar $T = 25\,°C$	[62]
GO + g-C_3N_4	Thermal oxidation + sonication	AAO	Vacuum filtration	700	650	39.2*	$\Delta P = 0$ bar $T = 25\,°C$	[97]
MOF (CuBDC) + GO	Sonication	α-Al_2O_3	Vacuum filtration	700	2870	95.1	$\Delta P = 0$ bar $T = 25\,°C$	[123]
g-C_3N_4	Chemical vapor deposition	AAO	Vacuum filtration	~128	1500	10	$\Delta P = 1$ bar $T = 250\,°C$	[96]
MOF ($Zn_2(bim)_4$)	N.A	α-Al_2O_3	Solvothermal	50	610	53.34*	$\Delta P = 0$ bar $T = 30\,°C$	[131]
MOF ($Zn_2(bim)_4$) + GO	N.A	α-Al_2O_3	Solvothermal	~200	418	89	$\Delta P = 0$ bar $T = 30\,°C$	[130]
MOF ($Co_2(bim)_4$)	N.A	α-Al_2O_3	Vapor phase transformation	~57	514	42.7	$\Delta P = 0$ bar $T = 30\,°C$	[129]
COFs (COF-LZU1 + ACOF-1)	N.A	α-Al_2O_3	Solvothermal	N.A	732	26.7*	$\Delta P = 0$ bar $T = 25\,°C$	[87]

Table 7.2 H_2/N_2 separation performance of 2D nanosheet-based membranes.

2D nanosheet	Nanosheet preparation method	Support	Membrane preparation method	Thickness (nm)	H_2 permeance (GPU)	Separation factor/ideal selectivity*	Test condition (trans-membrane pressure and temperature)	Refs.
GO	Chemical oxidization	α-Al_2O_3	Vacuum filtration	9	896	900	$\Delta P = 0$ bar $T = 20\,°C$	[77]
GO	Chemical oxidization	α-Al_2O_3	Hydrothermal self-assembly crosslink	350	2090	149	$\Delta P = 0.2$ bar $T = 25\,°C$	[137]
MOF ($Ni_3(HITP)_2$)	Soft-physical exfoliation	AAO	Vacuum filtration	50	568	8.5	$\Delta P = 0.2$ bar $T = 25\,°C$	[138]
COF (CTF-BTD) + GO	Wet-ball milling + ultrasonication	AAO	Vacuum filtration	140–260	655.6	18*	$\Delta P = 1$ bar $T = 150\,°C$	[136]
GO + g-C_3N_4	Thermal oxidation + sonication	AAO	Vacuum filtration	700	650	19.3*	$\Delta P = 0$ bar $T = 25\,°C$	[97]
g-C_3N_4	Chemical vapor deposition	AAO	Vacuum filtration	~128	1500	50	$\Delta P = 1$ bar $T = 250\,°C$	[96]
MOF ($Zn_2(bim)_4$) + GO	N.A	α-Al_2O_3	Solvothermal	~200	418	103	$\Delta P = 0$ bar $T = 30\,°C$	[130]
MOF (ZIF-L)	N.A	α-Al_2O_3	Solvothermal	N.A	711	177	$\Delta P = 0$ bar $T = 25\,°C$	[139]
MOF ($Zn_2(bim)_4$)	N.A	α-Al_2O_3	Solvothermal	50	610	66.55*	$\Delta P = 0$ bar $T = 30\,°C$	[131]
MOF ($Co_2(bim)_4$)	N.A	α-Al_2O_3	Vapor phase transformation	~57	514	38.4	$\Delta P = 0$ bar $T = 30\,°C$	[129]
COFs (COF-LZU1 + ACOF-1)	N.A	α-Al_2O_3	Solvothermal	N.A	731	83.9	$\Delta P = 0$ bar $T = 25\,°C$	[87]

Table 7.3 H_2/CH_4 separation performance of 2D nanosheet-based membranes.

2D Nanosheets	Nanosheet preparation method	Support	Membrane preparation method	Thickness (nm)	H_2 permeance (GPU)	Separation factor/ideal selectivity*	Test condition (trans-membrane pressure and temperature)	Refs.
MXene ($Ti_3C_2T_x$)	Sonication + HF etching	AAO	Vacuum filtration	N.A	956	324	$\Delta P = 0$ bar $T = 25\,°C$	[21]
GO	Chemical oxidization	α-Al_2O_3	Hydrothermal crosslink	350	1911	201	$\Delta P = 0.2$ bar $T = 25\,°C$	[137]
Graphene	Chemical vapor deposition	Porous tungsten	Spin coating + plasma treatment	80	3400	25.1	$\Delta P = 1$ bar $T = 150\,°C$	[140]
Graphene	Chemical vapor deposition	Porous tungsten	Spin coating	1	1224	25*	$\Delta P = 0$ bar $T = 150\,°C$	[60]
COF (CTF-BTD) + GO	Ball milling + ultrasonication	AAO	Vacuum filtration	140–260	655.6	26.1*	$\Delta P = 1$ bar $T = 150\,°C$	[136]
g-C_3N_4	Chemical vapor deposition	AAO	Vacuum filtration	~128	1500	60	$\Delta P = 1$ bar $T = 250\,°C$	[96]
MOF ($Zn_2(bim)_4$)	N.A	α-Al_2O_3	Solvothermal	50	610	89.75*	$\Delta P = 0$ bar $T = 30\,°C$	[131]
MOF ($Zn_2(bim)_4$) + GO	N.A	α-Al_2O_3	Solvothermal	~200	418	221	$\Delta P = 0$ bar $T = 30\,°C$	[130]
MOF ($Co_2(bim)_4$))	N.A	α-Al_2O_3	Vapor phase transformation	~57	514	51.3	$\Delta P = 0$ bar $T = 30\,°C$	[129]

Table 7.4 CO_2/N_2 separation performance of 2D nanosheet-based membranes.

2D nanosheets	Nanosheet preparation method	Support	Membrane preparation method	Thickness (nm)	CO_2 permeance (GPU)	Separation factor/ideal selectivity*	Test condition (trans-membrane pressure and temperature)	Refs.
MXene ($Ti_3C_2T_x$)	Sonication + HCl and LiF etching	α-Al_2O_3	Vacuum filtration	20	359	12.1*	$\Delta P = 0.5$ bar $T = 25\,°C$	[102]
MOF ($Cu(dhbc)_2(bpy)$) and GO	Interfacial reaction	α-Al_2O_3	Pressure assisted self-assembly	100	1144	22.8	$\Delta P = 0.4$ bar $T = 25\,°C$	[119]
GO	Chemical oxidization	AAO	Vacuum filtration	380	260	529	$\Delta P = 0.6$ bar $T = 0\,°C$	[143]
GO	Chemical oxidization	Polyethersulfone	Vacuum filtration	9	1015	680	$\Delta P = 0$ bar $T = 40\,°C$	[68]
Layered double hydroxide (MgAl–LDH)	Intercalating exfoliation	Poly(tetrafluoroethylene)	LbL + spray coating	~50	~7000	~86	$\Delta P = 0.27$ bar $T = 25\,°C$	[111]
GO–SILM	Sonication	AAO	Vacuum filtration	1050	68.5	382	$\Delta P = 0$ bar $T = 25\,°C$	[144]

Table 7.5 CO_2/CH_4 separation performance of 2D nanosheet-based membranes.

2D nanosheets	Nanosheet preparation method	Support	Membrane preparation method	Thickness (nm)	CO_2 permeance (GPU)	Separation factor/ideal selectivity*	Test condition (trans-membrane pressure and temperature)	Refs.
MXene ($Ti_3C_2T_x$)	Sonication + HCl and LiF etching	α-Al_2O_3	Vacuum filtration	20	359	15.3*	$\Delta P = 0.5$ bar $T = 25\,°C$	[102]
GO	Chemical oxidization	AAO	Vacuum filtration	380	260	262	$\Delta P = 0.6$ bar $T = 0\,°C$	[143]
MOF ($Cu(dhbc)_2(bpy)$) and GO	Interfacial reaction	α-Al_2O_3	Pressure-assisted assembly	100	1144	19.6	$\Delta P = 0.4$ bar $T = 25\,°C$	[119]
Layered double hydroxide (MgAl–LDH)	Intercalating exfoliation	Poly(tetrafluoroethylene)	LbL + spray coating	~50	~7000	~62	$\Delta P = 0.27$ bar $T = 25\,°C$	[111]
GO–PEGDA)	Sonication	PES	Vacuum filtration	110	220	75	$\Delta P = 15$ bar $T = 25\,°C$	[124]
GO–SILMs	Sonication	AAO	Vacuum filtration	1050	68.5	234	$\Delta P = 0$ bar $T = 25\,°C$	[144]

Table 7.6 *P*-xylene/*o*-xylene, He/H_2, He/N_2, *n*-butane/isobutane, H_2/C_3H_6, CO_2/H_2, H_2/C_3H_8 and H_2/C_3H_5 separation performance of 2D-materials membranes.

2D nanosheets	Nanosheet preparation method	Support	Membrane preparation method	Thickness (nm)	Permeance (GPU)	Separation factor/ideal selectivity*	Test condition (trans-membrane pressure & temperature)	Refs.
Zeolite (MFI)	Melt compounding	α-Al_2O_3	Vacuum filtration + secondary growth	200	*p*-xylene: 1045	*p*-xylene/ *o*-xylene: 40–70	$\Delta P = 4$ bar $T = 85–150\,°C$	[145]
Zeolite (MWW)	Melt compounding	α-Al_2O_3	Vacuum filtration + secondary growth	N.A.	He: 15.52	He/H_2: 3* He/N_2: 17*	$\Delta P = 4$ bar $T = 200\,°C$	[145]
Zeolite (MFI)	Melt compounding	α-Al_2O_3	Vacuum filtration + secondary growth	200	*p*-xylene: 1195	*p*-xylene/ *o*-xylene: 25–45	$\Delta P = 0$ bar $T = 150\,°C$	[146]
Zeolite (MFI)	Melt compounding	Polybenzimidazole	Vacuum filtration	N.A.	*n*-butane: 1045	*n*-butane/ isobutane: 5.4*	$\Delta P = 0$ bar $T = 25\,°C$	[147]
Zeolites (MFI)	Melt compounding	Si wafer	Vacuum filtration + secondary growth	<100	*n*-butane: 3791	*n*-butane/ isobutane: 15	$\Delta P = 0$ bar $T = 150\,°C$	[148]
MXene ($Ti_3C_2T_x$)	HF etching + sonication	AAO	Vacuum filtration	2000	H_2:1165 H_2: 1165	H_2/C_3H_5: 782 H_2/C_3H_8: 1102	$\Delta P = 0$ bar $T = 25\,°C$	[21]
Zeolite (MFI)	Seeded growth	Sintered silica fiber	Floating coating + secondary growth	500–1000	*p*-xylene: 657 *n*-butane: 866	*p*-xylene/ *o*-xylene: >10 000 *n*-butane/ isobutane: 24	$\Delta P = 0.5$ bar $T = 150\,°C$	[149]

GO	Sonication	α-Al_2O_3	Vacuum-spin + external force driven assembly	N.A.	H_2: 0.0011	H_2/C_3H_8: ~270	$\Delta P = 0.2$ bar $T = 25\,°C$	[62]
Layered double hydroxide (MgAl–LDH)	Intercalating exfoliation	Poly(tetrafluoroethylene)	LbL + spray coating	~50	CO_2: 7000	CO_2/H_2: 43	$\Delta P = 0.27$ bar $T = 25\,°C$	[111]
Zeolite (MFI)	Hydrothermal syntheses	Sintered silica fiber	Vacuum filtration	10–40	*p*-xylene: 1672	*p*-xylene/*o*-xylene: 1989	$\Delta P = 0.0053$ bar $T = 150\,°C$	[4]
GO + g-C_3N_4	Thermal oxidation + sonication	AAO	Vacuum filtration	700	H_2: 650	H_2/C_3H_8: 306 H_2/C_3H_6: 221	$\Delta P = 0$ bar $T = 25\,°C$	[97]
GO–SILMs	Sonication	AAO	Vacuum filtration	1050	CO_2: 68.5	CO_2/H_2: 24	$\Delta P = 0$ bar $T = 25\,°C$	[144]

7.5.3 Other Gas/Vapor Separations

Although lots of attention have been paid to the separation of H_2/CO_2, H_2/N_2, H_2/CH_4, CO_2/N_2, and CO_2/CH_4 due to their importance in the industries, source of clean energy and environmental situation, other types of separations can be accomplished by 2D-materials membranes. Recently, researchers have synthesized different kinds of 2D-materials membranes for *p*-xylene/*o*-xylene, *n*-butane/isobutane, H_2/C_3H_5 separations, alkanes, *and so on*. Table 7.6 illustrates a comprehensive review of recent gas/vapor separation systems by 2D-materials membranes.

7.6 Conclusions and Perspectives

In summary, this chapter reviews the most recent advancements in 2D-materials membranes, with an emphasis on membrane materials, membrane structural designs, membrane production techniques, unique membrane transport properties, and industrially appealing gas-separation processes. Nanoporous thin membranes have shown orders of magnitude more permeance than traditional polymeric membranes, enhancing existing membrane technology for large-scale and energy-efficient separation. Furthermore, due to their ultrathin thicknesses and size-selective interplanar spacing and/or in-plane slit-like holes, 2D lamellar membranes fabricated from diverse nanosheet building blocks have a low mass-transfer barrier and great separation performance. As a result, many 2D membranes exhibit both excellent selectivity and permeance or permeability, exceeding both micrometer-thick polycrystalline membranes and polymeric membranes in the literature. It is expected that more studies will be committed to this topic in the future, resulting in the creation of additional 2D membranes with improved separation performance. Despite significant advances over the last few years, 2D-materials membranes are still in the infancy level, with significant problems and constraints impeding their industrial application.

As responses to the current challenges, several perspectives linked to the ongoing development of 2D-materials membranes are emphasized in this chapter.

(1) Effective ways for obtaining sufficient quantities of high-quality building blocks to create large-area and defect-free membranes are urgently needed. This also pushes researchers to find universal solutions to collect 2D building blocks since the structural defects caused by the nanosheet preparation process (mechanical exfoliation) may result in a considerable loss of membrane selectivity.
(2) Scale-up the preparation of 2D-materials membranes and assemble into membrane modules are urgently required. Researchers should put effort into the invention of scaling-up methods for membrane preparation. In situ growth of the 2D materials on a hollow fiber, tube, or flexible flat plate is a good solution to realize the scaling-up and assemble into a module for commercial or industrial applications.

(3) Significant efforts should be applied to develop robust membranes with stable performance under realistic operating conditions. Currently, 2D-materials membranes are mostly tested in ideal conditions, e.g. zero trans-membrane pressure and room temperature. The technology development history of polymeric membranes reveals that membrane performance assessed in research laboratories under relatively perfect settings differs significantly from that found in the industry under rigorous conditions. Hence, fabricating robust membranes with high mechanical strength and high stability (e.g. chemical and thermal) is practically more important than selectivity and permeance.

References

1 Koros, W.J. and Zhang, C. (2017). Materials for next-generation molecularly selective synthetic membranes. *Nat. Mater.* 16 (3): 289–297.

2 Peng, Y. and Yang, W. (2020). 2D Metal-Organic framework materials for membrane-based separation. *Adv. Mater. Interfaces* 7 (1): 1901514.

3 Novoselov, K.S., Geim, A.K., Morozov, S.V. et al. (2004). Electric field effect in atomically thin carbon films. *Science* 306 (5696): 666–669.

4 Jeon, M.Y., Kim, D., Kumar, P. et al. (2017). Ultra-selective high-flux membranes from directly synthesized zeolite nanosheets. *Nature* 543 (7647): 690–694.

5 Min, B., Yang, S., Korde, A. et al. (2019). Continuous zeolite MFI membranes fabricated from 2D MFI nanosheets on ceramic hollow fibers. *Angew. Chem. Int. Ed.* 58 (24): 8201–8205.

6 Xu, L. and Sun, J. (2016). Recent advances in the synthesis and application of two-dimensional zeolites. *Adv. Energy Mater.* 6 (17): 1600441.

7 Shete, M., Kumar, M., Kim, D. et al. (2017). Nanoscale control of homoepitaxial growth on a two-dimensional zeolite. *Angew. Chem. Int. Ed.* 56 (2): 535–539.

8 Peng, L., Xu, Z., Liu, Z. et al. (2015). An iron-based green approach to 1-h production of single-layer graphene oxide. *Nat. Commun.* 6: 5716.

9 Janakiram, S., Espejo, J.L.M., Yu, X. et al. (2020). Facilitated transport membranes containing graphene oxide-based nanoplatelets for CO_2 separation: effect of 2D filler properties. *J. Membr. Sci.* 616: 118626.

10 Shi, F., Sun, J., Wang, J. et al. (2021). MXene versus graphene oxide: investigation on the effects of 2D nanosheets in mixed matrix membranes for CO_2 separation. *J. Membr. Sci.* 620: 118850.

11 Xie, A., Cui, J., Yang, J. et al. (2020). Graphene oxide/Fe (III)-based metal-organic framework membrane for enhanced water purification based on synergistic separation and photo-Fenton processes. *Appl. Catal., B* 2641: 18548.

12 Wang, Y., Li, L., Wei, Y. et al. (2017). Water transport with ultralow friction through partially exfoliated g-C_3N_4 nanosheet membranes with self-supporting spacers. *Angew. Chem. Int. Ed.* 56 (31): 8974–8980.

13 Fu, J., Xu, Q., Low, J. et al. (2019). Ultrathin 2D/2D WO_3/g-C_3N_4 step-scheme H_2-production photocatalyst. *Appl. Catal., B* 243: 556–565.

14 Lin, R., Li, L., Zhou, H. et al. (2018). Molecular sieving of ethylene from ethane using a rigid metal–organic framework. *Nat. Mater.* 17 (12): 1128–1133.

15 Kallo, M.T. and Lennox, M.J. (2020). Understanding CO_2/CH_4 separation in pristine and defective 2D MOF CuBDC nanosheets via nonequilibrium molecular dynamics. *Langmuir* 36 (45): 13591–13600.

16 Jiang, S., Shi, X., Zu, Y. et al. (2021). Interfacial growth of 2D MOF membranes via contra-diffusion for CO_2 separation. *Mater. Chem. Front.* 5 (13): 5150–5157.

17 Li, H., Gao, K., Mo, B. et al. (2021). Construction of porous 2D MOF nanosheets for rapid and selective adsorption of cationic dyes. *Dalton Trans.* 50 (9): 3348–3355.

18 Ying, Y., Tong, M., Ning, S. et al. (2020). Ultrathin two-dimensional membranes assembled by ionic covalent organic nanosheets with reduced apertures for gas separation. *J. Am. Chem. Soc.* 142 (9): 4472–4480.

19 Fan, H., Peng, M., Strauss, I. et al. (2020). High-flux vertically aligned 2D covalent organic framework membrane with enhanced hydrogen separation. *J. Am. Chem. Soc.* 142 (15): 6872–6877.

20 Li, J., Zhou, X., Wang, J. et al. (2019). Two-dimensional covalent organic frameworks (COFs) for membrane separation: a mini review. *Ind. Eng. Chem. Res.* 58 (34): 15394–15406.

21 Ding, L., Wei, Y., Li, L. et al. (2018). MXene molecular sieving membranes for highly efficient gas separation. *Nat. Commun.* 9: 1–7.

22 Li, J., Li, X., and Van der Bruggen, B. (2020). An MXene-based membrane for molecular separation. *Environ. Sci.: Nano* 7 (5): 1289–1304.

23 Ahmed, Z., Rehman, F., Ali, U. et al. (2021). Recent advances in MXene-based separation membranes. *ChemBioEng Rev.* 8 (2): 110–120.

24 Xing, Y., Akonkwa, G., Liu, Z. et al. (2020). Crumpled two-dimensional $Ti_3C_2T_x$ MXene lamellar membranes for solvent permeation and separation. *ACS Appl. Nano Mater.* 3 (2): 1526–1534.

25 Achari, A., Sahana, S., and Eswaramoorthy, M. (2016). High performance MoS_2 membranes: effects of thermally driven phase transition on CO_2 separation efficiency. *Energy Environ. Sci.* 9 (4): 1224–1228.

26 Su, Y., Liu, D., Yang, G. et al. (2020). Transition metal dichalcogenide (TMD) membranes with ultrasmall nanosheets for ultrafast molecule separation. *ACS Appl. Mater. Interfaces* 12 (40): 45453–45459.

27 Berean, K.J., Ou, J.Z., Daeneke, T. et al. (2015). 2D MoS_2 PDMS nanocomposites for NO_2 separation. *Small* 11 (38): 5035–5040.

28 Krasian, T., Punyodom, W., and Worajittiphon, P. (2019). A hybrid of 2D materials (MoS_2 and WS_2) as an effective performance enhancer for poly(lactic acid) fibrous mats in oil adsorption and oil/water separation. *Chem. Eng. J.* 369: 563–575.

29 Lu, G., Wu, T., Yuan, Q. et al. (2015). Synthesis of large single-crystal hexagonal boron nitride grains on Cu–Ni alloy. *Nat. Commun.* 6: 6160.

30 Song, L., Jia, H., Zhang, H. et al. (2020). Graphene-like h-BN/CdS 2D/3D heterostructure composite as an efficient photocatalyst for rapid removing rhodamine B and Cr (VI) in water. *Ceram. Int.* 46 (15): 24674–24681.

31 Zhang, Y., Shi, Q., Liu, Y. et al. (2015). Hexagonal boron nitride with designed nanopores as a high-efficiency membrane for separating gaseous hydrogen from methane. *J. Phys. Chem. C* 119 (34): 19826–19831.

32 Xu, R., Pang, W., Yu, J. et al. (2009). *Chemistry of Zeolites and Related Porous Materials: Synthesis and Structure*. Wiley.

33 Schulman, E., Wu, W., and Liu, D. (2020). Two-dimensional zeolite materials: structural and acidity properties. *Materials* 13 (8): 1822.

34 Roth, W.J., Nachtigall, P., Morris, R.E. et al. (2013). A family of zeolites with controlled pore size prepared using a top-down method. *Nat. Chem.* 5 (7): 628–633.

35 Choi, M., Na, K., Kim, J. et al. (2009). Stable single-unit-cell nanosheets of zeolite MFI as active and long-lived catalysts. *Nature* 461 (7261): 246–249.

36 Dakhchoune, M., Villalobos, L.F., Semino, R. et al. (2021). Gas-sieving zeolitic membranes fabricated by condensation of precursor nanosheets. *Nat. Mater.* 20 (3): 362–369.

37 Berry, V. (2013). Impermeability of graphene and its applications. *Carbon* 62: 1–10.

38 Yang, K., Wang, J., Chen, X. et al. (2018). Application of graphene-based materials in water purification: from the nanoscale to specific devices. *Environ. Sci.: Nano* 5 (6): 1264–1297.

39 Tarelho, J.P., dos Santos, M.P.S., Ferreira, J.A. et al. (2018). Graphene-based materials and structures for energy harvesting with fluids–a review. *Mater. Today* 21 (10): 1019–1041.

40 Cheng, Y., Pu, Y., and Zhao, D. (2020). Two-dimensional membranes: new paradigms for high-performance separation membranes. *Chem. Asian J.* 15 (15): 2241–2270.

41 Gadipelli, S. and Guo, Z.X. (2015). Graphene-based materials: synthesis and gas sorption, storage and separation. *Prog. Mater. Sci.* 69: 1–60.

42 Schrier, J. (2011). Fluorinated and nanoporous graphene materials as sorbents for gas separations. *ACS Appl. Mater. Interfaces* 3 (11): 4451–4458.

43 Koenig, S.P., Wang, L., Pellegrino, J. et al. (2012). Selective molecular sieving through porous graphene. *Nat. Nanotechnol.* 7 (11): 728–732.

44 Boutilier, M.S., Jang, D., Idrobo, J.-C. et al. (2017). Molecular sieving across centimeter-scale single-layer nanoporous graphene membranes. *ACS Nano* 11 (6): 5726–5736.

45 Mortazavi, B., Madjet, M.E., Shahrokhi, M. et al. (2019). Nanoporous graphene: a 2D semiconductor with anisotropic mechanical, optical and thermal conduction properties. *Carbon* 147: 377–384.

46 Zhang, L. and Jaroniec, M. (2020). Strategies for development of nanoporous materials with 2D building units. *Chem. Soc. Rev.* 49 (16): 6039–6055.

47 Jiang, D., Cooper, V.R., and Dai, S. (2009). Porous graphene as the ultimate membrane for gas separation. *Nano Lett.* 9 (12): 4019–4024.

48 Liu, H., Chen, Z., Dai, S. et al. (2015). Selectivity trend of gas separation through nanoporous graphene. *J. Solid State Chem.* 224: 2–6.

49 Tian, Z., Mahurin, S.M., Dai, S. et al. (2017). Ion-gated gas separation through porous graphene. *Nano Lett.* 17 (3): 1802–1807.

50 Emelianov, A., Kireev, D., Levin, D. et al. (2016). The effect of ultraviolet light on structural properties of exfoliated and CVD graphene. *Appl. Phys. Lett.* 109 (17): 173101.

51 Levin, D., Bobrinetskiy, I., Emelianov, A. et al. (2016). Surface functionalization of single-layer and multilayer graphene upon ultraviolet irradiation. *Semiconductors* 50 (13): 1738–1743.

52 Fischbein, M.D. and Drndić, M. (2008). Electron beam nanosculpting of suspended graphene sheets. *Appl. Phys. Lett.* 93 (11): 113107.

53 Zhou, Y., Jadwiszczak, J., Keane, D. et al. (2017). Programmable graphene doping via electron beam irradiation. *Nanoscale* 9 (25): 8657–8664.

54 Dyck, O., Yoon, M., Zhang, L. et al. (2020). Doping of Cr in graphene using electron beam manipulation for functional defect engineering. *ACS Appl. Nano Mater.* 3 (11): 10855–10863.

55 Surwade, S.P., Smirnov, S.N., Vlassiouk, I.V. et al. (2015). Water desalination using nanoporous single-layer graphene. *Nat. Nanotechnol.* 10 (5): 459–464.

56 Zhao, H., Fan, S., Chen, Y. et al. (2017). Oxygen plasma-treated graphene oxide surface functionalization for sensitivity enhancement of thin-film piezoelectric acoustic gas sensors. *ACS Appl. Mater. Interfaces* 9 (46): 40774–40781.

57 Celebi, K., Buchheim, J., Wyss, R.M. et al. (2014). Ultimate permeation across atomically thin porous graphene. *Science* 344 (6181): 289–292.

58 Lobo, D.E., Banerjee, P.C., Easton, C.D. et al. (2015). Miniaturized supercapacitors: focused ion beam reduced graphene oxide supercapacitors with enhanced performance metrics. *Adv. Energy Mater.* 5 (19): 1500665.

59 Wan, J., Huang, L., Wu, J. et al. (2018). Microwave combustion for rapidly synthesizing pore-size-controllable porous graphene. *Adv. Funct. Mater.* 28 (22): 1800382.

60 Huang, S., Dakhchoune, M., Luo, W. et al. (2018). Single-layer graphene membranes by crack-free transfer for gas mixture separation. *Nat. Commun.* 9: 2632.

61 Eda, G., Fanchini, G., and Chhowalla, M. (2008). Large-area ultrathin films of reduced graphene oxide as a transparent and flexible electronic material. *Nat. Nanotechnol.* 3 (5): 270–274.

62 Shen, J., Liu, G., Huang, K. et al. (2016). Subnanometer two-dimensional graphene oxide channels for ultrafast gas sieving. *ACS Nano* 10 (3): 3398–3409.

63 Hummers, W.S. Jr. and Offeman, R.E. (1958). Preparation of graphitic oxide. *J. Am. Chem. Soc.* 80 (6): 1339–1339.

64 Hung, W.-S., Tsou, C.-H., De Guzman, M. et al. (2014). Cross-linking with diamine monomers to prepare composite graphene oxide-framework membranes with varying *d*-spacing. *Chem. Mater.* 26 (9): 2983–2990.

65 Moon, I.K., Kim, J.I., Lee, H. et al. (2013). 2D graphene oxide nanosheets as an adhesive over-coating layer for flexible transparent conductive electrodes. *Sci. Rep.* 3: 1112.

66 Toda, K., Furue, R., and Hayami, S. (2015). Recent progress in applications of graphene oxide for gas sensing: a review. *Anal. Chim. Acta* 878: 43–53.

67 Chi, C., Wang, X., Peng, Y. et al. (2016). Facile preparation of graphene oxide membranes for gas separation. *Chem. Mater.* 28 (9): 2921–2927.

68 Zhou, F., Tien, H.N., Xu, W.L. et al. (2017). Ultrathin graphene oxide-based hollow fiber membranes with brush-like CO_2-philic agent for highly efficient CO_2 capture. *Nat. Commun.* 8: 2107.

69 Lyu, J., Wen, X., Kumar, U. et al. (2018). Separation and purification using GO and r-GO membranes. *RSC Adv.* 8 (41): 23130–23151.

70 Zhu, Y., Murali, S., Cai, W. et al. (2010). Graphene and graphene oxide: synthesis, properties, and applications. *Adv. Mater.* 22 (35): 3906–3924.

71 Gascho, J.L., Costa, S.F., Recco, A.A. et al. (2019). Graphene oxide films obtained by vacuum filtration: X-ray diffraction evidence of crystalline reorganization. *J. Nanomater.* 2019: 5963148.

72 Hu, M. and Mi, B. (2013). Enabling graphene oxide nanosheets as water separation membranes. *Environ. Sci. Technol.* 47 (8): 3715–3723.

73 Hu, M. and Mi, B. (2014). Layer-by-layer assembly of graphene oxide membranes via electrostatic interaction. *J. Membr. Sci.* 469: 80–87.

74 Kim, H.W., Yoon, H.W., Yoon, S.-M. et al. (2013). Selective gas transport through few-layered graphene and graphene oxide membranes. *Science* 342 (6154): 91–95.

75 Sun, P., Zhu, M., Wang, K. et al. (2013). Selective ion penetration of graphene oxide membranes. *ACS Nano* 7 (1): 428–437.

76 Xin, Z., Chen, X., Wang, Q. et al. (2013). Nanopolyhedrons and mesoporous supra-structures of zeolitic imidazolate framework with high adsorption performance. *Microporous Mesoporous Mater.* 169: 218–221.

77 Li, H., Song, Z., Zhang, X. et al. (2013). Ultrathin, molecular-sieving graphene oxide membranes for selective hydrogen separation. *Science* 342 (6154): 95–98.

78 Yaghi, O.M., Li, G., and Li, H. (1995). Selective binding and removal of guests in a microporous metal–organic framework. *Nature* 378 (6558): 703–706.

79 Zhou, H.-C., Long, J.R., and Yaghi, O.M. (2012). Introduction to metal–organic frameworks. *Chem. Rev.* 112 (2): 673–674.

80 Ahmad, N., Younus, H.A., Chughtai, A.H. et al. (2018). Synthesis of 2D MOF having potential for efficient dye adsorption and catalytic applications. *Catal. Sci. Technol.* 8 (16): 4010–4017.

81 Zhu, H. and Liu, D. (2019). The synthetic strategies of metal–organic framework membranes, films and 2D MOFs and their applications in devices. *J. Mater. Chem. A* 7 (37): 21004–21035.

82 Cheetham, A.K., Rao, C., and Feller, R.K. (2006). Structural diversity and chemical trends in hybrid inorganic–organic framework materials. *Chem. Commun.* 46: 4780–4795.

83 Peng, Y., Li, Y., Ban, Y. et al. (2014). Metal-organic framework nanosheets as building blocks for molecular sieving membranes. *Science* 346 (6215): 1356–1359.

84 Peng, Y., Li, Y., Ban, Y. et al. (2017). Two-dimensional metal–organic framework nanosheets for membrane-based gas separation. *Angew. Chem. Int. Ed.* 56 (33): 9757–9761.

85 Wang, X., Chi, C., Zhang, K. et al. (2017). Reversed thermo-switchable molecular sieving membranes composed of two-dimensional metal-organic nanosheets for gas separation. *Nat. Commun.* 8: 14460.

86 Diercks, C. and Yaghi, O. The atom, the molecule, and the covalent organic framework. *Science* 355 (6328): eaal1585.

87 Fan, H., Mundstock, A., Feldhoff, A. et al. (2018). Covalent organic framework–covalent organic framework bilayer membranes for highly selective gas separation. *J. Am. Chem. Soc.* 140 (32): 10094–10098.

88 Waller, P., Gándara, F., and Yaghi, O. (2015). A pyrene-based, fluorescent three-dimensional covalent organic framework. *Acc. Chem. Res.* 48: 3053–3063.

89 Li, Z., Feng, X., Zou, Y. et al. (2014). A 2D azine-linked covalent organic framework for gas storage applications. *Chem. Commun.* 50 (89): 13825–13828.

90 Huang, N., Chen, X., Krishna, R. et al. (2015). Two-dimensional covalent organic frameworks for carbon dioxide capture through channel-wall functionalization. *Angew. Chem. Int. Ed.* 54 (10): 2986–2990.

91 Feng, X., Ding, X., and Jiang, D. (2012). Covalent organic frameworks. *Chem. Soc. Rev.* 41 (18): 6010–6022.

92 Khattak, A.M., Ghazi, Z.A., Liang, B. et al. (2016). A redox-active 2D covalent organic framework with pyridine moieties capable of faradaic energy storage. *J. Mater. Chem. A* 4 (42): 16312–16317.

93 DeBlase, C.R. and Dichtel, W.R. (2016). Moving beyond boron: the emergence of new linkage chemistries in covalent organic frameworks. *Macromolecules* 49 (15): 5297–5305.

94 Li, Y., Chen, W., Gao, R. et al. (2019). 2D covalent organic frameworks with built-in amide active sites for efficient heterogeneous catalysis. *Chem. Commun.* 55 (96): 14538–14541.

95 Zhi, Y., Wang, Z., Zhang, H.L. et al. (2020). Recent progress in metal-free covalent organic frameworks as heterogeneous catalysts. *Small* 16 (24): 2001070.

96 Villalobos, L.F., Vahdat, M.T., Dakhchoune, M. et al. (2020). Large-scale synthesis of crystalline g-C_3N_4 nanosheets and high-temperature H_2 sieving from assembled films. *Scie. Adv.* 6 (4): eaay9851.

97 Zhou, Y., Zhang, Y., Xue, J. et al. (2021). Graphene oxide-modified g-C_3N_4 nanosheet membranes for efficient hydrogen purification. *Chem. Eng. J.* 420: 129574.

98 Fu, J., Yu, J., Jiang, C. et al. (2018). g-C3N4-based heterostructured photocatalysts. *Adv. Energy Mater.* 8 (3): 1701503.

99 Li, F., Qu, Y., and Zhao, M. (2015). Efficient helium separation of graphitic carbon nitride membrane. *Carbon* 95: 51–57.

100 Naguib, M., Mochalin, V.N., Barsoum, M.W. et al. (2014). 25th anniversary article: MXenes: a new family of two-dimensional materials. *Adv. Mater.* 26 (7): 992–1005.

101 Gogotsi, Y. and Anasori, B. (2019). The rise of MXenes, in. *ACS Nano* 13 (8): 8491–8494.

102 Shen, J., Liu, G., Ji, Y. et al. (2018). 2D MXene nanofilms with tunable gas transport channels. *Adv. Funct. Mater.* 28 (31): 1801511.

103 Ding, L., Wei, Y., Wang, Y. et al. (2017). A two-dimensional lamellar membrane: MXene nanosheet stacks. *Angew. Chem. Int. Ed.* 56 (7): 1825–1829.

104 Dou, H., Jiang, B., Xu, M. et al. (2019). Boron nitride membranes with a distinct nanoconfinement effect for efficient ethylene/ethane separation. *Angew. Chem. Int. Ed.* 58 (39): 13969–13975.

105 Ameen, A.W., Ji, J., Tamaddondar, M. et al. (2021). 2D boron nitride nanosheets in PIM-1 membranes for CO_2/CH_4 separation. *J. Membr. Sci.* 636: 119527.

106 Chen, C., Wang, J., Liu, D. et al. (2018). Functionalized boron nitride membranes with ultrafast solvent transport performance for molecular separation. *Nat. Commun.* 9: 1902.

107 Chen, D., Wang, W., Ying, W. et al. (2018). CO_2-philic WS_2 laminated membranes with a nanoconfined ionic liquid. *J. Mater. Chem. A* 6 (34): 16566–16573.

108 Wang, N., Huang, Z., Li, X. et al. (2018). Tuning molecular sieving channels of layered double hydroxides membrane with direct intercalation of amino acids. *J. Mater. Chem. A* 6 (35): 17148–17155.

109 Lu, P., Liu, Y., Zhou, T. et al. (2018). Recent advances in layered double hydroxides (LDHs) as two-dimensional membrane materials for gas and liquid separations. *J. Membr. Sci.* 567: 89–103.

110 Yu, N., Wang, L., Li, M. et al. (2015). Molybdenum disulfide as a highly efficient adsorbent for non-polar gases. *Phys. Chem. Chem. Phys.* 17 (17): 11700–11704.

111 Xu, X., Wang, J., Zhou, A. et al. (2021). High-efficiency CO_2 separation using hybrid LDH-polymer membranes. *Nat. Commun.* 12: 3069.

112 Yang, Y., Yang, X., Liang, L. et al. (2019). Large-area graphene-nanomesh/carbon-nanotube hybrid membranes for ionic and molecular nanofiltration. *Science* 364 (6445): 1057–1062.

113 Wang, S., Yang, L., He, G. et al. (2020). Two-dimensional nanochannel membranes for molecular and ionic separations. *Chem. Soc. Rev.* 49 (4): 1071–1089.

114 Tan, C., Cao, X., Wu, X.-J. et al. (2017). Recent advances in ultrathin two-dimensional nanomaterials. *Chem. Rev.* 117 (9): 6225–6331.

115 Wang, S., Mahalingam, D., Sutisna, B. et al. (2019). 2D-dual-spacing channel membranes for high performance organic solvent nanofiltration. *J. Mater. Chem. A* 7 (19): 11673–11682.

116 Ogino, I., Yokoyama, Y., Iwamura, S. et al. (2014). Exfoliation of graphite oxide in water without sonication: bridging length scales from nanosheets to macroscopic materials. *Chem. Mater.* 26 (10): 3334–3339.

117 Zhu, W., Gao, X., Li, Q. et al. (2016). Controlled gas exfoliation of boron nitride into few-layered nanosheets. *Angew. Chem. Int. Ed.* 55 (36): 10766–10770.

118 Rodenas, T., Luz, I., Prieto, G. et al. (2015). Metal–organic framework nanosheets in polymer composite materials for gas separation. *Nat. Mater.* 14 (1): 48–55.

119 Ying, Y., Zhang, Z., Peh, S.B. et al. (2021). Pressure-responsive two-dimensional metal–organic framework composite membranes for CO_2 separation. *Angew. Chem. Int. Ed.* 60 (20): 11318–11325.

120 Dong, R., Zhang, T., and Feng, X. (2018). Interface-assisted synthesis of 2D materials: trend and challenges. *Chem. Rev.* 118 (13): 6189–6235.

121 Kandambeth, S., Biswal, B., Chaudhari, H. et al. (2017). Selective molecular sieving in self-standing porous covalent-organic-framework membranes. *Adv. Mater.* 29: 1603945.

122 Akbari, A., Sheath, P., Martin, S.T. et al. (2016). Large-area graphene-based nanofiltration membranes by shear alignment of discotic nematic liquid crystals of graphene oxide. *Nat. Commun.* 7: 10891.

123 Yang, F., Wu, M., Wang, Y. et al. (2018). A GO-induced assembly strategy to repair MOF nanosheet-based membrane for efficient H_2/CO_2 separation. *ACS Appl. Mater. Interfaces* 11 (1): 990–997.

124 Wang, S., Xie, Y., He, G. et al. (2017). Graphene oxide membranes with heterogeneous nanodomains for efficient CO_2 separations. *Angew. Chem. Int. Ed.* 56 (45): 14246–14251.

125 Tsou, C.-H., An, Q.-F., Lo, S.-C. et al. (2015). Effect of microstructure of graphene oxide fabricated through different self-assembly techniques on 1-butanol dehydration. *J. Membr. Sci.* 477: 93–100.

126 Ibrahim, A.F. and Lin, Y. (2018). Synthesis of graphene oxide membranes on polyester substrate by spray coating for gas separation. *Chem. Eng. Sci.* 190: 312–319.

127 Liu, G., Jiang, Z., Cao, K. et al. (2017). Pervaporation performance comparison of hybrid membranes filled with two-dimensional ZIF-L nanosheets and zero-dimensional ZIF-8 nanoparticles. *J. Membr. Sci.* 523: 185–196.

128 Richardson, J.J., Björnmalm, M., and Caruso, F. (2015). Technology-driven layer-by-layer assembly of nanofilms. *Science* 348 (6233): aaa2491.

129 Nian, P., Liu, H., and Zhang, X. (2019). Bottom-up fabrication of two-dimensional Co-based zeolitic imidazolate framework tubular membranes consisting of nanosheets by vapor phase transformation of Co-based gel for H_2/CO_2 separation. *J. Membr. Sci.* 573: 200–209.

130 Li, Y., Liu, H., Wang, H. et al. (2018). GO-guided direct growth of highly oriented metal–organic framework nanosheet membranes for H_2/CO_2 separation. *Chem. Sci.* 9 (17): 4132–4141.

131 Li, Y., Lin, L., Tu, M. et al. (2018). Growth of ZnO self-converted 2D nanosheet zeolitic imidazolate framework membranes by an ammonia-assisted strategy. *Nano Res.* 11 (4): 1850–1860.

132 Yong, W.F., Li, F.Y., Chung, T.-S. et al. (2013). Highly permeable chemically modified PIM-1/Matrimid membranes for green hydrogen purification. *J. Mater. Chem. A* 1 (44): 13914–13925.

133 Bojdys, M.J., Müller, J.O., Antonietti, M. et al. (2008). Ionothermal synthesis of crystalline, condensed, graphitic carbon nitride. *Chem. Eur. J.* 14 (27): 8177–8182.

134 Wang, D., Wang, Z., Wang, L. et al. (2015). Ultrathin membranes of single-layered MoS_2 nanosheets for high-permeance hydrogen separation. *Nanoscale* 7 (42): 17649–17652.

135 Ying, Y., Liu, D., Ma, J. et al. (2016). A GO-assisted method for the preparation of ultrathin covalent organic framework membranes for gas separation. *J. Mater. Chem. A* 4 (35): 13444–13449.

136 Zhao, Y., Liu, P., Ying, Y. et al. (2021). Heating-driven assembly of covalent organic framework nanosheets for gas separation. *J. Membr. Sci.* 632: 119326.

137 Yang, J., Gong, D., Li, G. et al. (2018). Self-assembly of thiourea-crosslinked graphene oxide framework membranes toward separation of small molecules. *Adv. Mater.* 30 (16): 1705775.

138 Jiang, S., Shi, X., Sun, F. et al. (2020). Fabrication of crystalline microporous membrane from 2D MOF nanosheets for gas separation. *Chem. Asian J.* 15 (15): 2371–2378.

139 Zhong, Z., Yao, J., Chen, R. et al. (2015). Oriented two-dimensional zeolitic imidazolate framework-L membranes and their gas permeation properties. *J. Mater. Chem. A* 3 (30): 15715–15722.

140 Zhao, J., He, G., Huang, S. et al. (2019). Etching gas-sieving nanopores in single-layer graphene with an angstrom precision for high-performance gas mixture separation. *Sci. Adv.* 5: eaav1851.

141 Bui, M., Adjiman, C.S., Bardow, A. et al. (2018). Carbon capture and storage (CCS): the way forward. *Energy Environ. Sci.* 11 (5): 1062–1176.

142 Cheng, Y., Wang, Z., and Zhao, D. (2018). Mixed matrix membranes for natural gas upgrading: current status and opportunities. *Ind. Eng. Chem. Res.* 57 (12): 4139–4169.

143 Ying, W., Zhou, K., Hou, Q. et al. (2019). Selectively tuning gas transport through ionic liquid filled graphene oxide nanoslits using an electric field. *J. Mater. Chem. A* 7 (25): 15062–15067.

144 Ying, W., Cai, J., Zhou, K. et al. (2018). Ionic liquid selectively facilitates CO_2 transport through graphene oxide membrane. *ACS Nano* 12 (6): 5385–5393.

145 Varoon, K., Zhang, X., Elyassi, B. et al. (2011). Dispersible exfoliated zeolite nanosheets and their application as a selective membrane. *Science* 334 (6052): 72–75.

146 Agrawal, K.V., Topuz, B., Jiang, Z. et al. (2013). Solution-processable exfoliated zeolite nanosheets purified by density gradient centrifugation. *AIChE J.* 59 (9): 3458–3467.

147 Zhang, H., Xiao, Q., Guo, X. et al. (2016). Open-pore two-dimensional MFI zeolite nanosheets for the fabrication of hydrocarbon-isomer-selective membranes on porous polymer supports. *Angew. Chem. Int. Ed.* 55 (25): 7184–7187.

148 Agrawal, K., Topuz, B., Pham, T. et al. (2015). Oriented MFI membranes by gel-less secondary growth of sub-100 nm MFI-nanosheet seed layers. *Adv. Mater.* 27: 3243–3249.

149 Kim, D., Jeon, M.Y., Stottrup, B.L. et al. (2018). Para-Xylene ultra-selective zeolite MFI membranes fabricated from nanosheet monolayers at the air–water interface. *Angew. Chem. Int. Ed.* 130 (2): 489–494.

8

Layered Double Hydroxide Membranes for Versatile Separation Applications

Yi Liu

Dalian University of Technology, State Key Laboratory of Fine Chemicals, School of Chemical Engineering, Department of Chemical Technology, 2 Linggong Road, Dalian, 116024, P. R. China

8.1 Introduction on LDHs and LDH-Based Membranes

Hydrotalcite (HT) was first discovered by geologist Hochstetter in Sweden in 1842 [1]. In effect, HT represents a large mineral group of naturally occurring layered double hydroxides (LDHs), which could be expressed by the formula $[Mg_6Al_2(OH)_{16}]CO_3{\cdot}4H_2O$ in both rhombohedral and hexagonal forms, respectively [2]. LDHs, also commonly called HT-like compounds, are representative of anionic layered materials composed of layers of metal cations (M^{2+} and M^{3+}) with similar ionic radii (illustrated in Figure 8.1). These metal cations occupy the centers of edge-sharing octahedra, whose vertexes contain hydroxide anions that connect to form infinite positively charged brucite-like sheets [3], among which charge-compensating anions and solvation molecules are located. A general molecular formula of LDHs can be expressed as $\left[M^{2+}_{1-x}M^{3+}_{x}(OH)_2\right][A^{n-}]_{x/n}mH_2O$, where M^{2+} represents divalent metal ions (such as Mg^{2+}, Zn^{2+}, Cu^{2+}, Co^{2+}, or Ni^{2+}), M^{3+} represents trivalent metal ions (such as Al^{3+}, Fe^{3+}, Ga^{3+}, Cr^{3+}, or Mn^{3+}), and A^{n-} represents charge-compensating anions (such as $CO_3^{\ 2-}$, $NO_3^{\ -}$, $SO_4^{\ 2-}$, and $ClO_4^{\ -}$). Commonly the x value is in the range of 0.2–0.4 [4]. At present, LDHs have shown wide and promising application prospects in adsorption [5, 6], catalysts [7–9], polymeric modification [10–12], electrochemistry [13–15], and controlled drug delivery [16–18], which can be largely attributed to their unique structural and physicochemical features, such as uniform interlayer galleries, unusual thermal behavior, structural memory effect, anion-exchange ability, rich surface functional groups, high positive surface charge, easy delamination, and high-aspect ratios. All these features, in turn, render LDHs promising candidates for high-performing separation membranes [19–22].

It is noted, however, that research on LDH-based separation membranes has been rarely conducted until the year 2008, when T.T. Tsotsis et al. first employed the electrophoretic deposition (EPD) technique to deposit ball-milled commercial

Two-Dimensional-Materials-Based Membranes: Preparation, Characterization, and Applications, First Edition. Edited by Gongping Liu and Wanqin Jin.
© 2022 WILEY-VCH GmbH. Published 2022 by WILEY-VCH GmbH.

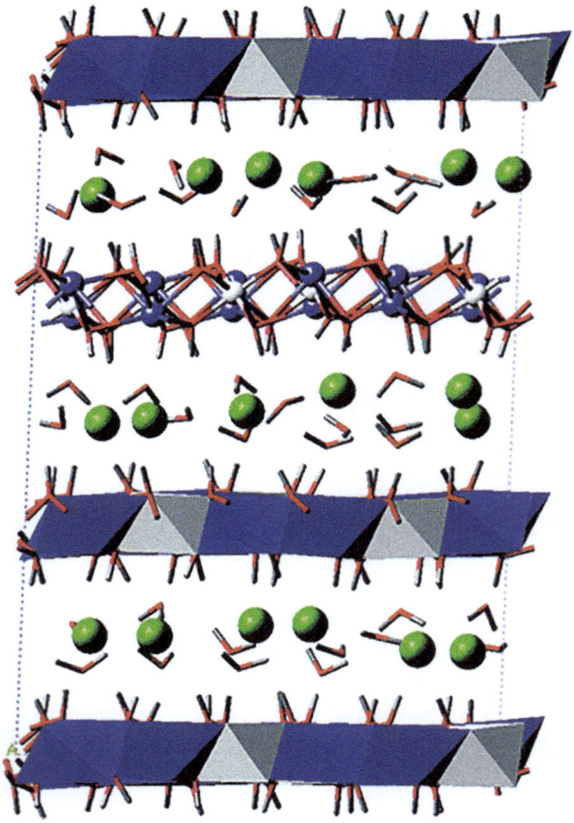

Figure 8.1 Crystal structure of Mg_2Al–Cl LDHs, $[Mg_2Al(OH)_6]Cl{\cdot}2H_2O$. Green balls = Cl^-, red sticks = O of water or OH, white sticks = H of water or OH, dark gray octahedra = Al of hydroxide layer, blue octahedra = Mg of hydroxide layer. Source: Cygan et al. [2]/with permission of Royal Society of Chemistry.

MgAl–CO_3 LDH powders on porous α-Al_2O_3 tubes [20]. Prepared LDH membranes exhibited preferred selectivity toward CO_2 with their CO_2/N_2 ideal selectivity being 40–70% higher than the corresponding Knudsen values, which was coincident with previous studies revealing their great promising applications in preferential CO_2 adsorption. In the follow-up work, a vacuum-suction approach was utilized instead of the EPD technique for the preparation of MgAl–CO_3 LDH membranes exhibiting a promising CO_2/N_2 selectivity (~35) after repairing these cracks and defects by silicone coating [22]. Nevertheless, the ideal H_2/CO_2 selectivity of pristine MgAl–CO_3 LDH membranes was only slightly higher than their Knudsen selectivity, thereby indicating that there existed substantial mesoscopic and microscopic defects derived from random stacking of LDH nanosheets (shown in Figure 8.2). Although there are still imperfections, this pioneering work, however, is still of great enlightening significance since it not only convinces the separation capability of LDH-based membranes but also establishes practical protocols for their facile preparation.

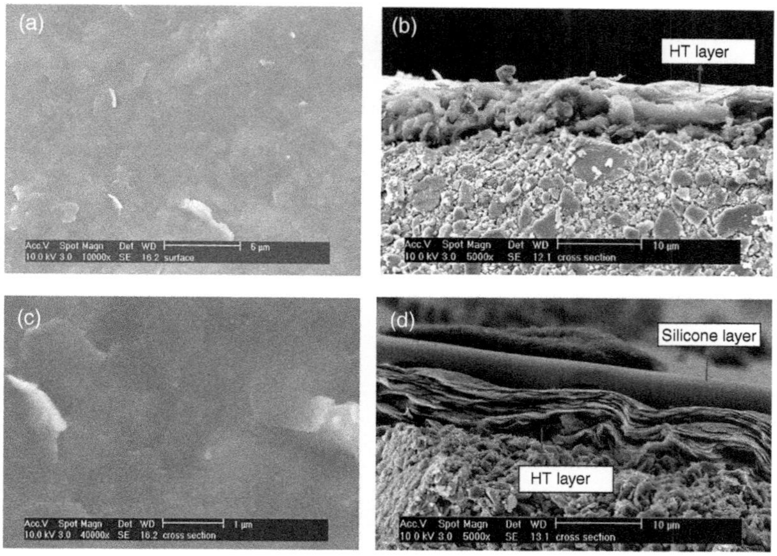

Figure 8.2 SEM images of (a, b) top and (c) the cross-section views of a vacuum-suction HT membrane. (d) The SEM image of (d) the silicone-coated HT membrane. Source: Kim et al. [22]/with permission from American Chemical Society.

8.2 Strategy for LDH-Based Membrane Preparation

In general, LDH-based membranes can be prepared by solution-based *in situ* growth, post-synthetic deposition, and blending with polymer precursor followed by *in situ* polymerization. It should be noted that prior to LDH-based membrane processing, exfoliation of LDH bulk crystals into unilamellar LDH nanosheets represents an important intermediate step, which leads to superior separation performances.

8.2.1 Solution-Based *In Situ* Growth

Solution-based *in situ* deposition refers to one-step synthesis of LDH membranes within alkaline aqueous solutions in the presence of mineralizing agents, such as urea, hexamethylenetetramine (HMT), $NH_3{\cdot}H_2O$, and NaOH. Compared with other methods, solution-based *in situ* deposition enables the formation of brucite-like sheets, their self-assembly into LDH crystallites, and firm attachment to the substrate surface in one step so that the synthetic procedure can be significantly simplified [23]. Moreover, with this method not only interlayer galleries between brucite-like sheets are highly uniform but also intercrystalline defects within LDH membranes can be minimized, thereby making them highly adaptable to precise molecular sieving.

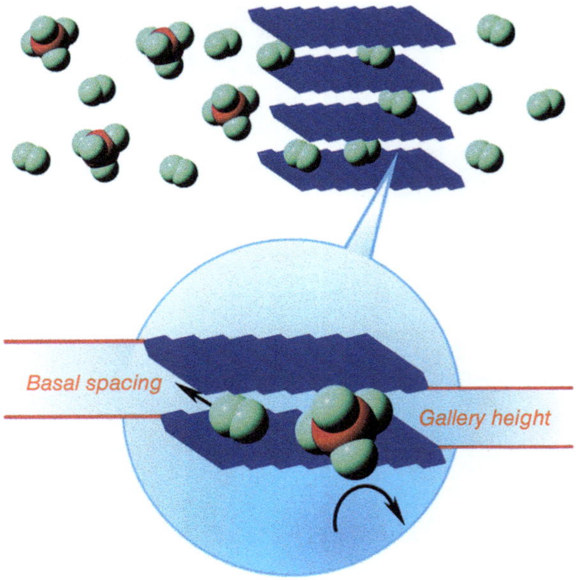

Figure 8.3 Schematic illustration of the concept of selective molecular sieving through interlayer galleries of LDHs. Source: Liu et al. [23]/with permission of Royal Society of Chemistry.

Y. Liu and J. Caro et al. pioneered *in situ* hydrothermal synthesis of highly H_2/CH_4 selective NiAl–CO_3 LDH membranes on porous γ-Al_2O_3 substrates with $NH_3 \cdot H_2O$ as the mineral agent and dissolved CO_2 as the precursor of charge-compensating anions (illustrated in Figure 8.3) [23]. Interestingly, it was noted that the concentration of CO_2 dissolved in the precursor solution exerted significant influence on its final microstructures [24]. Trace amounts of CO_2 led to the formation of loosely packed ab-oriented LDH membranes, while well-intergrown randomly oriented LDH membranes were formed within CO_2-saturated precursor solution. It was noted that although both ab- and randomly oriented LDH membranes exhibited clear size-based selectivity, randomly oriented LDH membranes showed much higher H_2/CH_4 selectivity due to decreased intercrystalline defects within prepared LDH membranes. In the follow-up work, different types of LDH membranes (including ZnAl–NO_3 [24] CoAl–CO_3 [25] and MgAl–Glycine [26]) were sequentially synthesized in the same manner.

It should be emphasized that although LDH membranes prepared via solution-based *in situ* growth exhibit certain superiority in microstructure and separation performance, the manufacturing process is highly demanding in terms of expertise and experience that, in turn, severely hinders their widespread applications.

8.2.2 Post-Synthetic Deposition

Post-synthetic deposition procedure can be described as follow—First, synthesis of LDH bulk crystals with desired chemical composition and microstructure via coprecipitation, hydrothermal synthesis, ion exchange, or calcination–reconstruction;

second, exfoliation of LDH bulk crystals for the purpose of obtaining monodispersed, unilamellar LDH nanosheets (this step is not mandatory); third, their deposition on substrates via vacuum deposition, dip coating, or EPD.

At present, diverse methods have been developed for the preparation of LDHs in powder form. Among them, the coprecipitation method employing $NaOH/NH_3 \cdot H_2O$ as mineral agents usually generates uniform LDH crystals with a wide grain size range but lower crystallinity [27], which facilitates selective anion-exchange intercalation or exfoliation. In contrast, the urea/HMT hydrothermal treatment method commonly leads to the formation of CO_3^{2-}-intercalated LDHs with larger grain size and higher crystallinity [28].

Further exfoliating LDH bulk crystals into single-layer nanosheets enables to not only maximize the surface area but also readjust the gallery height, which is quite advantageous for improving the separation performance of LDH-based membranes. In general conditions, once CO_3^{2-}-intercalated LDH bulk crystals are anion-exchanged to monovalent charge-compensating anions (typically NO_3^-), anion-exchanged LDH bulk crystals will be easily exfoliated by intercalating solvents (such as formamide) or intercalating anions (such as sodium dodecyl sulfate, SDS) as under vigorous stirring/ultrasonication (as illustrated in Figure 8.4) [29].

Finally, among versatile techniques developed for surface deposition of LDHs, vacuum filtration represents the most widely used technique due to its simplicity and competence in fine-tuning of the microstructure of deposited LDH layers.

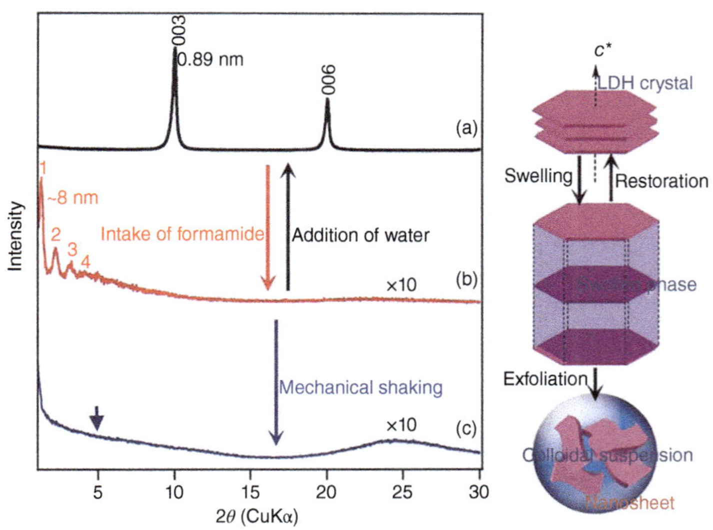

Figure 8.4 (Left panel) XRD patterns of (a) starting NO_3^--intercalated LDH, (b) after addition of formamide, and (c) colloidal aggregate centrifuged from the transparent LDH suspensions. The right panel shows a schematic illustration of the possible stages corresponding to the evolution of the XRD patterns. Source: Ma et al. [29]/with permission of Royal Society of Chemistry.

8.2.3 Blending with Polymers

Aiming at obtaining high-performing LDH-based MMMs, ideally, LDH fillers should be well-dispersed in the polymer matrix that, in turn, requires excellent mono-dispersity and uniformity of LDH fillers in the polymer precursor solution. However, owing to the intrinsic hydrophilic nature and high surface charge density of LDHs, in general, they cannot be well-dispersed in nonpolar solvents. In response to this issue, surface modification of LDH fillers with solvents [30], organosilicates [31], surfactants [32], oligomers [33], and intermediate polymers [34] prior to blending with polymer matrix has become indispensable for ensuring optimum compatibility between two phases.

8.3 Research Progress on LDH-Based Membranes

Exploration of potential applications of LDHs for membrane-based separation has become a hot research topic in recent decades due to their promising application prospects in efficient microfiltration, ultrafiltration (UF), nanofiltration (NF), reverse osmosis (RO), forward osmosis (FO), pervaporation, gas separation, and particulate matter (PM) capture. Herein we summarize major advances achieved in LDH-based separation membranes with emphasis on the elucidation of their structure–property–performance relationship. More specifically, progress made in LDH-based membrane research was sequentially introduced based on unique structural and physicochemical properties of LDHs, e.g. uniform interlayer galleries, high-aspect ratios, unusual thermal behavior, preferential adsorption properties, high surface charge, and intrinsic hydrophilicity.

8.3.1 Interlayer Gallery-Based Separation

Post-synthetic deposition procedure can be described as follow—First, synthesis of LDH bulk crystals with desired chemical composition and microstructure via coprecipitation, hydrothermal synthesis, ion-exchange, or calcination–reconstruction; second, exfoliation of LDH bulk crystals for the purpose of obtaining monodispersed, unilamellar LDH nanosheets (this step is not mandatory); third, their deposition on substrates via vacuum deposition, dip coating, or EPD.

8.3.1.1 Pristine Interlayer Gallery-Based Separation

As a representative of 2D materials, LDHs possess highly uniform interlayer galleries; moreover, by appropriate choice of corresponding brucite-like sheets and charge-compensating anions, the gallery height can be accurately tuned from nanometer to subnanometer scale, which endows them with unprecedented opportunities for precise molecular sieving.

Y. Liu and J. Caro et al. pioneered the preparation of well-intergrown NiAl–CO_3 LDH membranes on mesoporous γ-Al_2O_3 substrates by *in situ* hydrothermal growth [23, 24]. The XRD pattern indicated a gallery height of 0.31 nm, which was located

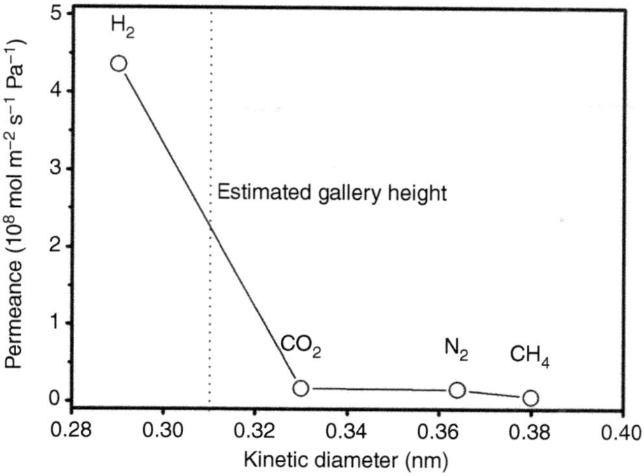

Figure 8.5 Permeances of the single gases (H_2, CO_2, N_2, and CH_4) of the NiAl–CO_3 LDH membrane at $\Delta P = 1$ bar, $T = 180\,°C$ as a function of molecular kinetic diameters. Source: Liu et al. [23]/with permission of Royal Society of Chemistry.

between kinetic diameters of H_2 (0.29 nm) and other gas molecules, such as CH_4 (0.38 nm). Correspondingly, separation factor (SF) of activated NiAl–CO_3 LDH membranes reached ~80 for the H_2/CH_4 gas pair, rendering them quite promising for application in H_2 purification (as shown in Figure 8.5). It was noted that in comparison with NiAl–CO_3 LDH membranes, ZnAl–NO_3 LDH membranes prepared in the same manner exhibited similar microstructure but inferior H_2/CH_4 selectivity (~13.7), which was ascribed to increased gallery height (0.42 nm), thereby allowing considerable permeation of CH_4 [24].

Since inevitably there existed an upper limit on the separation performance of a single-phase membrane, to overcome this limitation, Y. Liu and J. Caro et al. further prepared ZIF-8–ZnAl–NO_3 LDH composite membranes through facile solvothermal treatment of ZnAl–NO_3 LDH precursor membranes with 2-methylimidazole [35]. As shown in Figure 8.6, partial dissolution of the upper portion of ZnAl–NO_3 LDH precursor membranes resulted in reduced framework density and membrane thickness, leading to significantly enhanced H_2 permeance; meanwhile, since there existed two independent H_2-selective functional layers (i.e. ZIF-8 and ZnAl–NO_3 LDH), prepared ZIF-8–ZnAl–NO_3 LDH composite membranes exhibited enhanced H_2/CH_4 selectivity (c. 54.1), which was higher than pure ZIF-8 or ZnAl–NO_3 LDH membranes alone [36].

Besides gas separation, LDH-based membranes also hold great potential for pervaporation dehydration through combining the uniform interlayer gallery with intrinsic hydrophilicity. Recently, N.X. Wang prepared vertically aligned CoAl–CO_3 LDH membranes on tubular α-Al_2O_3 substrates through facile *in situ* hydrothermal growth. NH_4F was found to play a vital role in warranting vertical alignment of CoAl–CO_3 LDH nanosheets on substrates [25]. Owing to the 2.8 Å-sized interlayer gallery height and intrinsic hydrophilic nature of transport

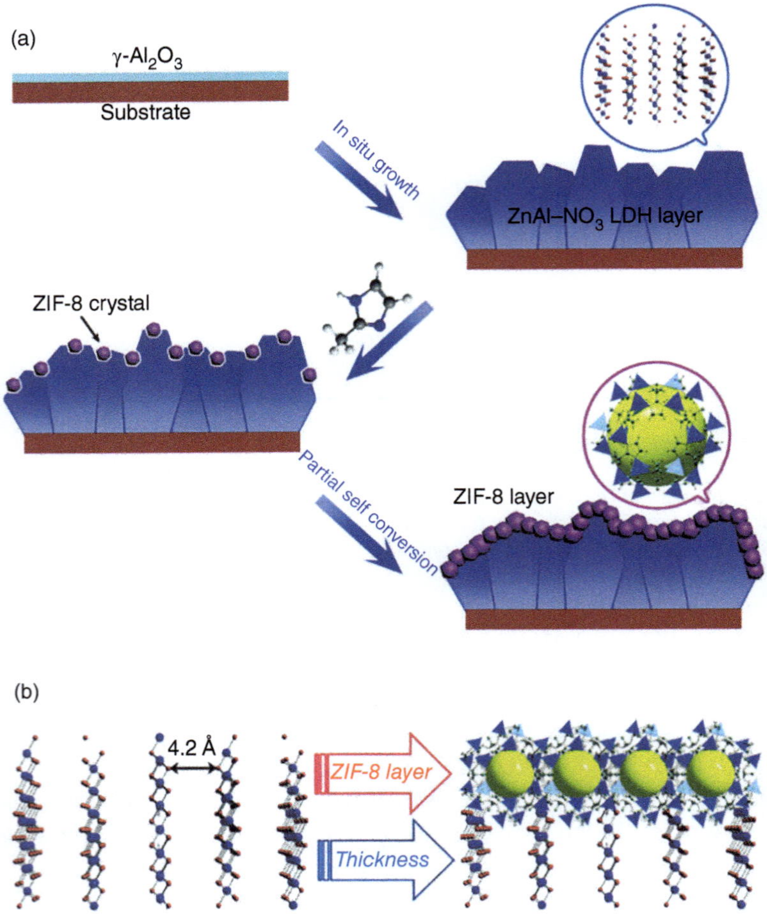

Figure 8.6 (a) Illustration of the preparation of the ZIF-8–ZnAl–NO_3 composite membrane on a γ-Al_2O_3-modified α-Al_2O_3 substrate via partial self-conversion method of the ZnAl–NO_3 LDH precursor layer. (b) Structural change of the ZnAl–NO_3 LDH membrane after the solvothermal treatment with 2-methylimidazole. Source: Liu et al. [35]/with permission of John Wiley & Sons.

channels, water molecules were enabled to preferentially permeate through the membrane, resulting in high and stable organic solvent dehydration performance.

Considering the importance of interlayer gallery environment on the separation performance of LDH membranes, Q.F. An et al. further prepared amino acid (e.g. glycine)-intercalated MgAl LDH membranes via facile *in situ* hydrothermal growth [26]. Compared with pristine MgAl–CO_3 LDH membranes, prepared MgAl–glycine LDH membranes exhibited remarkably enhanced water permeance while maintaining comparable rejection rate (98%) toward Eriochrome Black T, which was ascribed to enlarged gallery height (0.39 nm) and appropriate physical/chemical properties of LDH membranes through intercalation with glycine amino acid molecules. This study provided a facile strategy to tune the interlayer structure, and therefore, the separation performance of LDH membranes.

In addition to LDH membranes, very recently N.X. Wang and coworkers further prepared 300 nm-thick, vertically aligned layered α-$Co(OH)_2$ membranes on porous α-Al_2O_3 substrates [37]. As a structural analog of LDHs, α-$Co(OH)_2$-based membranes exhibited considerable fluxes in organic solvent nanofiltration (OSN) (methanol). The high flux was attributed to the vertically aligned separation channels, ultrathin membrane thickness, and high affinity of prepared membranes to methanol.

In contrast to inorganic substrates, recently, Z. Wang et al. prepared loose NiCo LDH/polyetherimide (PEI) hybrid layers on polyacrylonitrile (PAN) substrates through a solution-based method under mild reaction conditions (55 °C) [38]. 2D channels of LDH particles (gallery height: 5.0 Å), as well as open spaces on the Co^{2+}/PEI complex layer, significantly facilitated permeation of water molecules and ions through the membranes, resulting in enhanced water permeate flux. In addition, spontaneously formed ridge-and-valley surface structure improved the steric hindrance effect so that the macromolecule rejection was increased.

8.3.1.2 Regenerated Interlayer Gallery-Based Separation

Besides *in situ* generated interlayer galleries, several post-synthetic protocols, such as exfoliation–restacking, calcination–rehydration, and anion exchange with the desired anions, have been successfully employed to finely tune the interlayer structure of LDH-based membranes.

Recently Z.Y. Jiang et al. prepared ultrathin MgAl–CO_3 LDH membranes with the thickness of 8–70 nm through spin-coating-assisted self-assembly of exfoliated LDH nanosheets (shown in Figure 8.7) [39]. Relying on the intrinsic breathing effect of brucite-like nanosheets, CO_2 was able to transfer through interlayer galleries of LDH membranes in the form of CO_3^{2-} anions, which were subsequently released as CO_2 at the downstream side. Prepared LDH membranes displayed decent CO_2 permeance (150 GPU) and CO_2/CH_4 selectivity (~33), which were attributed to the contribution of breathing and size-sieving effects (gallery height ~3 Å).

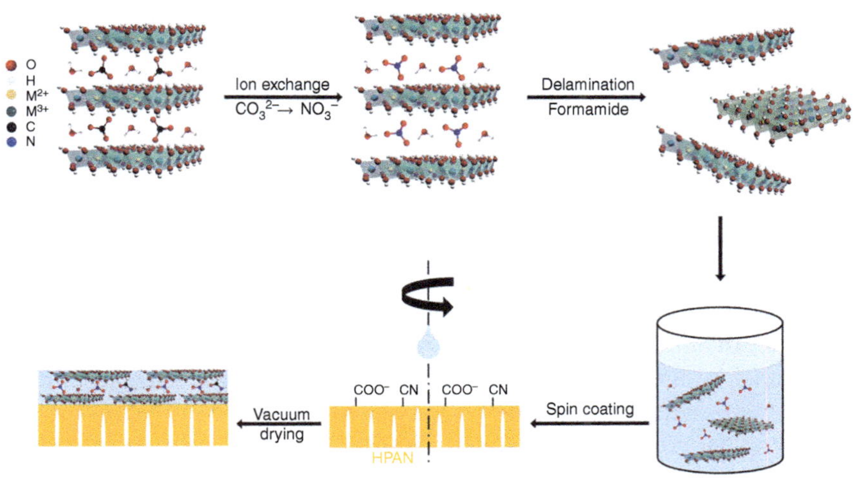

Figure 8.7 Schematic illustration of the preparation of LDH membranes with intrinsic breathing effect. Source: Liu et al. [39]/with permission of Elsevier.

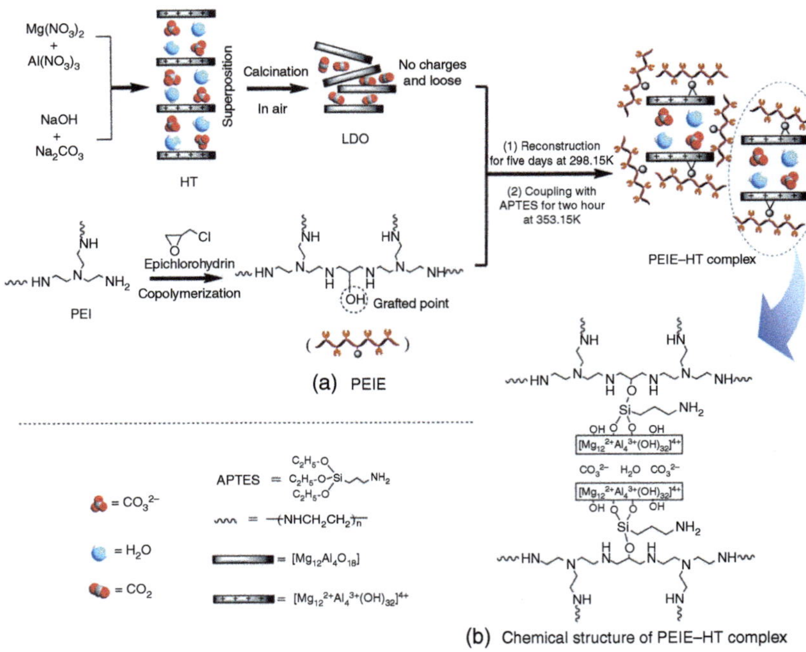

Figure 8.8 Reaction pathway for the formation of the PEIE–MgAl–CO_3 complex. Source: Liao et al. [40]/with permission of Royal Society of Chemistry.

In contrast, regulation of interlayer galleries of LDHs via calcination–rehydration enabled a more straightforward alternative to the common exfoliation–restacking method. For instance, Z. Wang prepared a new membrane by establishing high-speed CO_2 facilitated transport channels derived from MgAl–CO_3 LDH nanofillers in the fixed carrier membrane (schematically shown in Figure 8.8) [40]. It was interesting to note that instead of direct distribution, layered double oxides (LDOs), which derived from pre-calcined LDHs nanoparticles, were added to the polyethyleneimine (PEIE) and epichlorohydrin solution to *in situ* generate well-dispersed LDH nanofillers through the calcination–reconstruction step. Long-term exposure of LDOs in the presence of CO_2 warranted successful reconstruction of MgAl–CO_3 LDH nanofillers in PEIE solution. By adopting a similar method, a high-performing polyvinylamine–MgAl–CO_3 LDH membrane containing high-speed facilitated transport channels was successfully established for efficient CO_2 separation either [41].

LDHs with adjustable interlayer galleries also exhibited promising performances as OSN membranes. For instance, J.W. Chew et al. prepared a series of M^{2+}–Al LDH membranes exhibiting diverse physicochemical activities [42]. LDH nanosheets were prepared by urea hydrolysis followed by calcination and delamination in formamide. Subsequently, LDH nanosheets were stacked on nylon substrates via vacuum filtration. Prepared M^{2+}–Al LDH membranes not only exhibited superior solutes rejection but also rendered excellent NF permeances in both aqueous and organic solvents. Simulation results indicated that superior performance was derived from the preferential interfacial interactions between solvent molecules and LDH nanosheets.

Enlarging the gallery height via intercalation of bulky anions/anionic clusters represents an effective method to increase the water flux through LDH-based membranes in water treatment. Very recently, Q.F. An et al. prepared iron cage (Fe-cage)-intercalated CoAl LDH composite membranes on nanoporous PAN surface through vacuum-assisted self-assembly (shown in Figure 8.9) [43]. Fe-cages were intercalated into brucite-like sheets via solvothermal-assisted exfoliation of LDH nanosheets followed by restacking based on the electrostatic interaction. Owing to Fe-cage-induced improvement of the structural and chemical properties of the interlayer spacing, prepared CoAl–Fe-cage LDH membranes exhibited enhanced separation performance toward the removal of dyes and humic acid from aqueous solution.

Besides anionic metal clusters, small organic molecules could also be intercalated into interlayer galleries of LDH membranes to tailor the chemistry and structure of diffusion paths. S. Liang and coworkers fabricated superhydrophobic and oleophobic MgAl LDH membranes by *in situ* grafting interlayer galleries with alternately arranged sodium SDS and 3-aminopropyltriethoxysilane (APTS) chains during preparation [44]. Relying on the cross-linking between amine groups of APTS and carboxyl groups of poly(methacrylic acid), surface-tailored MgAl–SDS LDH nanosheets could be firmly attached to the membrane surface by simple dip coating. Long hydrophobic SDS and short hydrophilic APTS chains provided a superhydrophilic but oleophobic membrane surface. As a result, both wettability and hydrophilicity of the membrane were significantly enhanced. Prepared MgAl–SDS LDH membranes displayed higher water permeability without compromising their selectivity, which could be ascribed to increased electron donor component of surface energy, membrane surface roughness, and gallery height (~3.06 nm).

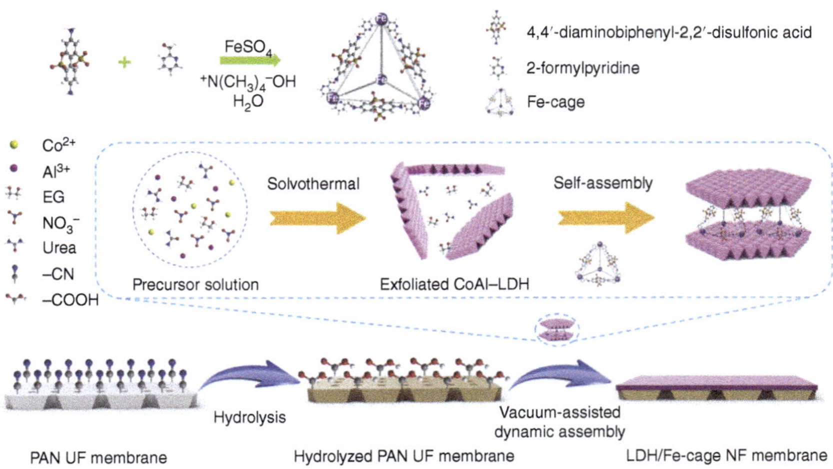

Figure 8.9 Schematic illustration of CoAl–Fe-cage LDH composite NF membrane preparation. Source: Wang et al. [43]/with permission of Elsevier.

8.3.2 Geometric Shape-Based Separation

Owing to their 2D layered structure, well-crystallized LDH crystals commonly exhibit a plate-like morphology with high-aspect ratios, rendering them potential candidates as geometric shape-based buffer layers for high-performing separation membranes.

8.3.2.1 Geometric Shape-Based Gas Separation

A typical example of effective utilization of geometric features of LDHs refers to their use as a "perch" to collect metal–organic framework (MOF) seeds while simultaneously preventing their detachment. Y. Liu and J. Caro et al. prepared vertically aligned MgAl–CO_3 LDH buffer layers on porous α-Al_2O_3 substrates by facile *in situ* hydrothermal growth [45]. Prepared LDH crystals exhibited typical plate-like morphology with even mutual distance. LDH buffer layer-modified substrates were then uniformly covered with nanosized ZIF-8 seeds by dip coating. Owing to the hindrance from adjacent MgAl–CO_3 LDH plates, trapped ZIF-8 seeds were firmly attached to substrates and would not easily peel off. ZIF-8 membranes exhibiting decent selectivity toward H_2/CO_2, H_2/N_2, and H_2/CH_4 gas pairs were obtained after secondary solvothermal growth.

Besides serving as the "perch" of ZIF-8 seeds, LDH buffer layers were further employed in assisting *in situ* solvothermal synthesis of well-intergrown ZIF-8 membranes (shown in Figure 8.10) [46]. Initially, uniform and vertically aligned ZnAl–CO_3 LDH buffer layers were *in situ* grown on γ-Al_2O_3 substrates following the van der Drift principle. Subsequently, well-intergrown ZIF-8 membranes were formed after *in situ* solvothermal growth, owing to the specific metal–imidazole interaction between the ZnAl–CO_3 LDH and ZIF-8 phases, as well as high-aspect ratios of ZnAl–CO_3 LDH plates. Gas permeation results indicated that selectivity toward H_2/CH_4 gas pair was comparable with that of other high-quality ZIF-8 membranes.

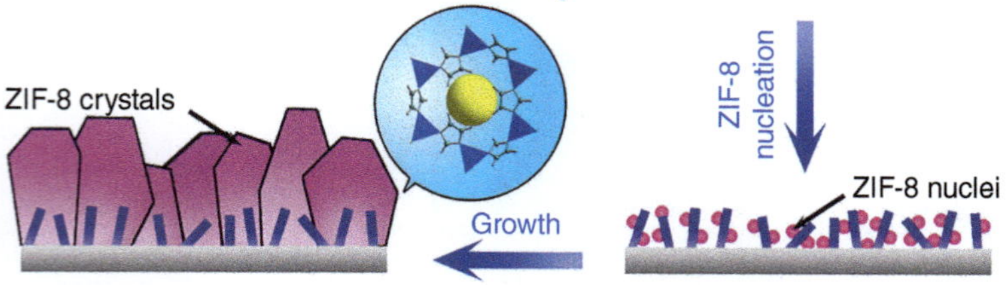

Figure 8.10 Schematic illustration of *in situ* solvothermal growth of ZIF-8 membrane on a ZnAl–CO_3 LDH buffer layer-modified γ-Al_2O_3 substrate. Source: Liu et al. [46]/with permission of American Chemical Society.

The follow-up work conducted by R. Matsuda and S. Kitagawa et al. further confirmed that introduction of CoAl–CO_3 and MgAl–CO_3 LDH buffer layers on substrates effectively promoted intergrowth of ZIF-7 films so that metal ions located in LDH buffer layers did not have to be the same as those residing in MOF films [47]. In addition, a variety of well-intergrown MOF films (such as ZIF-8 and HKUST-1), aside from ZIF-7, could also be grown on diverse LDH buffer layer-modified substrates, demonstrating the generality of this method.

Besides assisting in the synthesis of MOF membranes, Very recently, H.W. Fan and J. Caro et al. proposed to fabricate vertically aligned 2D covalent–organic framework (COF) layers via *in situ*-oriented growth on vertically aligned APTS-modified LDH buffer layers, which served as the template (shown in Figure 8.11) [48]. Owing to the in-plane growth of 2D COF–LZU1 layers along with the surface of CoAl–CO_3 LDH plates, straight 2D diffusion pathways (0.37 nm), perpendicular to the substrate surface, were spontaneously formed. Prepared COF–LZU1 membranes displayed ultrahigh H_2 permeance (~3600 GPU) as well as decent selectivity toward H_2/CO_2 (~31.6) and H_2/CH_4 (~29.5) gas pairs depending on their preferred orientation.

Besides COF membranes, introduction of LDH buffer layers also contributed to performance enhancement of Pd-based metallic membranes. Recently J.H. Huang and co-workers *in situ* synthesized high-density platelet-like LiAl–CO_3 LDH aggregates on the porous stainless steel (PSS) surface [49]. Surface modification with LDH buffer layers not only effectively reduced pore opening of PSS but also served as intermediate layers retarding Pd/Fe interdiffusion. Subsequently, thin Pd films were deposited on LDH buffer layer-modified PSS tubes by electroless plating. He-leaking test indicated that thin Pd layers on LDH-modified PSS substrates were free of defects. As a result, prepared Pd membranes exhibited excellent H_2 permeance, H_2/N_2 selectivity, and long-term operational stability.

Besides serving as buffer layers of a series of inorganic membranes, incorporation of LDH additives into polymer matrix further enabled us to tailor the domain

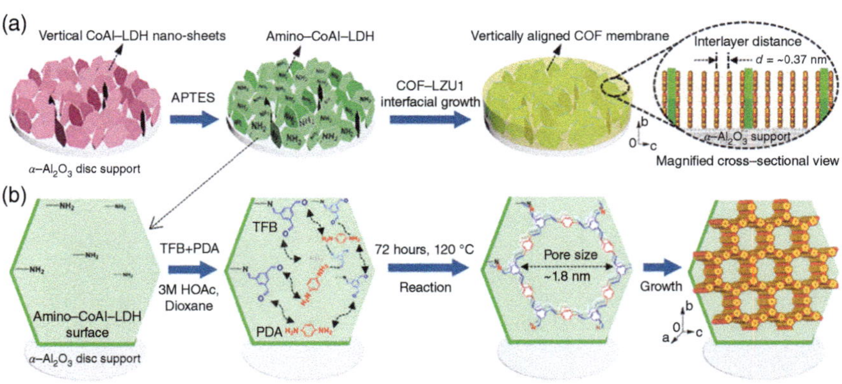

Figure 8.11 (a) Schematic illustration of the engineering of vertically aligned COF membranes. (b) Magnified scheme showing the growth of the 2D COF–LZU1 parallel to the amino-CoAl–CO_3 LDH nanosheet surface. Source: Fan et al. [48]/with permission of American Chemical Society.

structure and gas permeation behavior of pristine polymer membranes. Y.T. Zhang and J.D. Liu et al. fabricated MgAl–Cl LDH nanosheets via vigorous agitation and ultrasonication, respectively. Subsequently, poly(PEA–LDH–TMC)/polystyrene (PS) composite membranes were prepared via interfacial polymerization by adding LDH nanosheets into the aqueous phase of poly(ester-amide) (PEA). Prepared poly(PEA–LDH–TMC)/PS composite membrane exhibited significantly enhanced CO_2 permeability without significantly compromising CO_2/N_2 selectivity [50]. The increase in gas permeability was attributed to reduced crystallinity of the polymer and increased free volume caused by the addition of LDH nanosheets.

8.3.2.2 Geometric Shape-Based Liquid Separation

Both Wenzel and Cassie–Baxter theories figured out that enhancing surface roughness was essential for improving hydrophobicity/hydrophilicity of the corresponding flat surface. In this regard, incorporating LDHs into membranes was believed to be beneficial for high-efficiency liquid separation (in particular, water/oil separation) due to their high-aspect ratios as well as rich surface functional groups [51, 52].

Following the above principle, F. Li and co-workers prepared MgAl–CO_3 LDH microcrystal-coated commercial textile through a facile hydrothermal method followed by post-modification with sodium laurate (shown in Figure 8.12) [53]. Owing to the unique nest-like microstructure of LDH coatings as well as the low surface energy via post-modification, prepared textile became superhydrophobic and superoleophilic, which enabled effective separation of both light and heavy oils with high separation efficiency. In addition, they exhibited excellent selective oil adsorption capacity and reusability, making them competent "bags" for selective oil recovery from aqueous suspension.

In much the same manner, T. Zhang and F.X. Qiu et al. prepared ZnAl–CO_3 LDH/cellulose membranes by combining *in situ* growth and hydrophobic modification [54]. On the one hand, the rough surface was fabricated by *in situ* growth of ZnAl–CO_3 LDH nanosheets on AlOOH gel-modified cellulose fibers; on the other hand, the superhydrophobicity was endowed by grafting silane coupling agents. As a result, prepared membranes exhibited excellent separation properties

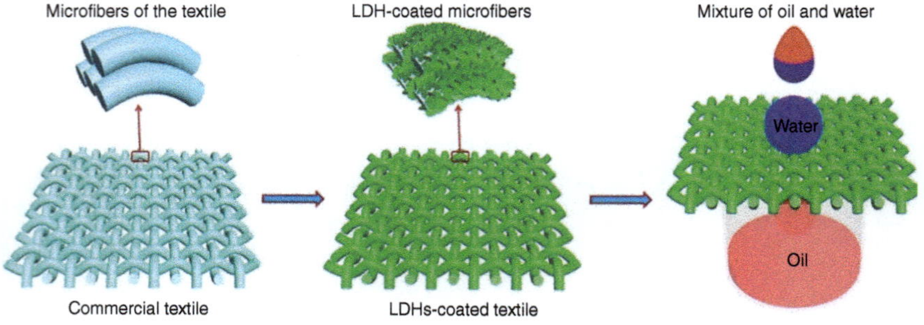

Figure 8.12 Schematic illustration of surface modification of commercial textile with LDHs and its application for oil/water separation. Source: Liu et al. [53]/with permission of American Chemical Society.

for surfactant-stabilized water-in-oil emulsions, high separation efficiency, great chemical durability, and excellent recyclability.

To simplify the membrane manufacturing process further, the above researchers developed a facile one-pot method to prepare superhydrophobic cellulose/LDH membranes in an open oil/water two-phase system so that the processes of hydrothermal reaction and hydrophobic modification could be perfectly integrated [55]. Two critical factors of superhydrophobic properties, i.e. surface roughness and low surface energy, could be accomplished in one step—Initially, AlOOH/filter papers were placed at the bottom of aqueous solution with oils floating on the water surface; subsequently, the aqueous solution containing Zn^{2+} ions, Cl^- anions, and mineral agents was heated to the boiling point. During this stage, Zn^{2+} ions reacted with AlOOH located on the surface of AlOOH/cellulose microfiber, thereby forming a rough surface comprising ZnAl–Cl LDH nanosheets. Upon complete evaporation of the aqueous solution, LDH-coated membranes were subsequently immersed in the oil phase and the whole reaction process entered the second stage for hydrophobic modification. During this stage, a self-assembled stearic acid layer was deposited on the surface of LDH nanosheets due to the coordination effect, ultimately forming a superhydrophobic surface. Prepared LDH/cellulose membranes were competent in effective separation of versatile oil/water mixtures.

In contrast, without being subjected to any post-treatment, J.D. Dai and co-workers reported the construction of grass-like structured composite membranes consisting of NiCo–CO_3 LDH arrays on polydopamine (PDA)-modified polyvinylidene fluoride (PVDF) membranes through facile *in situ* hydrothermal growth [56]. The surface morphology of NiCo–CO_3 LDH membranes could be easily tuned by hydrothermal reaction time. Prepared NiCo–CO_3 LDH composite membranes exhibited superhydrophilicity/underwater superoleophobicity rendering them competent for separating both surfactant-free and surfactant-stabilized oil/water emulsions by capillary effect under the driving of gravity.

In the follow-up study, Z.P. Zhou and co-workers prepared NiCo–CO_3 LDH (NC) membranes by *in situ* growth on PDA@PVDF substrates [57]. Obtained membranes exhibited superhydrophilicity/underwater superoleophobicity and high flux due to their high surface energy and sisal-like surface morphology. Prepared NC@PDA@PVDF hybrid membranes enabled effective separation of surfactant-free emulsion and surfactant-stabilized emulsion with the separation efficiency above 99.4%.

LDHs may further blend with other inorganic compounds with specific morphology to promote membrane separation performance. For instance, Y.J. Li and co-workers fabricated membranes with hierarchical architectures consisting of 1D and 2D materials for emulsified oil/water separation [58]. To verify the roles of 2D layered materials in oil/water separation, a facile layer-by-layer assembly method was developed to coat sepiolite (SeP)-grafted graphene oxide (GO) or MgAl–Cl LDH membranes on cellulose acetate (CA) substrates. Experimental results indicated that the hierarchical structure resulted in high permeate flux and excellent underwater superoleophobic properties. Moreover, SeP-grafted GO-coated membranes exhibited higher flux and slower permeability decline than SeP-grafted

MgAl–Cl LDH-coated membranes due to their discrepancy in surface roughness and surface charge.

8.3.2.3 Geometric Shape-Based Particulate Matter Capture

Besides their applications in gas and liquid separation, LDH-based membranes also held great promise for PM capture floating in air, relying on their morphological features. For instance, Y.F. Song and co-workers developed a "Seeds Embedded Epitaxial Growth" method for preparing PAN@LDH composite membranes with the "Mortise–Tenon" microstructure [59]. Owing to the steric effect, the inertance effect, and the electrostatic interactions between PM and MgAl–CO_3 LDH microparticles, prepared PAN@LDH composite membranes exhibited excellent removal efficiency for PM10, PM2.5, and PM0.3–0.5. In addition, spent membranes could be easily recovered while simultaneously maintaining high PM removal efficiency.

8.3.2.4 Geometric Shape-Based Sacrificing Templates

LDHs could also be employed as sacrificing templates to fabricate highly porous FO membranes to minimize the effect of the internal concentration polarization (ICP) in substrates. Very recently, Y.S. Li and co-workers incorporated MgAl–CO_3 LDH nanoparticles in the substrate using conventional non-solvent-induced phase inversion method followed by HCl washing, resulting in higher overall porosity and interconnected pore network with "sand rose" morphology (shown in Figure 8.13) [60]. Owing to high porosity and macropore structure in the pore-wall promoting connectivity of finger-like pores, as-obtained FO membranes exhibited higher water flux than commercial FO membranes.

8.3.3 Unusual Thermal Behavior-Based Separation

It is well recognized that in case bimetallic LDHs are subjected to calcination in appropriate environments, accompanied with the removal of surface functional groups, interlayer solvents and charge-compensating anions, the brucite-like sheets will be decomposed, thereby forming mixed metal oxides—LDOs, in which the crystalline phase is that of the oxide in larger proportion [61, 62]. Since the mean

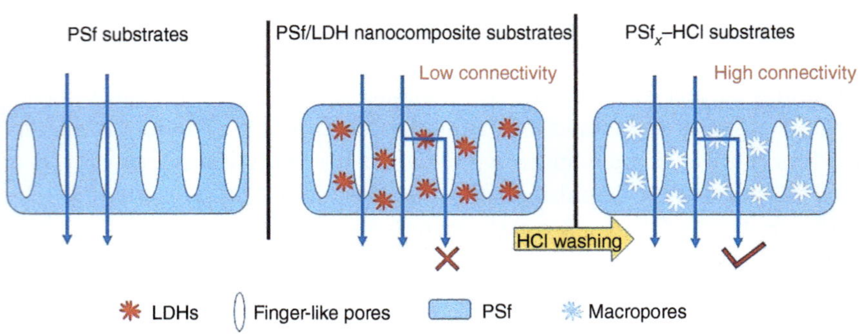

Figure 8.13 Schematic illustration of the function of TFC FO membranes based on the LDH-templates. Source: Lu et al. [60]/with permission of Elsevier.

size of these secondary nanoparticles is on the order of 10 nm, they are in general highly reactive resulting from the high surface free energy and dangling bond density; simultaneously, LDOs possess a strong chemical adsorption affinity toward acid gases (such as CO_2, SO_2, and NO_x) due to their intrinsic alkalinity. These unique properties endow them with promising applications in membrane-based separation.

Incorporation of LDOs exhibiting preferential adsorption toward acid gases represents a new concept for membrane performance enhancement. For instance, A.D. Wiheeb et al. fabricated HT–silica composite membranes through a sol–gel procedure followed by dip coating and calcination [63]. It was found that the CO_2 adsorption was notably improved when calcined HT was incorporated in silica membranes; simultaneously, the average pore size was increased. The CO_2/CH_4 separation performance of prepared HT–silica membranes not only easily surpassed the 2008 upper bound for glassy and polymeric membranes but also outperformed state-of-art microporous silica and ZIF-8 membranes. In their recent work, molecular transport mechanisms in HT–silica membranes were further elucidated. Research results implied that surface affinity derived from calcined HT additives contributed to the preferential diffusion of CO_2 through HT–silica membranes [64, 65].

In the follow-up work, the content of calcined HT on the separation performance of microporous HT–silica membranes was systematically investigated [66]. It was found that HT–silica membranes containing 15 vol% HT combined the molecular sieving behavior of pristine silica membranes with high CO_2 adsorption capacity of calcined HT components, resulting in concurrent increase in CO_2 permeability and CO_2 permselectivity. Nevertheless, excessive content of HT in the silica structure resulted in the loss of microporous nature of HT–silica membranes despite observable increase in affinity and permeability due to the increase in pore size. A further study of the effect of sintering temperature on membrane performance indicated that HT–silica membranes calcined at 500 °C exhibited the highest selectivity toward CO_2/H_2, CO_2/N_2, and CO_2/CH_4 gas pairs, while further increasing the sintering temperature inevitably led to better strength, improved microstructural integrity, and durability but lower gas selectivity [67].

Besides serving as selective adsorbents for acid gases, Y. Liu and J. Caro et al. pioneered the use of LDOs as metal sources of high-performing MOF membranes (shown in Figure 8.14) [68]. Initially, ZnO buffer layers were obtained by calcination of ZnAl–NO_3 LDH membranes. Subsequently, ZnO buffer layers, serving as the exclusive Zn^{2+} source of ZIF-8 membranes, were subjected to solvothermal treatment with a DMF solution containing 2-methylimidazole. Owing to rigidification of the framework caused by embedded ZnO buffer layers, prepared ZIF-8 membranes exhibited unprecedented gas selectivity for the H_2/CH_4 gas pair (~83.1).

Calcination of LDH membranes also altered their surface morphology by increasing surface roughness that, as mentioned above, may be utilized to enhance the separation performance of LDH-based membranes. Very recently N.X. Wang and co-workers synthesized CoAl–CO_3 LDH membranes composed of vertically aligned hexagonal-shaped LDH nanosheets by *in situ* hydrothermal growth on porous

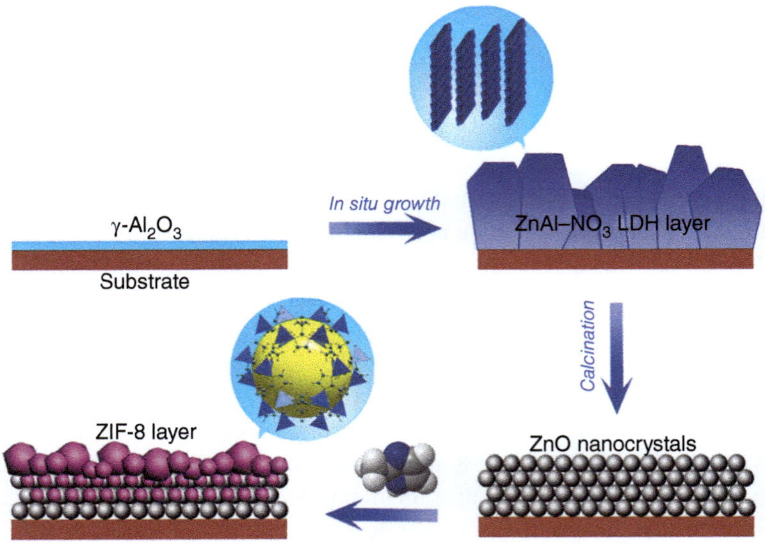

Figure 8.14 Schematic illustration of the fabrication of ZIF-8 membrane by partial conversion of ZnO buffer layer derived from ZnAl–NO_3 LDH membrane. Source: Liu et al. [68]/with permission of John Wiley & Sons.

α-Al_2O_3 tubes [69]. Obtained LDH membranes were then calcined to increase the surface roughness, reduce the surface free energy, and remove interlayer anions. As a result, calcined LDH membranes showed enhanced water stability and permeance for removal of dye molecules from water.

8.3.4 Photocatalytic Activity-Based Separation

Although being still rarely studied, LDHs may also act as high-performance photocatalytic membranes considering their intrinsic photocatalytic activity. F.P. Jiao and co-workers prepared $ZnWO_4$/NiAl–LDH/PVDF microfiltration membranes through a facile vacuum filtration method [70]. Owing to the considerable photocatalytic activity of $ZnWO_4$ under visible-light irradiation and fast photoelectron-hole separation rates resulting from the special heterostructure between $ZnWO_4$ and NiAl–LDHs, prepared $ZnWO_4$/NiAl–LDH/PVDF microfiltration membranes displayed a high degradation performance and membrane separation performance for the synergic removal of model wastewater pollutants. In particular, the high pollutant removal efficiency was ascribed to rapid adsorption of pollutants by loaded photocatalysts and the membrane interceptive role in the separation process. The issue of membrane fouling was mitigated significantly either.

8.3.5 Positive Surface Charge-Based Separation

Owing to the rich surface functionality and high positive surface charge of LDH nanofillers, LDH–polymer nanocomposite membranes may exhibit superior separation performances for fast and efficient removal of organic pollutants (such as oils, dyes, pigments, drugs, natural pollutants, and surfactants), heavy metal ions, and salts from water. In particular, the high positive surface charge of LDH nanofillers in polymer matrix endowed LDH–polymer nanocomposite membranes with positive zeta potential at the interface of two phases so that their rejection behavior toward charged solutes may be variant. In general, their rejection ratio toward positively charged solutes could be enhanced since the same charge repelled with each other.

8.3.5.1 Positive Surface Charge-Based Ultrafiltration

Y.T. Zhang and co-workers facilely exfoliated MgAl–NO_3 LDHs in dimethylacetamide [30]. The resulting colloidal solution was employed for fabrication of LDH/polysulfone (PSf) nanocomposite membranes via phase inversion method. Experimental results indicated that both LDH-based UF and NF membranes exhibited excellent separation performances. UF membranes exhibited higher rejection for poly(ethylene glycol) (PEG) 20 000 but lower protein adsorption amount for positively charged proteins; while NF membranes showed higher rejection for PEG 400 but a much lower value for $MgSO_4$. The above phenomena were attributed to the positively charged effect of prepared UF and NF membranes.

8.3.5.2 Positive Surface Charge-Based Nanofiltration

Y.T. Zhang and co-workers prepared charged nanosheets zwitterion–hydrotalcite (ZHT) by grafting sulfobetaine methacrylate (SBMA) on the surface of positively charged MgAl brucite-like sheets derived from MgAl–NO_3 LDHs via surface-initiated reverse atom transfer radical polymerization (RATRP) [33]. Subsequently, charged mosaic membranes were prepared by embedding different amounts of ZHT into polyethersulfone (PES) casting solution via non-solvent-induced phase separation (NIPS). Experimental results indicated that charged mosaic membranes displayed excellent dyes retention, superior salt permeation, and high water flux. Simultaneously, the retention of various salts remained quite low, which could be attributed to the decreased Donnan effect caused by charged-mosaic membranes (shown in Figure 8.15). In the same manner, Y.T. Zhang and co-workers prepared poly(ionic liquid) brushes-modified MgAl–NO_3 LDHs via RATRP method. Subsequently, novel positively charged loose LDH–PES membranes were prepared with the above NIPS method [34]. LDH–PES composite membranes exhibited facilitated transport of salts without sacrificing dyes rejections.

In a similar way, N. Yusof and coworkers fabricated CuAl–CO_3-modified LDH thin-film composite (TFC) membranes via interfacial polymerization for NF processes [71]. Owing to the positive surface charge intrinsic and intrinsic hydrophilicity of CuAl–CO_3 LDHs, incorporation of LDH nanofillers led to the formation of less negative and more hydrophilic membranes as compared to the pristine TFC membranes. The best-performing NF performances recorded an enhanced water flux and higher rejections toward various salts. In addition, their antifouling properties were improved.

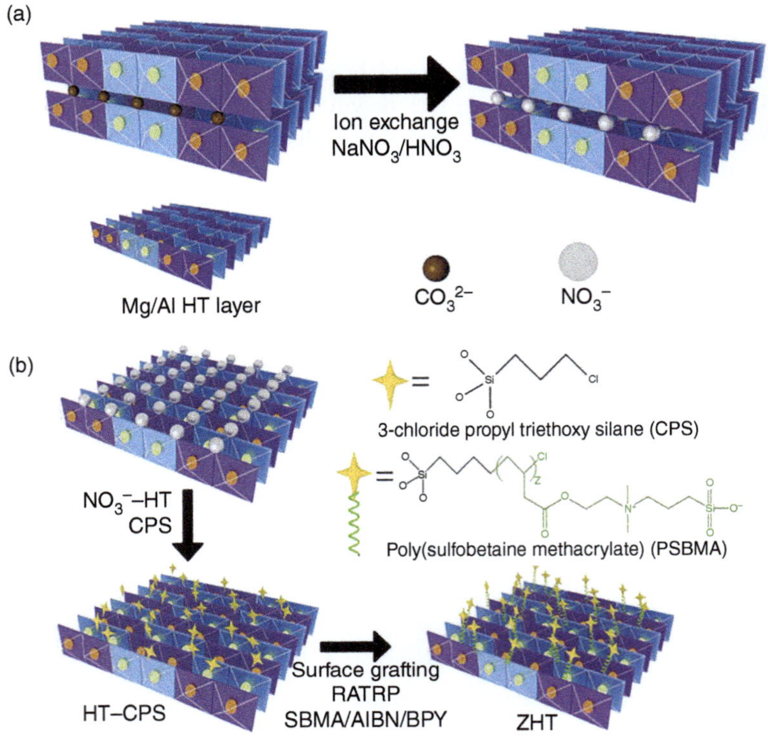

Figure 8.15 Schematic illustration of the preparation of ZHT via (a) anion-exchange of MgAl–NO_3 LDHs and (b) grafting of PSBMA via RATRP. Source: Wang et al. [33]/with permission of Elsevier.

Finally, electrostatic interaction and preferential adsorption of LDH nanofillers may jointly influence the performance of polymer membranes. Relying on the preferential adsorption of organic compounds on LDH nanofillers, A.M. Pandele and co-workers fabricated MgAl–SDS LDH/CA nanocomposite membranes for removal of two common drugs, i.e. diclofenac sodium (DS) and tetracycline (TC), from wastewater [32]. MgAl–SDS LDH nanofillers exhibited a high degree of exfoliation in the polymer matrix. Nanocomposite membranes fabricated under optimized conditions displayed not only an improved water permeability compared with pristine CA but also a ten-fold increase in adsorption capacity for DS, relying on strong electrostatic interactions between negatively charged drug molecules and positively charged MgAl brucite-like sheets. In comparison, the increase in adsorption capacity of TC was smaller, owing to hydrogen bonding interactions between TC molecules and MgAl–SDS LDH nanofillers.

8.3.5.3 Positive Surface Charge-Based Reverse Osmosis

L. Zhang and co-workers reported the use of MgAl–Cl LDH nanosheets as additives in the interfacial polymerization process to fabricate thin-film nanocomposite (TFN) RO membranes [72]. Physicochemical properties of TFN membranes were effectively adjusted by taking advantage of unique charged properties and hydrophilic nature of LDH nanosheets. LDH-filled TFN membranes exhibited enhanced hydrophilicity, and therefore, improved desalination performance. Moreover, the incorporation of LDH nanosheets resulted in a less negatively charged membrane surface so that significantly enhanced antifouling performances toward protein, cationic surfactant, and natural organic matter (NOM) foulants were realized.

8.3.5.4 Positive Surface Charge-Based Forward Osmosis

Relying on electrostatic interactions between LDH nanosheets and sulfonated polysulfone (sPSf), Y. Wang, Y.S. Li, and C.Y. Tang et al. firmly anchored sPSf in the PSf matrix with MgAl–NO_3 LDH nanosheets (shown in Figure 8.16) [73]. Resulting PSf/sPSf-LDH substrates exhibited higher mechanical strength, hydrophilicity, and water permeability in comparison with pristine PSf substrates. Moreover, compared with control TFC@PSf membranes, prepared LDH-loaded TFC membranes displayed increased water permeability, better permeability–selectivity balance, and reduced structural parameters, thereby leading to improved RO and FO performances.

8.3.5.5 Positive Surface Charge-Based Nanocomposite Membranes

Besides in pure form, relying on electrostatic interactions, positively charged LDH nanosheets are inclined to combine with negatively charged compounds (such as GO), serving as competent building blocks of LDH-based nanocomposite membranes. For instance, to overcome the intrinsic low-filtration flux and unstable performance of single-phase GO membranes, P.Y. Wang and co-workers proposed to self-assemble MgAl–CO_3 LDHs with GOs relying on the electrostatic interaction. Subsequently, GO–LDH composites were firmly deposited on PDA–CA substrates through vacuum filtration [74]. Relying on the double-layer channel structure and positive surface charge of LDH nanosheets, the stability and flux of prepared composite membranes were significantly improved compared with neat GO-based membranes.

LDH nanosheets may also contribute to solving several issues (such as swelling and disintegration) of GO membranes in water caused by hydration and electrostatic repulsion. Very recently, W.X. Fang and J. Jin et al. fabricated $Co(OH)_2$ nanosheets

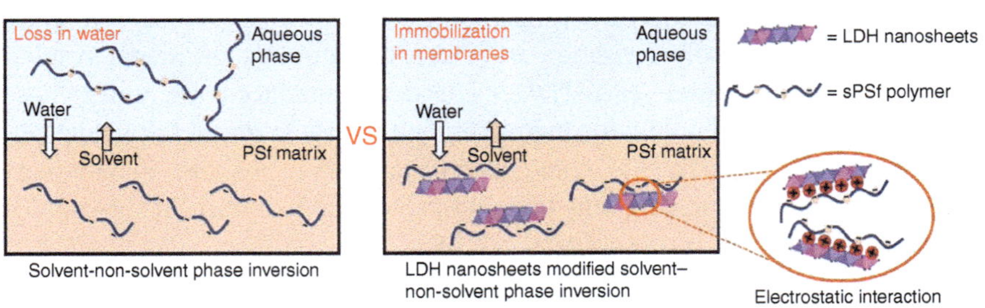

Figure 8.16 Comparison of the traditional solvent–non-solvent phase inversion and the LDH nanosheets modified solvent–non-solvent phase inversion. Source: Lu et al. [73]/with permission of Elsevier.

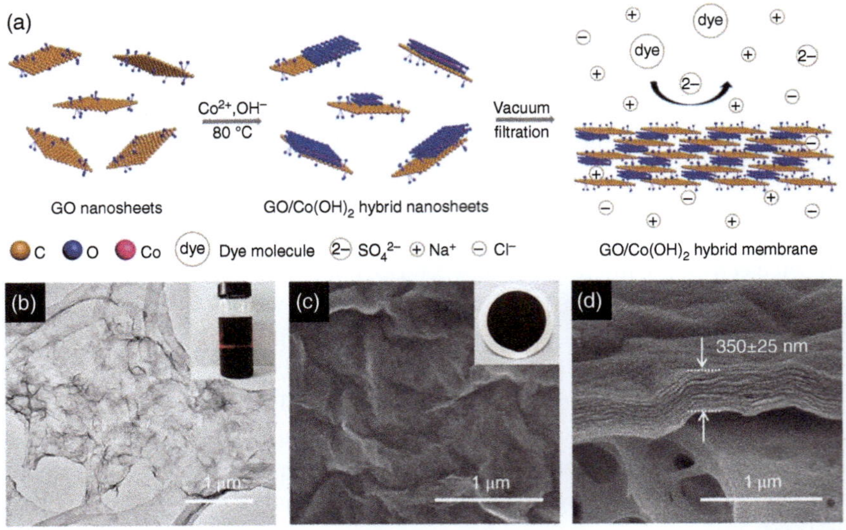

Figure 8.17 (a) Schematic showing the *in situ* growth of $Co(OH)_2$ nanosheets on the surface of GO nanosheets to form the $GO/Co(OH)_2$ hybrid nanosheets (GC), and the fabrication of GC hybrid membranes. (b) TEM image of GC hybrid nanosheets. (c) Top-view SEM image and (d) cross-sectional SEM image of GC hybrid membrane. The inserts in (b) and (c) are digital photographs of the uniform dispersion of GC nanosheets and the GC hybrid membrane, respectively. Source: Dong et al. [75]/with permission of Elsevier B. V.

by *in situ* growth on the surface of GO nanosheets, followed by coating obtained $GO/Co(OH)_2$ hybrid nanosheets on microporous substrates through vacuum suction to form $GO/Co(OH)_2$ hybrid membranes (shown in Figure 8.17) [75]. Owing to electrostatic and coordination interactions between GO and $Co(OH)_2$ nanosheets, the stability of $GO/Co(OH)_2$ hybrid membranes was greatly enhanced; in addition, prepared hybrid membranes survived in different pH environments, elevated transmembrane pressure, and long-term continuous crossflow test.

Besides directly serving as inorganic membranes, LDH–GO composites could act as competent nanofillers of polymer membranes. For instance, Y.T. Zhang and co-workers synthesized exfoliated MgAl–NO_3 LDH/GO (EHT/GO) hybrid nanosheets, which could be uniformly dispersed in PEI solution, via flocculation deposition [76]. Subsequently, ETH/GO–PEI thin-film selective layers were formed via interfacial polymerization on PES substrates. Experimental results indicated that incorporation of EHT/GO effectively enhanced the water flux. Moreover, the salt rejection of nanocomposite membranes followed the sequence: $MgCl_2 > NaCl > MgSO_4 > Na_2SO_4$, thereby confirming that prepared membranes were indeed positively charged, owing to the positively charged nature of MgAl–NO_3 LDH nanosheets.

8.3.6 Hydrophilicity-Based Water Treatment

Owing to the presence of substantial hydrophilic functional groups (i.e. hydroxide functional groups) on brucite-like sheets, LDH-based composite membranes are intrinsically highly hydrophilic, which is beneficial to improve water flux, rejection rate, fouling resistance, and long-term stability in water treatment.

8.3.6.1 Hydrophilicity-Based Microfiltration

S. Aber and co-workers fabricated microfiltration MMMs by embedding the Ag_3PO_4/ZnAlCu–CO_3 LDH nanocomposites into the PES matrix [77]. Experimental results implied that incorporation of Ag_3PO_4/ZnAlCu–CO_3 LDH nanocomposites increased the hydrophilicity, porosity, surface pore diameter, and antibacterial properties, thereby leading to enhanced antifouling properties. Compared with pristine PES membranes, there was an 18% increase in the pure water flux with a nanocomposite content of 0.5 wt%. Simultaneously, the flux recovery ratio of prepared antibiofouling membrane was increased from 50.9% to 62.3%. Later the influence of different additives and solvents on the separation performances was investigated further [78]. Among them, polyvinylpyrrolidone 29 000 and *N*-methyl-2-pyrrolidone (NMP) were considered as the optimum additives and solvents, respectively. Particularly, it was noted that, upon using NMP as the solvent, the pore geometry was changed from sponge-like to finger-like, resulting in remarkable reduction in diffusion resistance and therefore, improvement in permeability [79].

In the same manner, Z.X. Yu et al. constructed novel MgAl–CO_3 LDH/SeP composite membranes by layer-by-layer processing on CA substrates [80]. Owing to the excellent hydrophilicity and underwater superoleophobicity, prepared membranes exhibited an enhanced water flux while keeping a high dyes rejection. In addition, they exhibited a multifunctional characteristic for efficient removal of various oil/water emulsions and heavy metal ions in water due to the layer-by-layer sieving. Particularly, during the whole separation process prepared membranes exhibited excellent environmental and chemical stabilities.

8.3.6.2 Hydrophilicity-Based Ultrafiltration

A. Khataee and co-workers prepared MgAl–CO_3 LDH-filled PVDF UF membranes with the phase inversion method [81]. Owing to the presence of large amount of hydroxyl groups on the layers MgAl brucite-like sheets, prepared blended membranes exhibited enhanced hydrophilicity as well as increased porosity and average pore size. As a result, prepared blended membranes exhibited not only increased pure water flux but also improved antifouling performance during the whey protein concentration.

8.3.6.3 Hydrophilicity-Based Nanofiltration

N. Yusof and co-workers prepared CuAl–NO_3 LDH-incorporated polyamide (PA) TFN membranes via *in situ* polymerization [82]. In the presence of LDHs, not only did zeta potential became less negative but also the membrane surface became more hydrophilic and smoother. Prepared TFN membranes exhibited increased water flux (54.62 $l\,m^{-2}\,h^{-1}$ at 7 bar and 25 °C) and salt rejection (~96%, $MgSO_4$); simultaneously, the fouling resistance and pure water flux recovery of the membrane after bovine serum albumin filtration were enhanced.

In the same manner, S.J. Xia and co-workers prepared ZnAl–NO_3 LDH-modified PA NF membranes by interfacial polymerization [83]. Compared with raw PA membranes, LDH-incorporated PA layers became more hydrophilic and denser with more leaf-like folds, resulting in ~40.2% increase in water permeability and improved rejection toward NOM, as well as better antifouling properties. Simultaneously, the same authors prepared ZnAl–NO_3 LDH-modified PVDF UF membranes [84]. It was noted that owing to the presence of LDHs, small sponge-like and narrow finger-like pores were created. Combing with the significantly enhanced hydrophilicity, both water flux and antifouling performance of prepared membranes were greatly improved. In addition, they reported the fabrication of ZnAl–NO_3 LDH–PES UF membranes by a phase-inversion method [85]. It was observed that significant change in membrane morphology occurred and better hydrophilicity, water permeability, and antifouling ability were achieved for the PES/LDH membranes.

Besides in their pure form, Z.X. Yu and co-workers prepared LDH/GO/PVDF nanocomposite membranes via vacuum suction [86]. Relying on the intrinsic hydrophilicity of MgAl–CO_3 LDHs and their electrostatic interactions with GOs, the incorporation of LDHs significantly improved the hydrophilicity and mechanical stability of prepared membranes, resulting in excellent removal rates for Cu^{2+} and methylene blue over a long period of time in comparison with pristine GO/PVDF membranes.

8.3.6.4 Hydrophilicity-Based Reverse Osmosis

X.X. Tian et al. prepared TFN RO membranes by dispersing APTS-modified MgAl–CO_3 LDH nanoparticles in aqueous solution and incorporating these nanoparticles in PA layers during the interfacial polymerization process [31]. Owing to strong covalent interactions between nanoparticles and PA matrix, dispersibility of nanoparticles in aqueous solution and obtained membranes was improved, resulting in effective suppression of the aggregation, which was quite beneficial to enhance the compatibility between nanoparticles and the PA matrix. Compared with pristine RO membranes, TFN RO membranes incorporated with APTS-modified MgAl–CO_3 LDH nanoparticles exhibited better separation performances (in particular, selectivity) due to the introduction of high-speed water channels in APTS-modified MgAl–CO_3 LDH nanoparticles as well as their excellent compatibility with the PA matrix.

8.3.6.5 Hydrophilicity-Based Forward Osmosis

In recent decades, FO desalination process has been considered as a promising alternative to the conventional RO technology due to the low hydraulic pressure, high rejection to diverse contaminants, and enhanced antifouling properties. Several recent studies suggested that the incorporation of LDH additives into polymer membranes enabled considerable improvement of overall FO performances, at least, partially owing to the increased hydrophilicity.

Q. Wang and co-workers proposed the fabrication of novel TFN FO membranes based on MgAl–CO_3 LDH-blended UF substrates [87]. The addition of LDH nanoparticles endowed UF substrates with increased porosity, surface hydrophilicity, surface pore aperture, mechanical strength, and thermal stability. Prepared membranes exhibited increased water permeability with little compromise in reverse salt flux and alleviated ICP compared with pristine FO membranes.

Besides in pure LDH form, Q. Wang and co-workers further reported the use of LDH/GO hybrid as nanofillers of PSf substrates in the fabrication of TFN FO membranes [88]. Experimental results indicated that the addition of MgAl–CO_3 LDH/GO hybrid increased hydrophilicity, porosity, surface pore diameter, and mechanical strength of PSf substrates. As a result, obtained TFN membranes exhibited increased water permeability and salt rejection in comparison with pristine TFC membranes. Overall, the improvement in FO performance was ascribed to the lower structural parameters of the modified PSf substrates and the reduction of the ICP.

Moreover, to avoid the degradation of PA layers in TFC membranes due to chlorine-based chemical cleaning, Q. Wang and co-workers dip-coated MgAl–CO_3 LDH nanoparticles on TFC membranes by virtue of a PDA-induced immobilization process [89]. The accelerated chlorination experiments using 1000 ppm NaOCl indicated significantly enhanced chlorine-resistant performance of deposition-modified TFC membranes (the chlorine-resistant time prolonged ~96 times). MgAl–CO_3 LDH layers were served as a barrier to protect PA layers from chlorination. MgAl–CO_3 LDH-modified membranes exhibited a promising antifouling capability.

Y. Liu and Y.S. Li et al. further investigated the influence of the location of MgAl–CO_3 LDH additives on membrane degradation and fouling [90]. Four different FO membranes, i.e. TFC (PA membrane on PSf substrate), TFC–LDH (PA membrane on PSf substrate blended with LDHs), LDH@TFC (LDH post-modified PA membrane on PSf substrate), and LDH@TFC–LDH (LDH post-modified PA membrane on PSf substrate blended with LDHs), were prepared. Although the water flux of LDH@TFC–LDH was lower than TFC–LDH and the reverse salt flux was lower than LDH@TFC, the LDH@TFC–LDH was more hydrophilic and porous, which endowed the LDH@TFC–LDH with superior antifouling capacity and chlorine resistance without significantly sacrificing the water flux during the FO process.

In the follow-up work, A.A. Mungray and co-workers utilized the MgAl–CO_3 LDH–polyvinyl chloride composite as substrate material for the preparation of FO membranes [91]. It was found that LDH-containing substrates were more hydrophilic and porous while possessing lower value of structural parameter than the substrate in the absence of LDHs, resulting in a high water flux using deionized water as a feed solution and 1 M NaCl draw solution under an active layer facing feed solution and an active layer facing draw solution, respectively.

8.4 Summary and Outlook

LDHs, as representative of 2D layered compounds, possess an array of unique properties and features in terms of (i) uniform and adjustable interlayer galleries,

(ii) usual thermal behavior, (iii) memory effect, (iv) facile exfoliation, (v) plate-like morphology, (vi) high positive surface charge, (vii) rich surface functionality, (viii) anion exchangeability, (ix) hydrophilicity, and (x) low cost and facile synthesis. These properties and features have endowed them with unprecedented opportunities for applications in a variety of membrane-based separation processes, such as pervaporation, microfiltration, ultrafiltration, nanofiltration, reverse osmosis, forward osmosis, gas separation, and particulate matter capture. It should be noted that besides in their pure form, in most cases, LDHs could also serve as additives of various inorganic composite membranes and MMMs, thereby endowing them superior microstructures and enhanced separation performances; in addition, exfoliation of LDH bulk crystals further enabled us to tailor the membrane structure and function on a nanometer scale.

Nevertheless, some major issues concerning LDH-based separation membranes should be fully addressed before delving into practical applications. First, in-depth investigation of the surface and interface properties of LDH-based membranes is required since in many cases, the presence of nonselective intercrystalline defects and interfacial voids as the dominant diffusion paths will lead to significant deterioration of the separation performance of LDH-based membranes. Nevertheless, their in-depth structural elucidation remains rarely reported. Second, clear elucidation of the structure–performance relationship and precise manipulation of membrane microstructure. Since the separation performance of LDH-based membranes strongly depends on their microstructure (such as the preferred orientation, intercrystalline defects, interfacial voids, and thickness), the development of facile, inexpensive, and reproducible fabrication methods for accurate and deliberate control of their microstructure is still urgently required. Third, reproducible and reliable methods for the bulk production of LDH-based separation membranes should be established. Unfortunately, to date, there has been no published report on large-scale production of LDH-based separation membranes. To promote their commercial applications in the future, at least, pilot-scale production and application of LDH-based separation membranes should be carried out in the upcoming decades.

References

1 Khan, A.I. and O'Hare, D. (2002). Intercalation chemistry of layered double hydroxides: recent developments and applications. *J. Mater. Chem.* 12: 3191–3198.

2 Cygan, R.T., Greathouse, J.A., Heinz, H., and Kalinichev, A.G. (2009). Molecular models and simulations of layered materials. *J. Mater. Chem.* 19 (17): 2470–2481.

3 Wang, Q. and O'Hare, D. (2012). Recent advances in the synthesis and application of layered double hydroxide (LDH) nanosheets. *Chem. Rev.* 112 (7): 4124–4155.

4 Guo, X., Zhang, F., Evans, D.G., and Duan, X. (2010). Layered double hydroxide films: synthesis, properties and applications. *Chem. Commun.* 46 (29): 5197–5210.

5 Das, J., Patra, B.S., Baliarsingh, N., and Parida, K.M. (2006). Adsorption of phosphate by layered double hydroxides in aqueous solutions. *Appl. Clay Sci.* 32 (3–4): 252–260.

6 Lei, C.S., Zhu, X.F., Zhu, B.C. et al. (2017). Superb adsorption capacity of hierarchical calcined Ni/Mg/Al layered double hydroxides for Congo red and Cr(VI) ions. *J. Hazard. Mater.* 321: 801–811.
7 Sonf, F. and Hu, X.L. (2014). Exfoliation of layered double hydroxides for enhanced oxygen evoluation catalysis. *Nat. Commun.* 5: 4477.
8 Hu, X.L. and Song, F. (2014). Ultrathin cobalt-manganese layered double hydroxide is an efficient oxygen evolution catalyst. *J. Am. Chem. Soc.* 136 (47): 16481–16484.
9 Fan, G.L., Li, F., Evans, D.G., and Duan, X. (2014). Catalytic applications of layered double hydroxides: recent advances and perspectives. *Chem. Soc. Rev.* 43 (20): 7040–7066.
10 Leroux, F. and Besse, J.P. (2001). Polymer interleaved layered double hydroxide: a new emerging class of nanocomposites. *Chem. Mater.* 13 (10): 3507–3515.
11 Evans, D.G. and Duan, X. (2006). Preparation of layered double hydroxides and their applications as addtives in polymers, as precursors to magnetic materials and in biologiy and medicine. *Chem. Commun.* 42 (5): 485–496.
12 Gao, Y.S., Wu, J.W., Wang, Q. et al. (2014). Flame retardant polymer/layered double hydroxide nanocomposites. *J. Mater. Chem. A* 2 (29): 10996–11016.
13 Shao, M.F., Zhang, R.K., Li, Z.H. et al. (2015). Layered double hydroxides toward electrochemical energy storage and conversion: design, synthesis and applications. *Chem. Commun.* 51 (88): 15880–15893.
14 Wang, X., Sumboja, A., Lin, M.F. et al. (2012). Enhancing electrochemical reaction sites in nickel-cobalt layered double hydroxides on zinc tin oxide nanowires: a hybrid material for an asymmetric supercapacitor device. *Nanoscale* 4 (22): 7266–7272.
15 Li, Y., Zhang, L., Xiang, X. et al. (2014). Engineering of ZnCo-layered double hydroxide nanowalls twoard high-efficiency electrochemical water oxidation. *J. Mater. Chem. A* 2 (33): 13250–13258.
16 Choy, J.H., Jung, J.S., Oh, J.M. et al. (2004). Layered double hydroxide as an efficient drug reservior for folate derivatives. *Biomaterials* 25 (15): 3059–3064.
17 Li, B.X., He, J., Evans, D.G., and Duan, X. (2004). Inorganic layered double hydroxides as a drug delivery system-intercalation and in vitro release of fenbufen. *Appl. Clay Sci.* 27 (3–4): 99–207.
18 Rives, V., del Arco, M., and Martin, C. (2014). Intercalation of drugs in layered double hydroxides and their controlled release: a review. *Appl. Clay Sci.* 88–89: 239–269.
19 Lu, P., Liu, Y., Zhou, T.T. et al. (2018). Recent advances in layered double hydroxides (LDHs) as two-dimensional membrane materials for gas and liquid separations. *J. Membr. Sci.* 567: 89–103.
20 Kim, T.W., Sahimi, M., and Tsotsis, T.T. (2008). Preparation of hydrotalcite thin films using an electrophoretic technique. *Ind. Eng. Chem. Res.* 47 (23): 9127–9132.
21 Kim, T.W., Sahimi, M., and Tsotsis, T.T. (2009). The preparation and characterization of hydrotalcite micromembranes. *Chem. Eng. Sci.* 64 (7): 1585–1590.
22 Kim, T.W., Sahimi, M., and Tsotsis, T.T. (2009). The preparation and characterization of hydrotalcite thin films. *Ind. Eng. Chem. Res.* 48 (12): 5794–5801.

23 Liu, Y., Wang, N.Y., Cao, Z.W., and Caro, J. (2014). Molecular sieving through interlayer galleries. *J. Mater. Chem. A* 2 (5): 1235–1238.

24 Liu, Y., Wang, N.Y., and Caro, J. (2014). *In situ* formation of LDH membranes of different microstructures with molecular sieve gas selectivity. *J. Mater. Chem. A* 2 (16): 5716–5723.

25 Li, X.T., Wang, N.X., Huang, Z. et al. (2018). A vertically channeled lamellar membrane for molecular sieving of water from organic solvents. *J. Mater. Chem. A* 6 (37): 18095–18102.

26 Wang, N.X., Huang, Z., Li, X.T. et al. (2018). Tuning molecular sieving channels of layered double hydroxides membrane with direct intercalation of amino acids. *J. Mater. Chem. A* 6 (35): 17148–17155.

27 Yu, J., Wang, Q., O'Hare, D., and Sun, L. (2017). Preparation of two dimensional layered double hydroxide nanosheets and their applications. *Chem. Soc. Rev.* 46 (19): 5950–5974.

28 Hibino, T. and Ohya, H. (2009). Synthesis of crystalline layered double hydroxides: precipitation by using urea hydrolysis and subsequent hydrothermal reactions in aqueous solutions. *Appl. Clay Sci.* 45 (2): 123–132.

29 Ma, R.Z., Liu, Z.P., Li, L. et al. (2006). Exfoliating layered double hydroxides in formamide: a method to obtain positively charged nanosheets. *J. Mater. Chem.* 16 (39): 3809–3813.

30 Liu, J.D., Yu, L., and Zhang, Y.T. (2014). Fabrication and characterization of positively charged hybrid ultrafiltration and nanofiltration membranes via the in-situ exfoliation of Mg/Al hydrotalcite. *Desalination* 335 (1): 78–86.

31 Tian, X.X., Cao, Z., Wang, J. et al. (2020). Development of high-performance mixed matrix reverse osmosis membranes by incorporating aminosilane-modified hydrotalcite. *RSC Adv.* 10 (10): 5648–5655.

32 Raicopol, M.D., Andronescu, C., Voicu, S.I. et al. (2019). Cellulose acetate/layered double hydroxide adsorptive membranes for efficient removal of pharmaceutical environmental contaminants. *Carbohy. Polym.* 214: 204–212.

33 Wang, J., Zhang, Y.T., Zhu, J.Y. et al. (2016). Zwitterionic functionalized layered double hydroxides nanosheets for a novel charged mosaic membrane with high salt permeability. *J. Membr. Sci.* 510: 27–37.

34 Yu, L., Deng, J.M., Wang, H.X. et al. (2016). Improved salts transportation of a positively charged loose nanofiltration membrane by introduction of poly(ionic liquid) functionalized hydrotalcite nanosheets. *ACS Sust. Chem. Eng.* 4 (6): 3292–3304.

35 Liu, Y., Pan, J.H., Wang, N.Y. et al. (2015). Remarkably enhanced gas separation by partial conversion of a laminated membrane to metal-organic frameworks. *Angew. Chem. Int. Ed.* 54 (10): 3028–3032.

36 Bux, H., Liang, F.Y., Li, Y.S. et al. (2009). Zeolitic imidazolate framework membrane with molecular sieving properties by microwave-assisted solvothermal synthesis. *J. Am. Chem. Soc.* 131 (44): 16000–16001.

37 Liu, H.-X., Zhao, C., Wang, N.X. et al. (2018). Nanosheet α-$Co(OH)_2$ composite membranes with ultrathin separation layer for removing dyes from solvent with high flux. *Sep. Purif. Technol.* 207: 506–513.

38 Zhao, S., Zhu, H.T., Wang, Z. et al. (2018). A loose hybrid nanofiltration membrane fabricated via chelating-assisted in-situ growth of Co/Ni LDHs for dyre wastewater treatment. *Chem. Eng. J.* 353: 460–471.

39 Liu, Y.T., Wu, H., Min, L.F. et al. (2020). 2D layered double hydroxide membranes with intrinsic breathing effect toward CO_2 for efficient carbon capture. *J. Membr. Sci.* 598: 117663.

40 Liao, J.Y., Wang, Z., Gao, C.Y. et al. (2014). Fabrication of high-performance facilitated transport membranes for CO_2 separation. *Chem. Sci.* 5: 2843–2849.

41 Liao, J.Y., Wang, Z., Gao, C.Y. et al. (2015). A high performance PVAm-HT membrane containing high-speed facilitated transport channels for CO_2 separation. *J. Mater. Chem. A* 3 (7): 16746–16761.

42 Ang, E.H., Velioğlu, S., and Chew, J.W. (2019). Tunable affinity separation enables ultrafast solvent permeation through layered double hydroxide membranes. *J. Membr. Sci.* 591: 117318.

43 Wang, N.X., Li, Q., Li, X.T. et al. (2020). Vaccum-assisted assembly of iron cage intercalated layered double hydroxide composite membranes for water purification. *J. Membr. Sci.* 603: 118032.

44 Ma, Z.B., Zhang, S., Chen, G.H. et al. (2019). Superhydrophilic and oleophilic membrane functionalized with heterogeneously tailored two-dimensional layered double hydroxide nanosheets for antifouling. *J. Membr. Sci.* 577: 165–175.

45 Liu, Y., Wang, N.Y., Diestel, L. et al. (2014). MOF membrane synthesis in the confined spaced of a vertically aligned LDH network. *Chem. Commun.* 50 (32): 4225–4227.

46 Liu, Y., Wang, N.Y., Pan, J.H. et al. (2014). In-situ synthesis of MOF membranes on ZnAl-CO_3 LDH buffer layer-modified substrates. *J. Am. Chem. Soc.* 136 (41): 14353–14356.

47 Liu, Y., Matsuda, R., Kusaka, S. et al. (2018). Insights into inorganic buffer layer-assisted *in situ* fabrication of MOF films with controlled microstrucures. *CrystEngComm* 20 (43): 6995–7000.

48 Fan, H.W., Peng, M.H., Strauss, I. et al. (2020). High-flux vertically aligned 2D covalent organic framework membrane with enhanced hydrogen separation. *J. Am. Chem. Soc.* 142 (15): 6872–6877.

49 Chi, Y.-H., Uan, J.-Y., Lin, M.-C. et al. (2013). Preparation of a novel Pd/layered double hydroxide composite membrane for hydrogen filtration and characterization by thermal cycling. *Int. J. Hydrogen Energy* 38 (31): 13734–13741.

50 Zhang, Y.M., Wang, H.X., Zhang, Y.T. et al. (2017). Thin film composite membranes functionalized with montmorillonite and hydrotalcite nanosheets for CO_2/N_2 separation. *Sep. Purif. Technol.* 189: 128–137.

51 Li, X.M., Reinhoudt, D., and Crego-Calama, M. (2007). What do we need for a superhydrophobic surface? A review on the recent progress in the preparation of superhydrophobic surfaces. *Chem. Soc. Rev.* 36 (8): 1350–1368.

52 Zhang, X., Shi, F., Niu, J. et al. (2008). Superhydrophobic surfaces: from structural control to functional application. *J. Mater. Chem.* 18 (6): 621–633.

53 Liu, X.J., Ge, L., Li, W. et al. (2015). Layered double hydroxide functionalized textile for effective oil/water separation and selective oil adsorption. *ACS Appl. Mater. Interfaces* 7 (1): 791–800.

54 Yue, X.J., Zhang, T., Yang, D.Y. et al. (2018). Oil removal from oily water by a low-cost and durable flexible membrane made of layered double hydroxide nanosheet on cellulose support. *J. Clean. Prod.* 180: 307–315.

55 Yue, X.J., Li, J.X., Zhang, T. et al. (2017). In situ one-pot fabrication of durable superhydrophobic-superoleophilic cellulose/LDH membrane with hierarchical structure for efficiency oil/water separation. *Chem. Eng. J.* 328: 117–123.

56 Cui, J.Y., Zhou, Z.P., Xie, A.T. et al. (2019). Facile preparation of grass-like structured NiCo-LDH/PVDF composite membrane for efficient oil-water emulsion separation. *J. Membr. Sci.* 573: 226–233.

57 Cui, J.Y., Wang, Q.Q., Xie, A.T. et al. (2019). Construction of superhydrophilic and underwater superoleophilic membranes via in situ oriented NiCo-LDH growth for gravity-driven oil/water emulsion separation. *J. Taiwan Inst. Chem. Eng.* 104: 240–249.

58 Li, F., Gao, R.T., Wu, T., and Li, Y.J. (2017). Role of layered materials in emulsified oil/water separation and anti-fouling performance of modified cellulose acetate membranes with hierarchical structure. *J. Membr. Sci.* 543: 163–171.

59 Cheng, Y., Li, L.K., He, W.J. et al. (2020). Seeds embedded epitaxial growth strategy for PAN@LDH membrane with Mortise-Tenon structure as efficient adsorbent for particulate matter capture. *Appl. Catal., B* 263: 118312.

60 Lu, P., Li, W.J., Yang, S. et al. (2019). Layered double hydroxides (LDHs) as novel macropore-templates: the importance of porous substructures for forward osmosis desalination. *J. Membr. Sci.* 585: 175–183.

61 Valente, J.S., Hernandez-Cortez, J., Cantu, M.S. et al. (2010). Calcined layered double hydroxides Mg-Me-Al (Me: Cu, Fe, Ni, Zn) as bifunctional catalysts. *Catal. Today* 150 (3-4): 340–345.

62 Zhao, X.F., Zhang, F.Z., Xu, S.L. et al. (2010). From layered double hydroxides to ZnO-based mixed metal oxides by thermal deposition: transformation mechanism and UV-blocking properties of the product. *Chem. Mater.* 22 (13): 3933–3942.

63 Wiheeb, A.D., Kim, J., and Othman, M.R. (2006). Highly perm-selective micro-porous hydrotalcite-silica membrane for improved carbon dioxide-methane separation. *Sep. Sci. Technol.* 50 (11): 1701–1708.

64 Wiheeb, A.D., Ahmad, M.A., Murat, M.N. et al. (2014). Identification of molecular transport mechanisms in micro-porous hydrotalcite-silica membrane. *Transp. Porous. Med.* 104 (1): 133–144.

65 Wiheeb, A.D., Ahmad, M.A., Murat, M.N. et al. (2014). Predominant gas transport in microporous hydrotalcite-silica membrane. *Transp. Porous. Med.* 102 (1): 59–70.

66 Wiheeb, A.D., Ahmad, M.A., Murat, M.N. et al. (2014). The effect of hydrotalcite content in microporous composite membrane on gas permeability and permselectivity. *Sep. Sci. Technol.* 49 (9): 1309–1316.

67 Wiheeb, A.D., Ahmad, M.A., Murat, M.N. et al. (2014). The declining affinity of microporous hydrotalcite-silica membrane for carbon dioxide. *J. Porous Media* 17 (2): 159–167.

68 Liu, Y., Peng, Y., Wang, N. et al. (2015). Significantly enhanced separation using ZIF-8 membranes by partial conversion of calcined layered double hydroxide precursors. *ChemSusChem* 8 (21): 3582–3586.

69 Huang, Z., Wang, N.X., Li, X.T., and An, Q.-F. (2020). Calcination of layered double hydroxide membrane with enhanced nanofiltration performance. *J. Ind. Eng. Chem.* 89: 368–374.

70 Zhao, G.Q., Zou, J., Chen, X.Q. et al. (2020). Integration of microfiltration and visible-light-driven photocatalysis on a $ZnWO_4$ nanoparticle/nickel-aluminum-layered double hydroxide membrane for enhanced water purification. *Ind. Eng. Chem. Res.* 59 (14): 6479–6487.

71 Tajuddin, M.H., Yusof, N., Azelee, I.W. et al. (2019). Development of copper-aluminum layered double hydroxide in thin film nanocomposite nanofiltration membrane for water purification process. *Front. Chem.* 7 (2): 3.

72 Dong, H., Wu, L.G., Zhang, L. et al. (2015). Clay nanosheets as charged filler materials for high-performance and fouling-resistant thin film nanocomposite membranes. *J. Membr. Sci.* 494: 92–103.

73 Lu, P., Wang, Y., Wang, L. et al. (2020). Immobilization of sulfonated polysulfone via 2D LDH nanosheets during phase-inversion: a novel strategy towards greener membrane synthesis and enhanced desalination performance. *J. Membr. Sci.* 614: 118508.

74 Zhong, F., Wang, P.Q., He, Y. et al. (2019). Preparation of stable and superior flux GO/LDH/PDA-based nanofiltration membranes through electrostatic self-assembly for dye purification. *Polym. Adv. Technol.* 30 (7): 1644–1655.

75 Dong, Y.P., Lin, C., Gao, S.J. et al. (2020). Single-layered GO/LDH hybrid nanoporous membranes with improved stability for salt and organic molecules rejection. *J. Membr. Sci.* 607: 118184.

76 Wang, X., Wang, H.X., Wang, Y.M. et al. (2019). Hydrotalcite/graphene oxide hybrid nanosheets functionalized nanofiltration membrane for desalination. *Desalination* 451: 209–218.

77 Ghalamchi, L., Aber, S., Vatanpour, V., and Kian, M. (2019). Development of an antibacterial and visible photocatalytic nanocomposite microfiltration membrane incorporated by Ag_3PO_4/CuZnAl NLDH. *Sep. Purif. Technol.* 226: 218–231.

78 Oskoui, S.A., Vatanpour, V., and Khataee, A. (2019). Effect of different additives on the physicochemical properties and performance of NLDH/PVDF nanocomposite membrane. *Sep. Purif. Technol.* 209: 921–935.

79 Arefi-Oskoui, S., Khataee, A., and Vatanpour, V. (2017). Effect of solvent type on the physicochemical properties and performance of NLDH/PVDF nanocomposite ultrafiltration membranes. *Sep. Purif. Technol.* 184: 97–118.

80 Yu, Z.X., Li, X.H., Peng, Y.X. et al. (2019). MgAl-layered-double-hydroxide/sepiolite composite membrane for high-performance water treatment based on layer-by-layer hierarchical architectures. *Polymers* 11 (3): 525.

81 Arefi-Oskoui, S., Vatanpour, V., and Khataee, A. (2016). Development of a novel high-flux PVDF-based ultrafiltration membrane by embedding Mg-Al nanolayered double hydroxide. *J. Ind. Eng. Chem.* 41: 23–32.

82 Tajuddin, M.H., Yusof, N., Abdullah, N. et al. (2019). Incorporation of layered double hydroxide nanofillers in polyamide nanofiltration membrane for high performance of salts rejection. *J. Taiwan Inst. Chem. Eng.* 97: 1–11.

83 Zhao, Y., Li, N.N., and Xia, S.J. (2016). Polyamide nanofiltration membranes modified with Zn-Al layered double hydroxides for natural organic matter removal. *Comp. Sci. Technol.* 132: 84–92.

84 Zhao, Y., Li, N.N., Xu, B. et al. (2016). Preparation and characterization of a novel hydrophilic poly(vinylidene fluoride) filtration membrane incorporated with Zn-Al layered double hydroxides. *J. Ind. Eng. Chem.* 39: 37–47.

85 Zhao, Y., Li, N.N., Yuan, F.Z. et al. (2016). Preparation and characterization of hydrophilic and antifouling poly(ether sulfone) ultrafiltration membranes modified with Zn-Al layered double hydroxides. *J. Appl. Polym. Sci.* 133 (39): 43988.

86 Zeng, H.J., Yu, Z.X., Peng, Y.X., and Zhu, L.L. (2019). Environmentally friendly electrostatically driven self-assembled LDH/GO/PVDF composite membrane for water treatment. *Appl. Clay Sci.* 183: 105322.

87 Lu, P., Liang, S., Qiu, L. et al. (2016). Thin film nanocomposite forward osmosis membranes based on layered double hydroxide nanoparticles blended substrates. *J. Membr. Sci.* 504: 196–205.

88 Lu, P., Liang, S., Zhou, T.T. et al. (2016). Layered double hydroxide/graphene oxide hybrid incorporated polysulfone substrate for thin-film nanocomposite forward osmosis membranes. *RSC Adv.* 6 (61): 56599–56609.

89 Lu, P., Liang, S., Zhou, T.T. et al. (2017). Layered double hydroxide nanoparticle modified forward osmosis membranes via polydopamine immobilization with significantly enhanced chlorine and fouling resistance. *Desalination* 421: 99–109.

90 Lu, P., Li, W.J., Yang, S. et al. (2019). Layered double hydroxide-modified thin-film composite membranes with remarkably enhanced chlorine resistance and anti-fouling capacity. *Sep. Purif. Technol.* 220: 231–237.

91 Pardeshi, P.M., Mungray, A.K., and Mungray, A.A. (2017). Polyvinyl chloride and layered double hydroxide composite as a novel substrate material for the forward osmosis membrane. *Desalination* 421: 149–159.

9

MXene: A Novel Two-Dimensional Membrane Material for Molecular Separation

Guozhen Liu, Gongping Liu, and Wanqin Jin

Nanjing Tech University, College of Chemical Engineering, State Key Laboratory of Materials-Oriented Chemical Engineering, Department name: College of Chemical Engineering, 30 Puzhu South Road, Nanjing, 211816, P.R. China

9.1 Introduction

The discovery of graphene by the Geim group in 2004 has stimulated the research interests toward two-dimensional (2D) materials [1–5]. Unique physicochemical properties, including atomic thickness, micrometer lateral size, and high-aspect ratio, endow 2D materials as a potential candidate for membrane separation. In the past few years, various 2D materials, such as boron nitride (BN) [6], metal–organic framework (MOF) nanosheets [4], transition-metal dichalcogenide (TMD) [7], layered double hydroxide (LHD) [8], and zeolites [9], have been demonstrated to possess great prospect for development of high-performance membranes.

Recently, transition-metal carbides, carbonitrides, and nitrides, labeled as MXene, have been the latest addition to the 2D materials family [10, 11]. Generally, it is expressed using the formula $M_{n+1}X_nT_x$ ($n = 1, 2,$ or 3), where M represents an early transition metal (such as Ti, Zr, V, Nb, Cr, and Mo), A stands for the IIIA or IVA main group element, X is C and/or N elements, which was made by etching of the A layer from the MAX precursor phases in strong acid solution. The wet-etching process endows MXene nanosheets with abundant surface termination oxygen-containing groups (T_x), such as –OH and –O, which favor MXene with good hydrophilicity, versatile chemical tunability, and negative charge [11, 12]. After the first MXene material $Ti_3C_2T_x$ was reported, until now, more than 60 MXene materials have been discovered [13]. It should be noted that most other 2D materials, such as graphene oxide (GO), usually suffer from water stability. Nevertheless, owing to strong van der Waal force or hydrogen bonding, MXene could overcome the water swelling to maintain structural integrity [14, 15]. Furthermore, atomic-thickness MXene nanosheets with micrometer lateral size could be easily obtained via physical methods, such as sonication to be assembled as laminar structure similar to GO nanosheets. These natural physicochemical properties favor MXene to be a promising start for the membrane separation process.

Two-Dimensional-Materials-Based Membranes: Preparation, Characterization, and Applications, First Edition. Edited by Gongping Liu and Wanqin Jin.
© 2022 WILEY-VCH GmbH. Published 2022 by WILEY-VCH GmbH.

Recently, MXene has received widespread attention in membrane separation and few researchers have extensively discussed the status, prospects, and future development of MXene-based membranes [14, 16, 17]. However, research works on MXene-based membranes with outstanding separation efficiency were reported rapidly. Therefore, it is necessary to give a real-time review of MXene materials for membrane separation process. Hence, in this short review, we aim to briefly introduce the recent progress on MXene-based membrane for molecular sieving, including gas separation and water treatment. Besides, we also give a brief overview of synthesis and processing of MXenes and MXene-based membranes. In addition, we also further make prospects with an analysis of the challenges to overcome and an outlook for further research on MXene-based membranes for molecular separation. We hope that this short review will be useful for research scientists who are exploring the application of 2D materials on membrane separation.

9.2 Synthesis and Processing

9.2.1 Synthesis of Multilayered MXene Phases

Multilayered MXene powders are usually prepared from their precursor MAX phases. By taking the $Ti_3C_2T_x$ MXene as an example, the typical synthesis route of Ti_3AlC_2 MAX phase was reported by Naguib et al. [10]. To explain in detail, it was fabricated via ball-milling mixture of Ti_2AlC and TiC with a 1 : 1 molar ratio at 1350 °C for two hours under argon. Then, the resulting bricks were crushed by a mortar and pestle, followed by sieving through different mesh sieves. Due to that, the M—A is metal bond; it is difficult to remove the A layer to obtain MXene via simple mechanical shearing of the precursor MAX phases. Fortunately, the M—A bonds possess high chemical activity, which provides the possibility of selective etching A layer via chemical method to make MXene [13]. The pioneering work on MXene fabrication was reported by Yury Gogotsi and coworkers in 2011 [10], where hollow-fiber (HF) solution was employed as the etching agent and Ti_3AlC_2 MAX powders were immersed into concentrated HF solution, followed by repeated washing with deionized water using centrifugation; finally, $Ti_3C_2T_x$ MXene powders with accordion-like structure were obtained successfully. Furthermore, nearly all types of MXenes could be fabricated based on this method. However, HF solution with strong causticity is very dangerous; it is vitally important and necessary to develop less hazardous and milder etching agents.

Later, Halim et al. reported a milder and less hazardous etchant, NH_4HF_2. Meanwhile, during the etching process, species, such as NH_3 and NH_4^+, could be concomitantly intercalated into MAX phases to remove Al layer [18]. Such observation of intercalation of different species, including metal ions and organic molecules, was also reported by researchers. For example, Lukatskaya et al. intercalated various metal cations (Li^+, Na^+, Mg^{2+}, K^+, NH_4^+, and Al^{3+}) between $Ti_3C_2T_x$ MXene [19]. Besides, Wang and coworkers [20] developed a simple hydrothermal method to prepare $Ti_3C_2T_x$ MXene via exfoliating Al layer with NH_4F. In this work, effects of the ratio of reactants, reaction time, and temperature on the yields of MXene were investigated in detail. Finally, the optimal fabrication condition was that the reacted mixture with 5.0 g of NH_4F and 0.5 g of Ti_3AlC_2 in 60 ml of deionized water (DI) was maintained at 150 °C for 24 hours.

Recently, Xie et al. proposed a fluoride-free strategy via immersing the Ti_3AlC_2 powders into an aqueous NaOH solution at 80 °C for 100 hours followed by a H_2SO_4 hydrothermal treatment [21]. However, this method is time consuming and the transformation from Ti_3AlC_2 to $Ti_3C_2T_x$ is not complete. Besides, Li et al. improved this method to prepare high-purity $Ti_3C_2T_x$ [22]. They immersed 100 mg of Ti_3AlC_2 powders into 25 ml of NaOH (27.5 M) solution, which reacted at 270 °C for 12 hours. The resultant suspension was rinsed in DI water several times, followed by drying in a vacuum oven for 12 hours. Finally, $Ti_3C_2T_x$ powders with 92 wt% purity were successfully prepared. Although this method improves the preparation efficiency and the purity of products, the reaction condition is strict.

In addition, the in situ HF-containing strategy was developed. Ghidiu et al. reported a relatively safe and simple method to prepare high-quality MXene nanosheets via immersing Ti_3AlC_2 powders into a mixture of hydrochloric acid solution (HCl) and LiF solution [23]. Owing to its simplicity and safety, this method is widely adopted to prepare MXene powders. Nevertheless, it should be noted that the transformation of Ti_3AlC_2 to $Ti_3C_2T_x$ was not complete, and some original peaks belonging to Ti_3AlC_2 powders could be still found in the X-ray diffraction (XRD) patterns of Ti_3C_2 powders using the method, which is caused by the insufficient corrosivity of etching agents. Besides, new etching agents including FeF_3, NaF, KF, $NaHF_2$, KHF_2, and Na_4HF_2 were also developed to selective remove 'A' layer of MAX phases [20, 24–26].

9.2.2 Fabrication of Single MXene Flakes

Delamination of multilayer MXeneis powders is the key step to obtain single MXene nanosheets. Generally, the etching process with intercalation strategy endows strong M—A bonds replaced by weaker van der Waals and hydrogen bonds. Thus, it is possible to obtain single or few layers of MXene flakes via simple external force, such as sonication or handshaking. For example, Ghidiu et al. used HCl and LiF to etch Al atomic layer, where the Li^+ easily intercalated into the layers to auxiliary delaminate. Therefore, sonication was applied to exfoliate MXene powders to obtain high-quality nanosheets and yield single or double layers of nanosheets successfully. In addition, Lipatov et al. further improved the synthesis process with LiF/HCl etching method [27]. They named this process as minimally intensive layer delamination (MILD) method. The schematic diagram is shown in Figure 9.1a. In the modified process, the molar ratio of LiF to Ti_3AlC_2 was increased to 7.5 : 1 from 5 : 1, which could provide more Li^+ to intercalate into MAX phases to effectively facilitate removing the Al layer. Hence, manual shaking just could be employed to delaminate the MXene powders to achieve single nanosheets with larger lateral size and less defects compared with sonication (Figure 9.1b–e).

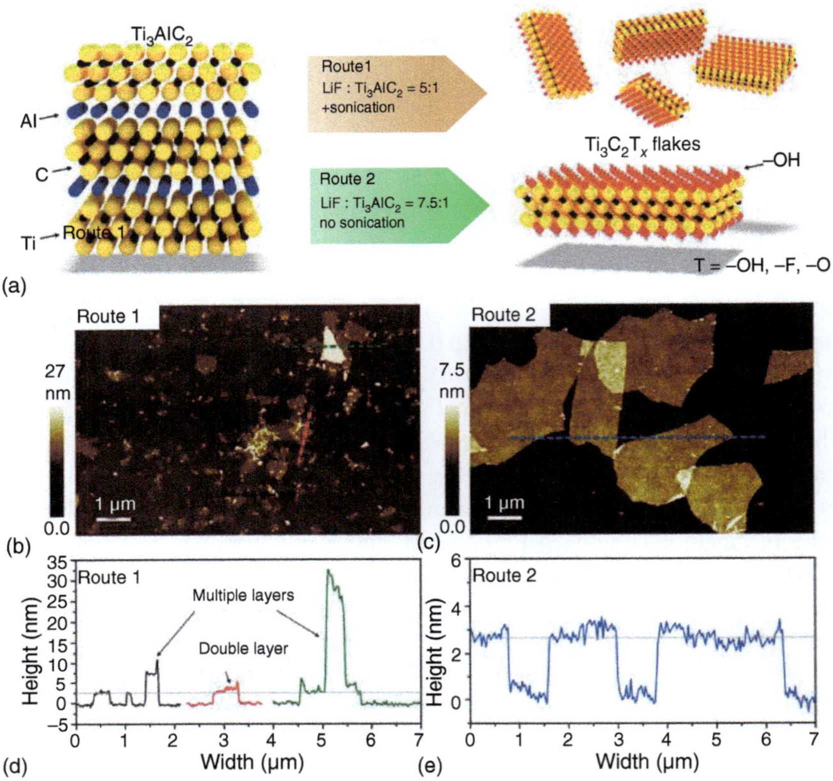

Figure 9.1 (a) Schematic diagram of the clay LiF/HCl method (Route 1) and the MILD method (Route 2); (b and c) AFM images of the $Ti_3C_2T_x$ flakes generated from Routes 1 and 2; (d and e) AFM height profiles measured along with the dashed in (b and c), respectively. Source: Lipatov et al. [27]/with permission of John Wiley & Sons.

Not all MXene powders prepared from different etching method can be exfoliated as single or few layer of flakes via simple bath sonication or handshaking. If there is no intercalation of species during the etching process, such as HF-etching process, a low yield of single/few-layered MXene would be achieved [28]. This phenomenon was first reported by Naguib et al. [10]. They applied sonication in methanol to delaminate the resulted $Ti_3C_2T_x$ powders generated from HF-etching. However, most of the MXene solution after sonication is about ten-layer thick and the yield of single-layered MXene flakes is very low. To resolve this problem to increase the concentration of single of few layered of MXene flakes, species, such as dimethyl sulfoxide (DMSO), can be as fillers to insert into the layer space of multilayered $Ti_3C_2T_x$ powders to exfoliate nanosheets with the assistance of mechanical stirring and sonication [29]. Besides, other intercalants, such as tetraalkylammonium hydroxide (TBAOH), isopropylamine, hydrazine, or urea, are also developed to expand MXene interlayer distance to facilitate the delamination of MXene [19, 25, 30, 31].

Almost all researchers applied two steps of etching and delamination to obtain single or few-layered MXene flakes. Recently, Feng et al. developed a one-step method with the assistance of ultrasonication to synthesize delaminated Ti_3C_2 MXene flakes [32]. They added the prepared Ti_3AlC_2 powders into aqueous HF (49%) solution and the whole system was sonicated at room temperature for four hours simultaneously, followed by centrifugation and shaking; finally, 46% of exfoliation rate was achieved. Although the etching reaction time was significantly reduced, high-quality atomic-thickness of MXene flakes could not obtain, the most exfoliated flakes with lateral size of 20–1000 nm in the suspension exhibited a thickness of less than 10 nm, which is considered as less than 10 MXene layers.

9.2.3 Surface Properties of MXene Flakes

After wet etching, the MXene surface was terminated with abundant feature groups, including hydroxyl (–OH), oxygen (–O), or fluorine (–F) groups, indicating hydrophilic nature, together with a highly negative ζ potential in the range of −39.5 to −63 mV [33–35], which may be lead to high stability of water-based solutions. This was verified by Maleski et al. [12]. They found that the resulted MXene colloidal solution in water suspension exhibited outstanding stability within 96 hours. However, there is a decrease in concentration and eventual sedimentation in the methanol, toluene, hexane, and 1,2-dichlorobenzene solution. The hydrophilic nature maybe endows MXene to possess potential application in water treatment. Meanwhile, to modify MXene to accomplish application diversification, surface engineering can be carried out to decorate MXene with negative charge with other functional materials based on electrostatic interaction and functional reaction [36–38]. However, it is critical to keep single MXene flakes stable for fabricating devices in various applications. Mashtalir et al. found that MXene flakes can be oxidized to form metal–oxide nanocrystals, such as TiO_2, in environments where oxygen and water are present [30]. The oxidation process generally starts from the edges of MXene flakes and then spreads to the entire surface through nucleation [39, 40]. Besides, the environment of oxygen-free degassed water or in dry air could effectively slow down the oxidation rate. Hence, to improve the lifetime of MXene flakes, MXene solution should be stored in hermetic inert gases-filled bottles at low temperatures. Meanwhile, another effective method is that the solution can also be made as membrane by filtration for storage and re-dispersing the membrane in solution if needed [11].

9.2.4 Preparation of MXene-Based Membranes

Owing to the ionization of carboxylic acid, the MXene nanosheets show a negative-charged behavior as confirmed by zeta potential [41]. Hence, the strong electrostatic repulsion interaction between MXene flakes makes sure that MXene nanosheets could homodisperse without overlapping for long time [42]. MXene-based membranes derived from nanosheets are mainly divided into two types—laminated membrane and composite membrane. The former is fabricated from MXene nanosheets via the flow-directed (i.e. filtration) method. This type of MXene membrane with layered structures possesses interlayer galleries to show extraordinary flexibility and promise practical utilization for molecular separation. The latter is formed by the integration of MXene with functional materials through covalent or noncovalent approaches. In this section, we discuss the fabrication method of MXene-based membranes in detail.

9.2.4.1 Drop-Coating

The detailed process to prepare membrane via drop-coating method is as follows—the suspension was drop coated on the surface of substrate (i.e. glass, organic or inorganic membrane, etc.) and finally, the membrane could be obtained after drying. Recently, Hao et al. have used drop-coating method to fabricate $Ti_3C_2T_x$ composite membranes for solvent transport [38], where $Ti_3C_2T_x$–PEI solution was drop cast upon the top of polyacrylonitrile (PAN) support at room temperature for 10 minutes and then the excess solution was removed, and then, trimesoyl chloride (TMC) was added to carry out the cross-linking reaction. Finally, the membrane was obtained after drying in an oven at 60 °C for two hours.

9.2.4.2 Spraying or Spinning Coating

The delaminated MXene solution was spin/spray coated on smooth substrate, such as glass, polyethylene terephthalate (PET) strips, to fabricate uniform membranes [11, 43]. For example, Zhang et al. pipetted MXene solution onto the substrate and it was fastened to the rotating plate of a spin coater. The membrane thickness could be controlled by adjusting the amount of added solution and/or the spinning speed [11].

9.2.4.3 Pressure-Assisted Filtration

In recent years, the most popular method to fabricate MXene-based membranes is pressure-assisted filtration. It is usually fabricated via filtrating MXene solution through microporous membranes, such as PAN, anodized aluminum oxide (AAO), nylon, and polyethersulfone (PES) [20, 44, 45]. The membrane thickness could be adjusted by varying the amount of MXene nanosheets. The vacuum-sucking force induces the compression of nanosheets paralleled the flow direction, leading to narrow interspace. Owing to the existence of interactions between nanosheets, including electrostatic repulsion, van der Waals attractive forces, and hydrogen bonding, the physiochemical properties of nanosheets remain unchanged [46].

9.3 MXene-Based Membranes for Molecular Separation

9.3.1 Liquid Separation

Studies have demonstrated that MXene with high water stability is hydrophilic, which can play an important role in water treatment [13, 34, 47]. In this section, we discuss in detail the following parts: (i) Desalination; (ii) organic solvent nanofiltration (OSN); (iii) pervaporation solvent dehydration; (iv) dyes and natural organic matter rejection; and (v) oil–water separation.

9.3.1.1 Desalination

It is well known that the water shortage has become a challenging problem, and seawater desalination using membrane technology is considered as an effective way to solve this water resource crisis [48, 49]. 2D materials-based membrane has been proved as a promising candidate for desalination [5, 50, 51]. Although MXene nanosheets were discovered in 2011, the pioneering work of applying MXene nanosheets for membrane separation was completed by Ren et al. in 2015 [47]. They first applied laminar MXene membrane to reject trivalent cation in solution, which was also used to compare with GO membrane. It can be seen that the assembled MXene/poly(vinylidene fluoride) (PVDF) membranes showed significantly lower permeation rated for multiple-charged ions than GO membrane; meanwhile, a better selectivity of MXene membrane for cations with radii smaller than 4.5 Å and charges between +2 and +4 was achieved. Ion intercalation could effectively adjust the layered space between 2D nanosheets [5, 19]. Thus, they also demonstrated that intercalated multivalent cations could narrow the channels between MXene membrane, while monovalent cations, such as Na^+, can enlarge it. A similar strategy was also reported by Wang group [52], where they designed non-swelling MXene membrane by intercalating Al^{3+} for ion separation using osmotic technology, which indicates long-term stable ion sieving toward small hydrated ions (Figure 9.2a). The Al^{3+} intercalation to the adjacent MXene layers could not only fix the interlayer distance but also suppress the swelling of MXene membrane in liquid water due to the strong interactions between Al^{3+} and oxygen-functional groups terminating at the MXene surface. Meanwhile, it was found that the Al^{3+} intercalation enhanced

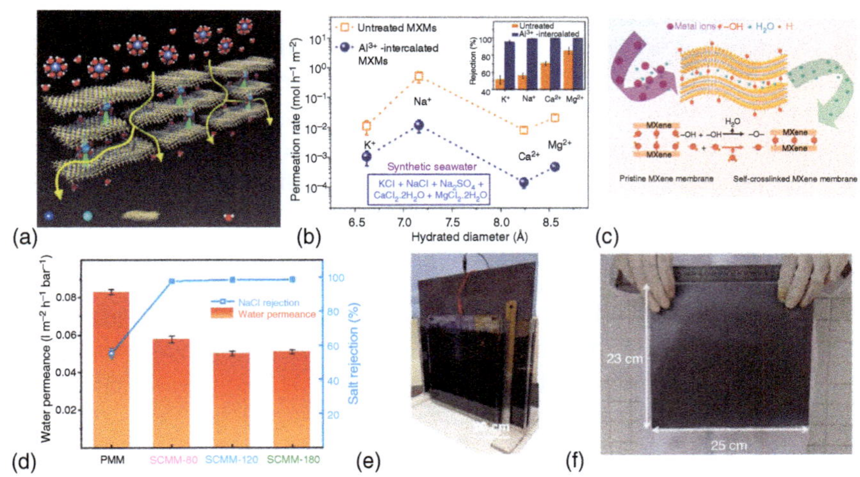

Figure 9.2 (a) Schematic of the effect of Al^{3+} intercalation on d-space; (b) The permeation rates of ions through MXene membranes. Source: Ding et al. [52]/with permission of Springer Nature. (c) Schematic of self-cross-linking process of the MXene membranes; (d) Water permeance and NaCl rejection of MXene-based membrane. Source: Lu et al. [53]/with permission of American Chemical Society. (e) Electrophoretic deposition device for large-area MXene membrane; (f) Digital photo of the EPD–MXene membrane with big area. Source: Deng et al. [54]/with permission from Elsevier B. V.

the permeation rate of MXene membrane for cationic components in the synthetic seawater by more than 50 times compared with un-intercalated MXene membrane, which effectively improve the rejection of multiple ionic components (Figure 9.2b). The developed Al^{3+}-intercalated MXene membrane demonstrated excellent rejection of ~89.5–99.6% for NaCl solution with fast water flux of ~1.1–8.5 $l\,m^{-2}\,h^{-1}$, which holds great promise on seawater desalination. Besides, they also proposed a new kind of self-cross-linked MXene membrane via a simple thermal treatment for monovalent ion exclusion [53]. The facile thermal treatment endows the formation of Ti—O—Ti bonds between adjacent MXene nanosheets utilizing the hydroxyl terminal groups via the reaction (–OH + –OH = –O– + H_2O) (Figure 9.2c), which could significantly improve the anti-swelling property. The self-cross-linked MXene membrane showed excellent salt rejection of 98.6% for NaCl solution with water permeance of ~0.05 $l\,h^{-1}\,m^{-2}\,bar^{-1}$ under cross-section filtration (Figure 9.2d). Although excellent salt rejection and water flux were achieved because of enhanced anti-swelling property and regular laminar structure, the external-pressure-driven desalination process remains to be demonstrated. Very recently, our group reported the fabrication of surface-charged MXene membranes by coating a polyelectrolyte (polyethylenimine, PEI) layer for nanofiltration desalination, achieving efficient rejection for divalent salts [55]. The PEI surface coating rationally tuned the surface-charge property and also enhanced the membrane hydrophilicity without sacrificial the laminar structure of MXene membrane. Owing to the synergic effect of electrostatic repulsion and size sieving, the resulting surface-charged-MXene membrane exhibited excellent water permeance and high salt rejection in nanofiltration process. Besides, Sun et al. also prepared MXene-derived membranes on α-Al_2O_3 tubular supports for ions rejection during nanofiltration process and the interlayer space was finely tuned via sintering-temperature regulation [56]. It was found that the interlayer spacing of MXene membranes decreased from 3.71 Å (60 °C) to 2.65 Å (450 °C). The MXene membrane with interlayer space of 2.68 Å at sintering 400 °C exhibits a relatively high ion-rejection rate of 99.2% for VO^{2+} with large water permeance of up to 11.5 $l\,m^{-2}\,h^{-1}\,bar^{-1}$. It should be noted that above laminar MXene membranes are normally prepared by vacuum filtration (VF) as the most common and straightforward approach. However, it is difficult to prepare defect-free and homogenous MXene membranes with large area by this method. Recently, Wang and coworkers have designed and fabricated MXene membranes larger than 500 cm^2 on commercial porous stainless steel mesh substrates within only several minutes by electrophoresis deposition (EPD) (Figure 9.2e, f) [54]. The EPD–MXene membranes with highly ordered 2D structure exhibited promising ions rejection ability of up to 99.5% (with permeation rates of the cations down to 10^{-3} $mol\,m^{-2}\,h^{-1}$) for the common light metal ions, such as Na^+ and Mg^{2+}, during forward osmosis (FO) process. Besides, brush coating can also be applied for large-scale preparation. Xie and coworkers fabricated a high-performance MXene-interlayered polyamide FO membrane with an area of $24 \times 8\,cm^2$ based on a combination of a facile and scalable brush coating of MXene on nylon substrates and the interfacial polymerization process [57]. The as-prepared FO membrane shows high water permeability of 31.8 $l\,m^{-2}\,h^{-1}$ and low specific salt flux of

$0.27\,g\,l^{-1}$ using $2.0\,mol\,l^{-1}$ sodium chloride as the draw solution. In addition to nanofiltration and osmosis process, pervaporation can also be used for desalination [45, 58]. Owing to abundant nanochannels and high hydrophilicity, ultrathin MXene membrane exhibited outstanding desalination performance with ultrahigh water flux of $85.4\,l\,m^{-2}\,h^{-1}$ and rejection of ~99.5 wt% for NaCl [45].

9.3.1.2 Organic Solvent Nanofiltration

OSN has been demonstrated as an efficient separation technique for aqueous solution treatment [59]. Unlike aqueous nanofiltration, which relies on steric and electrostatic interaction synergy, OSN is solely based on the physical size-exclusion effect and imposes stringent requirements that are hardly satisfied with current polymers [50, 60]. Generally, the traditional polymeric materials are limited by some drawbacks, including poor organic solvent and oxidation resistance and chemical and thermal stability [61]. Due to high stability in various organic solvents [12], MXene-based membrane could hypothetically provide outstanding organic permeation properties with high selectivity and stability. MXene nanosheets could also be, as fillers, incorporated into polymer matrix to improve the OSN performance. Wu et al. first applied MXene-based membranes for OSN [62], where MXene nanosheets were embedded into polymer matrix to facilitate the transfer of alcohol molecules. Results showed that 30% and 162% enhancements of alcohol flux for Ti_3C_2/PEI and Ti_3C_2/polydimethylsiloxane (PDMS) were achieved. Meanwhile, the rejection for molecular weight cutoff (MWCO) below 200 Da was dramatically improved, which shows that the trade-off between flux and rejection has been overcome. Besides, they also fabricated thin-film nanocomposite (TFN) membranes filled with functionalized $Ti_3C_2T_x$ MXene nanosheets with different groups via chemical modification for determining the solvent transport properties (Figure 9.3a) [63]. The introducing MXene nanosheets with different functional groups ($-NH_2$, $-COOR$, $-C_6H_6$, and $-C_{12}H_{26}$) effectively influenced the hydrophilicity of membrane, resulting in obvious enhancement of flux of PEI- and PDMS-based membranes for organic solvent, including isopropanol, ethyl acetate, toluene, and *n*-heptane, respectively (Figure 9.3b,c). Besides, Han et al. introduced MXene nanosheets into P84 polymer matrix to fabricate MXene-based mixed-matrix membranes, followed by cross-linking with triethylenetetramine (TETA) [66]. The developed membrane exhibits high flux ($268\,l\,m^{-2}\,h^{-1}$) and nearly 100% rejection for gentian violet (408) at 0.1 MPa and ambient temperature. Meanwhile, excellent solvent resistance to dimethylformamide (DMF), acetone, and methanol was achieved.

With micrometer lateral size and atomic thickness, the MXene nanosheets could also be stacked as laminate membranes; meanwhile, the regular and straight interlayer channels naturally formed between two adjacent MXene nanosheets, which permit precise molecular sieving. Wang et al. first reported the organic solvent behavior of laminar MXene membrane formed using flat double-layered flakes with ordered nanochannel structures [64]. Ultrahigh permeance of MXene membrane on six common solvents in industry was acquired, as shown in Figure 9.3d. The transport rate was related to molecular properties, as confirmed by their proposed model. Besides, dyes with different sizes were used to verify MXene membrane

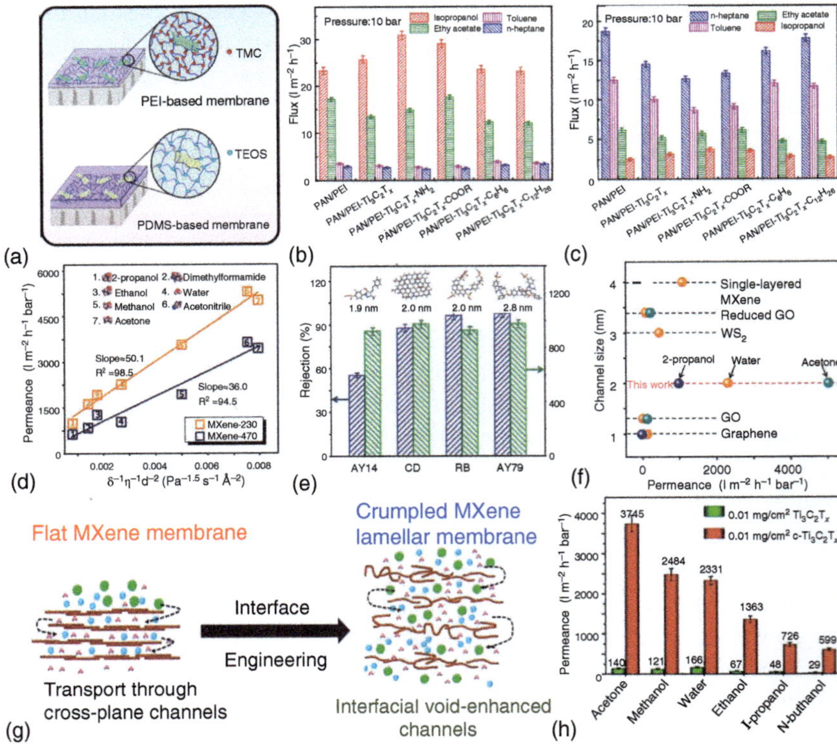

Figure 9.3 (a) Schematic diagrams of polymer-based membrane filled with MXene nanosheets; Various solvents flux of (b) PEI-based membranes and (c) PDMS-based membranes. Source: Hao et al. [63]/with permission of Elsevier. (d) The relation between solvent permeance and solvent property for MXene membranes; (e) Separation performance of MXene membrane on various molecules with different sizes; (f) Comparison of solvent permeance for lamellar membranes with different channel sizes. Source: Wang et al. [64]/with permission of John Wiley & Sons. (g) Schematic diagrams of transformation from flat MXene nanosheets to crumpled MXene nanosheets; (h) The solvent permeance of the $Ti_3C_2T_x$ and c-$Ti_3C_2T_x$ membranes with the same mass loading of 0.1 mg cm^{-2}. Source: Xing et al. [65]/with permission of American Chemical Society.

channel width of about 2 nm (Figure 9.3e). Due to that, irregular channels destroy the transport of bonded flow, other laminar membranes, such as GO and WS_2, with larger channels possess lower permeance (Figure 9.3f) [20, 67]. Besides, to further increase the solvent permeability, a new strategy on in situ enlarging interlayer channels via assembling the crumpled $Ti_3C_2T_x$ nanosheets (c-$Ti_3C_2T_x$) was proposed by Xing et al. [65]. The c-$Ti_3C_2T_x$ nanosheets were prepared by crumpling the flat nanosheets by using a facile freeze-drying method (Figure 9.3g). Due to the interfacial voids with enlarged interlayer space, the excellent permeation of c-$Ti_3C_2T_x$ membrane to different solvents at up to about 27 times (compared with noncrumpled $Ti_3C_2T_x$ membrane) was achieved (Figure 9.3h). Meanwhile, the trade-off between permeance and rejection has been efficiently addressed by regulating the thickness of the c-$Ti_3C_2T_x$ MXene membranes.

It was reported that regular GO membranes generally suffer from high rejection but low flux for dye solution and impermeability for organic solvents [68, 69], while it was different for MXene membrane. Thus, it is considered that the composite membrane with combination of MXene and GO could balance the separation performance with high solvent permeability and impressive molecules rejection, which was evidenced by Wei et al. [70]. Due to the hydrophilicity and 2D laminar structure of MXene and GO, excellent balanced separation performance of the GO/MXene (GM) composite membrane with high flux and rejection for both aqueous and organic solution was obtained. The resultant GM membranes exhibited ultrahigh flux of 21.02, 48.32, 25.03, 10.76, and 6.18 $l\,m^{-2}\,h^{-1}$ for water, acetone, methanol, ethanol, and isopropanol, respectively, and outstanding dyes molecular rejection of over 90%.

9.3.1.3 Pervaporation Solvent Dehydration

It is well known that dehydration of organic solvents is an important process in chemical industry and environment-friendly pervaporation technology has been widely used for solvent dehydration [71]. In addition to OSN, recently, MXene-based membranes have been used for solvent dehydration by pervaporation [72–75]. The first work of MXene-based membrane for organic solvent dehydration by pervaporation was reported by Xu et al., where they used single-layered 2D MXene nanosheets as filler to incorporate into the chitosan (CS) matrix for the organic solvent dehydration via pervaporation [74]. The optimal MXene/CS membrane with 3 wt% of nanosheets exhibited remarkable improvement of ~24% higher flux and ~250% separation factor for ethanol/water mixtures (Figure 9.4a), compared with

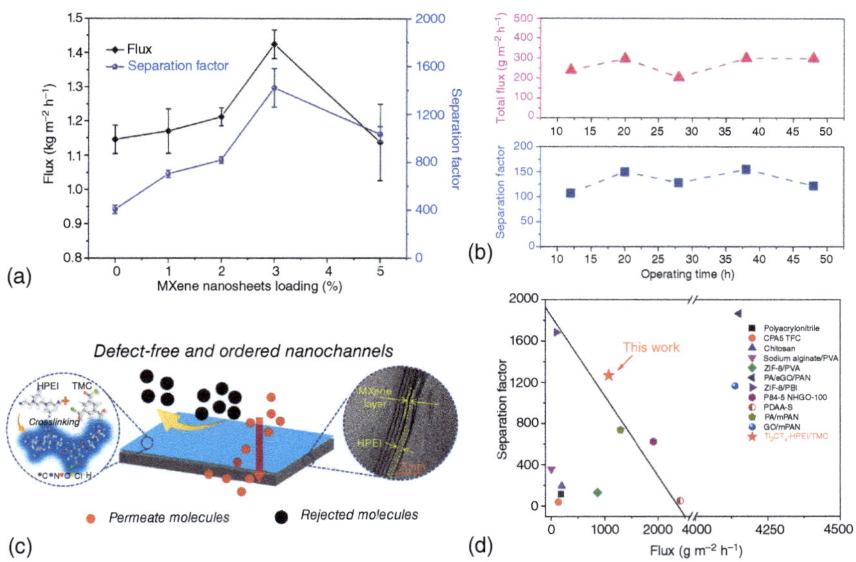

Figure 9.4 (a) Pervaporation performance of MXene/CS MMMs With different MXene loading for ethanol dehydration. Source: Xu et al. [74]/with permission of Elsevier. (b) Long-term stability of MXene membrane on separation for ethanol dehydration by pervaporation. Source: Wu et al. [73]/with permission of Elsevier. (c) Schematic of the defect-free and ordered MXene nanochannels; (d) Performance comparison of MXene–HPEI/TMC membrane on isopropanol dehydration by pervaporation with other membranes. Source: Liu et al. [75]/with permission of Royal Society of Chemistry.

the pristine CS membrane. In addition, due to abundant channels and enhanced selective transport of water molecules, the obtained MXene/CS mixed matrix membranes (MMMs) exhibited excellent separation performance with flux of ~ 1.4–1.5 kg m^{-2} h^{-1} and separation factor of 4898 and 906 for dehydration of ethyl acetate and dimethyl carbonate. However, the single-layered MXene nanosheets are irregularly distributed in membrane matrices, and the enhanced ability on separation performance is limited. To solve this problem, Li et al. presented an approach of incorporating multilayered MXene (mMXene) particles as porous fillers with sodium alginate (SA) to fabricate MMMs for ethanol dehydration [72]. The hydrophilic mMXene particles with uniform distribution in the SA matrix significantly enhanced the preferential sorption of water; the SA/mMXene MMMs showed ultrahigh separation factor of 9946 with a moderate flux of 505 g m^{-2} h^{-1}, which is superior to the performance of most reported membranes. Besides, the pristine MXene membrane with laminar structure could also be used for pervaporation organic solvent dehydration, which was reported by Wang group [73]. The prepared MXene membrane showed stable separation performance with a water/ethanol separation factor of 135.2 with a total flux of 263.4 g m^{-2} h^{-1} (Figure 9.4b). Nevertheless, the flux and separation factors are relatively low, which is due to the defects in the layered structure and the slight irregular stacking of nanosheets. Recently, many researches are devoted to regulate the stacking behavior and interlayer space of 2D nanosheets and seal membrane surface defects to improve permeability or selectivity [76, 77]. For example, our group proposed a new strategy of fabricating MXene membranes via introducing hyperbranched polyethylenimine (HPEI) between MXene nanosheets followed by interfacial polymerization with TMC on the assembled laminar membrane surface for organic solvent dehydration (Figure 9.4c) [75]. Due to the electrostatic interaction between positively charged HPEI and negatively charged MXene nanosheets, the ordered stacking structure was obtained. Meanwhile, less than 10 nm polymer generated by interfacial polymerization between HPEI and TMC, the defects on membrane surface was effectively sealed. The as-fabricated MXene-based membrane showed isopropanol dehydration performance with water content in the permeate side of more than 99 wt% and total flux of 1069 g m^{-2} h^{-1} that outperformed various types of membranes available in the literature (Figure 9.4d). In addition, to further improve the separation efficiency, it is desirable to develop a highly permeable and selective MXene membrane without the aid of a protective layer. Recently, we also have designed and fabricated MXene membranes without protective layer via functionalization by polyelectrolytes [78]. Because of the enhanced electrostatic interaction and hydrophilicity induced by the polyelectrolytes, the demonstrated poly(diallyldimethylammonium chloride) (PDDA)-modified MXene membrane showed outstanding separation performance with total flux of 1237 g m^{-2} h^{-1} and separation factor of 1932, which is superior to that of the MXene-based membrane with polymer layer to sealing potential defects.

9.3.1.4 Dyes and Natural Organic Matter Rejection

The breakthrough of MXene membrane on water purification by nanofiltration was reported by Wang group [20]. They prepared MXene membrane on rigid AAO substrate via vacuum filtration, followed by intercalating $Fe(OH)_3$ solution to create nanochannels, finally, the $Fe(OH)_3$ nanoparticles were removed via HCl treatment (Figure 9.5a). The MXene membrane with rough surface and typical lamellar structure (Figure 9.5b,c) showed excellent water permeance, which was higher than $1000\,l\,m^{-2}\,h^{-1}\,bar^{-1}$; meanwhile, the demonstrated MXene membrane achieved favorable rejection of >90% for molecules with size above 2.5 nm (Figure 9.5d). Besides, MXene nanosheets can also be deposited on the flexible flat substrate. Recently, the polymer PES and mixed cellulose ester (MCE) filters were used to support MXene nanosheets, and these resultant membranes exhibited high water flux and favorable rejection for dye molecules [44, 78]. However, the weak interface adhesive strength between MXene layer and substrate along with difficult to scale-up may hinder its practical application. Compared with flat-sheet substrates, hollow fiber possesses many advantages, including high packing density in membrane module, excellent chemical durability, and good solvent resistance, which could also be employed to support MXene membrane for molecular sieving. Sun et al. tried to obtain MXene–TiO_2 membranes on α-Al_2O_3 hollow-fiber (HF) supports by adding 2D MXene nanosheets into TiO_2 hydrosol using dip-coating method (Figure 9.5e) [79]. The optimal mesoporous MXene–TiO_2/HF membrane indicates high water flux of $102\,l\,m^{-2}\,h^{-1}\,bar^{-1}$ with a MWCO of 14854 Da (Figure 9.5f).

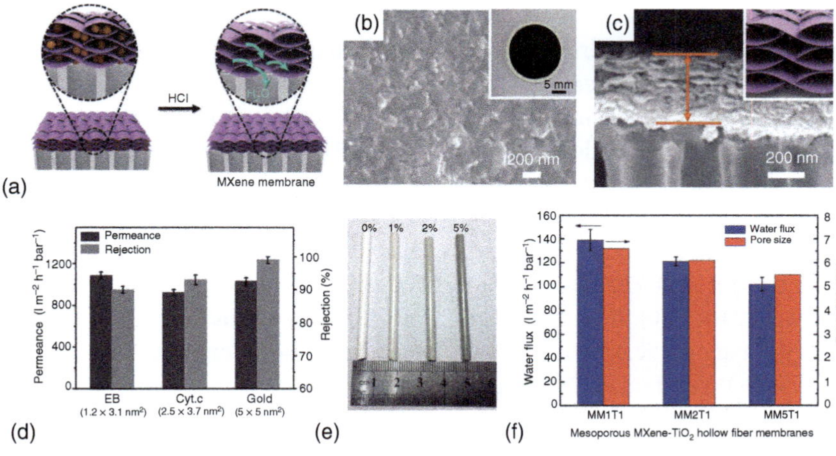

Figure 9.5 (a) Schematic diagram of MXene membrane; SEM images of MXene membrane, (b) surface and (c) cross section; (d) Separation performance of MXene membrane for dye rejection. Source: [20]. X. Xie et al., (2014)/with permission of John Wiley & Sons. (e) Photographs of the MXene–TiO_2 membranes with different MXenes contents; (f) Filtration performance of MXene–TiO_2/hollow-fiber membranes. Source: Sun et al. [79]/with permission of Institute of Physics Publishing (IOP).

We all know that the interlayer space of 2D nanosheets plays a dominant role in water permeability of 2D-based membranes [80, 81]. Strategy on creating nanochannels between 2D nanosheets via inserting 0D nanoparticles, 1D nanowires or carbon nanotubes, and 2D nanosheets has been widely studied to improve separation performance [82, 83]. Recently, metal nanoparticles, such as AgNPs, were introduced into MXene nanosheets, which is expected to improve the antifouling properties of the MXene membrane and increase the water flux [84]. The Ag@MXene membrane with 21% loading demonstrated enhanced water flux of ~420 l m^{-2} h^{-1} bar^{-1} (the pristine MXene membrane: ~118 l m^{-2} h^{-1} bar^{-1}) and improved bactericidal growth inhibition of over 99% for *E. coli* (~60% for pristine MXene membrane). In addition, Kang et al. incorporated the GO nanosheets into MXene nanosheets to enhance the dye rejection properties (Figure 9.6a) [85]. They found that the MXene/GO membrane showed a similar expansion of the interlayer spacing to 14.28 Å, where the thickness of an OH-terminated MXene single sheet around 9.2 Å, the effective interlayer spacing of MXene or composite membranes was around 5 Å, which displayed the high rejection for dye molecules with subnanometer size, but low rejection for salt (Figure 9.6b). Besides, to further enhance the water permeance, MXene could be as filler to insert into the layer space of other 2D materials nanosheets [86]. Liu et al. incorporated MXene nanosheets into GO to fabricate the GO/MXene membrane for dye and natural organic matter rejection (Figure 9.6c). Owing to the combined effects of molecular sieving, principally and electrostatic interaction, the developed membranes with high flux had an excellent removal efficiency of nearly 100% rejection rate (99.5% for NR: neutral red; MB: methylene blue; CV: crystal violet; BB: brilliants blue) for the small-molecule dyes and a natural organic matter found in raw water (humic acid [HA], bovine serum albumin [BSA]) (Figure 9.6d,e).

Reports have demonstrated that MXene nanosheets could be oxidized as TiO_2 nanoparticles under the environment with water and oxygen [11, 30]. Hence, probably MXene nanosheets could be in situ derived as 3D nanoparticles to enlarge the interlayer space of 2D-based membranes. Related research work was reported by Han et al. [87]. They added different ratios of MXene into GO nanosheets and fabricated GO/MXene membranes via filtration, followed by in situ oxidation via treating with treated with H_2O_2 (10%) for 12 hours to prepare the GO/TiO_2 membrane finally (Figure 9.6f). Owing to the lamellar structure of MXene nanosheets, the distribution of TiO_2 nanoparticles derived from the oxidation of MXene was consecutive among the GO membrane. The obtained GO/TiO_2 membrane achieved a highly improved water permeability up to 89.6 l m^{-2} h^{-1} bar^{-1}, which is 7.3 times that of pristine GO membrane and 2.4 times that of GO/TiO_2 membrane fabricated by traditional mixture methods. Although high-performance MXene membranes on natural organic matter rejection have achieved success, large-area MXene membranes on water treatment during nanofiltration process still remain high challenge. Very recently, a slot-die-coating method has been developed by Kim et al. for the preparation of large-area MXene membranes. The technique allows the fabrication of continuous and scalable coatings with a rapid coating speed of 6 mm s^{-1}. The thickness can be readily controlled from the nanometer scale to the micrometer scale, and the alignment of the nanosheet is enhanced by the shear force of the slot-die

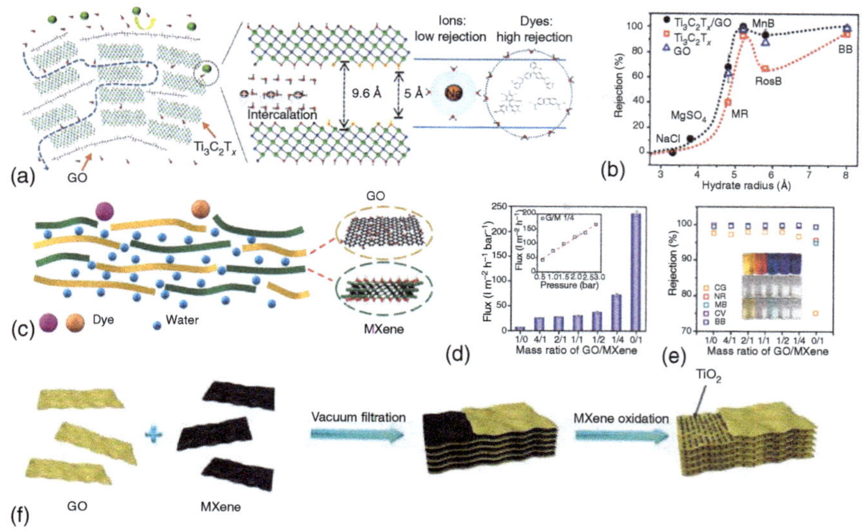

Figure 9.6 (a) Schematic diagrams of MXene–GO membrane on separation mechanism for hydrated ions and dye molecules; (b) Rejection performance of MXene–GO membrane. Source: Kang et al. [85]. Reproduced with permission of American Chemical Society. (c) Schematic illustration of GO/MXene lamellar membrane for separating dye molecule; (d) Flux of GO/MXene membranes with different ratio of MXene; (e) Separation performance of GO/MXene membranes for the various dyes molecules. Source: Liu et al. [86]/with permission of Elsevier B. V. (f) Schematic illustration of the fabrication of the GO–TiO_2 composite membrane derived from MXene nanosheets. Source: Han and Wu [87]/with permission of Royal Society of Chemistry.

head. Owing to the well-ordered interlayer structure, MXene membrane with diameter of 9 cm exhibited excellent nanofiltration performance with water permeance of 190 LMH bar^{-1} and a molecular weight cutoff of 269 Da [88].

9.3.1.5 Oil–Water Separation

The oil leakage and industrial oily wastewater are harming the environment and human health [89, 90]. Recently, more researchers focus on removing oil droplets from water using membranes derived from new materials, such as GO [91]. Owing to high hydrophilicity and atomic thickness, MXene nanosheets could also be assembled as laminar membrane for efficient oil/water separation [92, 93]. Wang group first developed ultrathin MXene membrane with 30 nm thickness for oil-in-water emulsions separation (Figure 9.7a) [92]. The resulted MXene membrane with high hydrophilicity and low-oil-adhesion force indicates outstanding oil/water emulsion separation efficiency up to 99.94% for toluene, pump oil, and soybean oil in water. Meanwhile, excellent recyclability with stable separation performance after 50 cycles test was obtained. In addition, for the salt-containing oily wastewater, higher oil rejection was obtained. Later, Li and coworkers explored the possibility of MXene membrane on oil/water separation in corrosive liquids, such as acidic, alkaline, and salty (Figure 9.7b) [93]. The MXene membrane with superhydrophilicity/underwater superoleophobicity could effectively separate

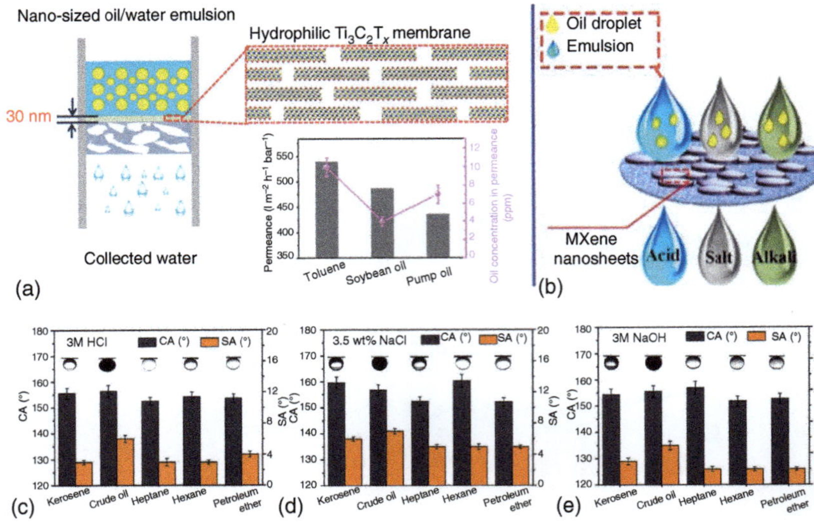

Figure 9.7 (a) Schematic diagrams of hydrophilic MXene membrane for efficient oil/water separation. Source: Li et al. [92]/with permission of Elsevier. (b) Schematic diagrams of ultrathin MXene membrane for effective separation of oil-in-water emulsions in acidic, alkaline, and salty environment; the MXene membrane with various types of oil contact angle and sliding angle in (c) 3 M HCl solutions, (d) 3.5 wt% NaCl, and (e) 3 M NaOH solutions. Source: Zhang et al. [93]/with permission of Elsevier.

diverse stabilized emulsions with high oil removal efficiency of >99.4% and high flux of up to 887 l m^{-2} h^{-1} bar^{-1}. Besides, it was found that the MXene membrane has excellent adaptability and anticorrosive ability to the harsh solution, including HCl, NaCl, and NaOH, which provides the possibility to separate oil/water mixture in harsh environments (Figure 9.7c–e). With the development of MXene membrane on oil/water separation, a new application of MXene is explored.

9.3.2 Gas Separation

Owing to high hydrophilicity and water stability, MXene-based membranes were mostly used for water treatment. However, with abundant transport channels and specific gas adsorption ability, MXene-based membranes can also be applied for gas mixture separation. The pioneering work of MXene membrane on gas separation was reported by Wang group [94]. They prepared freestanding MXene membrane with laminar structure by peeling it off from the substrate after drying (Figure 9.8a). The resultant MXene membrane shows high permeability for the small gas molecules (2164 Barrer for He and 2402 Barrer for H_2), which is much higher than that of the gases with bigger kinetic diameters, including CO_2, O_2, N_2, CH_4, C_3H_6, and C_3H_8, achieving high selectivity of small gas to big gas owing to the size-exclusion effect (Figure 9.8b). The ideal selectivity of the single-gas permeation of H_2/CO_2 is high up to 238.4 and that of the mixed-gas permeation was 166.6, which is much higher than the corresponding Knudsen coefficient of 4.7. This work

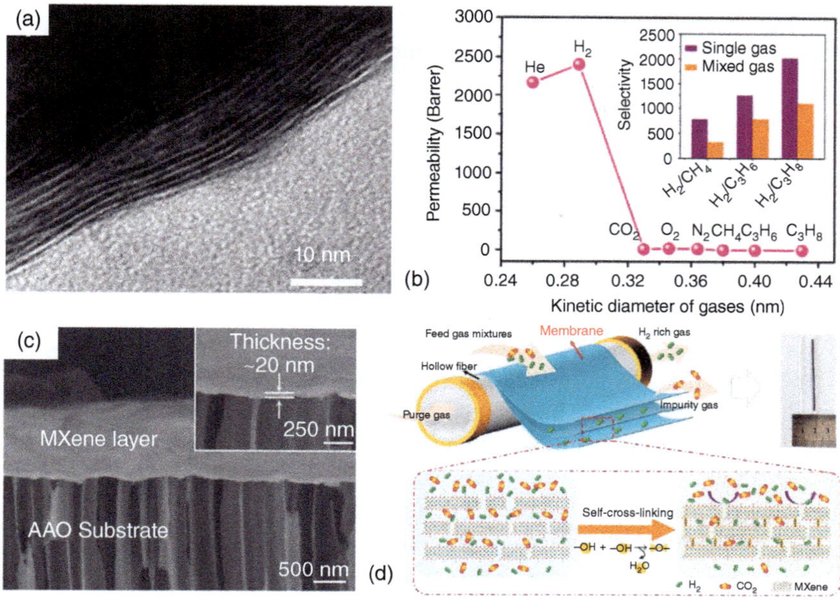

Figure 9.8 (a) Cross-sectional TEM image of MXene membrane with 2D channels; (b) Single-gas permeability through MXene membranes as a function of gas kinetic diameter (measured at 25 °C and 1 bar), insert is the selectivity of H_2 over other gases in both. Source: Ding et al. [94]/with permission from Springer Nature. (c) Cross-sectional SEM image of MXene nanofilm with 20 nm thickness. Source: Shen et al. [45]/with permission from John Wiley & Sons. (d) Schematic diagram of the self-cross-linked MXene hollow-fiber membrane for H_2/CO_2 separation. Source: Qu et al. [95]/with permission of Elsevier B. V.

opens the door of MXene membrane on gas separation. Although excellent high separation performance for H_2/CO_2 was achieved, the efficiency for CO_2 capture remains to be demonstrated.

Since the wet-etching process endows MXene nanosheets with abundant groups along with transport channels, it is possible to apply MXene membrane to separate target gas mixture, such as CO_2/N_2 and CO_2/CH_4, via versatile functionalities. Recently, our group has designed and engineered the MXene nanofilm with thickness of down to 20 nm for H_2- or CO_2-selective separation via chemical modification (Figure 9.8c) [45]. The pristine MXene nanofilm was fabricated via filtrating MXene solution through AAO substrate. Due to the well-stacking laminar structure and molecular sizing effect, the ultrathin 20 nm MXene nanofilm displays excellent gas separation performance of H_2/CO_2 with the H_2 permeance as high as 1584 GPU and the H_2/CO_2 selectivity of 27. To further explore the versatile potential of MXene nanofilms on gas separation, the gas transport behavior of nanofilms was tuned to realize CO_2 capture via introducing PEI and borate, followed by spin-coating polyether block amide (Pebax) for sealing some possible nonselective defects. Interestingly, the functional MXene nanofilm achieved an impressive CO_2 capture performance with the CO_2 permeance of 350 GPU and CO_2/CH_4 selectivity of 15.3. The separation mechanism of MXene nanofilms was further investigated, revealing the transformation from "diffusion-controlled" to "solution-controlled" channels after chemical tuning. Besides, Xu and coworkers applied a facile thermal treatment process to facilely adjust the interlayer space of developed self-cross-linked MXene hollow-fiber membranes (Figure 9.8d), where the dehydration reaction rate was visualized to illustrate the variation tendency of the interlayer space. As a result, the self-cross-linked hollow-fiber membranes exhibited H_2 permeance of 70.6 GPU and H_2/CO_2 selectivity of 30.3, with good operation stability for more than 120 hours [95].

Since MXene nanosheets exhibited high CO_2 adsorption property [45, 94], our group also incorporated MXene nanosheets as filler into Pebax matrix to fabricate mixed-matrix membrane for CO_2 capture [96]. The incorporated MXene nanosheets with preferential CO_2 sorption endow membrane abundant 2D channels, which facilitated the gas transport. The resulted MXene/Pebax membranes with MXene loading of 0.15 wt% exhibited ~81% higher CO_2 permeance and ~73.4% higher CO_2/N_2 selectivity, compared with the pristine pebax membrane. Besides, Shi et al. selected MXene or GO as typical nanosheet filler to be in pair with a non-2D filler, SiO_2 or halloysite nanotubes (HNTs) with Pebax as the polymer matrix to prepare mixed-matrix membranes for gas separation with the underlying synergy between two-dimensional (2D) nanosheets and a non-2D filler [96]. Based on the filler dispersion, nanosheet flexibility, and the tortuosity and connectivity of the surface diffusion pathways along with nanosheets, they found that the two 2D fillers found different preferential non-2D partners, where GO works better with HNTs than SiO_2, while MXene prefers SiO_2 to HNTs.

Although the above-developed MXene membranes performed excellent gas separation performance, they are tested at a low operation temperature (<150 °C). Fan et al. demonstrated and engineered an MXene laminar membrane on AAO substrate for hydrogen separation at high operation temperature with higher than 200 °C, [97] where the operation temperature was controlled by a furnace and the gas permeance was detected using the home-made device. The as-prepared MXene membrane indicates an excellent separation performance with the H_2 permeance of 2.0×10^{-7} mol m^{-2} s^{-1} Pa^{-1} and good selectivity of 41 for H_2/N_2 mixture at 320 °C. Meanwhile, the separation performance could be maintained stable after the long-term test for 200 hours. They owed this to the stable crystal structure and good thermal expansion matching with the substrate. Besides, the separation mechanism was also studied [66]. They found that molecular sieving plays a dominant role in controlling gas selectivity. This work displayed that the MXene can serve as the next-generation membrane material for gas separation in high-temperature environment.

At the end of this section, we introduce some works of MXene-based membranes on air purification [98]. Wang group incorporated 2D $Ti_3C_2T_x$ nanosheets into electrospun PAN to fabricate air filters for removing PM 2.5 (particulate matter with diameter ≤2.5 μm) [98]. The wet-etching process endows MXene nanosheets with plentiful surface-terminating groups, such as –OH, –F, –O, which will induce a strong interaction with PM 2.5 and bacteria, resulting in enhancing the removing efficacy of ~99.7% with a low-pressure drop of ~42 Pa. Meanwhile, the incorporated MXene nanosheets could restrain the propagation of bacteria (e.g. *E. coli* and *S. aureus*), which obviously improve air quality.

9.4 Conclusions and Perspective

In summary, 2D MXene has been demonstrated as a new star material in membrane separation for gas separation and water treatment. Meanwhile, plentiful oxygen-containing groups on MXene nanosheets and two-dimensional transport nanochannels contribute to precise molecular separation. In this minireview, the fabrication process of MXene nanosheets and MXene-based membranes has been briefly introduced. Besides, MXene nanosheets are used to be assembled as laminates or incorporated with polymers as composites. Meanwhile, we focus on the application of MXene-based membranes for water treatment, including desalination, organic solvent treatment, dyes and natural organic matter rejection, and oil–water separation and gas separation. Although considerable progress on MXene-based membrane has been achieved, there are some critical issues that still need to be addressed. First of all, as the starting point of future industry application, the acquisition of scale-up is imperative yet challenging, which could start with the preparation method. The spray coating with spinning process can be tried for large-area MXne membrane with less defects for separation. Besides, the transport model between the molecule transport behavior and the physicochemical property of MXene-based membranes should be proposed. Advanced characterization techniques along with simulation methods are of benefit to understand the effects of interlayer spaces, selective defects, and functional groups on transport behavior. Finally, MXene could be explored as promising membrane materials for some challenging research points, such as C2–C3 hydrocarbons and monatomic ion sieving.

References

1 Novoselov, K.S., Geim, A.K., Morozov, S.V. et al. (2004). Electric field effect in atomically thin carbon films. *Science* 306 (5696): 666–669.

2 Han, Y., Xu, Z., and Gao, C. (2013). Ultrathin graphene nanofiltration membrane for water purification. *Adv. Funct. Mater.* 23 (29): 3693–3700.

3 Lei, W., Mochalin, V.N., Liu, D. et al. (2015). Boron nitride colloidal solutions, ultralight aerogels and freestanding membranes through one-step exfoliation and functionalization. *Nat. Commun.* 6 (1): 8849.

4 Wang, Y., Li, L., Wei, Y. et al. (2017). Water transport with ultralow friction through partially exfoliated g-C_3N_4 nanosheet membranes with self-supporting spacers. *Angew. Chem. Int. Ed.* 56 (31): 8974–8980.

5 Chen, L., Shi, G., Shen, J. et al. (2017). Ion sieving in graphene oxide membranes via cationic control of interlayer spacing. *Nature* 550: 380–383.

6 Chen, C., Wang, J., Liu, D. et al. (2018). Functionalized boron nitride membranes with ultrafast solvent transport performance for molecular separation. *Nat. Commun.* 9 (1): 1902.

7 Ostwal, M., Shinde, D.B., Wang, X. et al. (2018). Graphene oxide – molybdenum disulfide hybrid membranes for hydrogen separation. *J. Membr. Sci.* 550: 145–154.

8 Liu, Y., Wang, N., and Caro, J. (2014). In situ formation of LDH membranes of different microstructures with molecular sieve gas selectivity. *J. Mater. Chem. A* 2 (16): 5716–5723.

9 Tsapatsis, M. (2014). 2-dimensional zeolites. *AIChE J.* 60 (7): 2374–2381.

10 Naguib, M., Kurtoglu, M., Presser, V. et al. (2011). Two-dimensional nanocrystals produced by exfoliation of Ti_3AlC_2. *Adv. Mater.* 23 (37): 4248–4253.

11 Zhang, C.J., Anasori, B., Seral-Ascaso, A. et al. (2017). Transparent, flexible, and conductive 2D titanium carbide (MXene) films with high volumetric capacitance. *Adv. Mater.* 29 (36): 1702678.

12 Maleski, K., Mochalin, V.N., and Gogotsi, Y. (2017). Dispersions of two-dimensional titanium carbide MXene in organic solvents. *Chem. Mater.* 29 (4): 1632–1640.

13 Anasori, B., Lukatskaya, M.R., and Gogotsi, Y. (2017). 2D metal carbides and nitrides (MXenes) for energy storage. *Nat. Rev. Mater.* 2 (2): 16098.

14 Huang, L., Ding, L., and Wang, H. (2021). MXene-based membranes for separation applications. *Small Sci.* 1: 2100013.

15 Yeh, C.-N., Raidongia, K., Shao, J. et al. (2015). On the origin of the stability of graphene oxide membranes in water. *Nat. Chem.* 7 (2): 166–170.

16 Karahan, H.E., Goh, K., Zhang, C. et al. (2020). MXene materials for designing advanced separation membranes. *Adv. Mater.* 32 (29): 1906697.

17 Li, J., Li, X., and Van der Bruggen, B. (2020). An MXene-based membrane for molecular separation. *Environ. Sci. Nano* 7 (5): 1289–1304.

18 Halim, J., Lukatskaya, M.R., Cook, K.M. et al. (2014). Transparent conductive two-dimensional titanium carbide epitaxial thin films. *Chem. Mater.* 26 (7): 2374–2381.

19 Lukatskaya, M.R., Mashtalir, O., Ren, C.E. et al. (2013). Cation intercalation and high volumetric capacitance of two-dimensional titanium carbide. *Science* 341 (6153): 1502–1505.

20 Ding, L., Wei, Y., Wang, Y. et al. (2017). A two-dimensional lamellar membrane: MXene nanosheet stacks. *Angew. Chem. Int. Ed.* 56 (7): 1825–1829.

21 Xie, X., Xue, Y., Li, L. et al. (2014). Surface Al leached Ti_3AlC_2 as a substitute for carbon for use as a catalyst support in a harsh corrosive electrochemical system. *Nanoscale* 6 (19): 11035–11040.

22 Li, T., Yao, L., Liu, Q. et al. (2018). Fluorine-free synthesis of high-purity $Ti_3C_2T_x$ (T = OH, O) via alkali treatment. *Angew. Chem. Int. Ed.* 57 (21): 6115–6119.

23 Ghidiu, M., Lukatskaya, M.R., Zhao, M.Q. et al. (2014). Conductive two-dimensional titanium carbide 'clay' with high volumetric capacitance. *Nature* 516 (7529): 78–81.

24 Feng, A., Yu, Y., Wang, Y. et al. (2017). Two-dimensional MXene Ti_3C_2 produced by exfoliation of Ti_3AlC_2. *Mater. Des.* 114: 161–166.

25 Feng, A., Yu, Y., Jiang, F. et al. (2017). Fabrication and thermal stability of NH_4HF_2-etched Ti_3C_2 MXene. *Ceram. Int.* 43 (8): 6322–6328.

26 Liu, F., Zhou, A., Chen, J. et al. (2017). Preparation of Ti_3C_2 and Ti_2C MXenes by fluoride salts etching and methane adsorptive properties. *Appl. Surf. Sci.* 416: 781–789.

27 Lipatov, A., Alhabeb, M., Lukatskaya, M.R. et al. (2016). Effect of synthesis on quality, electronic properties and environmental stability of individual monolayer Ti_3C_2 MXene flakes. *Adv. Electron. Mater.* 2 (12): 1600255.

28 Hong Ng, V.M., Huang, H., Zhou, K. et al. (2017). Recent progress in layered transition metal carbides and/or nitrides (MXenes) and their composites: synthesis and applications. *J. Mater. Chem. A* 5 (7): 3039–3068.

29 Mashtalir, O., Naguib, M., Mochalin, V.N. et al. (2013). Intercalation and delamination of layered carbides and carbonitrides. *Nat. Commun.* 4 (1): 1716.

30 Mashtalir, O., Cook, K.M., Mochalin, V.N. et al. (2014). Dye adsorption and decomposition on two-dimensional titanium carbide in aqueous media. *J. Mater. Chem. A* 2 (35): 14334.

31 Naguib, M., Unocic, R.R., Armstrong, B.L., and Nanda, J. (2015). Large-scale delamination of multi-layers transition metal carbides and carbonitrides "MXenes". *Dalton Trans.* 44 (20): 9353–9358.

32 Feng, W., Luo, H., Wang, Y. et al. (2018). Ultrasonic assisted etching and delaminating of Ti_3C_2 Mxene. *Ceram. Int.* 44 (6): 7084–7087.

33 Xie, X., Zhao, M.Q., Anasori, B. et al. (2016). Porous heterostructured MXene/carbon nanotube composite paper with high volumetric capacity for sodium-based energy storage devices. *Nano Energy* 26: 513–523.

34 Naguib, M., Mashtalir, O., Carle, J. et al. (2012). Two-dimensional transition metal carbides. *ACS Nano* 6 (2): 1322–1331.

35 Wang, Q.-W., Zhang, H.B., Liu, J. et al. (2019). Multifunctional and water-resistant MXene-decorated polyester textiles with outstanding electromagnetic interference shielding and joule heating performances. *Adv. Funct. Mater.* 29 (7): 1806819.

36 Xuan, J., Wang, Z., Chen, Y. et al. (2016). Organic-base-driven intercalation and delamination for the production of functionalized titanium carbide nanosheets with superior photothermal therapeutic performance. *Angew. Chem. Int. Ed.* 55 (47): 14569–14574.

37 Lin, H., Wang, Y., Gao, S. et al. (2017). Theranostic 2D tantalum carbide (MXene). *Adv. Mater.* 30 (4): 1703284.

38 Barankova, E., Tan, X., Villalobos, L.F. et al. (2017). A metal chelating porous polymeric support: the missing link for a defect-free metal–organic framework composite membrane. *Angew. Chem. Int. Ed.* 56 (11): 2965–2968.

39 Naguib, M., Mashtalir, O., Lukatskaya, M.R. et al. (2014). One-step synthesis of nanocrystalline transition metal oxides on thin sheets of disordered graphitic carbon by oxidation of MXenes. *Chem. Commun.* 50 (56): 7420.

40 Rakhi, R.B., Ahmed, B., Hedhili, M.N. et al. (2015). Effect of postetch annealing gas composition on the structural and electrochemical properties of Ti_2CT_x MXene electrodes for supercapacitor applications. *Chem. Mater.* 27 (15): 5314–5323.

41 Ying, Y., Liu, Y., Wang, X. et al. (2015). Two-dimensional titanium carbide for efficiently reductive removal of highly toxic chromium(VI) from water. *ACS Appl. Mater. Inter.* 7 (3): 1795–1803.

42 Huang, H., Ying, Y., and Peng, X. (2014). Graphene oxide nanosheet: an emerging star material for novel separation membranes. *J. Mater. Chem. A* 2 (34): 13772.

43 Sarycheva, A., Polemi, A., Liu, Y. et al. (2018). 2D titanium carbide (MXene) for wireless communication. *Sci. Adv.* 4 (9): eaau0920.

44 Han, R., Ma, X., Xie, Y. et al. (2017). Preparation of a new 2D MXene/PES composite membrane with excellent hydrophilicity and high flux. *RSC Adv.* 7 (89): 56204–56210.

45 Shen, J., Liu, G., Ji, Y. et al. (2018). 2D MXene nanofilms with tunable gas transport channels. *Adv. Funct. Mater.* 28 (31): 1801511.

46 Dikin, D.A., Stankovich, S., Zimney, E.J. et al. (2007). Preparation and characterization of graphene oxide paper. *Nature* 448 (7152): 457–460.

47 Ren, C.E., Hatzell, K.B., Alhabeb, M. et al. (2015). Charge-and size-selective ion sieving through $Ti_3C_2T_x$ MXene membranes. *J. Phys. Chem. Lett.* 6 (20): 4026–4031.

48 Van der Bruggen, B. and Vandecasteele, C. (2002). Distillation vs. membrane filtration: overview of process evolutions in seawater desalination. *Desalination* 143 (3): 207–218.

49 Elimelech, M. and Phillip, W.A. (2011). The future of seawater desalination: energy, technology, and the environment. *Science* 333 (6043): 712–717.

50 Zhang, M., Guan, K., Ji, Y. et al. (2019). Controllable ion transport by surface-charged graphene oxide membrane. *Nat. Commun.* 10 (1): 1–8.

51 Ries, L., Petit, E., Michel, T. et al. (2019). Enhanced sieving from exfoliated MoS_2 membranes via covalent functionalization. *Nat. Mater.* 18 (10): 1112–1117.

52 Ding, L., Li, L., Liu, Y. et al. (2020). Effective ion sieving with $Ti_3C_2T_x$ MXene membranes for production of drinking water from seawater. *Nat. Sustain.* 3: 296–302.

53 Lu, Z., Wei, Y., Deng, J. et al. (2019). Self-crosslinked MXene ($Ti_3C_2T_x$) membranes with good antiswelling property for monovalent metal ion exclusion. *ACS Nano* 13 (9): 10535–10544.

54 Deng, J., Lu, Z., Ding, L. et al. (2021). Fast electrophoretic preparation of large-area two-dimensional titanium carbide membranes for ion sieving. *Chem. Eng. J.* 408: 127806.

55 Meng, B., Liu, G., Mao, Y. et al. (2021). Fabrication of surface-charged MXene membrane and its application for water desalination. *J. Membr. Sci.* 623: 119076.

56 Sun, Y., Li, S., Zhuang, Y. et al. (2019). Adjustable interlayer spacing of ultrathin MXene-derived membranes for ion rejection. *J. Membr. Sci.* 591: 117350.

57 Wu, X., Ding, M., Xu, H. et al. (2020). Scalable $Ti_3C_2T_x$ MXene interlayered forward osmosis membranes for enhanced water purification and organic solvent recovery. *ACS Nano* 14 (7): 9125–9135.

58 Xu, K., Feng, B., Zhou, C., and Huang, A. (2016). Synthesis of highly stable graphene oxide membranes on polydopamine functionalized supports for seawater desalination. *Chem. Eng. Sci.* 146: 159–165.

59 Vandezande, P., Gevers, L.E., and Vankelecom, I.F. (2008). Solvent resistant nanofiltration: separating on a molecular level. *Chem. Soc. Rev.* 37 (2): 365–405.

60 Liang, B., Wang, H., Shi, X. et al. (2018). Microporous membranes comprising conjugated polymers with rigid backbones enable ultrafast organic-solvent nanofiltration. *Nat. Chem.* 10 (9): 961–967.

61 Karan, S., Jiang, Z., and Livingston, A.G. (2015). Sub–10 nm polyamide nanofilms with ultrafast solvent transport for molecular separation. *Science* 348 (6241): 1347–1351.

62 Wu, X., Hao, L., Zhang, J. et al. (2016). Polymer-$Ti_3C_2T_x$ composite membranes to overcome the trade-off in solvent resistant nanofiltration for alcohol-based system. *J. Membr. Sci.* 515: 175–188.

63 Hao, L., Zhang, H., Wu, X. et al. (2017). Novel thin-film nanocomposite membranes filled with multi-functional $Ti_3C_2T_x$ nanosheets for task-specific solvent transport. *Composites Part A* 100: 139–149.

64 Wang, J., Chen, P., Shi, B. et al. (2018). A regularly channeled lamellar membrane for unparalleled water and organics permeation. *Angew. Chem. Int. Ed.* 57 (23): 6814–6818.

65 Xing, Y., Akonkwa, G., Liu, Z. et al. (2020). Crumpled two-dimensional $Ti_3C_2T_x$ MXene lamellar membranes for solvent permeation and separation. *ACS Appl. Nano. Mater.* 3 (2): 1526–1534.

66 Jin, Y., Fan, Y., Meng, X. et al. (2019). Theoretical and experimental insights into the mechanism for gas separation through nanochannels in 2D laminar MXene membranes. *Processes* 7 (10): 751.

67 Huang, L., Chen, J., Gao, T. et al. (2016). Reduced graphene oxide membranes for ultrafast organic solvent nanofiltration. *Adv. Mater.* 28 (39): 8669–8674.

68 Liu, R., Arabale, G., Kim, J. et al. (2014). Graphene oxide membrane for liquid phase organic molecular separation. *Carbon* 77: 933–938.

69 Yang, Q., Su, Y., Chi, C. et al. (2017). Ultrathin graphene-based membrane with precise molecular sieving and ultrafast solvent permeation. *Nat. Mater.* 16 (12): 1198–1202.

70 Wei, S., Xie, Y., Xing, Y. et al. (2019). Two-dimensional graphene Oxide/MXene composite lamellar membranes for efficient solvent permeation and molecular separation. *J. Membr. Sci.* 582: 414–422.

71 Chapman, P.D., Oliveira, T., Livingston, A.G., and Li, K. (2008). Membranes for the dehydration of solvents by pervaporation. *J. Membr. Sci.* 318 (1-2): 5–37.

72 Li, S., Dai, J., Geng, X. et al. (2020). Highly selective sodium alginate mixed-matrix membrane incorporating multi-layered MXene for ethanol dehydration. *Sep. Purif. Technol.* 235: 116206.

73 Wu, Y., Ding, L., Lu, Z. et al. (2019). Two-dimensional MXene membrane for ethanol dehydration. *J. Membr. Sci.* 590: 117300.

74 Xu, Z., Liu, G., Ye, H. et al. (2018). Two-dimensional MXene incorporated chitosan mixed-matrix membranes for efficient solvent dehydration. *J. Membr. Sci.* 563: 625–632.

75 Liu, G., Shen, J., Ji, Y. et al. (2019). Two-dimensional Ti_2CT_x MXene membranes with integrated and ordered nanochannels for efficient solvent dehydration. *J. Mater. Chem. A* 7 (19): 12095–12104.

76 Shen, J., Liu, G., Huang, K. et al. (2016). Subnanometer two-dimensional graphene oxide channels for ultrafast gas sieving. *ACS Nano* 10 (3): 3398–3409.

77 Su, Y., Kravets, V., Wong, S. et al. (2014). Impermeable barrier films and protective coatings based on reduced graphene oxide. *Nat. Commun.* 5 (1): 1–5.

78 Zhang, S., Liao, S., Qi, F. et al. (2020). Direct deposition of two-dimensional MXene nanosheets on commercially available filter for fast and efficient dye removal. *J. Hazard. Mater.* 384: 121367.

79 Sun, Y., Xu, Z., Zhuang, Y. et al. (2018). Tunable dextran retention of MXene-TiO_2 mesoporous membranes by adjusting the 2D MXene content. *2D Materials* 5 (4): 045003.

80 Dervin, S., Dionysiou, D.D., and Pillai, S.C. (2016). 2D nanostructures for water purification: graphene and beyond. *Nanoscale* 8 (33): 15115–15131.

81 Liu, G., Jin, W., and Xu, N. (2015). Graphene-based membranes. *Chem. Soc. Rev.* 44 (15): 5016–5030.

82 Zhang, M., Guan, K., Shen, J. et al. (2017). Nanoparticles@rGO membrane enabling highly enhanced water permeability and structural stability with preserved selectivity. *AIChE J.* 63 (11): 5054–5063.

83 Han, Y., Jiang, Y., and Gao, C. (2015). High-flux graphene oxide nanofiltration membrane intercalated by carbon nanotubes. *ACS Appl. Mater. Inter.* 7 (15): 8147–8155.

84 Pandey, R.P., Rasool, K., Madhavan, V.E. et al. (2018). Ultrahigh-flux and fouling-resistant membranes based on layered silver/MXene ($Ti_3C_2T_x$) nanosheets. *J. Mater. Chem. A* 6 (8): 3522–3533.

85 Kang, K.M., Kim, D.W., Ren, C.E. et al. (2017). Selective molecular separation on $Ti_3C_2T_x$-graphene oxide membranes during pressure-driven filtration: comparison with graphene oxide and MXenes. *ACS Appl. Mater. Inter.* 9 (51): 44687–44694.

86 Liu, T., Liu, X., Graham, N. et al. (2020). Two-dimensional MXene incorporated graphene oxide composite membrane with enhanced water purification performance. *J. Membr. Sci.* 593: 117431.

87 Han, R. and Wu, P. (2019). High-performance graphene oxide nanofiltration membrane with continuous nanochannels prepared by the in situ oxidation of MXene. *J. Mater. Chem. A* 7 (11): 6475–6481.

88 Kim, J.H., Park, G.S., Kim, Y.J. et al. (2021). Large-area $Ti_3C_2T_x$-MXene coating: toward industrial-scale fabrication and molecular separation. *ACS Nano* 15 (5): 8860–8869.

89 Lo, Y. and Kang, D.-Y. (2016). Pseudopolymorphic seeding for the rational synthesis of hybrid membranes with a zeolitic imidazolate framework for enhanced molecular separation performance. *J. Mater. Chem. A* 4 (11): 4172–4179.

90 Peterson, C.H., Rice, S.D., Short, J.W. et al. (2003). Long-term ecosystem response to the Exxon Valdez oil spill. *Science* 302 (5653): 2082–2086.

91 Huang, Y., Li, H., Wang, L. et al. (2015). Ultrafiltration membranes with structure-optimized graphene-oxide coatings for antifouling oil/water separation. *Adv. Mater. Interfaces* 2 (2): 1400433.

92 Li, Z.K., Liu, Y., Li, L. et al. (2019). Ultra-thin titanium carbide (MXene) sheet membranes for high-efficient oil/water emulsions separation. *J. Membr. Sci.* 592: 117361.

93 Zhang, H., Wang, Z., Shen, Y. et al. (2020). Ultrathin 2D $Ti_3C_2T_x$ MXene membrane for effective separation of oil-in-water emulsions in acidic, alkaline, and salty environment. *J. Colloid Interface Sci.* 561: 861–869.

94 Ding, L., Wei, Y., Li, L. et al. (2018). MXene molecular sieving membranes for highly efficient gas separation. *Nat. Commun.* 9 (1): 155.

95 Qu, K., Dai, L., Xia, Y. et al. (2021). Self-crosslinked MXene hollow fiber membranes for H_2/CO_2 separation. *J. Membr. Sci.* 638: 119669.

96 Shi, F., Sun, J., Wang, J. et al. (2020). Exploration of the synergy between 2D nanosheets and a non-2D filler in mixed matrix membranes for gas separation. *Front. Chem.* 8: 58.

97 Fan, Y., Wei, L., Meng, X. et al. (2019). An unprecedented high-temperature-tolerance 2D laminar MXene membrane for ultrafast hydrogen sieving. *J. Membr. Sci.* 569: 117–123.

98 Gao, X., Li, Z.K., Xue, J. et al. (2019). Titanium carbide $Ti_3C_2T_x$ (MXene) enhanced PAN nanofiber membrane for air purification. *J. Membr. Sci.* 586: 162–169.

10

2D-Materials Mixed-Matrix Membranes

Ziting Zhu[1,2], Shaoyu Wang[1,2], Fusheng Pan[1,2], and Zhongyi Jiang[1,2]

[1] *Key Laboratory for Green Chemical Technology of Ministry of Education, School of Chemical Engineering and Technology, Tianjin University, Tianjin 300072, China*
[2] *Collaborative Innovation Center of Chemical Science and Engineering (Tianjin), Tianjin 300072, China*

10.1 Introduction

Molecular separation is a key process in chemical production to obtain qualified products. Due to the similar characteristics of the components to be separated, precise molecular separation is also a critical challenge in separation engineering [1]. As an emerging separation technology, membrane separation has the advantages of low energy consumption, easy scalability, and continuous operation. In the past few decades, it has experienced great growth and has been widely applied in gas separation [2–6] and liquid separation [7–10].

Materials play a key role in the separation performance of the membrane [11]. Since 1980s, polymers, such as cellulose (CA), polyethersulfone (PES), polyvinylidene fluoride (PVDF), and polytetrafluoroethylene (PTFE), have been widely used as membrane materials for industrial applications [12]. With vigorous development of membrane materials and fabrication methods, multiple polymer membranes, such as isotropic and anisotropic membranes [13, 14], symmetrical and asymmetric composite membranes [15, 16], are produced. However, conventional polymer membranes suffer from the "trade-off" effect between selectivity and permeability [17]. That means, polymer membranes with higher permeability generally have lower selectivity and *vice versa*, which severely limits the development and application of polymer membranes. One of the main reasons for the "trade-off" effect is a broad size distribution of free-volume elements in the polymer membranes [18].

Mixed-matrix membranes (MMMs) are usually composed of dispersed phase and continuous polymer matrix [19]. The superior processability and mechanical properties of the polymer, and the high permeability and/or selectivity of the dispersed phase can be combined to break through the limitation of the "trade-off" effect. The separation performance of the MMMs can be significantly improved by modulating

Two-Dimensional-Materials-Based Membranes: Preparation, Characterization, and Applications, First Edition. Edited by Gongping Liu and Wanqin Jin.
© 2022 WILEY-VCH GmbH. Published 2022 by WILEY-VCH GmbH.

the dispersed phase and/or the interface region [20]. Nanomaterials are one of the most widely used dispersed phase [21], including (i) zero-dimensional (0D) materials, such as quantum dots, hollow capsules, and microspheres, (ii) one-dimensional (1D) materials, such as nanotubes and nanofibers, (iii) two-dimensional (2D) materials, such as nanosheets, and (iv) three-dimensional (3D) materials, such as nanocages [22]. Among these materials, 2D nanomaterials with a high-aspect ratio (large lateral size and small thickness) are ideal building blocks for MMMs. The anisotropy structure drives the nanosheets to stack in parallel, forming a "brick and mortar" structure in the MMMs. During the process, interlayer pores and/or intralayer pores are introduced to the MMMs. The high-aspect ratio of the nanosheets facilitates the formation of sieving pores. They can achieve excellent performance enhancements at a low load. Rodenas et al. [23] compared the performance of MMMs with bulky-type mental organic frameworks (MOFs) (3D), nanosheet MOFs (2D), and nanodot MOFs (0D), and found that only the MMMs with 2D MOF nanosheets exhibited better CO_2/CH_4 selectivity than pure polymer membranes. Similar results were also observed in GO MMMs [24].

In addition to the polymer-based MMMs, 2D materials-based MMMs, where 2D materials serve as the continuous phase, have also attracted increasing attention in recent years. Unique properties of 2D materials, such as atomic thickness, molecular-sieving ability as well as excellent thermal and chemical stability, make them ideal building blocks for the new generation of molecular separation membranes [25–27]. 2D materials-based MMMs can break the limitation of "trade-off" effect and solve the issue of poor long-time stability [28]. In general, according to the structural difference of 2D materials, 2D-materials membranes could be divided into two categories. One is the membranes formed by porous 2D materials [29, 30], in which the transport of molecules mainly relies on the intrinsic pores and the functional structure around the rim [31]. The other one is the membranes assembled by the stacking of nonporous 2D materials [32–34], in which the molecular sieving is achieved by the interlayer pores. Materials with difunctional or multifunctional groups are introduced as the cross-linker to enhance the stability of the membranes, regulate the physical/chemical microenvironment and improve their separation performance.

In this chapter, the recent research progress on 2D-materials MMMs is summarized. MMMs with 2D materials as dispersed and continuous phases are discussed respectively. As shown in Figure 10.1, graphene oxide (GO), MOFs, and covalent organic frameworks (COFs) are the representative 2D materials, which are highlighted. The fabrication and applications of 2D-materials MMMs in precise molecular separation are elaborated. Finally, the existing challenges and perspectives on the future development of MMMs are discussed in detail, targeting at identifying the most promising strategies and directions for the engineering of 2D-materials MMMs.

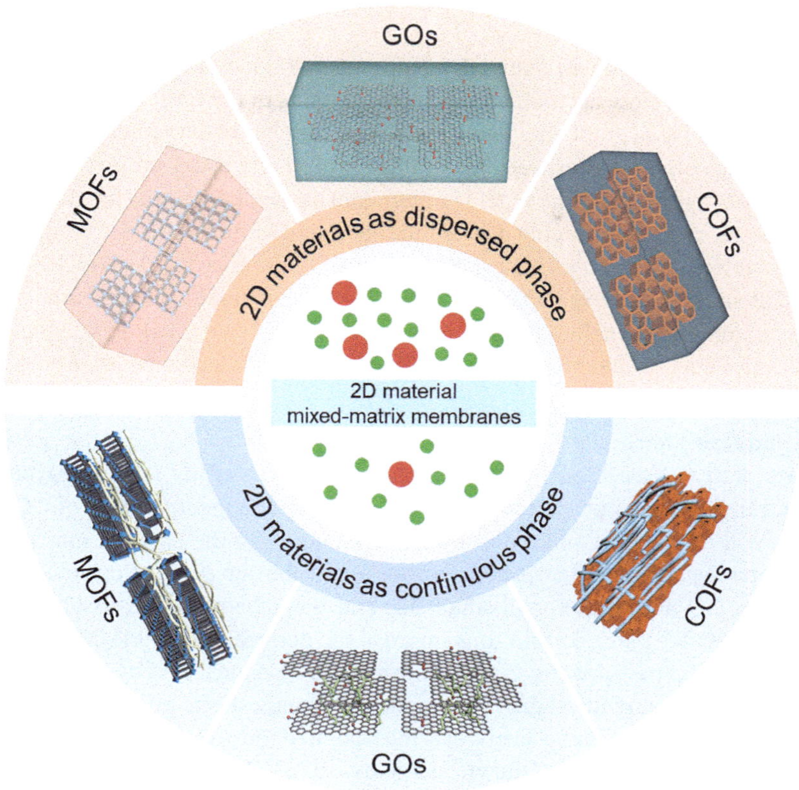

Figure 10.1 Schematic illustration of MMMs with 2D materials as dispersed phase and continuous phase. Source: Ziting Zhu, Shaoyu Wang, Fusheng Pan, Zhongyi Jiang.

10.2 Two-Dimensional Materials as Dispersed Phase of MMMs

Initially, MMMs use polymers as the continuous phase and other materials as the dispersed phase (commonly known as filler). There are three immediate challenges of MMMs—(i) the porosity of the filler, (ii) the interfacial compatibility between filler and polymers, and (iii) the loading of the filler. Due to their characteristics of high porosity (interlayer or intralayer), abundant functional groups, and great dispersibility, 2D materials, such as GO, MOFs, and COFs, have great potential to solve the above problems [35–38]. Thus, they have become one of the most widely used fillers in MMMs. Besides, the introduction of 2D materials will also endow MMMs with optimized structure and enhanced separation performance.

10.2.1 Graphene Oxide (GO)

Graphene, an atomic thickness material with hexagonal honeycomb lattice structure, is formed by the close accumulation of carbon atoms in the form of sp^2. GO is an oxidized derivative of graphene. There are a large number of oxygen-containing functional groups on GO nanosheets, including hydroxyl and epoxy groups on the surface and carboxyl groups at the edge. The abundant oxygen-containing functional groups make GO nanosheets highly hydrophilic and easy to functionalize. Owing to the unique molecular transport behavior between GO nanosheets [25, 39], GO is considered as an ideal dispersed phase of MMMs, which can modify the structure of MMMs in different ways, including increasing molecular transport channels, reducing nonselective defects, and introducing the functional site for facilitated transport.

10.2.1.1 Increasing Molecular Transport Channels

For MMMs with GO as filler, the sieving ability is enhanced by the interlayer spacing and the chemical properties of the GO surface. The incorporation of GO nanosheets will create more molecular transport channels in the polymer matrix, including interlayer spacing between nanosheets and the interfacial free-volume holes. Besides, GO nanosheets with small size and porous structure can shorten the path of molecular transport and thus improve the permeability of MMMs. Cao et al. [40] have prepared pristine graphene oxide (pGO) nanosheets by modified Hummer's method and reduced graphene oxide (rGO) nanosheets by a hydrothermal route. Two kinds of nanosheets are incorporated into a sodium alginate (SA) matrix to prepare MMMs (shown in Figure 10.2a). The free-volume holes around the SA–GO interface, the structural defects on the GO nanosheets, and the edge-to-edge slits between the GO nanosheets formed the interconnected channels. The separation performance of the MMMs was governed by the water channels, which can be adjusted by varying the GO content and structure. It can be observed from Figure 10.2b,c that rGO nanosheets possessed smaller sizes and more edges. Smaller nanosheet sizes, more structural defects, and fewer oxygen-containing groups facilitated the construction of water channels, while fewer negative charges and more non-oxidized regions further improved the permeability as well as the selectivity of the water channels.

Co-doption of GO nanosheets with other materials is an effective approach to increasing molecular transport channels in MMMs. Li et al. [41] mixed carbon nanotubes (CNTs) and GO into a polymer matrix, and prepared MMMs by solution-casting method. CNTs and GO nanosheets were uniformly dispersed in Matrimid matrix to construct an effective CO_2 transport channel, which synergistically improved the permeability and selectivity of the membrane. The smooth wall of CNTs allowed the target molecules to pass quickly, while GO nanosheets containing hydroxyl and carboxyl structure acted as a selective barrier. The flaky morphology of GO caused a strong steric effect and prevented the aggregation of CNTs. Therefore, the CNTs were dispersed between the GO nanosheet layers and were loosely wound in the polymer matrix. In the meantime, the attachment of CNTs can inhibit the aggregation of GO nanosheets in MMMs. This study also

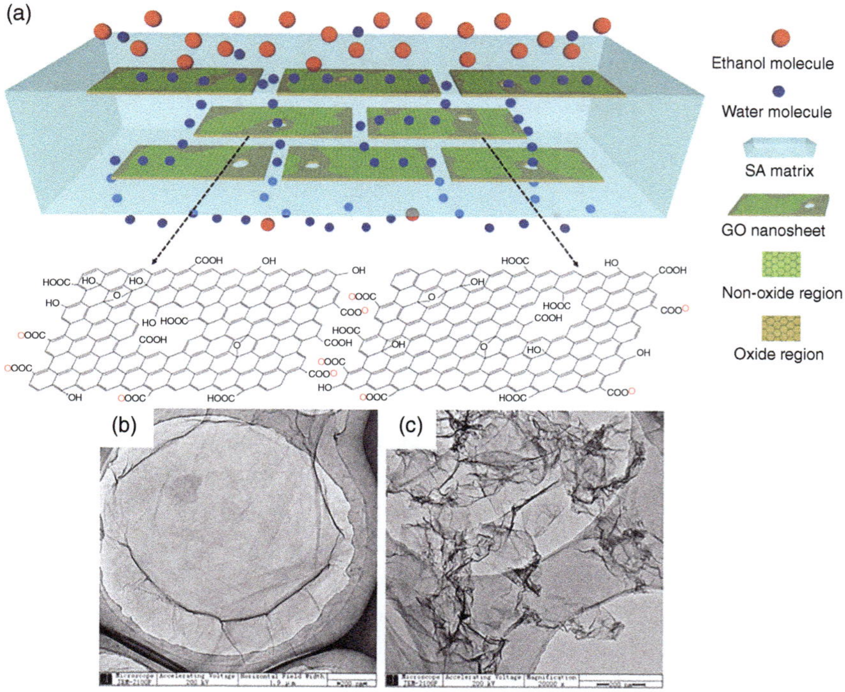

Figure 10.2 (a) Schematic illustration of mechanism for selective water permeation through the MMM with well-aligned GO nanosheets. TEM images of (b) pGO and (c) rGO. Source: Cao et al. [40]/with permission of Elsevier B. V.

proved that the MMMs with codoped CNTs and GO had better CO_2 separation performance than the MMMs doped with individual CNTs or GO.

10.2.1.2 Reducing Nonselective Defects

Nonselective defects usually originate from limited compatibility between dispersed phase and continuous phase, which will prevent the performance of MMMs from reaching the theoretical prediction. To enhance the interfacial compatibility, a variety of compounds, including macromolecules and small molecules, were modified onto GO nanosheets. Macromolecules can be introduced onto GO nanosheets for surface modification. Li et al. [42] filled GO nanosheets functionalized by polyethylene glycol and polyethyleneimine (PEG–PEI–GO) into the poly(ether-block-amide) (Pebax) matrix to prepare a high permeability and selectivity MMM (MP–MMM) as shown in Figure 10.3a). The improvement of membrane selectivity was manifested in the improvement of diffusion selectivity, solubility selectivity as well as reactivity selectivity. The mechanism of PEG–PEI–GO on membrane performance improvement is shown in Figure 10.3b. The high-aspect ratio of GO nanosheets generated a rigid interface between the polymer matrix and 2D nanomaterials. With the abundant primary, secondary, and tertiary amine groups on GO nanosheets modified by PEI, the interfacial compatibility between fillers and polymers was enhanced.

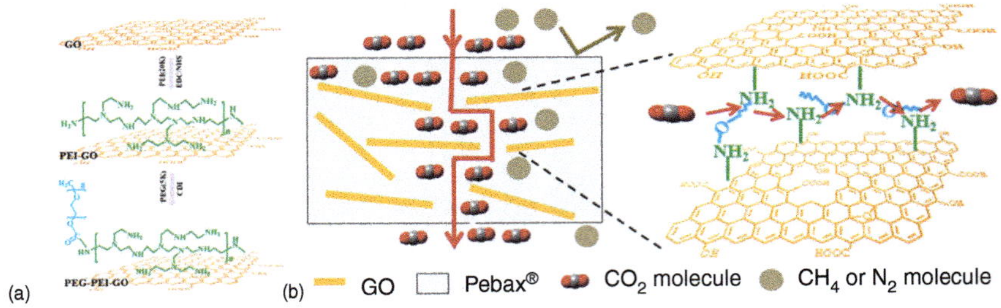

Figure 10.3 Schematic diagram of (a) the synthesis process of PEG–PEI–GO and (b) CO_2 transmission in MP–MMM. Source: Li et al. [42]/with permission of American Chemical Society.

Small molecules also can be grafted onto GO nanosheets for surface modification. Xin et al. [43] modified GO nanosheets with amino acid functionalization using dopamine (DA) and cysteine (Cys) and introduced amino acid-functionalized GO (GO–DA–Cys) into sulfonated polyether ketone (SPEEK) to prepare the MMMs. The amino acids containing carboxylic acid and primary amine groups enhanced the interfacial compatibility and increased the dissolution selectivity and reaction selectivity of the MMMs to CO_2 as well. The fabricated MMMs showed better separation performance than pure polymer membranes. In addition, the mechanical stability and thermal stability of the MMMs had also been greatly improved.

10.2.1.3 Introducing the Functional Sites for Facilitated Transport

The functionalized sites can enhance the transport of target molecules, thus improving the selectivity of MMMs. The incorporation of ions into GO nanosheets is a common method for functional modification of GO. Peng et al. [44] introduced zinc ions into GO nanosheets and prepared GO–DA–Zn^{2+} nanosheets through a dopamine-mediated complexation reaction. Then the prepared nanosheets were introduced into the Pebax matrix to fabricate MMMs for CO_2/CH_4 separation. Higher loading of Zn^{2+} (34.2 wt% of the nanosheets) can enhance the transmission of CO_2 based on the π-complexing mechanism, leading to the simultaneous improvement of CO_2 selectivity and permeability (Figure 10.4).

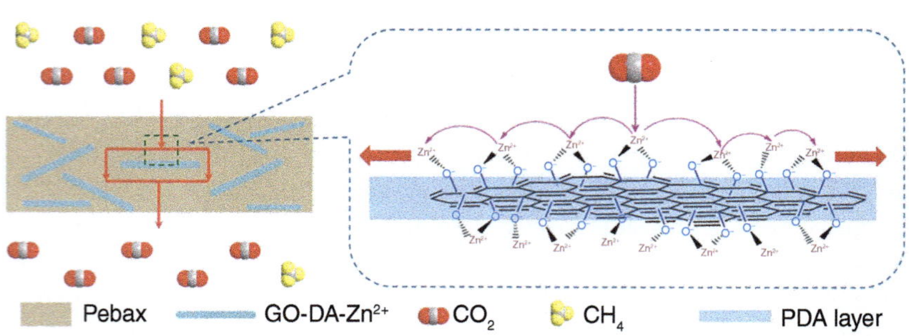

Figure 10.4 Schematic illustration of CO_2 transmission in Pebax/GO–DA–Zn^{2+} membrane. Source: Peng et al. [44]/with permission of Elsevier.

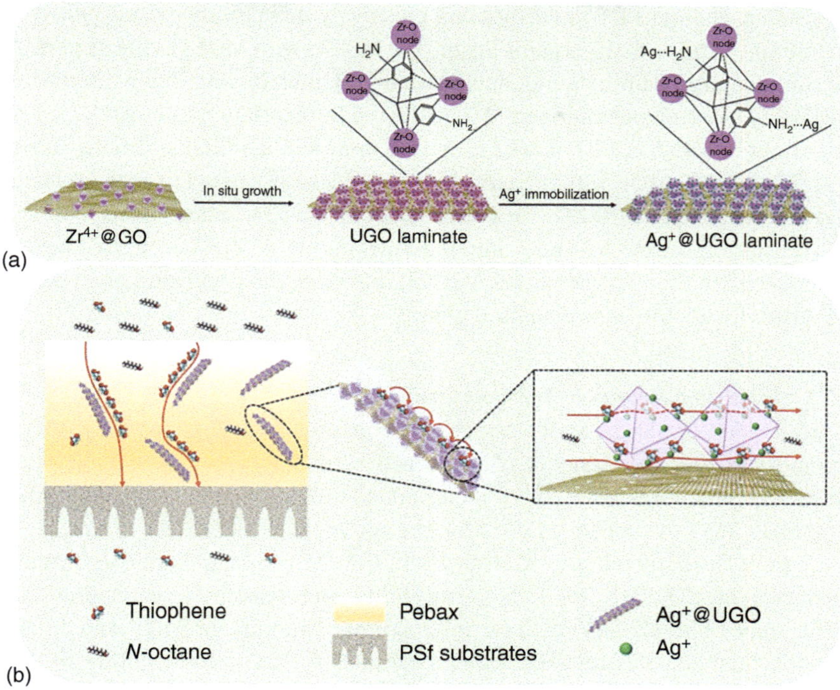

Figure 10.5 Schematic illustration of (a) the synthesis process of Ag+@UGO laminates and (b) molecules transport through Pebax-Ag^+@UGO-(X)/PSf membranes. Source: Zhang et al. [45]/with permission of Elsevier.

Zhang et al. [45] prepared a heterostructure laminate (UGO) by *in situ* growth of amine-functionalized zirconium funded metal–organic framework (UIO) on GO nanosheets and made it be a carrier to immobilize Ag^+. The prepared Ag^+@UGO (specific preparation process is shown in Figure 10.5a) were incorporated into the Pebax matrix for pervaporation desulfurization. Figure 10.5b shows the transport behavior of molecules in the Ag^+@UGO laminate membrane. The uniformly immobilized Ag^+ could reversibly interact with thiophene through a π-complexation reaction, which could promote the transport of thiophene, and increase the adsorption capacity and selectivity of thiophene. At the same time, UIO nanoparticles densely packed on GO nanosheets had better interparticle connectivity and could be used as a continuous molecular transport channel to enhance the migration efficiency of molecules and improve the permeability of the membrane.

10.2.2 Metal–Organic Frameworks (MOFs)

MOFs have the permanent porous crystal structure formed by metal ions and organic ligands connected by coordination bonds [46], representing an important class of crystalline molecular sieves. MOFs have the advantages of high porosity, high specific surface area, and regular pore structure, which is necessary to improve the separation performance of the membranes. Intrinsic porous structure provides low-resistance channels for molecules and effectively increase the permeability of the membrane. And with the organic ligand, the MMMs with MOFs as the dispersed phase can easily realize functional modification [47], thereby improving interfacial compatibility between polymer and MOFs. Zeolite imidazolate framework (ZIF) is an important subclass of MOFs, which are the most studied MOFs. Among these MOFs, ZIF-L is a 2D MOF with the same building blocks as ZIF-8. Liu et al. [48] had proved that when mixed 0D ZIF-8 nanoparticles and 2D ZIF-L nanosheets with SA for the separation of ethanol and water, the MMM with 2D ZIF-L exhibited better performance than that with 0D ZIF-8. We will introduce the significant progress in the MMMs with 2D MOFs as dispersed phase.

10.2.2.1 Increasing Molecular Transport Channels

The introduction of porous MOFs can create more molecular transfer channels in the membrane, and the existence of metal active centers can serve as facilitated transport carriers. Gou et al. [49] incorporated 2D $Cu(BPY)_2(OTF)_2$ into the Pebax@MHz 1657 to prepare MMMs for the separation of CO_2/CH_4. Compared with the pure Pebax membrane, MMMs with 2D MOF exhibited better separation performance of CO_2/CH_4 in both permeability and selectivity, exceeding the Robeson upper bound. The intrinsic microporous structure of MOF and inter-layer nanochannels constructed selective channels for CO_2, contributing to the excellent performance. The large number of affinity sites also increased the solubility of CO_2, achieving a synergistic improvement of permeability and selectivity (Figure 10.6).

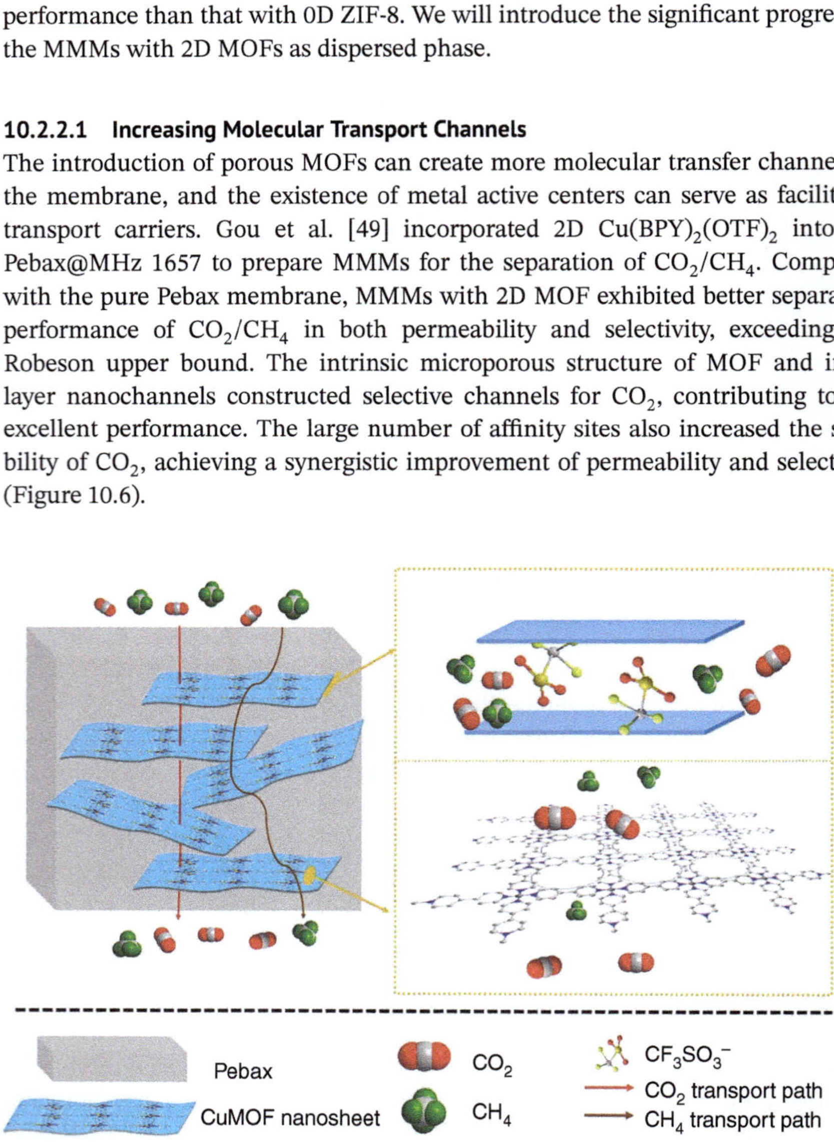

Figure 10.6 Schematic illustration of CO_2 transmission in Pebax/CuMOF membrane. Source: Gou et al. [49]/with permission of Elsevier.

10.2.2.2 Enhancing the Interfacial Compatibility Between Nanomaterials and Polymers

The introduction of fillers into polymer matrix always suffers from interfacial incompatibility, due to the affinity difference between them. However, MOF materials with specific physical/chemical properties can be used to improve compatibility and reduce interfacial defects, thus improving the selectivity of MMMs. Li et al. [50] chose a special hydrophobic MOF, Rho-[Zn(EIM)$_2$] (MAF-6), which had high thermal and chemical stability. MAF-6 and polydimethylsiloxane (PDMS) exhibited great compatibility, due to their similar affinity. Meanwhile, the hydrophobic MAF-6 showed a higher adsorption rate of ethanol than that of water, promoting selective transport in MMMs. The introduced MAF-6 inhibited the motion of PDMS, further improving the selectivity. But when the temperature increased, the separation factor declined, due to the weak interaction between PDMS chains and penetrants. It was found that when the loading amount of MAF-6 was 15 wt%, the MMM exhibited excellent separation performance, which exceed the highest level reported in the literatures at the time and overcome the "trade-off" effect.

10.2.3 Covalent Organic Frameworks (COFs)

COFs are a class of periodic crystalline framework material, consisting of organic monomers connected by covalent bonds [51, 52]. Since the discovery of COFs by Yaghi and coworkers in 2005 [53], a great variety of 2D COFs have been reported [54–56]. The intrinsic framework structures and tailorable organic linkers endowed 2D COFs with low density, large surface area, and facilely-tailored functionalities, which have attracted increasing interests in diverse fields, such as gas adsorption, functional devices, separation, and supercapacitors [57, 58]. Especially, 2D COFs are attracting particular attention in the field of molecular separation, due to their ordered and tunable pore structure and great stability. However, it is of great difficulty to fabricate defect-free and large area COFs membranes, because of the synthetic complexity and insolubility of COFs. Integrating COF nanosheets into the polymer matrix is a common approach to fabricate COF-based membranes. Compared with other fillers, such as GO and MOFs, the pure organic nature of COFs endows them with high compatibility with polymer matrixes, which facilitates the elimination of nonselective interface defects and agglomeration of oligomers. The orderly porous structures of COFs provide additional mass-transfer channels. The backbone of COFs contains abundant functional groups that can serve as sites for the introduction of specific functions by predesign or postsynthesis modifications. This makes COFs ideal carriers for functionalization. In addition, the distribution of COFs within the membrane can be controlled by modulating the membrane processing, which allows to form specific structures in the membranes. The MMMs combine the processable features of polymer materials and the unique performance of COFs, providing a universal and efficient method for fabricating continuous and defect-free membranes. Currently, nonsolvent-induce phase separation (NIPS) and interfacial polymerization (IP) are the most common methods to introduce COFs into the polymer matrix [37].

10.2.3.1 Increasing Molecule Transport Channels

In the past few years, various MMMs containing COFs have been reported. The simplest and most common method is to modify the COFs and introduce them into the polymer matrix. The introduced COFs will alter the properties of adjacent polymer phases, such as the packing, dynamics, or conformation of polymer chains near their surface. The well-distributed COFs alter the fractional free volume of polymer. The intrinsic porosity of COFs provides additional mass-transfer channels. Thus, the incorporation of COFs will significantly improve the overall transport properties. Inspired by the respiratory system in organisms, Yang et al. [59] reported MMMs with hierarchical pore structures, which was consisted of covalent organic nanosheets (CONs) and poly (ether sulfone) (PES). TpPa-1 and TpBD were synthesized by the Schiff base reactions and then were exfoliated into nanosheets. The prepared amide-modified CONs were introduced into PES matrices to prepare the MMMs. In the NIPS process, the CONs were concentrated on the membrane surface and generated hierarchical pore structures (Figure 10.7), which was advantageous for multiscale separation of membranes. The high COF content led to a more compact structure with micropores. Remarkably, the membrane-containing TpBD CONs (8 wt%) exhibited a high water/*n*-butanol separation factor of 2735 and remained robust for a long-term use up to about 150 hours. In addition to pervaporation, COF-based MMMs were believed as low carbon, efficient and sustainable method for gas separation. Zou et al. [60] used amide-modified COF nanosheets and Pebax to prepare MMMs with higher CO_2/N_2 gas selectivity. TpPa-1-nc synthesized by solid reaction was blended into Pebax to fabricate membranes. The uniform distribution of TpPa-1-nc within membranes was demonstrated by scanning electron microscopy (SEM) (Figure 10.8c). TpPa-1-nc disrupted the inherent packing of polymer chains and increased the accessible free volume, enhancing gas molecule diffusion. Due to the high polarizability and quadrupole moment, CO_2 was more likely to aggregate and permeate into the membranes. The uniform micropore and functional amide group of TpPa-1-nc realized the facilitated transport of CO_2,

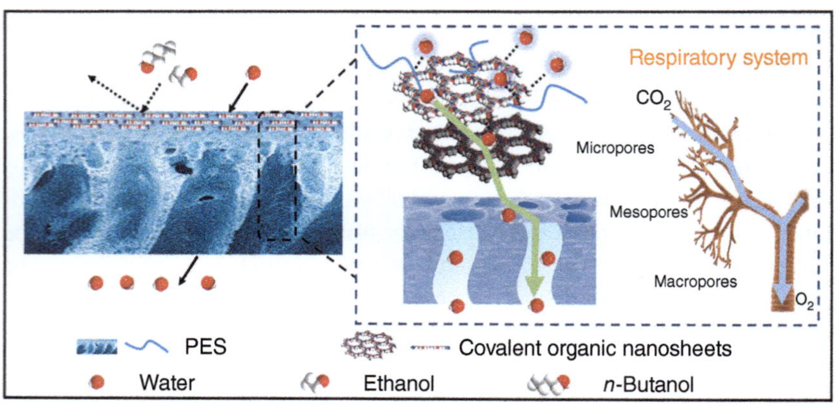

Figure 10.7 Schematic illustration of the molecule transport through the PES/TpPa-1 membrane. Source: Yang et al. [59]/with permission of Elsevier B. V.

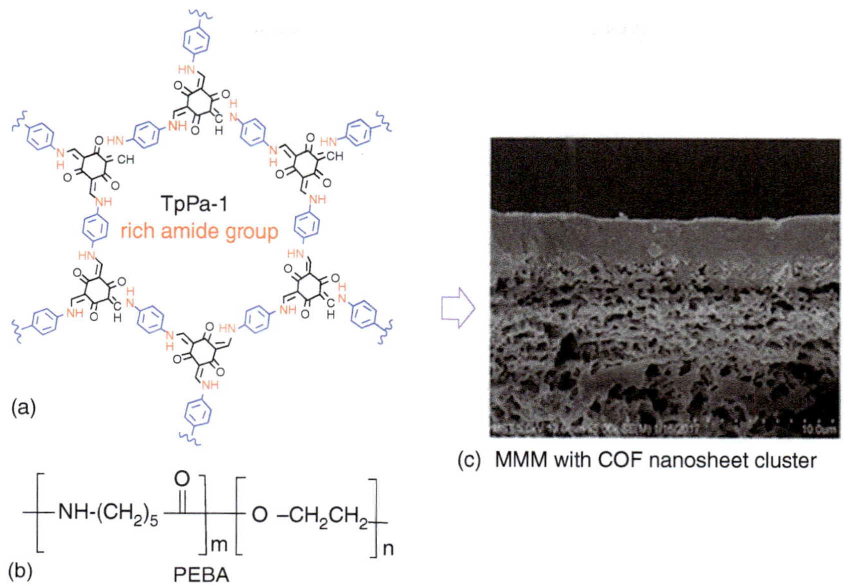

Figure 10.8 (a) TpPa-1 synthesized by Schiff base aldehyde–amine condensation (b) structure of Pebax (c) SEM image of cross section of TpPa-1-nc/ Pebax membrane. Source: Zou et al. [60]/with permission of American Chemical Society.

resulting in a much higher increase of CO_2 permeation than N_2. The MMM with 1 wt% COF nanosheets exhibited great separation factor (~72), with long-term stability of 120 hours.

10.2.3.2 Introducing Facilely-Tailored Functionality

The abundant organic monomers and diverse synthesis strategies enable the preparation of functionalized COFs. Through rational design, a variety of functional sites can be introduced on COF backbones to modulate the physical/chemical microenvironment and enhance the separation performance. Fan et al. [61] dispersed hydrazone-linked COF-42 in commercial polydimethylsiloxane (PDMS) to prepare MMMs. When used for the recovery of *n*-butanol from aqueous mixture, the pores inside COF-42 provided extra channels for both water and *n*-butanol. However, the spare space inside COF-42 could accumulate water through the H-bond network, resulting in a faster diffusion of *n*-butanol (Figure 10.9a). COF-X15, COF-X20, and COF-X21 were synthesized to understand the effect of functional groups and crystallinity of COFs, by replacing the vinyl groups of the COF-42 with 2-propyl, *n*-propyl, and *n*-butyl. As shown in Figure 10.9c, the membrane with stronger hydrophobic properties possessed a high separation factor because *n*-butanol molecules diffused more easily, while the transport of water molecules was blocked. The good compatibility, regular pore structures, decent chemical stability, and abundant binding sites make COFs ideal supports for functionalization in membranes. Inspired by the efficient facilitated diffusion via carrier proteins in biological membranes, Wang et al. [62] fabricated facilitated transport membranes based on

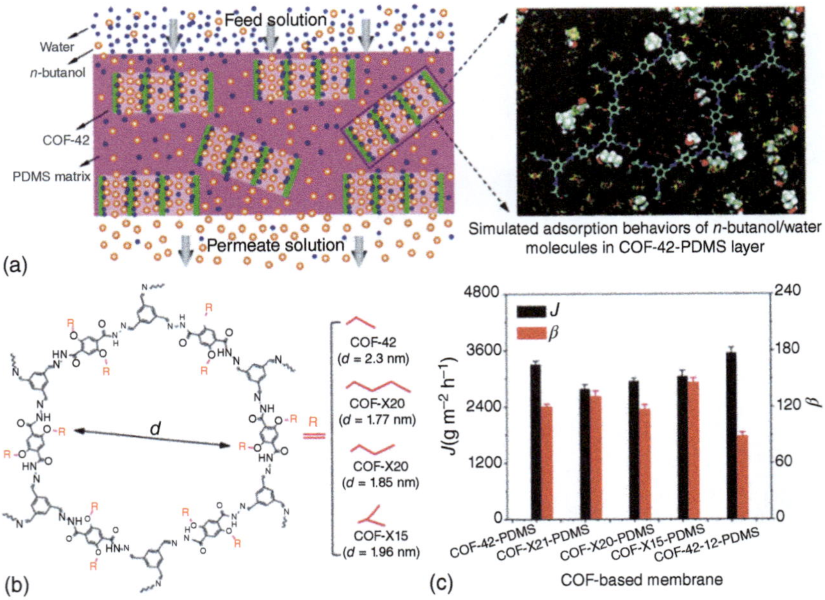

Figure 10.9 (a) Schematic illustration of COF-42/PDMS membrane for separating *n*-butanol from water. (b) Structure of COFs with different groups (R). (c) *N*-butanol flux and separation factor for different COF-based membranes in *n*-butanol/water mixture. Source: Fan et al. [61]/with permission of Royal Society of Chemistry.

CONs and transition-metal ions (Figure 10.10a). For the first time, crystalline CONs were used as supports to discover the microenvironment–performance relationship in facilitated transport membranes. The transition metals were immobilized on the CONs by coordination, forming Metal-ion@CONs. Then, the CONs were blended into Pebax to prepare facilitated transport membranes (Figure 10.10a). The images of energy dispersive spectrometer (EDS) mapping (Figure 10.10d) revealed good dispersion of metal ions and Ag^+@TpHz nanosheets in Pebax, which was attributed to the organic nature of CONs. Specific and reversible reactions between thiophene and Ag^+ enhanced the adsorption and diffusion of thiophene within the membranes. The ordered channels further reinforced the Ag^+-induced facilitated transport effect. The optimized membranes exhibited promising separation performance for gasoline desulfurization with permeation flux of 21.6 $kg\,m^{-2}\,h^{-1}$ and enrichment factor of 6.29, which was 22% and12% higher than that of Pebax-$AgNO_3$/PSf respectively.

10.2.3.3 Constructing Hierarchical Structures in MMMs

In addition to increasing mass-transfer channels and introducing functionality, the incorporation of COFs also have an impact on the membrane construction process, resulting in specific structures that enable intensified separation processes. The manipulation of COF distribution in polymer matrices has attracted great interest. In 2018, by mimicking the functionally graded biomaterials, such as seashells in nature, Yang et al. [63] proposed the concept of functionally graded membranes

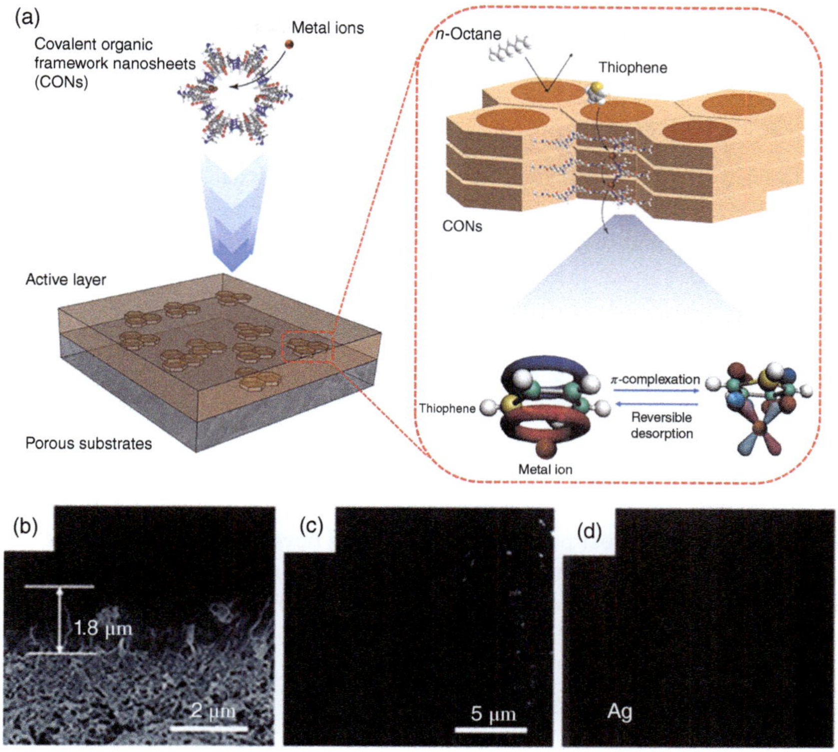

Figure 10.10 (a) Schematic illustration of channel-mediated facilitated transport membranes (b) Cross-sectional SEM images of Pebax-Ag^+@TpHz-0.4/PSf. (c) SEM surface images of Pebax -Ag^+@TpHz-0.4/PSf and (d) Elemental mapping of Ag in Pebax -Ag^+@TpHz-0.4/PSf. Source: Wang et al. [62]/with permission of Royal Society of Chemistry.

(COMs). The CONs were introduced into PES by NIPS method, during which surface segregation was employed to create the gradient distribution of CONs. By controlling both the amount of CON and the phase inversion temperature, the gradient distribution of CONs in the membrane can be tuned. An ultrahigh COF content up to 50.90 vol% on the membrane surface has been achieved. A high amount of CONs on the membrane surface could promote the adsorption of water molecules through hydrogen bonding. The dense structure of the surface imparted the molecular-sieving effect to the membrane, thus enhancing the water selectivity. After selective permeation through the COF-rich region, the porous structure of the main part of the membrane allowed rapid passage of water molecules with low transfer resistance. When used for water/ethanol separation, the functionally graded COMs exhibited a permeation flux of 2.48 kg m^{-2} h^{-1} and a high separation factor of 1430, with exceptional long-term operation stability up to 320 hours.

COF-based MMM approach showed plenty of advantages, including feasible operation and high generality to various COFs, which will garner increasing interest from the scientific community and industry. Further improving the compatibility of COFs through chemical structure modulation is crucial to fully exploit the potential of COF-based MMMs. Therefore, rational design and fabrication of 2D COF nanosheets are extremely important to maximize the advantages of COFs.

10.2.4 Other 2D Materials

Other ultrathin 2D materials, such as transition-metal dichalcogenides (TMDs), g-C_3N_4, and MXene have exhibited great potential as building blocks for MMMs. Their rationally tunable interlayer spacing and channel structure, tailored functionality, and large surface area can introduce surprising effects to mixed-matrix membranes.

10.2.4.1 Transition-Metal Dichalcogenides (TMDs)

TMDs nanosheets, such as MoS_2, $MoSe_2$, WTe_2, and WS_2, are an emerging class of membrane materials. Among them, MoS_2 is a representative 2D TMDs material. The high thermal stability, facile and scalable synthesis process, high rigidity, and many exceptional properties (e.g. electrical, antibacterial, and mechanical properties) endow MoS_2 with great potential in fabricating novel membranes with excellent separation performances [64]. In addition to pure MoS_2-based membranes, the MMMs approach greatly fulfills the advantages of MoS_2 for precise and rapid molecular separation. MoS_2-embedded MMMs have been used in water treatment, gas separation, and pervaporation [65–67].

Due to the high chemical stability and bearing few structural defects, MoS_2 nanosheets are used to construct facilitated transport pathways in MMMs. Pan et al. [67] embedded MoS_2 nanosheets into Pebax matrix to prepare facilitated transport membranes (Figure 10.11a). The reversible chemical interaction between MoS_2 and thiophene allowed MoS_2 nanosheets to serve as a carrier for facilitated transport. The large and regular surface of MoS_2 nanosheets provided a continuous channel for the transport of thiophene. The two effects synergistically enhanced the transport rate of thiophene within the membrane and improved the diffusion selectivity. When used to remove thiophene from the *n*-octane-thiophene mixture, the MMMs containing 4 wt% MoS_2 showed 22.27% higher permeability and 65.94% higher selectivity than these of Pebax membrane. Besides, MoS_2 exhibits excellent mechanical properties, and trace addition of MoS_2 to the polymer will remarkably improve the tensile strength of the MMMs [70]. When MoS_2 nanosheets were introduced into SA matrix, the load could be effectively transferred through the interfacial interaction between the nanosheets and polymer chains. The tensile strength and Young's modulus of SA-MoS_2-2 increased by 27% (from 104 to 132 MPa) and 57% (from 4.61 to 7.22 GPa), enabling the MMM to withstand fast feed washing over the surface. Moreover, the coordination between Mo atoms and oxygen-containing functional groups on SA polymeric chains improves the interfacial compatibility of these two materials. The swelling resistance, thermal stability, and tensile property of the MMM were considerably enhanced. The hydrophobic surface of MoS_2 nanosheets endowed water with faster transportation, which intensified the diffusion process of water in membranes. These features made the SA–MoS_2 MMMs practical for engineering applications.

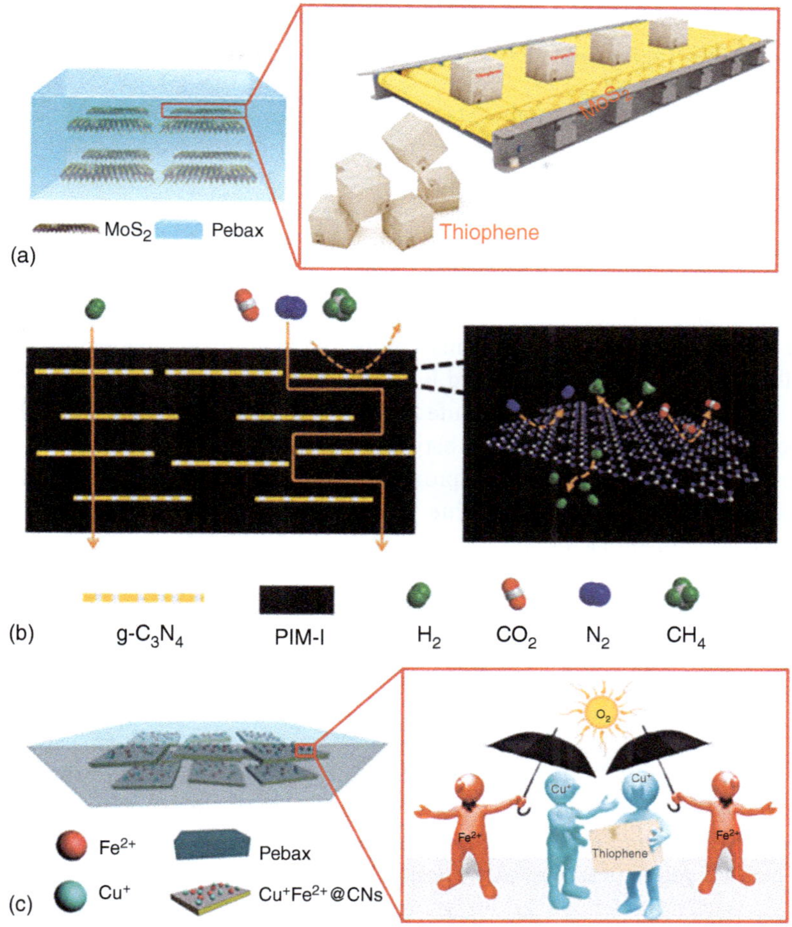

Figure 10.11 (a) Schematic illustration of structure and facilitated transport effect of the MoS_2 embedded MMMs. Source: Pan et al. [67]/with permission of Elsevier B.V. (b) Schematic illustration of molecular transport through PIM-1–g-C_3N_4 membrane. Source: Tian et al. [68]/with permission of Elsevier. (c) Schematic illustration of the Pebax-Cu^+Fe^{2+}@CNs-5/PSf membrane and the protective effect of the stabilizer Fe^{2+} on Cu^+. Source: Ding et al. [69]/with permission of Elsevier.

10.2.4.2 Graphitic Carbon Nitride (g-C_3N_4)

Graphite carbon nitride, generally known as g-C_3N_4, is a graphene-like layered material composed of C, N, and H impurities. It has been widely employed to construct functional membranes, due to its outstanding stability, intrinsic porous characteristics, and facile synthesis [71]. In 2016, Tian et al. [68] incorporated g-C_3N_4 nanosheets into polymers of intrinsic microporosity (PIM-1) to fabricate MMMs. The casting solution was prepared by dispersing g-C_3N_4 and PIM-1 in chloroform. Then, the solution was poured into Teflon dishes and evaporated to fabricate freestanding MMMs. The g-C_3N_4 nanosheets could efficiently modulate the pacing of PIM-1 chains, creating additional pathways at the interface between the 2D filler and the polymer matrix. Compared with pure PIM-1, the gas permeability of MMMs was greatly enhanced, especially at low g-C_3N_4 loadings. The sieving effect of g-C_3N_4 nanosheets increased the selectivity for H_2/CH_4 and H_2/N_2 (Figure 10.11b). Besides, the incorporation of g-C_3N_4 improved the mechanical properties and long-term performance of PIM-1. Similar to COFs, the facilely-tailored functionality and stable structure make g-C_3N_4 ideal platform for facilitated transport carriers. Ding et al. [69] fabricated facilitated transport MMMs by using g-C_3N_4 nanosheets as the supports to load Cu^+ and Fe^{2+}. The ultrathin CN nanosheet was covered with a polydopamine layer and metal ions were loaded on the layer through metal–organic coordination interaction. Finally, the nanosheets with facilitated metal ions were blended into Pebax matrix to prepare the MMMs. The metal ions were uniformly distributed within the membrane along with the nanosheets. The reversible interaction between Cu^+ and thiophene endowed the membranes with facilitated transport properties, greatly enhancing the selectivity for octane-thiophene model gasoline. When the content of Cu^+Fe^{2+}@CNs is 5 wt%, the enrichment factor and permeation flux of the MMM were 28.80% and 18.76% higher than these of pure Pebax membrane. Meanwhile, the stabilizer Fe^{2+} prevents the disproportionation and oxidation of Cu^+ (Figure 10.11c), significantly improving the long-term operation stability in the desulfurization process.

10.2.4.3 MXenes

MXenes are a new family of 2D materials with a layered hexagonal structure, prepared by selectively etching the A-layer in the MAX phase. They have been employed for many applications, due to their rich surface chemistry and unique physicochemical properties. Most of the research interest on MXene is concentrated on $Ti_3C_2T_x$, where T refers to the surface group (OH, O, or F). The polarity and abundant chemical sites of Mxenes endow the nanosheets with tailored functionalization and increase the compatibility at the filler/polymer interfaces [72]. Xu et al. [73] incorporated 2D $Ti_3C_2T_x$ MXene nanosheets into chitosan (CS) to fabricate MXene/CS MMMs for solvent dehydration. In the MMMs, the interlayer channels constructed of the MXene laminates greatly enhanced water transposition, while preventing ethanol, ethyl acetate, and dimethyl carbonate. Compared with the previous membrane, the permeation flux and separation factor of the MMMs were significantly improved.

In addition to the above materials, there are plenty of 2D materials that are used to fabricate MMMs, such as amphiphilic carbonaceous material (ACM) [74], boron nitride (BN) [75], and layered double hydroxide (LDH) [76]. The different properties of 2D materials might meet the needs of various applications. Due to their strong structural, physical, and/ or chemical similarities, it is an efficient method to fully utilize the lessons learned from each 2D material, when developing 2D material-based membranes.

10.3 Two-Dimensional Material as Continuous Phase of MMMs

In the previous part, MMMs with polymer were introduced as the continuous phase and 2D materials as the dispersed phase. Compared with pure polymer membranes, the separation performance of mixed-matrix membranes has been significantly improved, but they are still constrained by polymer materials. Recently, membranes with the 2D materials as the continuous phase and the polymer as the dispersed phase have been used for the separation of gas or liquid molecules [77–79], which can be regarded as another kind of MMMs. The 2D materials are used as the bulk of the membrane, and the polymer plays the role of cross-linker. The atomic thickness and unique physical/chemical properties of 2D materials contribute to the extraordinary separation performance of 2D materials-based MMMs, while the flexible polymers strengthen their mechanical property and processability and tailor the channel size.

10.3.1 Graphene Oxide (GO)

The pure GO membranes are often formed by tightly stacked nanosheets. The mass transport occurs in the interlayer channels between GO nanosheets, which can be easily tuned by the interlayer interactions (such as hydrogen bonds and π–π interactions) [80]. Due to the characteristics of rapid transfer of water molecules, GO membranes have been widely used in water treatment and pervaporation [81]. But, when used in aqueous environment, the swelling of the GO membranes will destroy the original interlayer spacing and decrease the separation performance of the membranes significantly. Therefore, a cross-linked structure is necessary to enhance the stability of GO membranes. The functional cross-linking of GO mainly includes two kinds—covalent cross-linking and noncovalent cross-linking. Covalent cross-linking is mainly achieved by covalent reaction between oxygen-containing groups on the surface of GO and other functional groups. Hung et al. [82] used diamine to covalently cross-link the oxygen-containing functional groups of GO, and the interlayer spacing of the GO membrane can be stabilized at 10.4 Å in the wet state. Noncovalent cross-linking structure is formed by the physical interaction between GO nanosheets and target cross-linkers, such as π–π interaction, electrostatic interaction, and van der Waals force. Xu et al. [83] utilized π–π interaction between GO nanosheets and 2D cationic porphyrin nanosheets to stabilize the structure of the GO membranes, and successfully controlled the interlayer spacing to about 1 nm, which enhanced the permeability while maintaining a good retention rate of target molecules.

The diamine molecules and 2D cationic porphyrin nanosheets mentioned above belong to small-molecule cross-linkers, polymers can be used as macromolecular cross-linkers to be introduced into GO membranes to prepare MMMs. The relative molecular mass of the polymers is generally above 20 000 g mol^{-1}. Polymer cross-linkers with a large number of functional groups increase the cross-linking efficiency and achieve the desired cross-linking effect with a small amount. With various sizes of polymers, the interlayer spacing of GO-based membrane can be preciously controlled. The incorporation of polymers will also introduce functional sites.

10.3.1.1 Controlling Interlayer Spacing

Polymers can directly cross-link unmodified GO to prepare MMMs. Zhang et al. [7] grafted polydopamine (PDA) onto GO nanosheets, and then prepared MMMs by vacuum-assisted self-assembly. The prepared membranes showed excellent selectivity and permeability in ethanol/water separation, overcoming the upper bond. DA reacted with oxygen-containing functional groups on the GO nanosheets to form "raised points," which increased the surface roughness of the GO nanosheets and expanded the interlayer spacing. The cross-linked structure also increased the stability of the MMMs. The degree of polymerization of PDA increased with the reaction time. By adjusting the dosage and reaction time of PDA, the physical and chemical microenvironment of GO could be easily regulated (Figure 10.12).

With the development of supramolecular chemistry, supramolecular polymers with specific structures and properties [84] can be used as cross-linkers for GO-based MMMs. Song et al. [85] took primary amine-terminated polyamidoamine (PAMAM) dendrimers with medium relative molecular mass and multiple periphery amine groups as cross-linkers to prepare GO-based MMM for the separation of butanol/water and desalination (Figure 10.13). The channel size could be continuously and accurately tuned by varying the molecular mass of branched PAMAM dendrimers. The GO nanosheets were firmly connected by the multiple interactions between the oxygen-containing functional groups on the GO nanosheets and the amine groups of PAMAM. The cross-linked structure formed an interlayer channel for mass transport with great stability. When used for butanol/water separation and pervaporation desalination, it is the MMMs that exhibited excellent separation performance surpassing the upper bound of most membranes.

10.3.1.2 Modulating the Physical/Chemical Microenvironment

The introduced polymers will significantly change the physical/chemical microenvironment of the MMMs. By rational design, the cross-linkers could also further

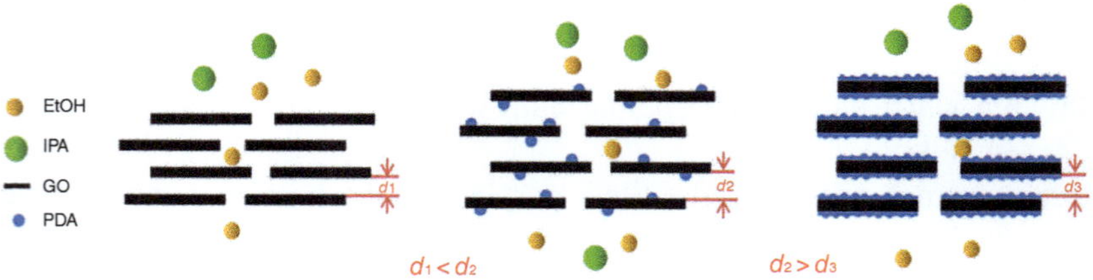

Figure 10.12 Structural illustration of GO and PDA-g-GO membranes with different loading of PDA. Source: Zhang et al. [7]. Reproduced with permission. Copyright 2018, John Wiley & Sons.

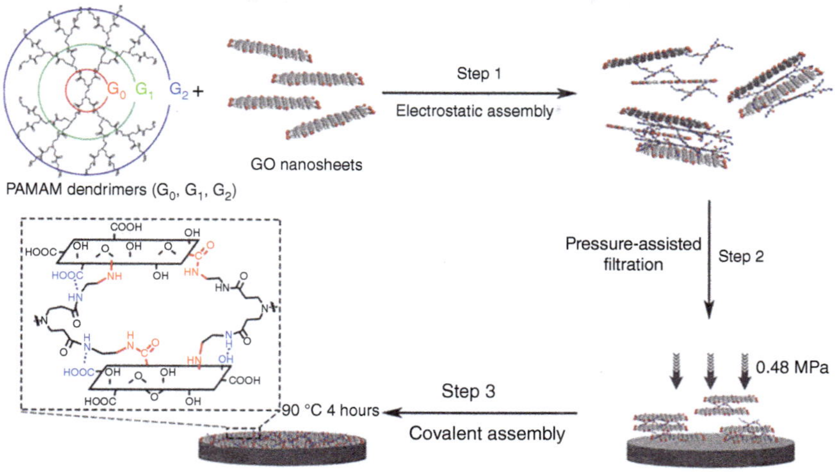

Figure 10.13 Schematic illustration of the assembly process of GO–PAMAM/PTFE membranes. Source: Song et al. [85]. Reproduced with permission. Copyright 2019, Royal Society of Chemistry.

improve the performance of GO membranes and construct more stable structures. Lim et al. [86] introduced tannic acid (TA) to the surface of GO and then reacted with hyperbranched polyethyleneimine (PEI) via the catechol groups of TA under weakly alkaline conditions (Figure 10.14). The amine groups were covalently cross-linked, and PEI was introduced to prepare MMM for ion separation. The prepared MMMs exhibited excellent ion rejection and water flux. Polymers can also be incorporated to modify the channel of membranes. Pan et al. [87] introduced polyvinylamine (PVAm) as a cross-linker into the GO membrane to obtain a structure-stable membrane. Then, PVAm was used as a mineralizer to induce the biomimetic mineralization to *in situ* generate silica as the second cross-linker to further regulate and stabilize the size of channels (Figure 10.15). The prepared dual-cross-linked MMMs were used for separation of butanol/water and showed superior performance overcoming the upper limit.

10.3.2 Metal–Organic Framework (MOF)

MOFs, also known as porous coordination polymers, are a class of crystalline porous materials. Due to their high internal surface area, tunable pore size, facilely-tailored functionality, and structural flexibility, MOFs have promising potential in a variety of applications [88, 89], especially in membrane separation [36, 90]. The flexible and adjustable structural design allows the pore sizes of MOFs to be precisely positioned between the molecular diameters of the mixtures to be separated, potentially achieving highly selective molecular sieving. The electron-deficient unsaturated sites can reversibly interact with electron-rich molecules, such as CO_2 and hydrocarbons. This reversible interaction will enhance adsorption selectivity. Through rational design or postsynthetic modification, a variety of functional groups are introduced into

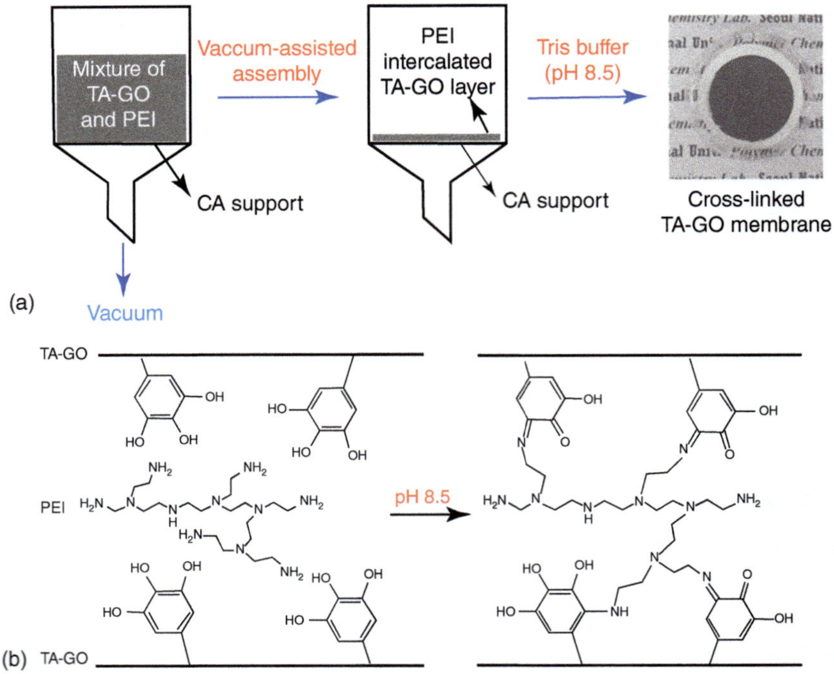

Figure 10.14 Schematic illustrations for (a) the preparation of the cross-linked TA–GO membrane and (b) the formation of the covalently-bonded interlayer bridges between the adjacent TA–GO sheets. Source: Lim et al. [86]/with permission of Elsevier B.V.

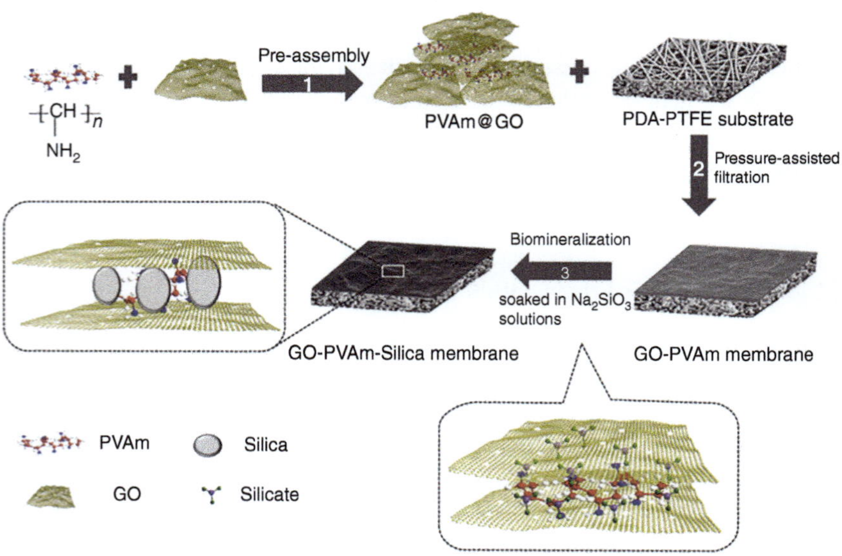

Figure 10.15 The preparation of GO–PVAm-silica membranes. Source: Pan et al. [87]/with permission of Elsevier.

MOF backbones. Based on this, the physical/chemical microenvironment within the membranes can easily be modulated to intensify the transport of target molecule. However, pure MOF membranes are mechanically fragile, which makes them susceptible to cracking during synthesis and processing. The unsuitable orientation, lattice geometry, and grain size will lead to nonselective crack or grain boundary defects. Generally, the thickness of MOF membranes varies from several micrometers to tens of micrometers. As the membrane thickness increases, the mass-transfer resistance will increase and the permeability will decrease. This increases the difficulty of overcoming "trade-off" effects. Since MOFs are comprised of coordination bonds, their stability is susceptible to humidity, temperature fluctuations, and PH. MMMs through incorporating polymers into MOF nanosheets matrix have become a new trend in the field of MOF membranes [91]. Unsaturated metal sites and functional groups of organic ligands in MOFs provide bonding sites for polymers. The flexible polymers reinforce the backbones of MOFs and increase their stability in different environments while maintaining their high crystallinity. Besides, the polymers will enhance the processability of MOF membranes and fill the defects, facilitating the preparation of ultrathin membranes. The tethered polymer can partition the pores of MOFs and provide special features.

10.3.2.1 Enhancing Processability and Stability of MOFs

The flexible polymers can regulate the assembly of MOFs nanosheets, enhancing the mechanical properties and processability. The cross-linked structure connected by high-strength bonds reduces the effects of environmental fluctuations and intensifies the stability of the framework. Li et al. [92] regulated the assembly of 2D $Zn_2(Bim)_4$ nanosheets via ionic bonds between polycations and the peripheral hydroxy groups of the nanosheets to prepare membranes. The attached polymer improved the dispersion stability of MOF nanosheets. The large-aspect ratio and cross-linked structure drove MOF nanosheets to stack in orientation parallel to the plane of the support, reducing crack, and grain boundary defects. Based on this novel, 2D MOF functional layer, a crack-free, ultrathin (~37 nm) membrane with aligned channels was prepared (Figure 10.16a). The cross-linked structure enhanced stability of MOFs in aqueous environment, which is crucial for long-term use in nanofiltration. Zhang et al. [94] fabricated carboxymethyl cellulose (CMC) cross-linked ZIF-L membrane for gas separation. The complex particles were prepared by adding a certain amount of ZIF-L nanosheets into the 2 wt% CMC suspension. Owing to the anion charge repulsion effect between the complex particles, ZIF-L nanosheets were well dispersed in aqueous solutions. The resulting CMC–ZIF-L solution was cast onto a stainless steel substrate to acquire MMMs. The highly dense structural CMC reduced nonselective gas permeation, increasing the selectivity of ZIF-L membranes.

10.3.2.2 Modulating the Physical/Chemical Microenvironment

When introduced into MOF matrix, the polymer chain will increase the layer spacing and reduce the intrinsic pore size of MOF nanosheets. The smaller pores

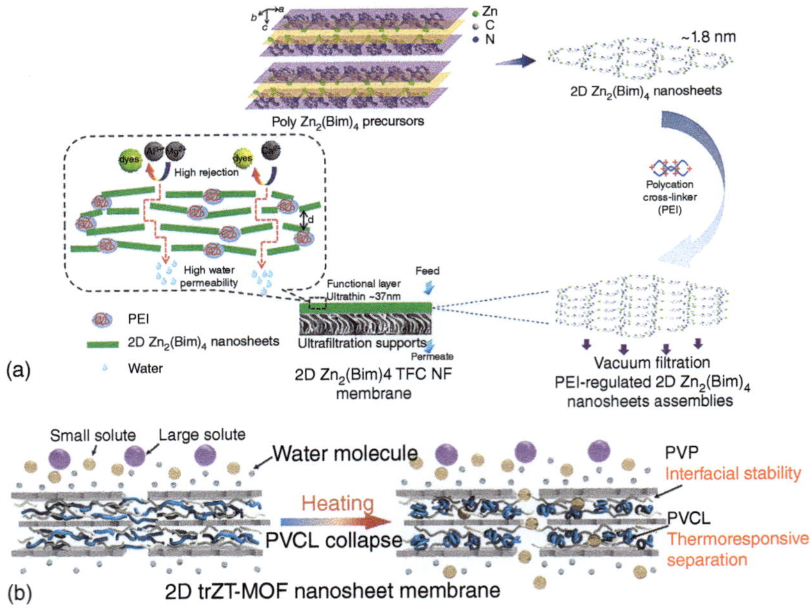

Figure 10.16 (a) Schematic illustration of the fabrication of the 2D $Zn_2(Bim)_4$ TFC NF membrane. Source: Li et al. [92]/with permission of Elsevier. (b) Schematic illustration of the "smart" membrane separation performance of the trZT-MOF nanosheet membrane with thermoresponsive nanochannels. Source: Jia et al. [93]/with permission of Springer Nature.

facilitate molecular sieve, while larger layer spacing allows for rapid molecular transportation. Besides, the group of polymers also modulate the chemical microenvironment in the membranes. Jia et al. [93] prepared a stable 2D MOF MMM via twin-chain stabilizing strategy. Polyvinylpyrrolidone (PVP) and poly(*N*-vinyl caprolactam) (PVCL) were attached to trZT–MOF surface via the strong interactions between carbonyl and Zn^{2+}. This structure confined the growth of MOFs along with the *z*-axial, formatting ultrathin trZT–MOF nanosheets with large lateral sizes. Meanwhile, the polymers eliminated the stacking defects and regulated layer spacing. The intrinsic porosity of MOF nanosheet and interlayer nanochannels contributed to the high permeance. When heated, the PVCL chains would collapse, increasing the size of nanochannels. The LCST phase transition of PVCL chains endowed the trZT–MOF nanosheet membrane with sensitive and reversible thermoresponsive separation properties (Figure 10.16b).

10.3.3 Covalent Organic Frameworks (COFs)

As highly ordered framework polymers, COFs are considered to be promising high-performance membrane materials. Due to the periodic framework pore structures, COFs are unconstrained by the rigidity and spacing of polymer chains and circumvent the "trade-off" effect and physical aging issues. Long-range structural order of COFs reduce mass-transfer resistance and enable rapid molecular transport, resulting in high permeability. The nanosheets contain high density, homogeneous, and tailorable functional groups [95, 96]. Thus, COFs can achieve facilitated transport mechanism to obtain high separation selectivity. But the inevitable defects of the structure and insolubility in various solvents limit the processability of COFs. The pore size of COFs is larger than 0.6 nm, which is much larger than most gas and liquid molecules. Incorporating polymers into COFs is a feasible method to solve the above challenges. Functional groups on COF nanosheets provide reaction sites for linking polymers. Flexible polymers will strengthen the mechanical property and processability of MMMs. The sheltering effect of polymers will reduce the effective pore size of MMMs, benefiting molecular sieve.

10.3.3.1 Regulating the Physical/Chemical Microenvironment

Flexible polymers can be immobilized on the surface of COF nanosheets by chemical bonding, which can effectively tune the structure of the MMMs. Meanwhile, the functional groups of the anchored polymers will significantly alter the chemical microenvironment of the MMMs. In 2019, Yang et al. [78] proposed mixed-dimension COFs membranes for the first time (Figure 10.17a). 2D COF nanosheets and 1D cellulose nanofibers (CNFs) were used as building blocks to fabricate the MMMs. Negatively charged CNFs and positively charged $TpTG_{Cl}$ nanosheets were mixed in the solution and tightly bound via electrostatic interactions (Figure 10.17f). Then, the MMMs were fabricated by a vacuum-assisted self-assembly method on a polyacrylonitrile (PAN) substrate. The thickness of the membrane can be tuned from dozens of nanometers to a few microns by varying the volume of the filtrate. The transmission electron microscope (TEM) images (Figure 10.17b–e) showed that the planar $TpTG_{Cl}$ was covered with dense

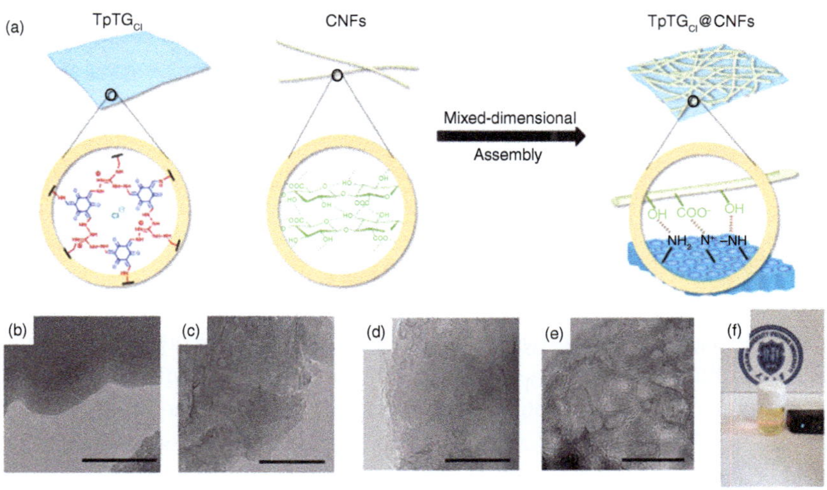

Figure 10.17 Mixed-dimensional assembly of $TpTG_{Cl}$ and CNFs. (a) A schematic illustration showing the assembly process and the interactions. High magnification TEM images of (b) pure $TpTG_{Cl}$, (c) $TpTG_{Cl}$@CNFs-1, (d) $TpTG_{Cl}$@CNFs-3, (e) $TpTG_{Cl}$@CNFs-5 (scale bar, 100 nm), and (f) Tyndall scattering effect in the aqueous solution of $TpTG_{Cl}$@CNFs-X nanocomposites. Source: Yang et al. [78]/with permission of Springer Nature.

networks of 1D CNFs. The high water absorption of CNFs endowed the MMMs with super hydrophilic surface, while the ordered channels of $TpTG_{Cl}$ conferred fast water diffusion to the MMMs. The synergy of two effects significantly reduced the water permeation resistance and intensified the water transportation. Meanwhile, the pore size was reduced from 1.3 nm to 0.45–1.0 nm by the sheltering effect of CNFs, endowing membranes with precise molecular-sieving ability. When used for alcohol dehydration, dye rejection, or salt rejection, the repeated concentrating-sieving process through $TpTG_{Cl}$@CNFs-X nanocomposites achieved a high permeation flux along with a high water concentration in permeate. Along with the mixed-dimensional strategy, a series of COF membranes were prepared, including TpPa-1, TpBD, TpHZ, and TpBD(OH). Compared with the corresponding pristine COF membrane, the mechanical strength and stability of these membranes have been significantly improved, owing to the multiple interactions between the 2D nanosheets and 1D nanofibers.

10.3.3.2 Modulating Crystallinity, Porosity, Mechanical Properties

In "bottom-up" approach, polymers can be used as building blocks attached in the framework of COF to prepare COF-based MMMs. Compared to small molecules, polymers have a larger volume and more functional groups, resulting in different behavior during the reaction. Wang et al. [77] reported polymer-covalent-organic-frameworks (polyCOFs), which are assembled from customized linear polymers and COF monomers. In this study, a PEG-lated linear polymer, containing 2,5-diethoxyterephthalohydrazide (DTH) moieties, was designed and synthesized firstly. Then, a three-component condensation reaction of 1,3,5-triformylbenzene (TB) with DTH and PEG-lated linear polymer occurred via an interfacial polymerization, forming the polyCOF-42 membranes (Figure 10.18a). The PEGs in polymer building blocks improved the water solubility of the monomers, which facilitated the formation of membranes. The large volume slowed down the diffusion of polymer monomers in solution and decreases the reaction rate at the interface, improving the crystallinity of COFs. The introduced linear polymer enhanced the processability and mechanical performance (Figure 10.18b), while COFs maintained high crystallinity and porosity. Notably, the COF pore sizes were tailored from 2.4 to 1.3 nm, enhancing molecular-sieving effect (Figure 10.18d). The crystallinity, porosity, mechanical properties, and thickness of polyCOF membranes can be readily adjusted by changing the molar ratio of polymer reactants and DTH, and kinds of polymers. However, due to the high polydispersity index (PDI) of polymers, small monomers are necessary to obtain membranes with high crystallinity, which serve as co-building blocks. To address these issues, Guo et al. [97] used DTH dimer with PEG bridging linkers as the monomer of COF-42 to prepare membranes. The freestanding and uniform PEG–COF-42 membranes were harvested through a two-component condensation of TB and DTH dimer (Figure 10.18e), which was simpler and more controllable. The properties of crystalline membranes were heavily dependent on the length of PEG linkers.

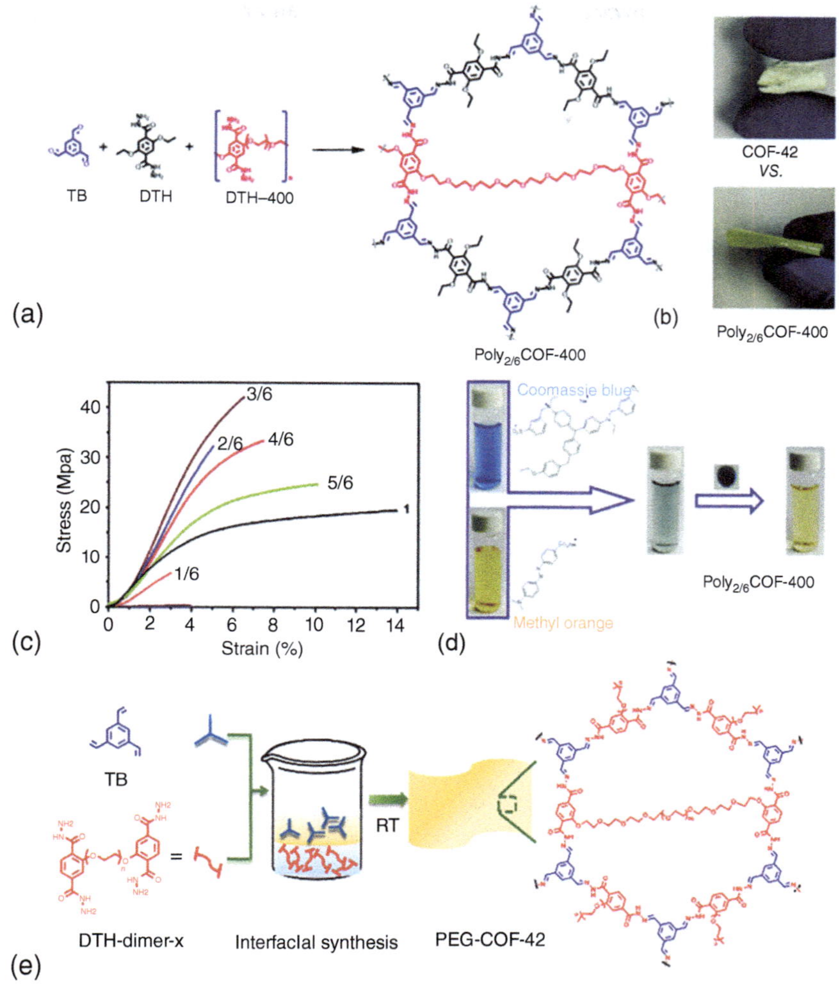

Figure 10.18 (a) Strategy to fabricate polyCOF membranes via a three-component condensation. (b) Comparison of the mechanical properties of COF-42 versus poly2/6COF-42. (c) Stress–strain curves of polyxCOF-42 membranes with different polymer contents ($x = 1/6, 2/6, 3/6, 4/6, 5/6, 1$) compared with the COF-42 membrane. (d) Separation of Coomassie brilliant blue R-250 and methyl orange dye solution by filtration with a poly2/6COF-42 membrane. Source: Wang et al. [77]/with permission of Royal Society of Chemistry. (e) Illustration of the approach to fabricate PEG–COF-42 membrane using interfacial synthesis. Source: Guo et al. [97]/with permission of American Chemical Society.

10.4 Conclusion and Outlook

Due to the combination of robust-processing properties of polymer materials and high separation performance of 2D materials, MMMs are one of the most attractive membranes with high application potential. The rapid development of 2D materials and their MMMs have attracted more and more interest in precise molecular separation. The readily adjustable interlayer and/or intralayer pore size, great compatibility with polymer, exceptional stability, and abundant functional groups make 2D materials ideal building blocks to fabricate MMMs. In this chapter, the structural and separation performance of MMMs are focused using 2D materials as dispersed phases or continuous phases.

When the 2D materials serve as dispersed phase, they can be uniformly dispersed in polymer matrix. The incorporated 2D materials can modify the neighbor polymer chains and introduce specific functions, significantly improving the overall separation performance of MMMs. Compared with isotropic fillers, MMMs containing thin nanosheets can achieve great separation performance at lower loading. The interaction between 2D materials and polymer chains can strengthen the mechanical properties and long-term stability. Besides, the assembled nanosheets with interlayer channels or intrinsic pores provide extra size exclusion and molecular transport channels, improving the selectivity and permeability. The modification of functional groups of 2D materials can obviously intensify specific separation processes. The decent stability and abundant binding sites endow 2D materials with huge potential as supports for various carriers in membranes, such as transition-metal ions. Furthermore, the precise distribution of 2D materials in membranes and regulation of the membrane structure can be achieved by adjusting the conditions of membrane processing.

When the 2D materials serve as continuous phase, novel fabrication methods are emerging, such as mixed-dimension membrane. The abundant functional groups on 2D materials provide binding sites for polymer. Compared to van der Waals forces, covalent bonding or coordination interaction enables a more stable cross-linked structure and enhances the compatibility between the two phases. This novel strategy solves the challenge between separation performance and mechanical strength in 2D material-based membranes and adapts them to precise molecular separations. The ordered channels of 2D materials are conducive to efficient molecular sieving and break the constraints of "trade-off" effect. The flexible polymer enhances its mechanical properties and reduces processing difficulty for 2D-material membranes. Polymers with special properties can also introduce tunable functionality to the membranes. Overall, these advanced 2D materials lead to new research trends in MMMs, including continuous phase–dispersed phase interface design, membrane structure regulation, and ultrathin MMM preparation, to construct MMMs with excellent separation performance and economical attractivity for industrial molecular separations.

Despite the remarkable progress achieved, some crucial challenges remain for future development and application of 2D-material MMMs.

Firstly, the separation performance of polymer-based MMMs is mainly determined by the polymers. Although a variety of MMMs containing 2D materials have been prepared and their performance is much higher than that of pure polymer membranes, it is far from the separation performance of pure 2D-material membranes. Limited by the precipitation and agglomeration of 2D-material dispersed phase, the loading of most 2D materials in the MMMs cannot reach particularly high, which cannot fully utilize the excellent performance of 2D materials. It is still a huge challenge to achieve especially high loading and uniform dispersion of 2D materials in the membrane to promote their separation performances.

Secondly, the interfacial interactions between 2D materials and polymers are difficult to be manipulated with high precision. The 2D materials will alter the packing dynamic and conformation of polymer near them, which will impart great effect on molecular transport. The unfavorable orientation and packing of 2D materials or insufficient interfacial compatibility usually result in nonselective voids and defects, which will severely reduce selectivity. On the other hand, the polymer chains could probably block the pore of 2D materials, lowering the permeability. It is of great importance to improve the interaction between 2D materials and polymer for practical usage. Most 2D materials have abundant functional groups, but the introduction or modification of functional groups is still necessary, which can improve the compatibility and change chemical affinities to penetrants. 2D materials-based MMMs become a new trend, such as mixed-dimensional membranes and polyCOF membranes. They consist of 2D nanosheets and polymers linked to the nanosheets via covalent or coordination bonds. They can more precisely modulate the interactions between polymers and 2D materials, alleviating the formation of nonselective defects and the blockage of intrinsic pores. However, synthesis of MMMs with high crystallinity remains challenging, especially for presynthesis strategies. The preformed structures will hinder subsequent reactions and lead to difficulties in the uniform distribution of polymers in MMMs. Therefore, the further development of 2D materials with controlled structure and properties is an important challenge to fabricate MMMs with high performance.

Thirdly, the separation performance of 2D-material MMMs cannot be theoretically predicted. For conventional polymer membranes, there are a series of mass transport models to explain the transfer phenomenon. But for the emerging 2D-material MMMs, the model that can adequately describe the separation process has rarely been reported. Because the structure of the MMMs is difficult to be fully characterized, and small molecules show anomalous mass-transfer behaviors in subnanometer channels. More experimental explorations are urgently needed for the further understanding of the relationship between the structure and performance of MMMs. The experimental data and computational studies should be combined. More computational studies, including machine learning and artificial Intelligence, need to be performed in the design and discovery of new 2D-material MMMs.

Finally, many novel 2D materials and 2D-material MMMs are expected to emerge as the development of material science and chemical engineering. In the past few years, various 2D materials have been spawned by the rapid development of chemistry and materials science. However, only part of them has been applied to fabricate MMMs in molecular separation. More MMMs constructed by existing 2D materials should be attempted. It is of great importance to explore MMMs with novel physical or chemical structures.

MMMs constructed of 2D materials and polymer represent a new field of membrane research. It can be expected that their remarkable new properties and multifunctional nature could overcome the permeability–selectivity "trade-off" limitation and endow them with great prospects for many applications, relevant to gas separation, pervaporation, water treatment, etc.

References

1 Sholl, D.S. and Lively, R.P. (2016). Seven chemical separations to change the world. *Nature* 532 (7600): 435–437.

2 Venna, S.R. and Carreon, M.A. (2010). Highly permeable zeolite imidazolate framework-8 membranes for CO_2/CH_4 separation. *J. Am. Chem. Soc.* 132 (1): 76–78.

3 Li, Y., Liang, F., Bux, H. et al. (2010). Zeolitic imidazolate framework ZIF-7 based molecular sieve membrane for hydrogen separation. *J. Membr. Sci.* 354 (1–2): 48–54.

4 Comesaa-Gándara, B., JieChen, G.C. et al. (2019). Redefining the Robeson upper bounds for CO_2/CH_4 and CO_2/N_2 separations using a series of ultrapermeable benzotriptycene-based polymers of intrinsic microporosity. *Energy Environ. Sci.* 12: 2733–2740.

5 Zhang, C. and Koros, W.J. (2017). Ultraselective carbon molecular sieve membranes with tailored synergistic sorption selective properties. *Adv. Mater.* 29 (33): 1701631.1–1701631.6.

6 Zhang, Q., Li, H., Chen, S. et al. (2020). Mixed-matrix membranes with soluble porous organic molecular cage for highly efficient C_3H_6/C_3H_8 separation. *J. Membr. Sci.* 611: 118288.

7 Zhang, W., Jia, J., Qiu, Y. et al. (2018). Polydopamine-grafted graphene oxide composite membranes with adjustable nanochannels and separation performance. *Adv. Mater. Interfaces* 5: 1701386.

8 Sun, H., Magnuson, Z., He, W. et al. (2020). PEG@ZIF-8/PVDF nanocomposite membrane for efficient pervaporation desulfurization via a layer-by-layer technology. *ACS Appl. Mater. Interfaces* 12: 20664–20671.

9 Liu, G., Shen, J., Liu, Q. et al. (2017). Ultrathin two-dimensional MXene membrane for pervaporation desalination. *J. Membr. Sci.* 548–558.

10 Singh, R. (2014). *Membrane Technology and Engineering for Water Purification: Application, Systems Design and Operation*, 2e. Elsevier.

11 Koros, W.J. and Zhang, C. (2017). Materials for next-generation molecularly selective synthetic membranes. *Nat. Mater.* 16 (3): 289–297.

12 Keating, J.J., Imbrogno, J., and Belfort, G. (2016). Polymer brushes for membrane separations: a review. *ACS Appl. Mater. Interfaces* 8 (42): 28383–28399.

13 Matsuyama, H., Yuasa, M., Kitamura, Y. et al. (2000). Structure control of anisotropic and asymmetric polypropylene membrane prepared by thermally induced phase separation. *J. Membr. Sci.* 179 (1–2): 91–100.

14 Plisko, T.V., Penkova, A.V., Burts, K.S. et al. (2019). Effect of Pluronic F127 on porous and dense membrane structure formation via non-solvent induced and evaporation induced phase separation. *J. Membr. Sci.* 580: 336–349.

15 Tsai, H.A., Chung, L.H., Lee, K.R. et al. (2013). The preparation of polyamide/polyacrylonitrile composite hollow fiber membranes for pervaporation. *Appl. Mech. Mater.* 377: 246–249.

16 Koros, W.J. (2002). Gas separation membranes: needs for combined materials science and processing approaches. *Macromol. Symp.* 188 (1): 13–22.

17 Yin, J. and Deng, B. (2015). Polymer-matrix nanocomposite membranes for water treatment. *J. Membr. Sci.* 479: 256–275.

18 Park, H.B., Jung, C.H., Lee, Y.M. et al. (2007). Polymers with cavities tuned for fast selective transport of small molecules and ions. *Science* 318 (5848): 254–258.

19 Park, H.B., Kamcev, J., Robeson, L.M. et al. (2017). Maximizing the right stuff: the trade-off between membrane permeability and selectivity. *Science* 356 (6343): 1138–1148.

20 Chung, T.S., Lan, Y.J., Yi, L. et al. (2007). Mixed matrix membranes (MMMs) comprising organic polymers with dispersed inorganic fillers for gas separation. *Prog. Polym. Sci.* 32 (4): 483–507.

21 Ozturk, B. and Ozen, H.A. (2021). The importance of nanomaterials in gas separation. *Mater. Today Proc.* 42: 1547–1549.

22 Furukawa, S., Reboul, J., Diring, S. et al. (2014). Structuring of metal–organic frameworks at the mesoscopic/macroscopic scale. *Chem. Soc. Rev.* 43 (16): 5700–5734.

23 Rodenas, T., Luz, I., Prieto, G. et al. (2015). Metal-organic framework nanosheets in polymer composite materials for gas separation. *Nat. Mater.* 14 (1): 48–55.

24 Jie, S., Liu, G., Kang, H. et al. (2015). Membranes with fast and selective gas-transport channels of laminar graphene oxide for efficient CO_2 capture. *Angew. Chem. Int. Ed.* 54 (2): 578–582.

25 Nair, R.R., Wu, H.A., Jayaram, P.N. et al. (2012). Unimpeded permeation of water through helium-leak-tight graphene-based membranes. *Science* 335 (6067): 442–444.

26 Kim, H.W., Yoon, H.W., Yoon, S.-M. et al. (2013). Selective gas transport through few-layered graphene and graphene oxide membranes. *Science* 342: 91–95.

27 Hang, L., Song, Z., Zhang, X. et al. (2013). Ultrathin, molecular-sieving graphene oxide membranes for selective hydrogen separation. *Science* 342 (6154): 95–98.

28 Mo, G.H., Adam, F., and Park, H.B. (2019). Two-dimensional nanoporous materials: membrane platform for gas and liquid separations. *2D Mater.* 6 (4): 042002.

29 Bachman, J.E., Smith, Z.P., Li, T. et al. (2016). Enhanced ethylene separation and plasticization resistance in polymer membranes incorporating metal-organic framework nanocrystals. *Nat. Mater.* 15: 845–849.

30 August, D.P., Dryfe, R., Haigh, S.J. et al. (2020). Self-assembly of a layered two-dimensional molecularly woven fabric. *Nature* 588: 429–435.

31 Wang, L., Boutilier, M., Kidambi, P.R. et al. (2017). Fundamental transport mechanisms, fabrication and potential applications of nanoporous atomically thin membranes. *Nat. Nanotechnol.* 12 (6): 509–522.

32 Huang, X., Qi, X., Boey, F. et al. (2012). Graphene-based composites. *Chem. Soc. Rev.* 41 (2): 666–686.

33 Yang, L., Yzb, D., Xza, C. et al. (2021). MoS_2-based membranes in water treatment and purification. *Chem. Eng. J.* 422: 130082.

34 Ding, L., Li, L., Liu, Y. et al. (2020). MXene membranes for production of drinking water from seawater. *Nat. Sustain.* 3: 296–302.

35 Tan, P., Sun, J., Hu, Y. et al. (2015). Adsorption of Cu^{2+}, Cd^{2+} and Ni^{2+} from aqueous single metal solutions on graphene oxide membranes. *J. Hazard. Mater.* 297: 251–260.

36 Qian, Q., Asinger, P.A., Lee, M.J. et al. (2020). MOF-based membranes for gas separations. *Chem. Rev.* 120 (16): 8161–8266.

37 Yuan, S., Li, X., Zhu, J. et al. (2019). Covalent organic frameworks for membrane separation. *Chem. Soc. Rev.* 48: 2665–2681.

38 Zimmerman, C.M., Singh, A., and Koros, W.J. (1997). Tailoring mixed matrix composite membranes for gas separations. *J. Membr. Sci.* 137 (1): 145–154.

39 Boukhvalov, D.W., Katsnelson, M.I., and Son, Y.W. (2013). Origin of anomalous water permeation through graphene oxide membrane. *Nano Lett.* 13: 3930–3935.

40 Cao, K., Jiang, Z., Zhao, J. et al. (2014). Enhanced water permeation through sodium alginate membranes by incorporating graphene oxides. *J. Membr. Sci.* 469 (11): 272–283.

41 Li, X., Ma, L., Zhang, H. et al. (2015). Synergistic effect of combining carbon nanotubes and graphene oxide in mixed matrix membranes for efficient CO_2 separation. *J. Membr. Sci.* 479: 1–10.

42 Li, X., Cheng, Y., Zhang, H. et al. (2015). Efficient CO_2 capture by functionalized graphene oxide nanosheets as fillers to fabricate multi-permselective mixed matrix membranes. *ACS Appl. Mater. Interfaces* 7 (9): 5528–5537.

43 Xin, Q., Li, Z., Li, C. et al. (2015). Enhancing the CO_2 separation performance of composite membranes by the incorporation of amino acid-functionalized graphene oxide. *J. Mater. Chem. A* 3: 6629–6641.

44 Peng, D., Wang, S., Tian, Z. et al. (2017). Facilitated transport membranes by incorporating graphene nanosheets with high zinc ion loading for enhanced CO_2 separation. *J. Membr. Sci.* 522: 351–362.

45 Zhang, Y., Song, J., Pan, F. et al. (2020). Constructing high-efficiency facilitated transport pathways via embedding heterostructured Ag+@MOF/GO laminates into membranes for pervaporative desulfurization. *Sep. Purif. Technol.* 245: 116858.

46 Yaghi, O.,.M. (2016). Reticular chemistry-construction, properties, and precision peactions of frameworks. *J. Am. Chem. Soc.* 138: 15507–15509.

47 Lv, Y., Ma, L., Svec, F. et al. (2019). Engineering of filler/polymer interface in metal-organic framework-based mixed-matrix membranes to enhance gas separation. *Chem. Asian J.* 14 (20): 3502–3514.

48 Liu, G., Jiang, Z., Cao, K. et al. (2017). Pervaporation performance comparison of hybrid membranes filled with two-dimensional ZIF-L nanosheets and zero-dimensional ZIF-8 nanoparticles. *J. Membr. Sci.* 523: 185–196.

49 Gou, M., Zhu, W., Sun, Y. et al. (2020). Introducing two-dimensional metal-organic frameworks with axial coordination anion into Pebax for CO_2/CH_4 separation. *Sep. Purif. Technol.* 259 (1–2): 118107.

50 Li, Q., Cheng, L., Shen, J. et al. (2017). Improved ethanol recovery through mixed-matrix membrane with hydrophobic MAF-6 as filler. *Sep. Purif. Technol.* 178: 105–112.

51 Yaghi, O.M. (2016). Reticular chemistry-construction, properties, and precision reactions of frameworks. *J. Am. Chem. Soc.* 138 (48): 15507–15509.

52 Yaghi, O.M. (2019). Reticular chemistry in all dimensions. *ACS Cent. Sci.* 5 (8): 1295–1300.

53 Cote, A.P., Benin, A.I., Ockwig, N.W. et al. (2005). Porous, crystalline, covalent organic frameworks. *Science* 310 (5751): 1166–1170.

54 Evans, A.M., Castano, I., Brumberg, A. et al. (2019). Emissive single-crystalline boroxine-linked colloidal covalent organic frameworks. *J. Am. Chem. Soc.* 141 (50): 19728–19735.

55 Ma, T.Q., Kapustin, E.A., Yin, S.X. et al. (2018). Single-crystal X-ray diffraction structures of covalent organic frameworks. *Science* 361 (6397): 48–52.

56 Li, Y., Wang, C., Ma, S.J. et al. (2019). Fabrication of hydrazone-linked covalent organic frameworks using alkyl amine as building block for high adsorption capacity of metal ions. *ACS Appl. Mater. Interfaces* 11 (12): 11706–11714.

57 Geng, K.Y., He, T., Liu, R.Y. et al. (2020). Covalent organic frameworks: design, synthesis, and functions. *Chem. Rev.* 120 (16): 8814–8933.

58 Li, X., Yadav, P., and Loh, K.P. (2020). Function-oriented synthesis of two-dimensional (2D) covalent organic frameworks-from 3D solids to 2D sheets. *Chem. Soc. Rev.* 49 (14): 4835–4866.

59 Yang, H., Wu, H., Xu, Z. et al. (2018). Hierarchical pore architectures from 2D covalent organic nanosheets for efficient water/alcohol separation. *J. Membr. Sci.* 561: 79–88.

60 Zou, C.C., Li, Q.Q., Hua, Y.Y. et al. (2017). Mechanical synthesis of COF nanosheet cluster and its mixed matrix membrane for efficient CO_2 removal. *ACS Appl. Mater. Interfaces* 9 (34): 29093–29100.

61 Fan, H.W., Xie, Y.F., Li, J.C. et al. (2018). Ultra-high selectivity COF-based membranes for biobutanol production. *J. Mater. Chem. A* 6 (36): 17602–17611.

62 Wang, M.D., Pan, F.S., Yang, H. et al. (2019). Constructing channel-mediated facilitated transport membranes by incorporating covalent organic framework nanosheets with tunable microenvironments. *J. Mater. Chem. A* 7 (16): 9912–9923.

63 Yang, H., Wu, H., Yao, Z.Q. et al. (2018). Functionally graded membranes from nanoporous covalent organic frameworks for highly selective water permeation. *J. Mater. Chem. A* 6 (2): 583–591.

64 Wang, Z.Y. and Mi, B.X. (2017). Environmental applications of 2D molybdenum disulfide (MoS_2) nanosheets. *Environ. Sci. Technol.* 51 (15): 8229–8244.

65 Jiang, S.D., Koh, A., Chong, K.H. et al. (2019). Opening organic solvent pathways by molybdenum disulfide in mixed matrix membranes for molecular separation. *J. Membr. Sci.* 585: 60–66.

66 Shen, Y., Wang, H., Xiang, Z. et al. (2016). MoS_2 nanosheets functionalized mixed matrix membrane for enhanced CO_2 capture via surface drop-coating method. *ACS Appl. Mater. Interfaces* 8 (35): 23371–23378.

67 Pan, F.S., Ding, H., Li, W.D. et al. (2018). Constructing facilitated transport pathway in hybrid membranes by incorporating MoS_2 nanosheets. *J. Membr. Sci.* 545: 29–37.

68 Tian, Z.Z., Wang, S.F., Wang, Y.T. et al. (2016). Enhanced gas separation performance of mixed matrix membranes from graphitic carbon nitride nanosheets and polymers of intrinsic microporosity. *J. Membr. Sci.* 514: 15–24.

69 Ding, H., Pan, F.S., Mulalic, E. et al. (2017). Enhanced desulfurization performance and stability of Pebax membrane by incorporating Cu^+ and Fe^{2+} ions co-impregnated carbon nitride. *J. Membr. Sci.* 526: 94–105.

70 Song, Y.M., Jiang, Z.Y., Gao, B.X. et al. (2018). Embedding hydrophobic MoS_2 nanosheets within hydrophilic sodium alginate membrane for enhanced ethanol dehydration. *Chem. Eng. Sci.* 185: 231–242.

71 Wang, Y., Gao, B.Y., Yue, Q.Y. et al. (2020). Graphitic carbon nitride (g-C3N4)-based membranes for advanced separation. *J. Mater. Chem. A* 8 (37): 19133–19155.

72 Karahan, H.E., Goh, K., Zhang, C.F. et al. (2020). MXene materials for designing advanced separation membranes. *Adv. Mater.* 32 (29): 1906697.

73 Xu, Z., Liu, G.Z., Ye, H. et al. (2018). Two-dimensional MXene incorporated chitosan mixed-matrix membranes for efficient solvent dehydration. *J. Membr. Sci.* 563: 625–632.

74 Gao, B.X., Jiang, Z.Y., Liu, G.H. et al. (2016). Enhanced pervaporative performance of hybrid membrane by incorporating amphiphilic carbonaceous material. *J. Membr. Sci.* 520: 951–963.

75 Wang, Y.Q., Low, Z.X., Kim, S. et al. (2018). Functionalized boron nitride nanosheets: a thermally rearranged polymer nanocomposite membrane for hydrogen separation. *Angew. Chem. Int. Ed.* 57 (49): 16056–16061.

76 Zhang, N., Wu, H., Li, F.C. et al. (2018). Heterostructured filler in mixed matrix membranes to coordinate physical and chemical selectivities for enhanced CO_2 separation. *J. Membr. Sci.* 567: 272–280.

77 Wang, Z.F., Yu, Q., Huang, Y.B. et al. (2019). PolyCOFs: a new class of freestanding responsive covalent organic framework membranes with high mechanical performance. *ACS Cent. Sci.* 5 (8): 1352–1359.

78 Yang, H., Yang, L.X., Wang, H.J. et al. (2019). Covalent organic framework membranes through a mixed-dimensional assembly for molecular separations. *Nat. Commun.* 10 (2101): 1–10.

79 Ang, H.X. and Hong, L. (2017). Polycationic polymer-regulated assembling of 2D MOF nanosheets for high-performance nanofiltration. *ACS Appl. Mater. Interfaces* 9 (33): 28079–28088.

80 Hu, R., He, Y., Huang, M. et al. (2018). Strong adhesion of graphene oxide coating on polymer separation membranes. *Langmuir* 34: 10569–10579.

81 Yeh, C.N., Raidongia, K., Shao, J. et al. (2014). On the origin of the stability of graphene oxide membranes in water. *Nat. Chem.* 7 (2): 166–170.

82 Hung, W.S., Tsou, C.H., Guzman, M.D. et al. (2014). Cross-linking with diamine monomers to prepare composite graphene oxide-framework membranes with varying d-spacing. *Chem. Mater.* 26 (9): 2983–2990.

83 Xu, X.L., Lin, F.W., Du, Y. et al. (2016). Graphene oxide nanofiltration membranes stabilized by cationic porphyrin for high salt rejection. *ACS Appl. Mater. Interfaces* 8 (20): 12588–12593.

84 Aida, T., Meijer, E.W., and Stupp, S.I. (2012). Functional supramolecular polymers. *Science* 335 (6070): 813–817.

85 Song, Y., Li, R., Pan, F. et al. (2019). Ultrapermeable graphene oxide membranes with tunable interlayer distances via vein-like supramolecular dendrimers. *J. Mater. Chem. A* 7: 18642–18652.

86 Lim, M.Y., Choi, Y.S., Kim, J. et al. (2017). Cross-linked graphene oxide membrane having high ion selectivity and antibacterial activity prepared using tannic acid-functionalized graphene oxide and polyethyleneimine. *J. Membr. Sci.* 521: 1–9.

87 Pan, F., Li, Y., Song, Y. et al. (2019). Graphene oxide membranes with fixed interlayer distance via dual crosslinkers for efficient liquid molecular separations. *J. Membr. Sci.* 595: 117486.

88 Li, Y.Z., Fu, Z.H., and Xu, G. (2019). Metal-organic framework nanosheets: preparation and applications. *Coord. Chem. Rev.* 388: 79–106.

89 Ji, Z., Wang, H.Z., Canossa, S. et al. (2020). Pore chemistry of metal-organic frameworks. *Adv. Funct. Mater.* 30 (41): 2000238.

90 Denny, M.S., Moreton, J.C., Benz, L. et al. (2016). Metal-organic frameworks for membrane-based separations. *Nat. Rev. Mater.* 1 (12): 1–17.

91 Jin, F.Z., Liu, J.J., Chen, Y. et al. (2021). Tethering flexible polymers to crystalline porous materials: a win–win hybridization approach. *Angew. Chem. Int. Ed.* 60 (26): 14222–14235.

92 Li, T., Ren, Y., Wu, D.W. et al. (2020). A novel water-stable two-dimensional zeolitic imidazolate frameworks thin-film composite membrane for enhancements in water permeability and nanofiltration performance. *Chemosphere* 261: 127717.

93 Jia, W., Wu, B.H., Sun, S.T. et al. (2020). Interfacially stable MOF nanosheet membrane with tailored nanochannels for ultrafast and thermo-responsive nanofiltration. *Nano Res.* 13 (11): 2973–2978.

94 Zhang, F., Dou, J., and Zhang, H. (2018). Mixed membranes comprising carboxymethyl cellulose (as capping agent and gas barrier matrix) and nanoporous ZIF-L nanosheets for gas separation applications. *Polymer* 10 (1340): 1–15.

95 Ding, S.Y. and Wang, W. (2013). Covalent organic frameworks (COFs): from design to applications. *Chem. Soc. Rev.* 42 (2): 548–568.

96 Segura, J.L., Royuela, S., and Ramos, M.M. (2019). Post-synthetic modification of covalent organic frameworks. *Chem. Soc. Rev.* 48 (14): 3903–3945.

97 Guo, X.X., Mao, T.H., Wang, Z.F. et al. (2020). Fabrication of photoresponsive crystalline artificial muscles based on pegylated covalent organic framework membranes. *ACS Cent. Sci.* 6 (5): 787–794.

11

Transport Mechanism of 2D Membranes

Tae Hoon Lee and Ho Bum Park

Hanyang University, Department of Energy Engineering, Wangsimni-ro 222, Seoul 04763, Republic of Korea

11.1 Introduction

Before discussing the main topic of this chapter, it is worthwhile reviewing the fundamental mass-transport theories in conventional polymeric membranes which are part of the basis of the transport phenomena in two-dimensional (2D) membranes. A membrane is defined as a thin and permselective barrier that allows the permeation of a target species while rejecting the others in a homogeneous mixture [1, 2]. A driving force is exerted over the boundary of feed and permeate sides separated by a membrane having a certain thickness (l), which initiates the transport of a particle or molecule driven by the potential gradient (ΔX). The extent of this driving force (F_D) is described as follows [2]:

$$\text{Driving force } (F_D) = \frac{\Delta X}{l}\left[\frac{\text{N}}{\text{mol}}\right] \tag{11.1}$$

In general, the chemical potential difference ($\Delta\mu$, controlled by external pressure or concentration of solution) and the electrical potential difference (ΔE, controlled by an electric field) are two main factors that govern the mass transport through a membrane in common applications [3, 4]. Other possible driving forces include temperature gradient, magnetic fields, gravity, and centrifugal fields [2]. These forces are not necessarily to be independent of each other and sometimes offer new opportunities to control the mass-transport properties by combining more than two forces (e.g. electrochemical potential and electromagnetic potential) [5, 6].

A constant flow from a high potential to a low potential will be established after reaching a steady state if the driving force is kept unchanged. At this equilibrium, the mass flux (J), defined as the movement of mass across a unit area at a given time (unit: [mass/time·area]), is proportional to the driving force as follows [2]:

$$\text{Flux } (J) = \text{Proportionality factor } (A) \times \text{Driving force } (F_D) \tag{11.2}$$

This relation is analogous to Fick's First law that depicts the mass transport when a concentration gradient is applied to the membrane [7]. Indeed, the relation in

Two-Dimensional-Materials-Based Membranes: Preparation, Characterization, and Applications, First Edition. Edited by Gongping Liu and Wanqin Jin.
© 2022 WILEY-VCH GmbH. Published 2022 by WILEY-VCH GmbH.

Eq. (11.2) is widely adopted in most of the membrane systems regardless of the size and charge of permeants, which will be discussed in Section 11.2.

The existence of the proportionality factor (A) in Eq. (11.2) indicates that the structural properties of a membrane may largely determine its mass-transport properties from both perspectives of kinetics (e.g. penetrant mobilities) and thermodynamics (e.g. membrane-penetrant interactions). Typical membrane structures and the corresponding transport mechanisms are illustrated in Figure 11.1a, roughly classified by their pore size (porous/nonporous membranes) and electrical charge (positively/negatively charged or neutral membranes) [1, 2]. Indeed, industrial applications of these membranes follow the physicochemical structures of membranes installed into the unit modules (Figure 11.1b). For example, membranes can be applied to diverse liquid separation processes, such as osmotic processes (reverse osmosis [RO] or forward osmosis [FO], <10 Å, e.g. water/monovalent ion separation [8]), nanofiltration (10–100 Å, e.g. water/divalent ion separation [9] and organic solvent nanofiltration [OSN] [10]), ultrafiltration (100–1000 Å, e.g. separation of proteins or viruses [11, 12]), and microfiltration (>1000 Å, e.g. removal of fungi or microplastics [13]) given that the pore size of membrane falls into the right range required for the target separation application. The classification of membranes by pore size is also available for other phases (gas/vapor and ion), and thus it is not surprising that a membrane material commonly used in gas separation can be

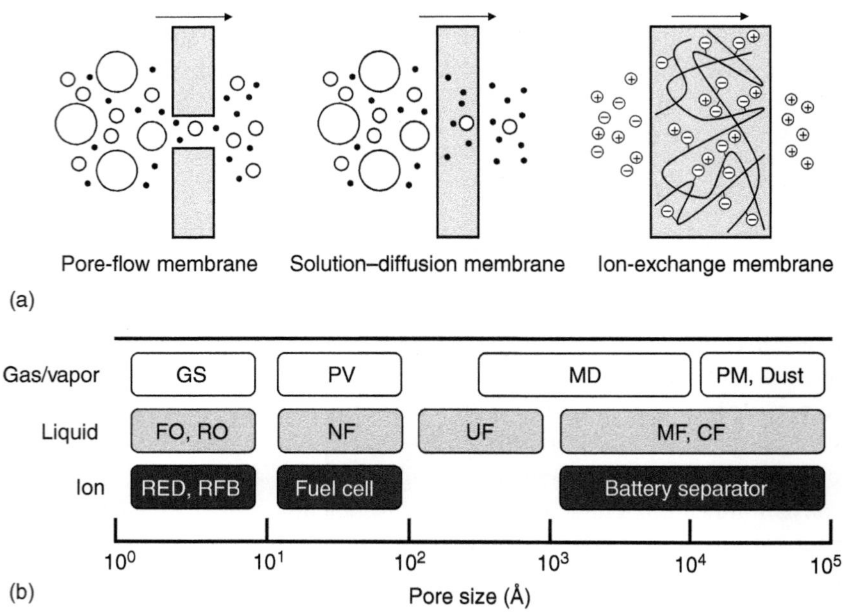

Figure 11.1 Schematic illustrations of (a) mass transport through membranes based on different transport mechanisms and (b) representative membrane processes and applications at a given pore size range. GS: Gas separation, PV: Pervaporation, MD: Membrane distillation, PM: Particulate matter, FO: Forward osmosis, RO: Reverse osmosis, NF: Nanofiltration, UF: Ultrafiltration, MF: Microfiltration, CF: Conventional filtration, RED: Reverse electrodialysis, RFB: Redox flow battery.

utilized in RO desalination or electrolytes for a fuel cell which all share the same pore size range (1–10 Å). Besides the pore size of membranes, the morphology of membranes (e.g. pore size distribution and tortuosity), the chemistry of pore walls (e.g. chemical affinity and wettability), and membrane configuration (e.g. integrally skinned asymmetric [ISA] membranes and thin-film composite [TFC] membranes) are crucial factors that should be considered for the design of optimal membranes showing both high flux and high selectivity [4, 6].

In this context, it is of the utmost importance to tailor the membrane structure and geometries to be appropriate for the application of interest. For instance, cylindrical pores perpendicular to the membrane surface offer a faster transport pathway for the desired species than that for randomly oriented pores [12]. The resistance to mass transport can also be minimized by reducing the membrane thickness [14, 15]. Hence, an ideal membrane should be fabricated as thin as possible, and its porous characteristics (pore size, porosity, and pore size distribution) are needed to be optimized to improve the efficiency of the separation process. However, these aspects have been very challenging to be precisely control in polymer membranes because of their broad molecular weight distribution and intrinsically fluctuating nature of polymer chains [4]. Thus, polymeric membrane materials generally suffer from poor separation performances ruled by the permeability–selectivity trade-off relation, that is, a membrane with high permeability exhibits a low selectivity and vice versa [16]. The flexible backbones of polymeric materials can also intensify physical aging by elapsed time as well as penetrant-induced plasticization, which is unfavorable for guaranteeing the long-term stability of polymer membranes [17]. Furthermore, insufficient chemical and thermal stabilities of polymer membranes often limit their applications under harsh operating conditions [1, 10, 18]. To overcome these issues, researchers have dedicated themselves to developing better membrane materials, such as microporous polymers (thermally rearranged [TR] polymers [19] and polymers of intrinsic microporosity [PIMs] [20]), zeolites [21], silica [22], carbon molecular sieves (CMS) [23, 24], metal–organic frameworks (MOFs) [25], carbon nanotubes (CNTs) [26], and mixed-matrix membranes (MMMs) consisting of dispersed nanofillers and continuous polymer matrix [27, 28]. These advanced membrane materials have displayed excellent separation performance owing to the fine-tuned membrane structures. Nevertheless, thinning the developed membrane materials less than 1 μm to maximize the flux and productivity of the separation process (see Eq. (11.2)) has been rarely reported and remains an overlooked challenge to date [4, 24, 25, 28].

Recently, the rise of 2D materials [29], such as graphene, graphene oxide (GO), hexagonal boron nitride (hBN), molybdenum disulfide (MoS_2), and transition metal carbide (Mxene), has been facilitated a major breakthrough in membrane technology [3, 4, 6, 30, 31]. While it has been believed that the pristine 2D materials are generally impermeable barriers not allowing transport of any molecules [32, 33], the remarkable advancements in pore engineering in 2D materials pioneered themselves as a next-generation membrane material demonstrated by ultrafast and selective mass-transport properties together with outstanding mechanical robustness, thermal and chemical stability [34–36]. Most importantly, the atomic thicknesses of these materials enable facile fabrication of ultrathin membranes to significantly boost the membrane flux by maximizing the applied driving force to the membrane system—at least from a theoretical aspect—which makes them an ideal platform to design high-performance membranes in a wide range of membrane processes [4, 37]. Of particular interest, 2D membranes are highly efficient in emerging membrane applications that require precise separation of molecular mixtures whose size is less than 10 Å, including gas separation, osmotic processes, and energy conversion and generation [38]. Moreover, the separation abilities of 2D membranes can be further improved by various strategies, including (i) controlling the defect density in the basal plane, (ii) stacking the 2D sheets to fabricate the layered structure, and (iii) introducing functional groups to provide chemical (or electrostatic) affinity to the desired molecules [37].

In addition to these promising potentials, theoretical studies and lots of experimental evidence have revealed that there exist unique mass-transport properties in 2D membranes, which far deviate from those in conventional polymer membranes [3, 39, 40]. In this regard, understanding the transport mechanisms in 2D membranes and the origins behind such phenomena is essential, which may guide the rational design of advanced 2D membranes and their future applications. This chapter aims to provide readers with a brief overview of the current knowledge of mass-transport mechanisms through 2D membranes. Fundamentals of mass transport in conventional membranes were first introduced by classical theories, which are divided into porous membranes (pore-flow model), nonporous membranes (solution–diffusion model), charged membranes (Donnan effects, and extended Nernst–Plank equation), and trade-off relation that limits their separation performance are also discussed. Next, the nanofluidic transport under confined dimensions is highlighted and how this interesting phenomenon can be beneficial to control the molecular transport properties in 2D membranes. Finally, the structural effects of 2D membranes in their transport mechanisms are discussed with several important examples, which include nanoporous atomically thin 2D membranes, staked 2D membranes with laminar structure, and 2D materials-embedded MMMs.

11.2 Fundamentals of Mass Transport Through Membranes

Controlling the transport of mixed components is the most important ability of membranes. In general, the experimentally obtained mass flux J (Eq. (11.2)) of species A or B is normalized by either membrane area or thickness to estimate the actual productivity and selectivity of a membrane as follows:

$$\text{Permeance } (Q) = \frac{\text{Mass, volume, or molar flow rate across membrane}}{\text{Membrane area} \times \text{Pressure difference}} = \frac{J}{\Delta P} \tag{11.3}$$

$$\text{Permeability } (P) = Q \times \text{membrane thickness} = \frac{J}{\Delta P} \times l \tag{11.4}$$

$$\text{Selectivity } (A/B) = \frac{\text{Permeance or Permeability of species } A}{\text{Permeance or Permeability of species } B} \tag{11.5}$$

$$\text{Rejection} = 1 - \frac{\text{Solute concentration in permeate side}}{\text{Solute concentration in feed side}} \tag{11.6}$$

Although there are different equations and terminologies to express the mass-transport properties of membranes depending on the state of penetrants and applications [2], the above equations are the representative figure of merits of a membrane material [3]. As discussed previously, mass transport through membranes is mostly influenced by the structure of membranes (e.g. pore size and porosity), the chemical affinity between membranes and permeants, and an electrostatic charge of membranes. Hence, the consideration for mass-transport phenomena in membranes needs to start from distinguishing the membrane material as a porous/nonporous or charged/neutral membrane.

11.2.1 Transport Mechanism in Porous Membranes

Figure 11.2 shows the possible mechanisms of mass transport through porous membranes. In porous membranes, the mass flux is determined by the pore size of membranes and the size of permeants. Additionally, pore affinity (e.g. electrostatic interactions, chemical affinity, and wettability) may influence mass transport. For porous membranes used in microfiltration or ultrafiltration processes, these membranes display a pore size range within 10 nm to 10 μm, and the mass transport in this regime is characterized by the classical convective flow theory [41]. The simplest way to consider the model prediction of these membranes is assuming that most of the pores are cylindrical and perpendicular to the membrane surface. If the length of all cylindrical pores is equal to the membrane thickness (l) with same

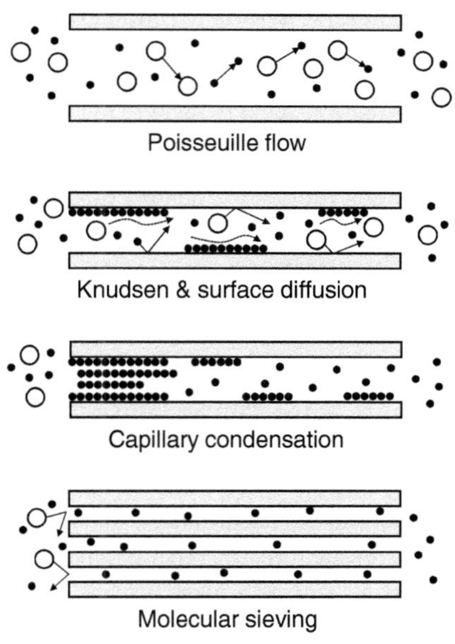

Figure 11.2 Schematic representation of diffusion mechanisms in porous membranes.

pore radius (r), the mass flux can be described by the following Hagen–Poiseuille equation [42]:

$$J = \frac{\varepsilon r^2}{8\eta\tau}\frac{\Delta P}{l} \tag{11.7}$$

where ε is the surface porosity, which is the product of the ratio of effective pore area to the membrane area A and the number of pores (n_p) ($\varepsilon = n_p \cdot \pi r^2/A$), η is the viscosity of the permeating flow, τ is the pore tortuosity (equal to unity for completely cylindrical and perpendicular pores), and ΔP is the pressure difference across the membrane [2]. This relation can be successfully used to analyze the mass-transport properties in porous anodic alumina oxide (AAO) membranes or track-etched polycarbonate membranes which serve as an ideal framework to investigate the Poiseuille flow by satisfying the above assumptions [43–46]. However, in general, common phase inversion membranes for industrial use represent a significantly different flux from the prediction by Eq. (11.7) due to the asymmetric and randomly oriented pores, requiring modification of the Hagen relation with some adjustable parameters [2, 47, 48].

For gas transport, if the pore size is much larger than the size of a molecule but smaller than its mean free path (i.e. the average distance transversed by a moving molecule between collisions), the collisions between the gas molecules and the pore wall become more pronounced than their self-collisions. The mean free path (λ) is affected by the pressure (P) and temperature (T), given by:

$$\lambda = \frac{k_B T}{\sqrt{2}\pi d_{gas}^2 P} \tag{11.8}$$

where k_B is Boltzmann constant and d_{gas} is the diameter of a gas molecule. Equation 11.8 shows that the mean free path increases as decreasing pressure. For example, the mean free path of an oxygen molecule is ~70 Å at 10 bar and 70 μm at 10 mbar. Consequently, the effects of molecule-wall collisions will be dominant in gas transport through common nanofiltration or ultrafiltration membranes at a high feed pressure above 1 bar [2]. On the other hand, the molecules are so close to each other in liquids that the mean free path in the liquid components is typically less than a few Å and thus Knudsen diffusion is negligible.

The gas flux governed by Knudsen diffusion is given by [49, 50]:

$$J = \frac{\varepsilon D_k}{RT}\frac{\Delta P}{l} \tag{11.9}$$

$$D_k = \frac{2r}{3}\sqrt{\frac{8RT}{\pi M_w}} \tag{11.10}$$

where D_k is the Knudsen diffusion coefficient, R is the ideal gas constant, and M_w is the molecular weight of a gas molecule. The Knudsen flow is generally observed in ultrafiltration membranes (pore size: 10–100 nm) that are used as substrates for gas-separation membranes [51]. Here, the separation of the gas mixtures occurs based on the ratio of the square root of the molecular weight of each gas species. The selectivity from Knudsen diffusion is typically insufficient to meet the requirements of the separation processes [1].

As a special case of the Knudsen flow regime, when there is a strong interaction between the condensable gases and the pore walls, surface adsorption, and diffusion can further contribute to the gas diffusion through a small-pore-diameter membrane [1]. This phenomenon has been observed in amorphous carbons and microporous glasses [52, 53]. The transport of sorbed gas molecules can be boosted by the preferred diffusion based on a hopping-type mechanism [38]. For example, Shelekhin et al. observed that the surface diffusion of gases through a microporous glass is accelerated as the condensability of gas increases [53]. Such a strong interaction between the condensable gases and the pore walls can even restrict the permeation of noncondensable gases, which is called capillary condensation [1]. This effect becomes more noticeable when evaluating a gas mixture instead of a single-component gas. Barrer and coworkers first noticed the blocking of hydrogen gas with low condensability by the capillary condensation of other components, such as highly condensable sulfur dioxide inside the membrane pores. This effect became more prominent by increasing the partial pressure of sulfur dioxide [54]. In this case, the membrane could completely block the light gas (hydrogen) while allowing the permeation of larger gas molecules (sulfur dioxide), inversing the general trend of gas permeability expected from the size-dependent diffusion of gas molecules.

When the pore size falls into the microporous range (<10 Å), the membranes separate the close-sized molecules by molecular-sieving mechanism based on the steric hindrance of the transport of larger molecules [24]. Here, the diffusion of species i is not only affected by the activation energy of diffusional jump (E_{D_i}, enthalpic activation) in the average diffusive jump length (λ_{D_i}) but the geometric constraints by the shape of micropores (S_{D_i}, entropic activation), and the diffusion selectivity between species i and j in molecular-sieving regime can be expressed as [25]:

$$\frac{D_i}{D_j} = \left[\frac{\lambda_{D_i}{}^2}{\lambda_{D_j}{}^2}\right]\left[\exp\left(-\frac{E_{D_i} - E_{D_j}}{RT}\right)\right]\left[\exp\left(-\frac{S_{D_i} - S_{D_j}}{R}\right)\right] \tag{11.11}$$

In Eq. (11.11), the first exponent considers energetic selectivity governed by size difference in molecules while the second considers entropic selectivity related to the effects of shape difference [24, 25]. The molecular-sieving mechanism is rarely found in polymeric membranes because their chains keep fluctuating and thus their indefinite free volume (i.e. pore size) is not enough to provide molecular-sieving ability. Inorganic membranes, such as CMS and MOF materials, typically exhibit remarkably high selectivity compared to polymer membranes due to the improved molecular-sieving effects by their finely defined pore size distribution in the microporous range (<20 Å) or even ultramicroporous range (<7 Å) [23, 25].

11.2.2 Transport Mechanism in Nonporous Membranes

The major difference between the definition of porous and nonporous membranes is whether the pore size can be regarded as fixed or not [38]. For nonporous membranes, particularly for polymers with low free volume, the free-volume elements (usually envisioned as pores) fluctuate continuously by the thermal motion of the polymer chains [28]. Thus, it is expected that the pores in dense polymer membranes repeatedly appear and disappear while the permeants move across the membrane [55]. The solution–diffusion model is widely acknowledged to explain the molecular transport in dense, nonporous membranes. This mechanism depicts that molecules permeate through the nonporous membrane by several sequential steps, as illustrated in Figure 11.3. First, the permeants approach the membrane surface from the feed side and dissolve into the membrane matrix. Subsequently, the sorbed molecule diffuses across the membrane driven by the concentration gradient and then finally desorbs at the permeate side. The permeability coefficient of species i (P_i) is decoupled into diffusivity coefficient (D_i) and solubility coefficient (S_i) as follows [56]:

$$\text{Permeability}(P_i) = \text{Diffusivity}(D_i) \times \text{Solubility}(S_i) \tag{11.12}$$

Importantly, several assumptions should be considered to make the solution–diffusion model valid [56, 57]—(i) the membrane material is uniform and defect-free, (ii) the chemical potential gradient is continuous, and (iii) the applied pressure across the membrane is constant at the highest value—these aspects are taken into account in Figure 11.3.

Solubility is a thermodynamic parameter and is defined as the amount of sorbed penetrants into the membrane at the equilibrium condition. The solubility of gases in rubbery polymers is proportional to the applied pressure, which is consistent with Henry's law. On the other hand, sorption in glassy polymers is described by the dual-mode sorption model, accounting for the contribution of Langmuir sorption into the excess free volume in a nonequilibrium state together with the Henry sorption [56].

Diffusivity is a kinetic parameter in the solution–diffusion model, the measure of how fast the permeants is transported from the upstream side (e.g. higher pressure)

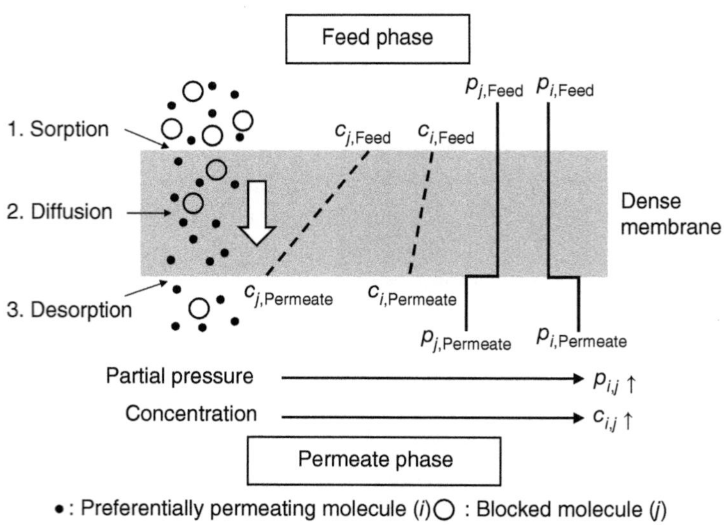

Figure 11.3 Mass transport through a nonporous membrane based on solution–diffusion mechanism.

to the downstream side (e.g. lower pressure) by a concentration gradient. This phenomenon was first reported by Fick back in 1855, which is now called Fick's law of diffusion as follows [7]:

$$J_i = -D_i \frac{dc_i}{dx} \tag{11.13}$$

where J_i is the flux, dc_i/dx is the concentration gradient, and D_i is the diffusivity of component i. The minus sign means that the direction of diffusion is down the concentration gradient. Diffusivity is affected by the geometry and size of the penetrants, which is influenced by the free-volume distributions and functional groups of membranes [2, 24].

Based on the above relations, the selectivity of compound i to j can also be decoupled into diffusivity selectivity and solubility selectivity, given by:

$$\alpha_{i/j} = \frac{P_i}{P_j} = \frac{D_i}{D_j} \times \frac{S_i}{S_j} \tag{11.14}$$

Namely, the selectivity is determined by the ratio of molecular size (i.e. mobility in membranes) and the ratio of permeant condensability (i.e. specific interaction between the permeant and the membrane) [28]. For polymeric membranes, both diffusivity and solubility depend on the physicochemical properties of membranes, such as microstructures, states (glassy or rubbery), and free-volume elements [4]. In some cases, the diffusivity is concentration-dependent, especially when treating the highly condensable gases (e.g. CO_2 and hydrocarbons) and liquids which are enough to swell the original polymer matrix, and thus the diffusivity increases with increasing concentration of permeants [1, 2]. For example, increasing water solubility in the membrane leads to a significant expansion in free volume by excessive swelling of the polymer chains, accompanying a dramatic increase in both water diffusivity and permeability. This phenomenon is usually called plasticization, which can be controlled by tuning the cross-linking degree and the number of fixed charge groups in polymer membranes [58–62].

Although the solution–diffusion mechanism is very simple and effective to explain the molecular transport in flexible polymers with low free volume, this model is often invalid for several classes of novel membrane materials, such as TR polymers [19], PIMs [63], and microporous inorganic membranes (e.g. CMS and MOFs) [24] which possess relatively fixed microporosity capable of sharp molecular sieving due to the rigid backbone structures [38]. To better clarify the mass transport in these materials, it is necessary to properly integrate different transport mechanisms in accordance with the various properties of membrane materials.

11.2.3 Transport Mechanism in Charged Membranes

Charged membranes (or ion-exchange membranes, IEMs) containing fixed charge groups are another important class of membrane materials in some applications where the electric field is exerted as a driving force, including electrodialysis (ED), reverse electrodialysis (RED), membrane capacitive deionization (MCDI), fuel cells, and flow batteries [1, 64, 65]. These membranes provide selective transport of ionic mixtures by excluding the similarly charged ions (co-ions) while allowing the oppositely charged ions (counter-ions). Based on the charged characteristic of the fixed ionic groups, the IEMs are generally classified into cation-exchange membranes (CEM, negatively charged groups), anion-exchange membranes (AEM, positively charged groups), amphoteric membranes, and bipolar membranes [64]. Since an IEM contains a high concentration of fixed charged groups, typically ~1–3 meq/g, the membrane is prone to lose mechanical stability by excessive swelling by solvent molecules when immersed in an electrolyte solution [1, 66]. Therefore, common IEM membranes have both hydrophilic, charged functional groups for ion conduction and hydrophobic backbone for mechanical robustness [67]. This structural basis of IEMs results in the microphase separation between the fixed charge groups and hydrophobic domains, which has been typically observed in the perfluorosulfonic acid polymer membranes (e.g. Nafion®, the most investigated CEM) [68].

Ion selectivity of IEMs was first explained by Donnan in 1911 [69]. The effects of co-ion exclusion (i.e. Donnan effects or Donnan exclusion) by charged membranes can be described based on the equilibrium thermodynamics which allows the electric potential at the membrane-solution interface and the activity of ions in the solution and membrane as follows:

$$\varphi_{\text{Don}} = \varphi^m - \varphi^s = \frac{RT}{z_i F} \ln \frac{a_i^s}{a_i^m} \tag{11.15}$$

where φ_{Don} is the Donnan potential, the potential difference between the membrane (φ^m) and the solution (φ^s) potentials; R is the gas constant; T is the absolute temperature; F is the Faraday constant; z_i, a_i^s, and a_i^m refer to the valence and activity of ion i in the solution and membrane, respectively [64]. Figure 11.4a–c shows the charge distribution of co-ions and counter-ions between the membrane and bulk solutions coupled with the corresponding Donnan potentials at the CEM and AEM [64]. Equation 11.15 leads to two important conclusions—(i) the higher fixed charge density will accompany the lower amount of co-ion sorption in the membrane matrix and (ii) the co-ion concentration in the membrane matrix will increase as increasing the external electrolyte concentration. The Donnan exclusion provides fundamental rationales for elucidating the ion selectivity of charged membranes up to now. However, the basic assumption behind the Donnan theory—the membrane phase is regarded as a solution containing uniformly distributed fixed charges—is often not valid due to the heterogeneous microstructure of most IEMs, and some deviations between the experimental co-ion solubility and the theoretical value have been found [64]. Recently, Kamcev et al. integrated the Donnan theory and Manning's counter-ion condensation theory to compromise such discrepancies, demonstrating the excellent agreement between the calculated values by their approach and experimentally determined data on co-ion sorption in IEMs [65, 70].

Ion transport in IEMs is generally driven by the combination of electrical potential difference and concentration difference. In addition, a convective term is often considered in pressure-driven filtration processes by using IEMs. By assuming the ideal conditions (i.e. extremely dilute solution) and the absence of coupling

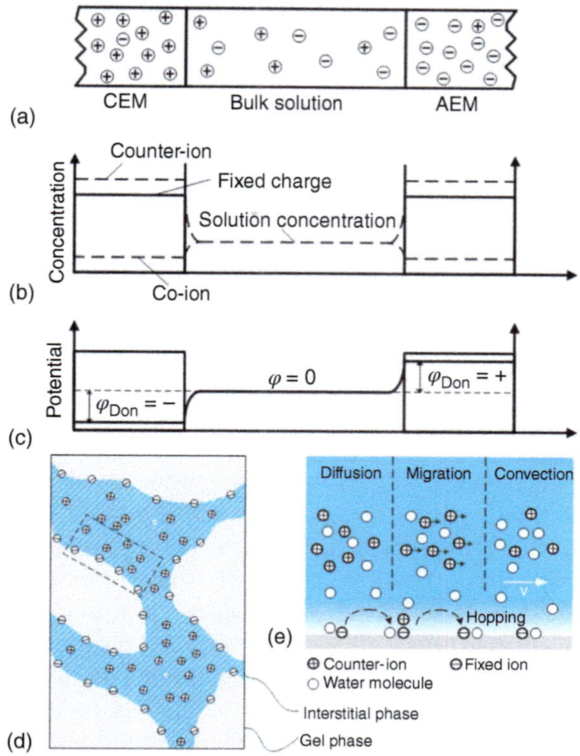

Figure 11.4 Schematic illustrations of (a, b) the distribution of the co-ions and counter-ions between the membrane and the bulk solution, (c) the corresponding Donnan potential at the CEM and AEM. (d) Two-phase structure model consisting of gel phase swollen by interstitial-free water phase, and (e) possible mechanisms for ion transport in IEMs, including diffusion, electromigration, convection, and hopping of counter-ions. Source: Luo et al. [64]/with permission of Elsevier.

phenomena, the extended Nernst–Planck equation describes the quantitative ionic flux (J_i) of these three forces governing the ion transport in IEMs, given by [2]:

$$J_i = vC_i - D_i\frac{dC_i}{dx} - \frac{z_iFC_iD_i}{RT}\frac{d\varphi}{dx} \tag{11.16}$$

where v is the fluid velocity of the solution, C_i, D_i, and z_i are the concentration, diffusivity, and charge number of ion species i, respectively, and x is the distance variable across the membrane. The first term represents the convective transport of ions overlapped by the electroosmotic solvent transfer, the second term describes the concentration gradient-driven ion diffusion, and the third term represents the migration of ions induced by the electric potential gradient. This relation is particularly convenient since ionic transport properties are usually decoupled into terms of the parameters found in Eq. (11.16) [64]. If necessary, the Maxwell–Stefan approach to mass transfer can be used to describe the diffusion of multicomponent ionic mixtures [71].

As observed in the porous and nonporous membranes, water and ion-transport properties in IEMs are strongly influenced by the physical and chemical structures of a membrane, such as fixed charge density, water content, specific interaction between ionic species and membranes, and microphase-separated morphology [64, 67, 68]. Figure 11.4d illustrates a two-phase model of the microphase-separated IEMs which include the swollen polymer gel phase (grey part) and the interstitial phases (blue part) [64]. The gel phase consists of the hydrophilic fixed charged groups coupled with the hydrophobic backbone chains, while the interstitial phase is the microvoid between the gel phases originated from the interstices, pores, and structural defects. The interstitial phase is regarded as the space filled with the electroneutral bulk solution when the membrane is hydrated with ionic solutions and contributes to the co-ion transport through IEMs. The void size is related to the degree of hydration and swelling ratio of the membrane [68]. Freeman and coworkers revealed the importance of sorbed water content in the membrane (expressed as the volume fraction of water to the membrane (ϕ_w)) in the ideal Donnan effects and the corresponding co-ion sorption [72–74]. The results indicate that the increase in ϕ_w in the sulfonated aromatic hydrocarbon polymers is responsible for the heterogeneous fixed charge distribution in IEMs [74].

The real mechanisms for ion transport are rather complicated compared to the fundamental theories discussed above. H^+ and OH^- transports in IEMs are the representative unique cases commonly observed in fuel cell applications [75]. Here, in addition to the three major mechanisms (convection, diffusion, and migration), the surface site "hopping" of H^+ ion is likely to contribute to the transport of protons (Figure 11.4e) [64]. When an IEM contacts with water, the fixed charged groups are surrounded by water molecules which hinder the possibility of electrostatic interaction between the counter-ions and the fixed groups [76]. Thus, the transport of counter-ions may occur by the hopping mechanism depending on the distance between the closely located fixed groups and the ion-exchange capacity of IEMs. For proton hopping, the Grotthuss mechanism explains the transport of H^+ ions from one hydronium ion to the nearby water molecule via rearrangement of hydrogen bond network, while the H^+ ion transport as an individual ion is a small extent because of the hydration shell [38, 75]. Likewise, OH^- ion transport follows a similar but slightly different mechanism accompanying the addition of another water molecule and the local recombination of the hydrogen bond network [77].

11.2.4 Permeability–Selectivity Trade-Off for Polymers

Ideally, a membrane should exhibit both high permeability and high selectivity to improve the efficiency of the separation process. However, it has been extremely challenging to achieve simultaneous increases in permeability and selectivity, which is known as a trade-off relationship [4, 16]. That is, nature appears to create membrane materials having high permeability and low selectivity, or vice versa.

A representative example of the trade-off relationship is found in polymeric gas-separation membranes. In 1991, Robeson first reported the trade-off between permeability and selectivity by empirical relation based on the database of gas transport membranes [78], which was revisited in 2008 and now called as Robeson upper bound [16]. In this relation, polymers possessing the highest selectivity between species i and j ($\alpha_{i,j}$) at a given permeability of species i (P_i) locate close to a line, defined as the upper bound, in compliance with the following relation:

$$\alpha_{i,j} = \beta_{i,j} P_i^{-\lambda_{i,j}} \tag{11.17}$$

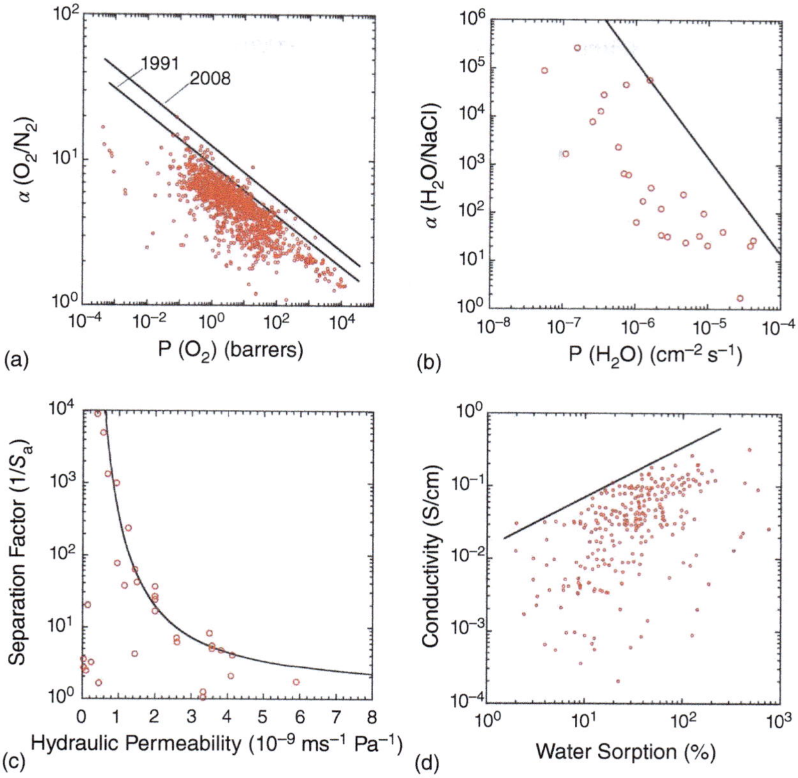

Figure 11.5 Representative upper-bound relations in polymer membranes for different applications. (a) Gas separation – O_2/N_2 separation, (b) desalination – water/NaCl separation, (c) ultrafiltration – water/protein (Bovine Serum Albumin, BSA) separation, (d) polymer electrolyte – ionic conductivity for membranes at a given level of water uptake. Source: Park et al. [4]/with permission of American Association for the Advancement of Science.

where $\beta_{i,j}$ and $\lambda_{i,j}$ are variables depending on the gas pair. An example of O_2/N_2 separation membranes for air separation is represented in Figure 11.5a. Clearly, a more permeable material is prone to display a lower selectivity and vice versa. The upper-bound behavior was revealed to be universal for all gas pairs [16].

Freeman established a theoretical expression to analyze the upper bound in gas-separation membranes [79]. In this model, the slope $\lambda_{i,j}$ of upper bound, depends on the ratio of diameters of gas molecules, is given by:

$$\lambda_{i,j} = \left(\frac{d_j}{d_i}\right)^2 - 1 \tag{11.18}$$

where d_j and d_i are the kinetic diameters of the larger and smaller gases, respectively. The front factor in Eq. (11.17), $\beta_{i,j}$, relates to the gas solubility is represented as follows:

$$\beta_{i,j} = \frac{S_i}{S_j} S_i^{\lambda_{i,j}} \exp\left(-\lambda_{i,j}\left(b - f\frac{1-a}{RT}\right)\right) \tag{11.19}$$

where S_i and S_j are the solubility coefficient of gas i and j, respectively, a and b are adjustable constants, and f is related to the average chain-to-chain distance and chain stiffness of polymer membranes [4]. There are several assumptions needed to satisfy this model—(i) gas transport is governed by solution–diffusion mechanism, (ii) gas diffusion follows the Arrhenius equation (i.e. activated process), (iii) the activation energy of diffusion, E_D, depends on the square of molecular size, (iv) there exists a universal and linear relation between E_D and the front factor of the Arrhenius equation, and (v) gas solubility relies on the condensability of gas species.

The permeability–selectivity upper-bound limitation does not necessarily occur in gas separation. It has been recognized that the upper bounds also appear in other synthetic membrane-based separation processes, such as desalination (water/NaCl separation [80], Figure 11.5b), porous ultrafiltration membranes (water/protein separation [12], Figure 11.5c), and polymer electrolyte membranes (proton conductivity at a given level of water sorption [81], Figure 11.5d), and many others [4]. Although the fundamental transport mechanisms significantly vary depending on target applications and membrane materials (e.g. porous/nonporous membranes and charged/neutral membranes), the universal existence of the upper bound suggests the generality of the trade-off behavior in all classes of synthetic membranes. Thus, the upper-bound theory provides the membrane researchers with a useful guideline of the current status and directions for future developments of membrane materials. However, care must be taken not to much rely on these plots since most of the upper bound is based on the separation performances based on single component, not a mixture, and do not consider the effects of surface modification, the number of different layers, composites, and scalability of given membrane materials [4, 82]. Ultimately, a comprehensive understanding of material design to membrane processing, module development, and process control is recommended toward the successful application of developed membrane materials [17, 82].

11.3 Nanofluidic Transport Through Confined Dimensions

For decades, the prevalence of trade-off limitation in polymer membranes has rendered researchers seeking novel membrane materials whose separation mechanism is distinctive from conventional theories. An important outcome of these attempts is the discovery of carbon nanofluidic phenomena that explain the anomalously fast transport of water molecules inside graphitic nanochannels [83]. After that, nanofluidic transport has drawn intensive attention in fundamental and practical importance in both basic science and engineering fields [39]. Generally, nanofluidics covers the mass transport through confined dimensions within a length scale of nanochannel from a few Å to ~100 nm. In this regime, the surface characteristics of channel walls (e.g. surface charge and channel size) largely contribute to the transport of molecules and lead to various interesting transport phenomena. Note that nanofluidic transport is distinguished from transport through typical polymer

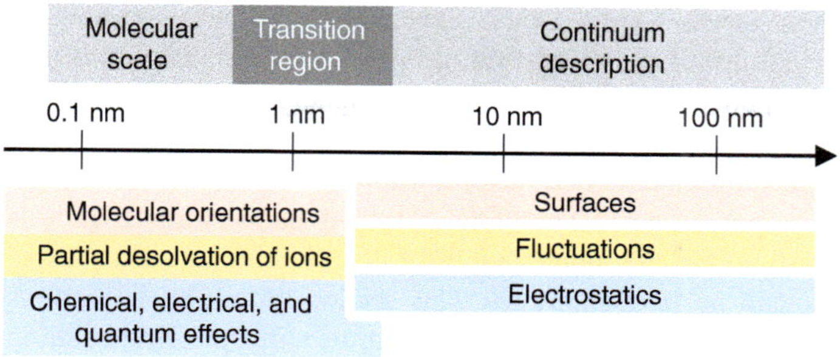

Figure 11.6 Transitions in mass-transport properties from bulk state to nanofluidics.

membranes having much smaller and nonpermanent pores, which can be best described by the solution–diffusion model as discussed previously.

As the terminology "nanofluidics" implies that something unprecedented may occur for molecular transport confined by the nanometric size of channels, it is necessary to define when the breakdown of the conventional theories on bulk flow emerges. Depending on the length scale of the host channel, Figure 11.6 separates the transport regime of nanofluidics into (i) classical continuum description (e.g. Navier–Stokes equations of hydrodynamics) is still applicable and bulk effects, such as surface properties of channels, thermal fluctuations, and electrostatics, are observed, (ii) mass transport follows intermediate behavior between continuum and discrete modeling (transition region, ~1 nm), and (iii) discrete particle effects at subnanometer scale mostly govern the molecular transport [39, 40]. The length scales associated with the bulk effects have provided key fundamentals in this field of nanofluidics for a decade. However, a recent review by Faucher et al. pointed out that there are critical knowledge gaps in mass transport through "single-digit" nanopores (<10 nm) largely because the current understanding is established mainly based on the macroscopic fluidics and continuum electrostatics [84]. For example, Tunuguntla et al. studied anomalous water and ion-transport properties through 0.8-nm-diameter and ~10-nm-long CNT porins which only contain ~30 water molecules, suggesting the importance of discrete particle effects with a statistical mechanics description [85]. In addition, when the dimension of the channel becomes down to the scale of permeating molecules or their surrounding solvents, typically below 1 nm, the physicochemical properties of fluids may represent significant deviations from predictions by common intuition from classical hydrodynamics due to the complicated interplay of short- and long-range interactions between molecules, which is now called nanoconfinement effects [86]. The dimensions of the confinement depending on the confining material are also responsible for tuning the nanoconfinement effects and the corresponding transport properties in such architectures. Overall, integrating the knowledge from the past to the present is essential to understand the nanofluidic phenomena in various forms of confined dimensions toward right connection between fundamental theories and practical applications.

With the recent advancements in nanomaterials and nanofabrication technology, artificial nanofluidic systems are not limited to purely curiosity-driven interests and have become more accessible and even scalable, flourishing numerous applications in biotechnology, tribology, energy storage and conversion, environmental fields, and membrane separations [87]. To date, tremendous efforts have been dedicated to the understanding of the nanofluidic phenomena and engineering the surface chemistry and confinement effects of nanochannels to exploit their unique transport behaviors, especially in 2D materials-enabled nanochannels. In this regard, it is essential to review the fundamentals of nanofluidics before exploring the transport mechanism in 2D membranes. This section introduces (i) various confinement architectures for nanofluidic systems and the corresponding building block materials, (ii) fundamentals of nanofluidic transport based on classical and current theories, and (iii) the roles of molecular confinement in mass transport through subnanometer channels.

11.3.1 Confinement Architectures for Artificial Nanofluidic Systems

Designing nanochannels in a confined geometry is the first step toward the development of nanofluidic systems. In general, the confinement architectures are classified by dimensions of scaffold materials as follows—(i) zero-dimensional (0D) nanopores, (ii) one-dimensional (1D) nanotubes, and (iii) slit-like two-dimensional (2D) nanochannels, respectively (Figure 11.7). Based on these frameworks, controlling the structures of scaffold materials (e.g. morphology, length or thickness,

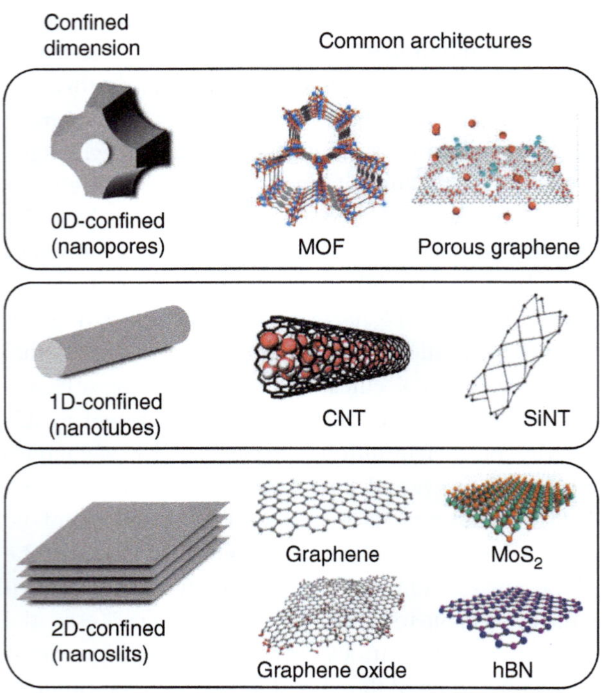

Figure 11.7 Classification of confinement architectures and the corresponding representative materials to exert nanoconfinement effects.

and pore size distribution) and surface chemistry (e.g. electrostatic charges and functional groups with molecular recognition) of confinement architectures are commonly performed strategies to investigate the nanoconfinement effects and improve their separation abilities [40, 86].

Nanofluidics in 0D-confined channels can be built by materials consisting of nanopores whose diameter and length are in the same range, typically processed by "bottom–up" or "top–down" approaches. The former case uses molecular precursors to form nanoporous materials, such as zeolites, MOFs, COFs, and metal–organic cages (MOCs) [25]. These crystalline materials are suitable for nanofluidic studies because of their relatively fixed pore size and size distribution in the subnanometer range compared to those in amorphous microporous materials, such as CMSs, PIMs, TR polymers, and porous aromatic frameworks (PAFs). In the "top–down" case, initially, the nanopores on membranes were fabricated by drilling through ceramic precursors such as silicon nitride (SiN) or silicon carbide (SiC) [39]. These materials exhibited limitations by the difficulties in controlling pore size distribution and thickness down to subnanometer. Recently, a monolayer of 2D materials, such as graphene, hBN, and MoS_2, has intrigued considerable attention as an excellent candidate to replace the silicon derivatives and serve as a versatile platform benefiting from its atomic thickness to minimize the mass-transport resistance. The nanopores in the 2D basal plane can be created in the range from a few Å to tens of nm by top–down perforation approaches, including ion bombardment, chemical or plasma etching, and defect engineering during the earlier synthesis step [3].

Nanotubes are defined as cylindrical channels having inner diameter D and length L, where $L \gg D$ with a typical aspect ratio (L/D) of 20–1000 [39]. CNTs are the most investigated nanomaterial in 1D-confined nanofluidics by showing extraordinary mass-transport properties due to the molecularly smooth graphitic surfaces which allow ultrafast molecular transport [88]. Both the diameter and length of CNTs determine their separation abilities. CNTs are classified by the number of walls, including single-walled carbon nanotubes (SWCNTs), few-walled carbon nanotubes (FWCNTs), and multi-walled carbon nanotubes (MWCNTs) which are synthesized by arc-discharge method or chemical vapor deposition (CVD) [89]. Furthermore, the entrance and exterior surface of CNTs can be modified by postfunctionalization to preferentially draw the target molecules. Other types of 1D architecture include boron nitride nanotubes (BNNTs), silicon nanotubes (SiNTs), TiO_2 nanotubes, and organic nanotubes [40]. One of the major challenges of the 1D-confined nanofluidics is the very difficult assembling of the nanotube arrays on a practically meaningful scale without any leakages between the outer surfaces of nanotubes [90].

A slit-like channel formed by layer-by-layer stacking of 2D materials has been an excellent playground for understanding the 2D-confined nanofluidics over the past decade [91]. In 2012, Nair et al. first reported an ultrafast diffusion of water vapor molecules in a freestanding graphene oxide (GO) film with a thickness of a few micrometers while other molecules like ethanol, hexane, and even helium hardly permeate with a negligible amount [32]. Inspired by this pioneering work, GO laminate has been intensively explored to date, initially as the part of the extension of anomalous molecular transport in graphitic channels observed in CNTs, and now becomes one of the most viable membrane materials utilizing nanofluidic transport due to their ease of postfunctionalization and excellent scalability for practical applications [92]. However, the nanochannels in GO membranes are prone to swell when immersed in aqueous solutions owing to the presence of hydrophilic oxygen-functional groups as spacers between graphitic basal planes [93]. The limited channel size down to 5–10 Å is another concern in GO membranes which may not be effective for precise molecular sieving of close-sized molecular mixtures [94]. To overcome these disadvantages, other novel 2D materials, such as reduced graphene oxide (rGO), solvent-exfoliated graphene, hBN, MoS_2, and Mxenes, are currently gaining much attention and often exhibit excellent separation performances ascribed to their unique surface properties and finely tuned nanochannels [30, 31].

11.3.2 Continuum Modeling of Nanofluidic Transport in Confined Channels

Although recent studies have demonstrated the discrete nature of fluids on the atomic scale in both theoretical and experimental ways, continuum description based on bulk hydrodynamics still serves as a cornerstone of explaining transport phenomena in nanofluidic systems [87]. Indeed, the experimental results predicted by this approach are usually close to the predictions from discrete modeling, while a drastic change in transport behavior is observed for more severe confinements—this will be discussed in Section 11.3.5. Bocquet and Charlaix stated that the continuum modeling may be valid down to the ~1 nm of nanochannel size for water transport at room temperature by defining the molecular timescale required for momentum to diffuse across the system with a critical channel size [95]. Notably, a nanometric characteristic length scale is widely adopted in common nanomaterials comprising confinement architectures, as shown in Figure 11.7. This section presents fundamental transport theories based on continuum modeling in nanochannels.

The Navier–Stokes equation, developed in the nineteenth century, is surprisingly robust even at describing nanofluidics which is far below the human scale—though it might have not been the original scope of this historic finding [95]. This classical theory is well established in common disciplines, but it cannot be overlooked in the emerging interdisciplinary field of nanofluidics, and thus is summarized here. For a laminar flow of an incompressible fluid at constant viscosity, a compact form of the Navier–Stokes equation in the single vector equation is given by [96]:

$$\rho\frac{\partial v}{\partial t} = f - \nabla P + \eta\nabla^2 v \tag{11.20}$$

where ρ is the density of the fluid, v is the flow velocity, t is the time, and η is the dynamic viscosity of the fluid, respectively. The left-hand side of the equation indicates mass and acceleration of fluid which equals zero for steady-state flows. The parameters in righthand side indicate a body force applied to the unit system (f, e.g. gravitational force, electric field, and buoyancy force), the pressure gradient (∇P), and diffusion-like viscosity term ($\eta\nabla^2 v$), respectively.

Given that the length scale is very low (<10 nm) for water flow in nanofluidic devices, the Reynolds number is low (i.e. $Re \ll 1$) at fluid velocities less than 10 m s^{-1} and Stokes flow where viscous forces overwhelm advective inertial forces can be safely used to describe the flow in a nanochannel [39, 95, 97]:

$$\nabla P = f + \eta \nabla^2 v \tag{11.21}$$

The above equation, called the Stokes equation, is a linearized form of the Navier–Stokes equation, and thus relatively simple to be solved by a number of common methods for linear differential equations.

Stokes flow is often analyzed with no-slip boundary conditions where the velocity of the fluid is presumed to vanish at an interface between the fluid and the channel wall. However, as the nanochannel size decreases more, the contribution of interfacial dynamics will become more pronounced in the fluid transport. A representative factor of this consideration is the friction force, which is proportional to the velocity of fluid according to the linear response theory. For a fluid moving in the direction of x axis with slip velocity v_x in parallel with a surface of normal z axis, the friction force per unit area is $\sigma_{xz} = \lambda v_x$, where σ the stress tensor and λ the coefficient of friction per unit area. For a Newtonian fluid, the friction coefficient is given by $\lambda = \eta \partial_z$, and we have the Navier boundary condition as [83]:

$$v_x = b \left.\frac{\partial v_x}{\partial z}\right|_{\text{wall}} \tag{11.22}$$

where b is the Navier length or slip length ($b = \eta/\lambda$).

Intuitively, the slip length is defined as the depth inside the channel wall where a linear extrapolation of fluid velocity profile vanishes. On the other hand, it also serves as an indicator to gauge the molecular interaction between the fluid and the channel, providing more degree of freedom to design nanofluidic systems. Depending on the extent of interfacial interaction (i.e. the friction force), two extreme conditions can be envisioned; (i) the slip length can be zero at no-slip boundary condition ($\lambda \to \infty$) and (ii) the slip length converges to infinity at perfect slip ($\lambda \to 0$) (Figure 11.8a). Larger slip lengths can come from lower fluid–solid interaction (e.g. water transport through hydrophobic graphitic walls), which can be exploited to boost the transport rate of molecules in pressure-driven flows [95, 97].

Considering the typical slip length within several tens or hundreds of nm in nanofluidic systems [88], control of the slip effects is expected to largely influence the transport phenomena in confined dimensions. Water transport through the inner diameter of CNTs is the most investigated example to showcase the effects of the slip length [98]. Basically, the volumetric flow rate in a cylindrical tube ($Q_{\text{H-P}}$) can be described by the no-slip Hagen–Poiseuille relation (Eq. (11.7)). However, the underlying assumption of this formula—the channel length L is much higher than the size of the channel (i.e. $L \gg R$ if the channel size is given by radius R) and thus the effect of channel entrance is negligible—may not be applicable in a nanoscale channel due to the deviations from a macroscopic perspective [39]. The entrance effects can be appreciated by deeming a flow through an infinitely thin nanopore. When the geometry of the channel mouth is a circular nanopore with a radius of R,

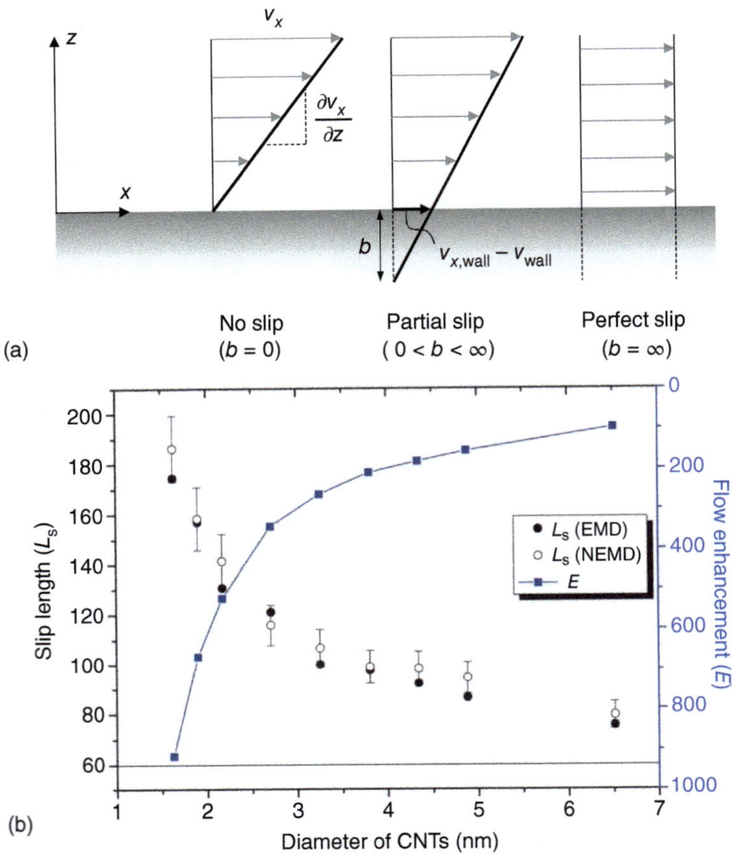

Figure 11.8 (a) Three cases of slip flow through a stationary surface with the slip length. (b) The slip length (represented as L_s here) predicted using the equilibrium molecular dynamics (EMD) and nonequilibrium molecular dynamics (NEMD) simulations and flow enhancement (E) against the diameter of CNTs. The black line at 60 (±6) is the slip length of water on a graphene surface. Panel b – Source: Kannam et al. [88]/with permission of American Institute of Physics.

Sampson's equation can be used to express the flow rate through the pore entrance (Q_p) as [39]:

$$Q_p = \frac{R^3}{3\eta}\Delta P \tag{11.23}$$

where ΔP is pressure drop. By taking into account the entrance process (Eq. (11.23)), the corrected flow rate (Q_c) can be corrected as follows:

$$Q_c = \frac{\Delta P}{R_{H-P} + R_p} = \frac{\pi R^4}{8\eta L}\frac{\Delta P}{1 + \frac{3\pi}{8}\frac{R}{L}} \tag{11.24}$$

where R_{H-P} and R_p are hydrodynamic resistances of the channel and the pore, respectively [39, 99]. Next, it is necessary to combine the contribution of the slip flow in the 1D-confined CNTs. In the case of a positive slip length where the water intermolecular interaction is larger than the water–CNT interaction [100], the volumetric slip flow (Q_s) is accelerated by slippery effects, and the corresponding flow enhancement (E) is defined as follows [83, 88]:

$$E = \frac{Q_s}{Q_c} = \left(1 + \frac{4b}{R}\right) \tag{11.25}$$

where R is the inner radius of CNT. Equation 11.25 implies that the ratio of the slip length to the channel size will determine how much the slippery effects affect the nanofluidic transport. That is, the molecular transport in the smallest channel, of size $R \ll b$, will approach a perfect-slip flow where entrance effects should be taken into account [39]. Kannam et al. further validated this relation by calculating the flow enhancement using equilibrium molecular dynamics (EMD) and nonequilibrium molecular dynamics (NEMD) simulations [88]. As the diameter of the CNT increases, the slip length gradually decreases, and the flow-enhancement approaches unity for a CNT with a micrometer-sized diameter (Figure 11.8b). The results match well with experimental data on nanofluidic water transport in CNTs [101, 102]. In addition to the 1D-confined nanotube case, slip flow in 2D-confined graphene nanochannels has also been investigated based on the planer Poiseuille flow in both computational and experimental approaches [98, 103].

11.3.3 Mechanisms of Nanofluidic Transport in Atomically Thin Nanopores

The theories on slippery effects are mainly established to describe the nanofluidics in 1D- or 2D-confined channels. On the other hand, for molecular transport across atomically thin nanopores (0D-confined), Sampson's equation (Eq. (11.23)) can be used for the free effusion behavior where the transport is governed by intermolecular collisions while Knudsen flow (Eq. (11.9)) will be more dominant upon reducing the Knudsen number, defined as $K_n = \lambda/L$, where λ is the mean free path (Eq. (11.8)) and L is the size of nanopore [39]. Convincing results have demonstrated that these approaches are effective for 0D-confined nanofluidics studies, mostly performed by nanoporous monolayer graphene as a versatile platform. Many theoretical studies coupled with some of the remarkable experimental evidence have reported exceptionally high selectivity achieved by nanoporous atomically thin membranes (NATMs) [3], which cannot be predicted from either Sampson's equation or even Knudsen flow. In these cases, additional factors in molecular scale should be taken into account, such as surface adsorption, penetrant-pore interaction, and conformational entropy (shape effects). Density-functional theory (DFT) calculations and molecular dynamics simulations are widely adopted tools to calculate the transport rate of molecules across a nanopore, further validated by many experimental efforts recently [30, 91].

Karnik and coworkers proposed the effectiveness of the Arrhenius approach to approximate the energy barrier for gas transport in NATMs based on the ratio of Arrhenius factors [3]. Depending on the differences in the diameter of the gas molecule (d_m) and the nanopore (d_p), gas transport across a nanopore is divided

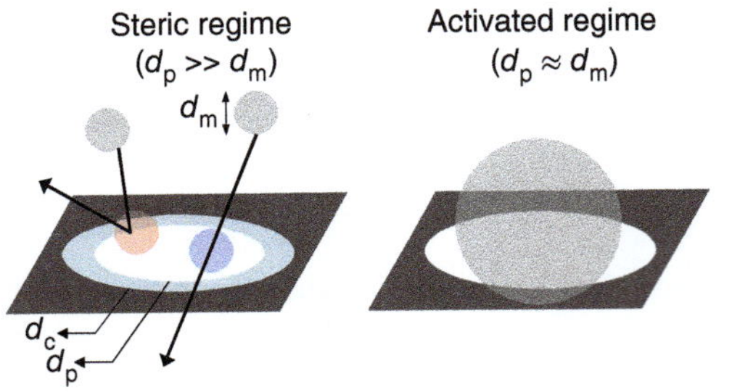

Figure 11.9 Transport model and mechanism of gas transport through porous monolayer graphene. Here, carbon centers located at d_c with pore diameter defined as $d_p = d_c - d_{vdw}/\sqrt{2}$, where d_{vdw} is the van der Waals diameter of a carbon atom.

into two scenarios; (i) if $d_p > d_m$, gas transport is governed by steric hindrance and follows Sampson's equation, given by an effective area $A_{eff} = \pi/4(d_p - d_m)^2$ which is smaller than the area of nanopore $A_{pore} = (\pi/4)d_p^{\,2}$ and (ii) if $d_p \leq d_m$, the attractive/repulsive interactions between the penetrant and the pore present an energetic barrier to the transport in the nanopore (Figure 11.9). The former is referred to as a steric regime that typically displays a low selectivity limited by the differences in molecular size. The latter is called an activated regime where a high selectivity can be achieved depending on the activation energy difference. Here, the molecule–pore interaction can be roughly estimated by the Lennard–Jones potential, given that the pore consists of fixed atoms, and the barrier height (E) is written as [3]:

$$E \approx \frac{\pi d_c}{a} 4\varepsilon \left[\left(\frac{\sigma}{\frac{d_c}{2}} \right)^{12} - \left(\frac{\sigma}{\frac{d_c}{2}} \right)^{6} \right] \approx \frac{4\pi\varepsilon d_c}{a} \left(\frac{\sigma}{\frac{d_c}{2}} \right)^{12} \tag{11.26}$$

where $d_c/2$ is defined as the center-to-center distance between the pore and the atom constituting the pore, a is the distance between the adjacent atoms on the pore edge, ε and σ are the Lennard–Jones parameters, respectively. $\pi d_c/a$ represents the average number of atoms on the pore edge. Given that the kinetic energies of all crossing gas molecules exceed the energy barrier of the pore (Eq. (11.26)), an Arrhenius-type expression describes the effective pore size in this regime:

$$\frac{A_{eff}}{A_{pore}} \approx \frac{1}{2} \text{erfc} \left(\sqrt{\frac{E}{k_B T}} \right) \tag{11.27}$$

It was found that the effective pore areas calculated from the above two scenarios agree well with the simulation results in the literature, which exhibits two distinct sections divided into steric regime (i.e. larger pore sizes) with high permeance and low selectivity and activated regime (i.e. smaller pore sizes) with low permeance and high selectivity [3]. Notably, this simple approach based on the steric exclusion and the activation theory results in the same conclusion from Robeson's upper-bound theory that predicts a trade-off relation between permeability and selectivity. Beyond the gas transport, recent studies have also demonstrated the activation behavior of liquid and ion transport in atomically thin nanopores, supporting its universal applicability regardless of the states of penetrants on a molecular scale [104, 105].

11.3.4 Effects of Electrical Double Layer in Nanofluidic Ion Transport

The previous subsections 11.3.2 and 11.3.3 discussed the transport mechanisms in nanofluidic systems with no consideration of the electrostatic properties of nanochannels. Now, the influences of nanoscale confinement are explored on charged channels, particularly with the role of the electrical double layer (EDL) in ionic distribution and transport.

Since almost all materials constituting channel walls intrinsically carry fixed surface charges, an oppositely charged region of counter-ions is formed in the liquid-wall interface to maintain electroneutrality [87]. This screening region is called the EDL because of the presence of both the surface charge of walls and the corresponding counter-ion region. The EDL was initially treated as a simple capacitor with fixed counter-ions, called the Helmholtz model in 1879. However, all the counter-ions are not necessarily to be bound to the wall surface, and the thermal (Brownian) motion of ions rather results in the ionic distribution along with the distance from the wall. In 1910, Gouy and Chapman proposed that the counter-ion distribution in EDL can be postulated by the Poisson's equation (for the potential distribution) coupled with the Boltzmann distribution (for the distribution of ions), given by [99]:

$$-\nabla^2\psi = \frac{e}{\varepsilon}\sum_i c_i z_i \exp\left[\frac{-z_i e\psi}{k_B T}\right] \tag{11.28}$$

where $\nabla^2\psi$ indicates the divergence of the electrostatic potential gradient, which can be cast in Cartesian coordinates for parallel flat walls with respect to the perpendicular coordinate y (i.e. $d\nabla^2\psi/dy^2$), e is the unit charge, ε is the permittivity of the liquid; c_i and z_i are the concentration and the charge number of the ionic species i, ψ is the electrostatic potential, k_B is Boltzmann's constant, and T is the absolute temperature, respectively. Assuming low potentials (i.e. $ze\psi \ll kT$), the length of an EDL is denoted as the Debye length (λ_D). This important parameter is regarded as the distance from the charged wall surface where the potential has been exponentially decayed, which can be written as [97]:

$$\lambda_D = \sqrt{\frac{\varepsilon k_B T}{\sum_i N_A c_i z_i^2 e^2}} \tag{11.29}$$

where N_A is Avogadro's number. The Debye length offers basic criteria for assessing the nanofluidic ion transport in confined channels, typically 1–100 nm in aqueous solutions [106]. Unfortunately, neither the Helmholtz nor the Gouy–Chapman model provides satisfactory agreement over a large range of potential. This is because the assumptions that the solvent could be deemed as a structureless dielectric with constant permittivity and the electrolyte ions could be regarded as point charges may not be valid depending on the ionic concentration or the electric potential. In 1924, Stern combined these ideas and introduced a model of the electric double layer comprised of some of the charges in a fixed layer and some of the charges in a diffuse layer, as illustrated in Figure 11.10 [99]. The outermost layer is composed of mobile co-ions and counter-ions at the location of the slip plane relative to a

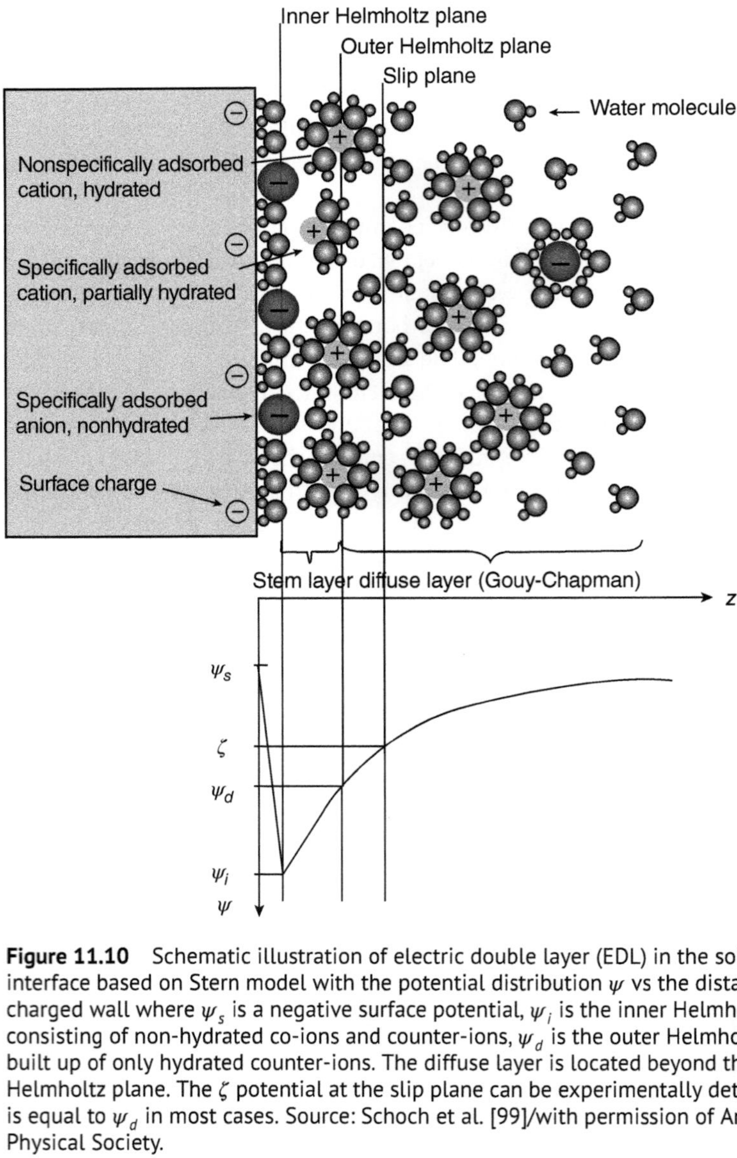

Figure 11.10 Schematic illustration of electric double layer (EDL) in the solid-electrolyte interface based on Stern model with the potential distribution ψ vs the distance z from the charged wall where ψ_s is a negative surface potential, ψ_i is the inner Helmholtz plane layer consisting of non-hydrated co-ions and counter-ions, ψ_d is the outer Helmholtz plane layer built up of only hydrated counter-ions. The diffuse layer is located beyond the outer Helmholtz plane. The ζ potential at the slip plane can be experimentally determined, which is equal to ψ_d in most cases. Source: Schoch et al. [99]/with permission of American Physical Society.

point in the bulk solution away from the solid-electrolyte interface, bearing the zeta potential (ζ). In concentrated solutions, the contribution of the static, Helmholtz portion is much greater than that of the diffuse part of the double layer, and vice versa.

For nanofluidic systems where the wall-to-wall distance (h) is in the order of Debye length, the diffuse layer from one charged wall surface can influence the next wall depending on the geometries of nanochannel and electrolyte composition. If the channel height is higher than Debye length ($h \gg \lambda_D$, Figure 11.11a), the diffuse layer

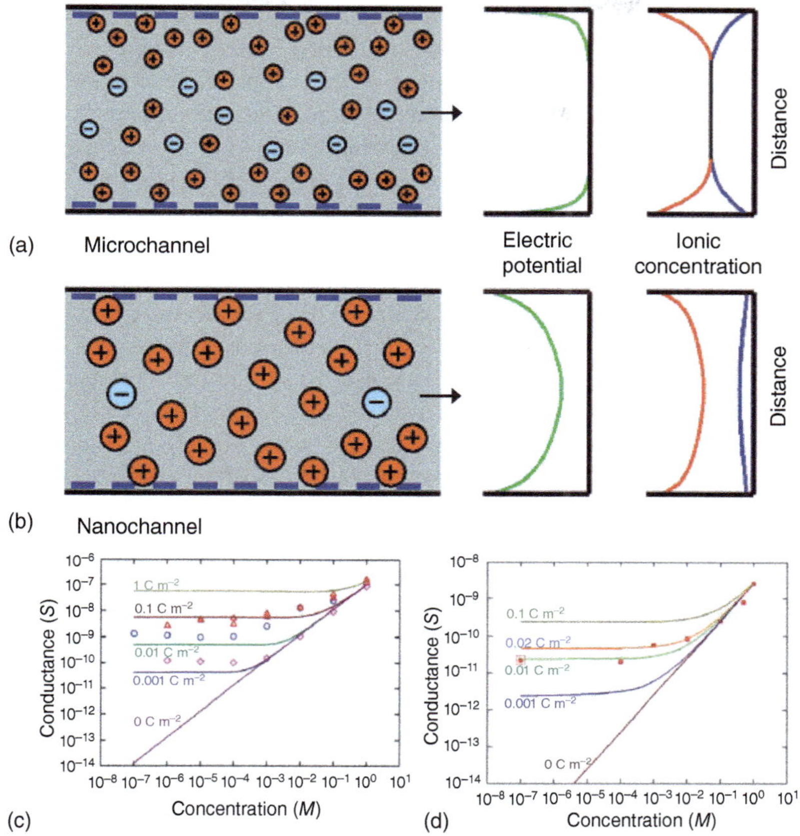

Figure 11.11 Effects of surface charge in charge distribution, electric potential, and ionic concentration for (a) microchannels and (b) nanochannels. The cation concentration (orange) and anion concentration (blue) in the microchannel is equal to those in bulk solution. In a nanochannel, the cation concentration is much higher than the anion concentration. The electrical conductance of the (c) nanochannel and the (d) nanotube devices deviates from bulk solution ($0\,\text{C m}^{-2}$). Source: Karnik et al. [106]/with permission of American Chemical Society.

of two walls will not meet and the resultant ionic current (I) can be described as the sum of a bulk term and surface term by EDL [39]:

$$I = 2w\frac{e^2D}{k_BT}E(\rho_s h + \Sigma) \tag{11.30}$$

where D is the diffusion coefficient, w is the width of nanochannel, E is the applied electric field, and $\rho_s h$ and Σ account for the number of charge carriers in the bulk and at surfaces, respectively. In contrast, If the channel height is lower than Debye length ($h \ll \lambda_D$, Figure 11.11b), the contribution from bulk solution becomes absent and all the quantities can be treated uniform across the channel. This regime is called the Debye overlap displaying ionic current–voltage relation as follows [39]:

$$I_{ep} = 2w\frac{e^2D}{k_BT}E\sqrt{(\rho_s h)^2 + \Sigma^2} \tag{11.31}$$

Equations 11.30 and 11.31 indicate that ion enrichment and exclusion effects are more noticeable in nanofluidic systems, potentially presenting the nanochannels with molecular sieving, much like membranes. In microchannels, the Debye length is generally much smaller than the channel dimensions. Thus, the bulk portion of the fluid is not affected by the EDL and direct manipulation of ion transport in the microchannel is impossible. On the other hand, in nanochannels whose scale is comparable to or even smaller than the Debye length, the superimposed diffuse layer of counter-ions between the two wall surfaces enables direct ionic manipulation based on Donnan effects by controlling electrostatic variables, such as surface charge, electrolyte concentration, and field effect [95, 99]. For example, Yang and coworkers fabricated two types of nanofluidic transistors based on metal–oxide, one is a two-dimensional nanochannel with a height of 30–40 nm and the other is a one-dimensional nanotube with an inner diameter of 10–100 nm [106]. The experimental ionic conductance of the negatively charged nanochannel and nanotube devices agrees well with the theoretical predictions, indicating that the ion transport is governed by surface charge of wall material due to the overlap of Debye regime with a high cation/anion selectivity by Donnan exclusion behavior (Figure 11.11c,d). In a concentrated electrolyte solution, however, the Debye length would become comparable with the channel size, and thus the selectivity sill decrease drastically (i.e. charge screening effects) [107].

11.3.5 Various Confinement Effects in Nanofluidic Transport at the Subnanometer Scale

When a nanochannel size becomes atomically thin less than 1 nm, the predictions from continuum modeling may lead to errors and deviate from the real situations. This is because the interactions between individual molecules will govern the fluid properties due to the discrete nature of molecules. The consideration of discrete modeling can be performed by molecular dynamics simulations that account for the electrostatic and van der Waals interactions in nanofluidic systems. For more information on this topic, the readers are suggested to follow the excellent reviews [39, 95, 99]. Meanwhile, with the recent advancements in the material platforms for Ångström-sized channels, such as 0D nanopores, 1D nanotubes, and 2D nanochannels, there are a number of interesting phenomena resulting from unprecedented confinement effects at the subnanometer scale, which have been evidenced by both theoretical and experimental studies for more than a decade [40]. In general, if a nanospace is dimensionally confined in scale to the size or shape of its penetrant molecules, the physicochemical properties of fluid may experience drastic changes due to the more complicated intermolecular forces and molecule–channel interactions from both short- and long-range forces. Herein, several important and relatively common nanoconfinement effects are introduced.

11.3.5.1 Molecular Rearrangement

In a subnanometer scale, the confined molecules are likely to undergo molecular orientation and rearrangement due to the limited space available for molecular interactions as it was in the bulk phase. For example, the extraordinary water transport in carbon nanofluidics is generally attributed to the extremely smooth graphitic walls constituting the channel that repels the confined water molecules and makes them exhibit different phases, orientation, and density compared with bulk water [83, 108]. Bocquet et al. studied the effects of interfacial friction between water molecules and graphitic surfaces in various topologies, such as the type of CNTs (armchair/zigzag), the location of water (inner/outer surface), and the dimension of graphitic surface (1D/2D) (Figure 11.12a). The results revealed that the spatial rearrangement of water molecules is related to the curvature of a confined system that influences the friction coefficient. This work indicates there may be a significant difference in molecular rearrangement depending on the confinement dimension despite the use of the same base material. Recently, Geim and coworkers observed that the 2D-layered water in graphene nanocapillaries form a "square ice," not existing in the conventional phase of water, originated from the vanished interfacial interaction between water and surface inside the hydrophobic confinement architecture [112]. Such unique molecular structures under nanoscale confinement are not solely limited to the water, and thus the precise control of such spatial orientations is expected to not only govern the separation efficiency but tune the chemical reactivity of nanofluidic systems [86].

11.3.5.2 Partial Dehydration or Desolvation

It has been recognized that ions likely diffuse through a membrane as dissociated, hydrated ions due to the favorable interactions between charged ions and polar water molecules. Thus, the effective diameter of an ion is generally thought of as being the diameter of the fully hydrated ion [72]. On the other hand, when the confinement system is reduced to a scale comparable to the diameter of hydrated ions, the hydration shell that is already established may be frustrated, leading to partial dehydration of the surrounding water molecules. The extent of dehydration is determined by the confined space, the electrokinetics, and the interplay among the ion-solvent-channel ternary system. A good example of this phenomenon is the rapid transport of K^+ ions across cell membranes while completely excluding Na^+ ions, potentially ascribed to the fully dehydrated nature of K^+ ions unlike the partially dehydrated Na^+ ions in the membrane channel, which maximizes both permeability and selectivity of K^+ ions [113]. A similar result could be achieved in a nanoporous graphene membrane that demonstrates dehydration as a universal mechanism to govern the ion selectivity in an atomically thin pore [114]. Lee et al. also reported that polymer nanocomposite membranes containing nanoporous ZIF-8 nanoparticles exhibit anomalous ion-transport properties compared to those of common polymer membranes by showing that the permeability enhancement rather follows in order of the hydrated diameter of ions, called "reverse selectivity" by the partial dehydration of penetrating ions (Figure 11.12b) [110]. Furthermore, an artificial nanochannel of ~3.4 Å in height was fabricated by introducing a monolayer graphene spacer between graphite or hBN crystals. In this 2D nanofluidic device, no permeation of Na^+ and Cl^- was detected while only proton (H^+) with the smallest size could transport through the monolayer water built inside the

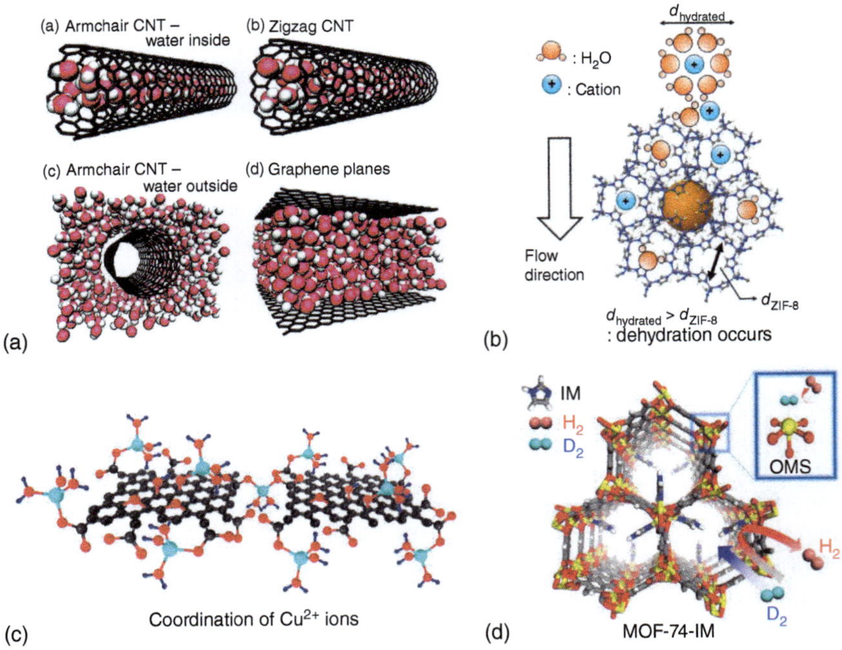

Figure 11.12 Typical confinement effects in mass transport behaviors. (a) Nanoconfinement-induced rearrangement of water molecules through carbon nanotube with different geometries and topologies. (b) Ion transport through the ZIF-8 cavity is smaller than its hydrated state accompanied by the partial dehydration phenomenon. (c) Coulombic interaction between GO nanocapillaries and Cu^{2+} ions. (d) Quantum-sieving effects in MOF-74 due to strong open metal sites. Panel a – Source: Falk et al. [109]/with permission of American Chemical Society. Panel b – Source: Lee et al. [110]. Reproduced with permission of American Chemical Society. Panel c – Source: Sun et al. [36]. Reproduced with permission of American Chemical Society. Panel d – Source: Kim et al. [111]. Reproduced with permission of American Chemical Society.

capillaries [115]. These examples suggest controlling the extent of ion dehydration may contribute to improving the selectivity of desired ion species and realizing the high-performance nanofluidic devices for various applications, which are not limited to the aqueous solutions and can also be applied to other organic solvents and electrolytes [86].

11.3.5.3 Electrical Effects

In addition to the common driving forces, such as concentration gradient and pressure difference, electrical effects can be exploited to improve the transport rate of molecules, particularly for ion transport, by changing the electrostatic interactions between the permeants and the nanoconfined channels. A similar conclusion was introduced in Section 11.3.4. That is, these effects would be more pronounced in the regime where the channel size is close to or even lower than its Debye length, resulting in the overlapping of EDL and thus the co-ions would be repelled by the Donnan exclusion mechanism. The subnanometer channel will also undergo the partial or complete dehydration of ions that induces the rectified and nonlinear characteristics of current–voltage experiments. Accordingly, electric potential significantly influences the ion transport within nanoconfined, charged channels by changing the distribution and concentration of ions [5]. Furthermore, the electrostatic interaction is affected by the coordination bonding with transition metals. Sun et al. reported that the strong coordination between heavy metal ions and functional groups in GO membranes governed the selective ion transport through the GO nanocapillaries, supported by the ion adsorption experiments (Figure 11.12c) [36]. Likewise, tuning the surface charge of the channels is a simple strategy to improve the ion selectivity by increased electrostatic attraction or repulsion. Jin and coworkers coated positively or negatively polymer electrolytes onto GO membranes to control their surface charge, and the resultant highly charged GO membranes exhibited excellent potential in nanofiltration application with their remarkably high water permeance together with monovalent/divalent ion selectivity [116].

11.3.5.4 Quantum Effects

As an extreme case, if the confinement scale approaches the de Broglie wavelength of atoms (λ_i), e.g. 1.45 Å for proton and 1.02 Å for deuteron atom at 300 K [40], quantum effects may govern the mass transport and have been paid much attention for hydrogen isotope separation [117]. For example, there are two major mechanisms to separate D_2/H_2 mixture by nanoporous materials—(i) direct capture of heavier isotope gas molecules that possess a shorter λ_i and a higher diffusivity at cryogenic temperatures via kinetic quantum-sieving (KQS) effects and (ii) introduction of strong adsorption sites preferential to heavier isotopes that provide chemical affinity quantum-sieving (CAQS) effects. Kim et al. demonstrated that the D_2/H_2 separation efficiency can be significantly boosted by simultaneously utilizing these two quantum-sieving effects together, which could be realized in 1D-confined nanochannel in MOF-74 functionalized with imidazole groups that selectively adsorb heavier isotopes (Figure 11.12d) [111]. Quantum confinement also occurs in intralattice spaces of 2D materials which have been regarded as an impermeable barrier to any other molecules [32]. A counterintuitive result was reported by Hu et al. revealing that the proton transport is actually possible through a perfect monolayer of graphene or hBN depending on their electron-density distribution [118]. A follow-up study by another group showed the single-oriented few-layered hBN with AA-stacked structure could efficiently block the H_2 gas crossover while allowing the fast proton transport, emphasizing the usefulness of tailoring quantum effects in nanochannels in practical applications [119].

11.4 Unique Mass-Transport Properties in 2D Membranes: Structural Aspects

The above discussions may explain the reason why many researchers have recently dedicated themselves to seeking a thinner membrane material governed by new transport mechanisms, especially for molecular separation applications. Undoubtedly, 2D materials are just the right platform to develop next-generation membranes

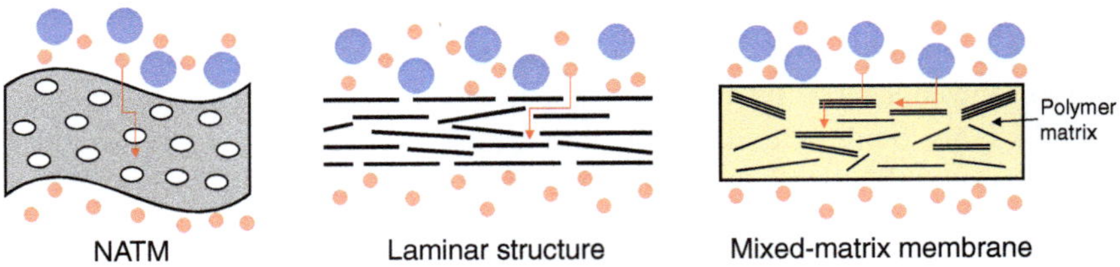

Figure 11.13 Typical structures of 2D material-based membranes. Here, NATM stands for nanoporous atomically thin membranes.

due to their atomic thickness, nanofluidic transport properties, and excellent physicochemical stability [37]. As illustrated in Figure 11.13, there are three main categories of 2D materials-based membranes—(i) NATMs, (ii) staked 2D membranes with laminar structure, and (iii) 2D materials-embedded MMMs [91]. For all of these membranes, basically, molecular transport is achieved through nanopores in the 2D basal plane and/or nanochannel between the staked 2D layers. That is, the molecular transport through 2D membranes is strongly influenced by the membrane morphologies or geometries, such as pore size and density in the basal plane, the number of layers, flake size, stacked manner, and dispersion quality, which can be controlled by various synthesis and functionalization methods [92]. This section accounts for unique mass-transport properties in 2D material-based membranes, with particular emphasis on the role of structural properties, and discusses how these mechanisms could enable the 2D membranes to display excellent separation abilities of molecular mixtures far superior to those of conventional membrane materials. Since the well-established principles of transport mechanisms, such as solution–diffusion model, molecular sieving, and nanofluidic phenomena, are still applicable to 2D materials-based membranes, reviewing several essential examples of each 2D membrane structure reported to date are focused.

11.4.1 Nanoporous Atomically Thin 2D Membranes (NATMs)

NATMs consist of a monolayer of a 2D material with in-plane nanopores formed by intrinsic defects during the synthesis stage or artificial pore generation. Not surprisingly, this form of 2D membranes is pursued as an ideal structure to maximize the flux of the membrane system, provided that the pore size or distribution is enough to allow selective transport of the molecular mixture [3]. Transport in NATMs can be satisfactorily explained by the mechanisms introduced in Section 11.3.3, including steric effects and activated transport models. In particular, the selectivity of NATMs largely depends on their pore size distribution since transport is more pronounced in a larger pore which offers a lower resistance. Thus, for fast and selective transport in NATMs, it is critical to control uniform pore size distribution as well as the appropriate pore size in 2D materials close to that of desired species. Furthermore, functional groups and surface charges at the pore rim can also affect the separation performance of NATMs through specific interactions [30].

A number of theoretical studies have reported that ultrahigh selectivities (several orders of magnitude higher than conventional membranes) can be achieved in graphene and other 2D materials-based NATMs given that the pore size is in the activated regime [3]. Despite the promising potential predicted by simulation studies, for gas separation, most of the NATMs exhibited a low gas selectivity governed by the Knudsen regime, particularly when the membrane size is enlarged to a centimeter-scale for practical interest. Boutilier et al. suggested that an optimal support layer should be chosen by considering the intrinsic defects in macroscopic areas of graphene, which is matched to the resistance of selective pores in graphene membrane to mitigate nonselective leakage while maintaining gas permeance [120]. To minimize the defects during membrane fabrication, AAO support was mechanically polished and then monolayer graphene grown by CVD was transferred to the AAO support. Subsequently, ion bombardment was applied to the graphene layer to control the pore density. Notably, the prepared graphene membrane showed a higher He/SF_6 selectivity compared to the corresponding Knudsen selectivity, reflecting the contribution of molecular-sieving-based gas transport (Figure 11.14a). Based on the resistance model (Figure 11.14b), considering the areal fraction of tear-free graphene layer (γ) and resistance of selective pores (R_s) and AAO support (R_{AAO}), a permeance model was proposed to predict how much the selective pores contribute to the gas transport based on the mean (μ) and standard deviation (σ) of a log-normal pore size distribution (Figure 11.14c) together with the steric exclusion model (Figure 11.14d). Ultimately, the experimental data is proven to be consistent with the model prediction and the normalized selectivity to remove the portion of Knudsen diffusion demonstrated its molecular-sieving nature (Figure 11.14e,f). Moreover, the extracted gas-separation performance of graphene layer even exceeded the Robeson limit for H_2/CO_2 separation, proving the excellent potential of graphene-based NATMs for gas separation.

The transport of water molecules is largely influenced by the pore size of NATMs by combining hydrogen bonding and osmotic effects. A defect-engineering approach was reported by Jang et al. introducing a defect-healing step in macroscale via interfacial polymerization of two monomer phases sandwiching a graphene layer [121]. They found that the experimental water and solute transport properties through the graphene NATM, whose pore size was controlled by ion bombardment and plasma etching, could be theoretically predicted from a model based on the no-slip hydrodynamics, including contributions from Sampson flow and osmotic pressure. Beyond the aqueous solutions, Cheng et al. have recently demonstrated that the graphene NATMs follow the molecular size-dependent transport properties of various organic solvents and solutes, exhibiting an excellent separation performance in OSN processes [105]. The authors explained the subcontinuum molecular geometry realized by the delicate fabrication of graphene membranes is responsible for the selective transport of liquid molecules, deviating from the viscous-dominated transport in dense membranes.

Electrostatic interactions are indispensable to explain the ion-transport properties across NATMs. The pore rim can be functionalized with charged groups, and both steric exclusion and Donnan effects will coexist and affect ion selectivity. Cheng et al. prepared a centimeter-scale graphene NATM via a size-selective defect-sealing strategy using polyhedral-oligomeric-silsesquioxane as a filler material [122]. In this

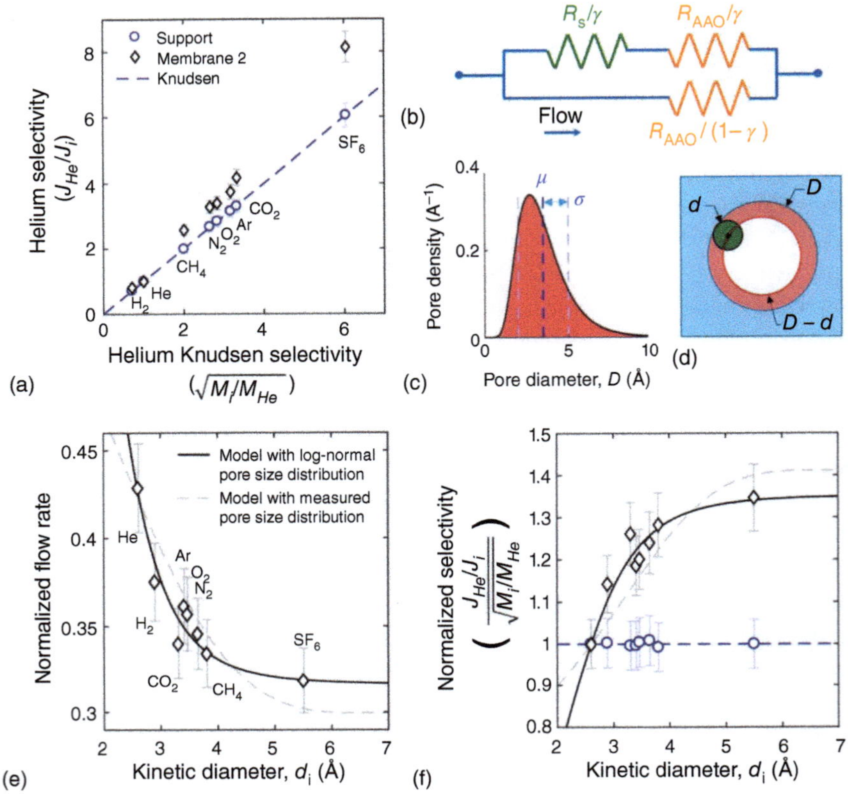

Figure 11.14 Gas transport properties of graphene-based NATM coated onto porous AAO support. (a) Relative selectivity to Helium plotted with the corresponding Knudsen selectivity. Schematic illustrations of (b) resistance model, (c) nominal pore size distribution, (d) effective pore size (D-d) when a gas molecule with a kinetic diameter of d permeate through pore diameter D. (e) Normalized gas flow rates compared to that of AAO support before graphene transfer, compared with the fitted models. (f) Normalized selectivity to the corresponding Knudsen selectivity depending on the kinetic diameter of gas molecules. Source: Boutilier et al. [120]/with permission of American Chemical Society.

system, A sharp solute exclusion cut-off was observed with ~23 times higher water permeance than commercial nanofiltration membranes. Notably, a very high monovalent ion rejection could be achieved due to the steric effects by the exceptionally small nanopore size down to 0.5 nm compared to the tested monovalent ions (K^+: 0.662 nm, Na^+: 0.716 nm, and Cl^-: 0.664 nm, respectively). On the other hand, if the hydrated ion diameter exceeds the size of a nanopore in NATM, one should consider that an energetic barrier to ion transport may occur for its dehydration that depends on the properties of ions (e.g. size and charge number) and electric polarizability of 2D materials constituting NATMs (Figure 11.15a) [123, 124]. Fu et al. demonstrated that ion dehydration serves as the main mechanism for selective ion transport through graphene NATMs, evidenced by both experimental results and theoretical calculations [123]. It was found that the monovalent cations could more easily pass through the graphene nanopore than the divalent cations due to their

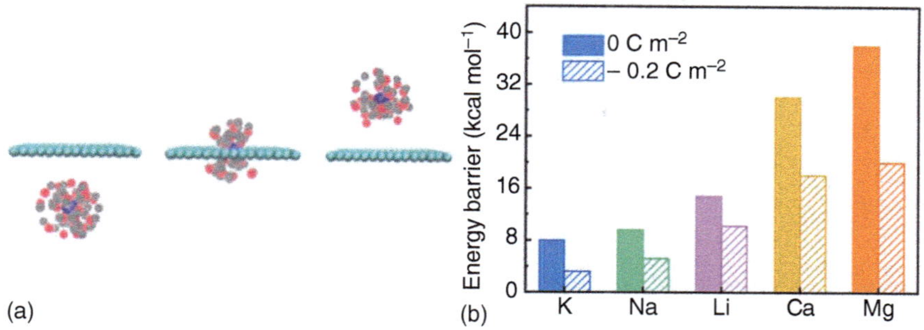

Figure 11.15 (a) Schematic illustration of ion-transport properties through nanoporous graphene monolayer via dehydration. (b) Free-energy barrier for cations to permeate through the graphene nanopore calculated by molecular dynamics simulation. Source: Fu et al. [123]/with permission of American Chemical Society.

smaller energy barriers for dehydration that can be further reduced by granting negative charges (Figure 11.15b).

Unlike any other ions, proton shows distinct transport behaviors. That is, it has been unveiled that through-plane proton transport is actually possible in intralattice of 2D materials depending on the activation energy barrier governed by quantum oscillations [118]. Mogg et al. discovered a perfect Nernst selectivity toward protons in a monolayer of exfoliated graphene and hBN crystal [125]. Intriguingly, although the hBN membrane showed no microscopic defects, confirmed by the gas leakage test, only the proton transport could be achieved while no permeation of counter-ions was detectable. The results also support the activation models applied to explaining the anomalous proton transport through high-quality 2D materials and indicate more detailed transport theories should be developed to utilize this unique transport mechanism in proton selective membranes.

11.4.2 Staked 2D Membranes with Laminar Structure

Until now, 2D materials-based NATM is still in its infancy due to the very laborious fabrication protocols and the difficulties in defect control on an atomic scale. To circumvent these challenges in NATMs, researchers have proposed a new type of 2D membrane, that is, assembling the dispersed nanosheets into a stacked structure with interlayer nanochannels for diffusion pathway, designated as 2D laminar membranes [32, 34, 35]. These membranes can be prepared using various coating methods, such as vacuum filtration, dip coating, spin coating, spray coating, and bar coating (Figure 11.16a), some of which are industrially scalable [126]. The physicochemical properties of 2D laminar membranes are highly related to their fabrication method used [37, 126]. Generally, a slower assembly of nanosheets will result in more defect-free and dense membranes, and vice versa [127]. Thus, rational choice of coating conditions is desired to achieve desired separation abilities and scalability of resultant membranes.

It is widely accepted that molecular transport through 2D laminar membranes occurs in the structural defects on the basal plane of 2D materials at an initial stage,

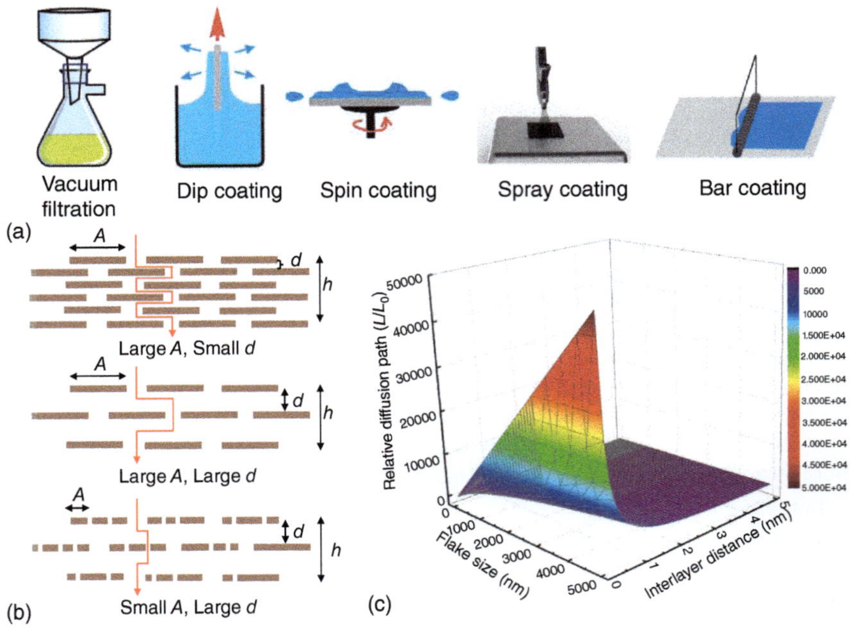

Figure 11.16 (a) Preparation methods of 2D laminar membranes. (b) Schematic illustration of diffusion path of molecules depending on flake size (*A*) and interlayer distance (*d*) at a given membrane thickness (*h*). (c) A plot of relative diffusion path (L/L_0) with variables of flake size and interlayer distance.

and subsequently, the molecules diffuse across their interlayer channels. Hence, the manipulation of the size and functionalities of pore entrance and nanochannels is expected to improve the separation performance of the laminar membranes. From a structural point of view, the stacked structure of 2D laminar membranes makes the effective diffusion path (L_{eff}) long and tortuous, which can be simply described as follows [32, 128]:

$$L_{eff} = A \times \frac{h}{d} \tag{11.32}$$

where A is the average flake size of 2D nanosheets, h is the membrane thickness, and d is the interlayer distance, respectively. Namely, at a given membrane thickness, a faster molecular transport can be obtained when the laminar membrane possesses a smaller flake size with a larger interlayer distance that provides a less tortuous diffusion pathway, as depicted in Figure 11.16b. To quantify these effects, the relative diffusion path (L/L_0) is calculated, the ratio of effective diffusion path of the modified 2D laminar membrane to that of control membrane with a fixed flake size (A_0) and interlayer distance (d_0) at a constant membrane thickness, given by:

$$L/L_0 = \frac{A}{A_0} \times \frac{d_0}{d} \tag{11.33}$$

Notably, Figure 11.16c captures the importance of structural parameters in 2D laminar membranes, which indicates that the diffusion path becomes much lengthier up to four orders of magnitude at a typical range of flake size (a few μm) and interlayer distance (~1 nm) of interest. Therefore, it is necessary to develop effective strategies to reduce the tortuosity of laminar membranes toward the high productivity of membrane systems [30, 129, 130]. Meanwhile, one should also consider that an excessive reduction in flake size or excessive enlargement of interlayer distance may cause insufficient packing of 2D sheets and thereby accompanies a loss in selectivity or failure to form membranes due to the lack of mechanical robustness [131, 132].

2D laminar membranes have gained much attention after a pioneer work reported by Nair et al. in 2012 [32]. The authors discovered the micrometer-thick, freestanding GO membranes are perfectly impermeable to various organic solvents and even small gases, including helium while allowing ultrafast transport of water vapor which is four to five orders of magnitude higher than that of small gases. The extraordinary water vapor permeation rate is ascribed to the unimpeded water flow through the hydrophobic sp^2 hybrid carbon regions, while oxygen-functional groups in sp^3 carbon regions act as the spacer to allow the intercalation of water molecules inside the nanochannels of the GO membrane.

In 2013, a counterintuitive work reported by Kim et al. demonstrated that the GO membrane has excellent potential to separate small gas molecules, and its separation performance can be tuned by controlling the stacking manner of GO sheets [34]. They fabricated two types of ultrathin (3–10 nm) GO membranes by coating the aqueous GO solution on the porous polyethersulfone (PES) support. One was prepared by contacting the PES support with GO dispersion, followed by spin coating, resulting in a less interlocked GO membrane due to the preferential face-to-face stacking of GO sheets (Figure 11.17a). This membrane showed a typical Knudsen diffusion whose permeance order follows $H_2 > He > CH_4 > N_2 > O_2 > CO_2$, while a high H_2/CO_2 selectivity (~30) is totally unexpected from the corresponding Knudsen selectivity (~4.7) (Figure 11.17b). This was attributed to the CO_2-philic carboxylic acid groups at the edges of the nanopores created by less interlocked GO sheets, strongly binding with CO_2 molecules and thus leading to low CO_2 permeance. At a hydrated state, the presence of water molecules solely improved the dissolution of CO_2 molecules rather than the noncondensable small gases, leading to 50 times higher CO_2 permeance than that of GO membrane in a dry state. The other type of GO membrane was prepared by directly dropping a given amount of GO dispersion during the spin-coating process, resulting in a highly interlocked GO membrane governed by capillary interactions between GO sheets (Figure 11.17c). In this case, the GO membrane displayed a gas permeance order of $CO_2 > He > H_2 > CH_4 > O_2 > N_2$, significantly deviating from the former GO membrane, and CO_2/N_2 selectivity of ~20 was achieved (Figure 11.17d). Here, the densely packed GO layers are responsible for the selective CO_2 transport by a water-enhanced separation mechanism that preferentially sorbs and transports CO_2 than other gases, evidenced by its relative humidity (RH)-dependent CO_2 separation efficiency.

The above two important examples by Nair et al. and Kim et al. showing opposite transport behavior of GO-based laminar membranes, may arouse some questions.

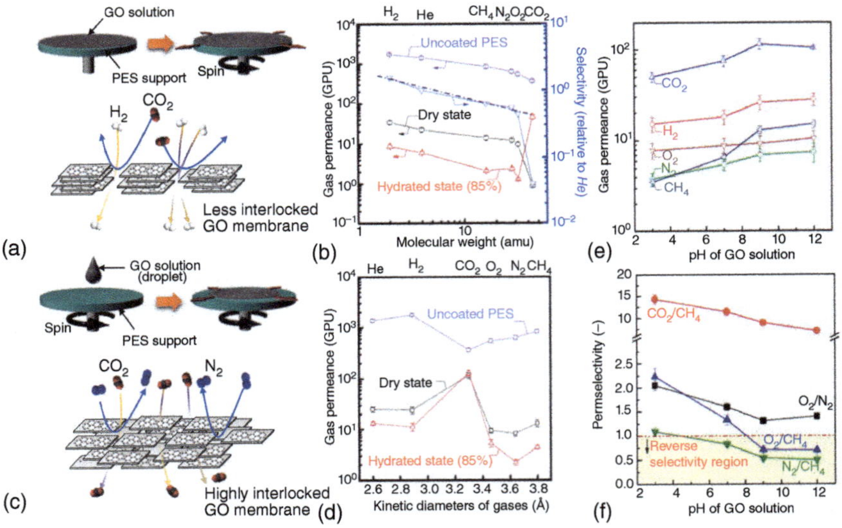

Figure 11.17 (a) Schematic illustration of gas transport pathway and (b) the corresponding separation performances of less interlocked GO membranes (dashed line indicates the ideal Knudsen selectivity). (c) Schematic illustration of gas transport pathway and (d) the corresponding separation performances of highly interlocked GO membranes. Gas permeation data. (e) Gas permeance and (f) permselectivity of the GO membranes as a function of the pH of the GO dispersion. Panel a–d. Source: Kim et al. [34]/with permission of American Association for the Advancement of Science. Panel e–f. Source: Roh et al. [133]/with permission of American Chemical Society.

For example, what makes the difference between them? While the answer is still on the debate, these results indicate that the molecular transport in 2D laminar membranes may vary to a significant extent depending on the flake size, thickness, stacking manner, degree of functionalization, amount of intercalated water, and even testing conditions [37, 91]. As a follow-up study, Kim and coworkers attempted to further control the tortuosity and interlayer distance of GO membranes by adjusting the pH of GO dispersion before the spin-coating process [133]. A higher solution pH resulted in a more pronounced repulsion between the GO sheets, confirmed by more increased negative zeta potential, enlarging the interlayer distance of the resultant GO membranes. Consequently, the gas permeance of the GO membrane was enhanced by increasing the pH of the GO solution due to the reduced tortuosity for gas diffusion (Figure 11.17e). Meanwhile, the relative permeance enhancement shows considerable differences depending on the tested gas molecule, especially for CH_4, resulting in a reverse selectivity (<1) for N_2/CH_4 pair (Figure 11.17f). The authors attributed this behavior to the highest kinetic diameter of CH_4 among the other gases that possibly make CH_4 is more sensitive to the enlargement of channel size in GO membranes. Similarly, Huang et al. reported that the creation of nanopores in GO sheets by hydrogen peroxide treatment could significantly improve the molecular-sieving abilities of the GO membrane by offering a shorter diffusion path [129].

In general, liquid and ion transport in 2D laminar membrane shares the same principles adopted to transport mechanism in gas separation, apart from the fact that the interlayer spacing can be expanded significantly depending on the extent of the solvent-sheet interactions. Joshi and coworkers first investigated the permeation of water, organic solutes, and ions with different charges through freestanding GO membranes [118]. Ultrafast water and ion permeation was observed, showing several times higher diffusion coefficients than those of bulk fluid, which may be explained by a capillary-like high pressure exerted inside the GO nanocapillaries. They found that sharp molecular sieving of ions or molecules was possible, governed by the physical size of penetrants with a cutoff radius of 4.5 Å. The results were ascribed to the pore size of the hydrated GO membrane (~9–10 Å), calculated by taking into account the effective thickness of the graphene layer (3.4 Å), which is perfectly fitted for discriminating the molecules or ions with a radius above 4.5 Å. However, the assumptions in this study are too simplified to consider the complex chemistry and structural aspects of the real GO membrane and the calculation of effective length based on the total tortuous path is not practical to estimate the separation performance of membranes [134]. To more clarify the ambiguous transport mechanism in GO membranes, Cho et al. revisited the water and ion-transport properties in GO membranes within the framework of the classical solution–diffusion model. A very high water sorption coefficient (0.93) (or partition coefficient) of the GO membrane was found, possibly ascribed to its abundant hydrophilic functional groups, which is much higher than that of common hydrophilic polymers used in water purification (less than 0.6). This allows a significant amount of ions to be sorbed which follows the size of hydrated ions and thus affects the ion permeability of the swollen GO membrane (Figure 11.18a). Likewise, the GO membrane exhibited higher salt sorption coefficients than those of hydrophilic polymers, such as a hydrogel consisting of cross-linked polyethylene glycol (XLPEGDA) (Figure 11.18b). Despite the high water and ion sorption in the GO membrane, the highly tortuous structure causes the GO membrane to possess a very low diffusion coefficient and relatively poor water/NaCl separation performance compared to other polymeric membrane materials (Figure 11.18c). Nevertheless, the authors emphasized that there are many opportunities to improve the performance given the excellent thin-film manufacturing capabilities of GO membranes and many physical and chemical modification methods to control the tortuous diffusion path. In this context, Chen et al. fabricated reduced holey GO membranes having controlled surface porosity as well as the interlayer distance of nanochannels (Figure 11.18d) [130]. By adjusting these structural parameters, both water permeance and salt rejection were considerably enhanced (Figure 11.18e), which reaches the level of the commercial nanofiltration membranes based on polyamide chemistry. Again, this work demonstrates the excellent potential of GO membranes for water purification if their structures are optimized for the favorable permeation of desired species.

Beyond the graphene-based materials, there are plenty of 2D material candidates, such as hBN, Mxene, MoS_2, MOF, and COF nanosheets, which serve as a next-generation platform toward advanced 2D laminar membranes. Notably, these materials often overcome the limitations of graphene laminar membranes based

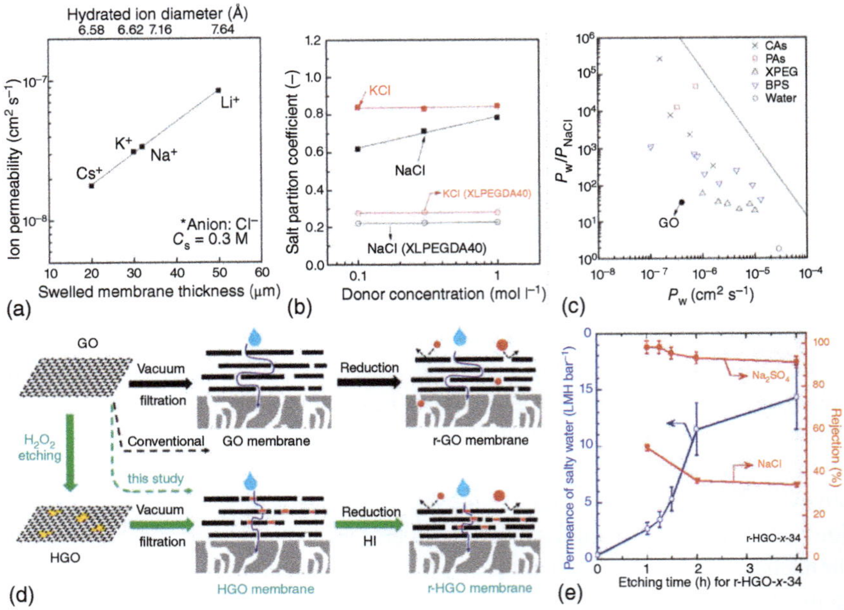

Figure 11.18 (a–c) Fundamental water and ion-transport properties of freestanding GO membranes. (a) Effects of hydrated ion diameter in membrane thickness and corresponding ion permeability, (b) effects of salt concentration in salt sorption (or partition) coefficient, and (c) water/NaCl separation performance plotted with upper bound. (d–e) Nanofiltration by reduced holey graphene oxide membranes. (d) Schematic illustration of the preparation of reduced holey GO membranes (designated as r-HGO-*x*-*y*, where *x* is the etching time for pore generation and *y* is the concentration of hydroiodic acid (HI) for reduction) and (e) nanofiltration performance of prepared r-HGO membranes depending on the etching time. Panel a–c. Source: Cho et al. [93]/with permission of Elsevier Ltd. Panel d–e. Source: Chen et al. [130]/with permission of American Chemical Society.

on the different physicochemical properties coupled with the unique structure that influence the separation mechanisms, which includes—(i) a narrower interlayer distance that maximizes selectivity [135, 136], (ii) resistance to water-induced swelling to improve ion selectivity and mechanical robustness [137–139], and (iii) intrinsically nanoporous basal plane that directly allows the fast transport in through-plane direction [140–142].

11.4.3 2D Materials-Embedded Mixed-Matrix Membranes (MMMs)

Although 2D materials-based NATMs and laminar membranes have exhibited interesting transport properties and promising potential as a novel, high-performance membrane class, their scale-up fabrication is still very challenging, and thus more intensive studies are required to resolve this issue at present. Alternatively, incorporating 2D sheets into a polymeric matrix to produce MMMs is a simple and rational strategy to exploit the unique transport properties of 2D materials as well as the good film processability of polymers [4]. Compared to typical 3D filler materials, at the same filler concentration, 2D fillers offer a much higher interfacial area with polymer matrix due to their high-aspect ratio and thereby more efficiently influence the separation performance of MMMs depending on the orientation and dispersion quality, the filler-matrix interactions, pore density, and surface chemistry of 2D fillers [30]. These effects will not be considered here since the transport mechanism in 2D membranes are focused, not the general reviews on MMMs, and readers who are interested in more details on MMMs are referred to the following chapters in this book.

Theoretical predictions of the gas permeability of MMMs provide useful guidance to not only rational matching of filler and matrix materials but also estimating the intrinsic separation performance of fillers based on the "back-calculation" method [24]. To date, different modeling approaches have been developed, which can be roughly classified by the morphologies of MMMs, such as ideal and nonideal MMMs. An ideal MMM can be regarded as a binary phase system that consists of dispersed fillers and polymer matrices without any interfacial defects or matrix rigidification. On the other hand, a nonideal MMM should be considered as a multiphase system comprising more than three components, which follows the kinds of filler–matrix interactions. These nonideal morphologies include interfacial voids, sieve-in-a-cage, rigidified polymer chains surrounding the dispersed fillers, and pore blockage by fillers [25].

Current models for transport through MMMs are mostly based on the resistance model approach and the effective medium approach, which are analogous to the conventional theories for explaining the thermal or electrical conductivity of composite materials. These approaches can be used to calculate the effective permeability of the MMM (P_{MMM}) using the known information of constituting phases, such as the dispersed filler at a given volume fraction (ϕ_{f}) and the continuous matrix, denoted as f and c, respectively [143].

The resistant model approach is originated from Ohm's law for a series/parallel array of resistors and Fick's law for mass flux through a membrane. The model states that the permeability (P_i) is inversely proportional to the transport resistance (R_i) of given material i, which can be expressed as:

$$R_i = \frac{l}{P_i A} \tag{11.34}$$

where l is the membrane thickness and A is the effective area of the membrane, respectively. The simplest resistance model considers a two-phase laminated composite which comprises alternating sheets of matrix and filler material assembled into the series or parallel connection to the flow direction (Figure 11.19a).

For MMMs with series configuration:

$$\frac{1}{R_{\mathrm{MMM}}} = \frac{1}{R_{\mathrm{c}}} + \frac{1}{R_{\mathrm{f}}} \tag{11.35}$$

$$P_{\mathrm{MMM}} = \frac{P_{\mathrm{c}} P_{\mathrm{f}}}{P_{\mathrm{c}} \phi_{\mathrm{f}} + P_{\mathrm{f}}(1 - \phi_{\mathrm{f}})} \tag{11.36}$$

For MMMs with parallel configuration:

$$R_{\mathrm{MMM}} = R_{\mathrm{c}} + R_{\mathrm{f}} \tag{11.37}$$

$$P_{\mathrm{MMM}} = P_{\mathrm{f}} \phi_{\mathrm{f}} + P_{\mathrm{c}}(1 - \phi_{\mathrm{f}}) \tag{11.38}$$

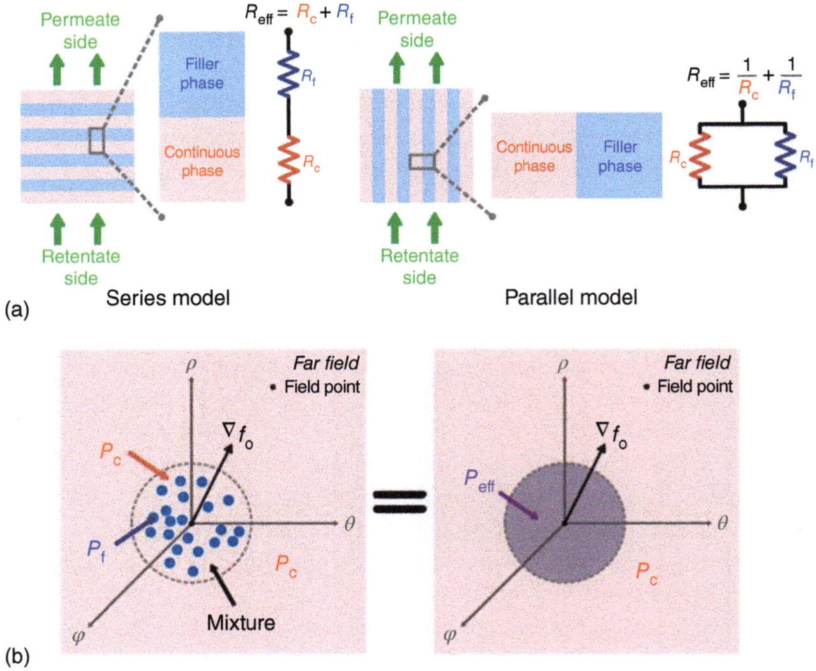

Figure 11.19 Schematic illustration of (a) resistance model approach assuming a series/parallel multilayer composite and (b) effective medium approach based on Maxwell's theory. A composite sphere consisting of spherical fillers (phase f) in a matrix (phase c) is assumed to be equal to an infinite matrix of phase c. Source: Monsalve-Bravo et al. [143]/MDPI/CC BY-4.0.

Thus, these two-phase resistance models can be used to infer the orientation or connectivity of the dispersed fillers inside the polymer matrix [27]. Note that the series model in Eq. (11.36) provides the lower limit for the permeability of a penetrant through an ideal MMM, while the parallel model in Eq. (11.38) provides that of the upper limit.

The main framework behind the effective medium approach is substituting a given composite system with an imaginary homogeneous medium having equivalent properties of the composite. These models generally consider randomly distributed spherical fillers, which differs from the resistance models that assume regular distributions of platelet or cubic fillers. In 1873, Maxwell proposed a way to calculate the electrical conductivity of a composite consisting of infinitely diluted, homogeneously distributed spherical fillers and an infinite matrix. The calculation is based on the assumption that the far-field potential of filler is identical to that of a homogeneous sphere with the volume of the original composite, as described in Figure 11.19b. Wagner and Sillars further extended the Maxwell model to spheroids. The permeability of a MMM can be expressed by the Maxwell–Wagner–Sillars model [144]:

$$P_{\mathrm{MMM}} = P_{\mathrm{c}} \left[\frac{nP_{\mathrm{f}} + (1-n)P_{\mathrm{c}} - (1-n)\phi_{\mathrm{f}}(P_{\mathrm{c}} - P_{\mathrm{f}})}{nP_{\mathrm{f}} + (1-n)P_{\mathrm{c}} + n\phi_{\mathrm{f}}(P_{\mathrm{c}} - P_{\mathrm{f}})} \right] \tag{11.39}$$

where n is the particle shape factor within the range of $0 < n < 1$. The two extremes of Eq. (11.39) lead to a parallel bilayer model (Eq. (11.38)) when n is zero or a series bilayer model (Eq. (11.36)) when n is unity, respectively. For a limit of $n = 1/3$, Eq. (11.39) is reduced to the following equation, which is originally developed by Maxwell to consider a dilute dispersion of spherical fillers in a composite material:

$$P_{\mathrm{MMM}} = P_{\mathrm{c}} \left[\frac{P_{\mathrm{f}} + 2P_{\mathrm{c}} - 2\phi_{\mathrm{f}}(P_{\mathrm{c}} - P_{\mathrm{f}})}{P_{\mathrm{f}} + 2P_{\mathrm{c}} + \phi_{\mathrm{f}}(P_{\mathrm{c}} - P_{\mathrm{f}})} \right] \tag{11.40}$$

Equation 11.40 is called the Maxwell model and has been widely adopted in the theoretical and experimental studies on the permeability of MMMs. The Maxwell model is applicable when ϕ_{f} is less than about 0.2 since the assumption behind this model states that the streamlines around the fillers are not affected by the existence of neighboring ones [144]. The Maxwell model is effective to predict the permeability of MMM possessing an ideal morphology without any defects and changes in the separation performance of each matrix and filler phase. For nonideal cases, this model must be corrected to account for the morphological transitions, such as interfacial defects, rigidification of polymer chains, and pore blockage.

The above models have been widely used in permeability studies in MMMs containing sphere-like 3D fillers. When a filler is not spherical and exhibits different structural properties, such as aspect ratio, dimension, and agglomeration of embedded fillers, especially for 2D fillers, modification of the original models is necessary to account for these parameters. The Lewis–Nielsen model, originally developed for calculating an elastic modulus of the composite system, is the most popular method to consider the packing-related effects of fillers in MMMs, given by [143]:

$$P_{\mathrm{MMM}} = P_{\mathrm{c}} \left[\frac{1 + 2\phi_{\mathrm{f}}(b-1)/(b+2)}{1 - \psi\phi_{\mathrm{f}}(b-1)/(b+2)} \right] \tag{11.41}$$

$$\psi = 1 + \left(\frac{1 - \phi_{\mathrm{m}}}{\phi_{\mathrm{m}}^2} \right) \phi_{\mathrm{f}} \tag{11.42}$$

where b is the permeability ratio of filler to a continuous matrix ($P_{\mathrm{f}}/P_{\mathrm{c}}$), ϕ_{m} is the maximum volume fraction of packed fillers and is known to be equal to 0.64 for randomly packed uniform spheres. ϕ_{m} relies on the particle shape, particle size distribution, and degree of aggregation of embedded fillers over the range of $0 < \phi_{\mathrm{f}} < \phi_{\mathrm{m}}$. Note that the Lewis–Nielsen model (Eq. (11.41)) approaches the Maxwell model (Eq. (11.40)) when $\phi_{\mathrm{m}} \to 1$.

Another model proposed by Cussler is applied to a semidilute dispersion of high-aspect ratio platelets with a staggered array, expressed as follows:

$$P_{\mathrm{MMM}} = P_{\mathrm{c}} \frac{1}{1 - \phi_{\mathrm{f}} + 1 \Big/ \left(\frac{P_{\mathrm{f}}}{\phi_{\mathrm{f}} P_{\mathrm{c}}} + 4 \frac{1 - \phi_{\mathrm{f}}}{\alpha^2 \phi_{\mathrm{f}}^2} \right)} \tag{11.43}$$

where α is the aspect ratio of dispersed filler.

For a composite system containing impermeable filler material in barrier applications, the modified Nielsen model satisfactorily predicts the permeability of MMM at very dilute concentrations by considering both aspect ratio and alignment of embedded filler, written as [145]:

$$P_{\mathrm{MMM}} = P_{\mathrm{c}} \frac{1 - \phi_{\mathrm{f}}}{1 + \frac{\alpha}{3}\left(S + \frac{1}{2}\right)\phi_{\mathrm{f}}} \tag{11.44}$$

$$S = \frac{1}{2}\langle 3\cos^2\theta - 1\rangle \tag{11.45}$$

where S represents the extent of filler alignment inside the composite when θ is the angle between the unit vectors of preferred orientation (n) and the sheet normal (p) directions. Here, $S = 0$ for randomly distributed fillers and $S = 1$ for perfectly aligned fillers perpendicular to the mass-flow direction. Figure 11.20 illustrates the effects of orientation of 2D filler by changing the length (L) of platelets at a fixed width (W) and volume fraction, showing that P_{MMM} significantly reduces as increasing the aspect ratio and horizontal orientation of embedded filler [145]. By comparing with experimental permeability, Eq. (11.44) provides useful information on the packed morphologies and alignment properties of 2D fillers inside the MMM. These barrier

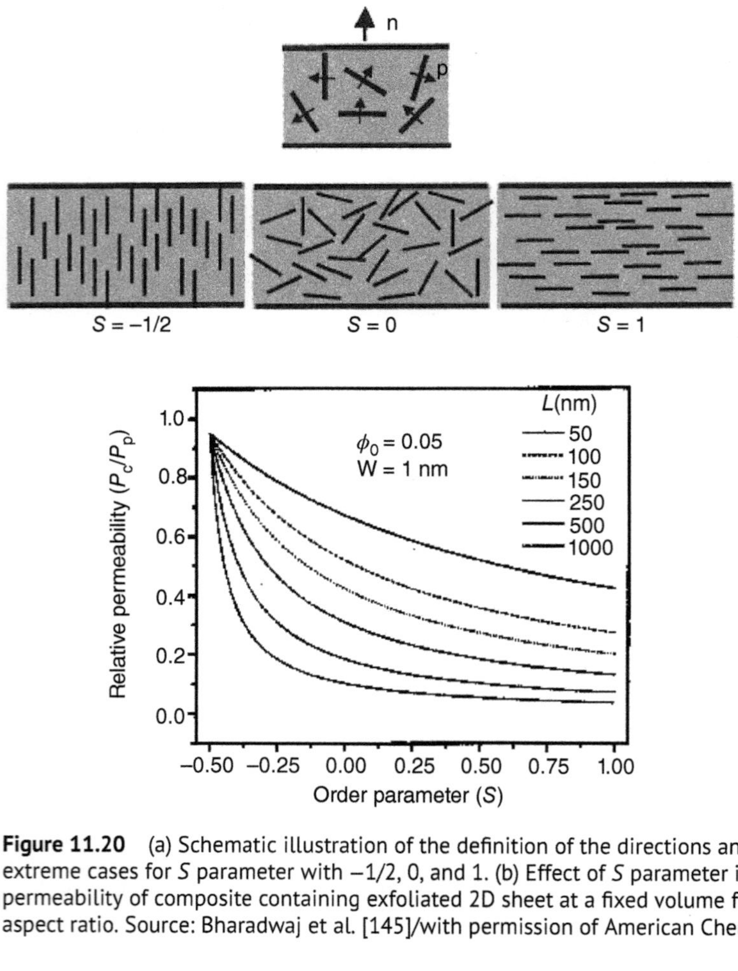

Figure 11.20 (a) Schematic illustration of the definition of the directions and three extreme cases for S parameter with −1/2, 0, and 1. (b) Effect of S parameter in relative permeability of composite containing exfoliated 2D sheet at a fixed volume fraction and aspect ratio. Source: Bharadwaj et al. [145]/with permission of American Chemical Society.

properties have been demonstrated by many studies on MMMs containing 2D silicates, clays, GO, and exfoliated graphite incorporated into various polymer matrices [33]. On the other hand, recent studies have shown that porous 2D materials having intrinsic pores onto the basal planes, including porous graphene oxide (PGO), and nanosheets of zeolites, MOFs, and COFs, are more likely to improve the permeability by reducing the tortuosity and to increase the selectivity due to the molecular-sieving effects by the surface nanopores [30, 31, 91].

11.5 Summary and Outlook

In this chapter, transport phenomena in 2D materials-based membranes have been reviewed, starting from the classical transport theories to the current knowledge accumulated by many theoretical and experimental studies. Mass transport in membranes can be largely classified by pore size and electrostatic charge, explained by the pore-flow model, the solution–diffusion model, and the effects related to electrochemical potential (e.g. Donnan effects). To overcome the limitations governed by the trade-off relations in conventional polymer membranes, tremendous efforts have been dedicated to seeking novel membrane materials and new transport mechanisms, mostly explored in a nanofluidic system established by 0D, 1D, and 2D-confined architectures. Continuum modeling by conventional hydrodynamics (e.g. the Navier–Stokes equation) can be used to satisfactorily estimate the nanofluidic phenomena in confined 1D nanochannels or 2D nanoslits, exhibiting channel-size-dependent transport properties which are affected by the extent of slip flow. Depending on the pore size, transport through 0D nanopores can be described by classical Sampson's flow model, steric exclusion, and activated transport theory. Ion transport through the confined dimension is highly dependent on the size and charge density of the nanochannel, which determines the ion selectivity that can be explained by electrical double-layer theories and the Debye length overlapping regime. Beyond the continuum descriptions, there are diverse confinement effects in nanofluidics transport, including discrete modeling, molecular rearrangement, dehydration of ions, electrical effects, and quantum effects, which significantly change the mass-transport properties in nanofluidic systems compared to those in the bulk state. Finally, the unique transport properties in 2D membranes are explored by classifying them into NATMs, laminar membranes, and MMMs. Transport mechanisms in each 2D membrane configuration are discussed with several important examples, with the emphasis on their morphological aspects, such as pore density, pore size, aspect ratio, tortuosity effects, stacked manner, and dispersion quality. If these structural properties are precisely controlled, 2D membranes often show far superior separation performances compared to those of conventional membrane materials by overcoming the trade-off limitations due to their unique mass-transport properties. We believe that the contents described here may help the readers to understand the transport mechanism of 2D materials-based membranes toward rational design of next-generation membranes for real applications.

References

1 Baker, R.W. (2004). *Membrane Technology and Applications*. Wiley.
2 Mulder, M. (1996). *Basic Principles of Membrane Technology*. Kluwer Academic Publishers.
3 Wang, L.D., Boutilier, M.S.H., Kidambi, P.R. et al. (2017). Fundamental transport mechanisms, fabrication and potential applications of nanoporous atomically thin membranes. *Nat. Nanotechnol.* 12 (6): 509–522.
4 Park, H.B., Kamcev, J., Robeson, L.M. et al. (2017). Maximizing the right stuff: the trade-off between membrane permeability and selectivity. *Science* 356 (6346), eaab0530.
5 Sun, P.Z., Zheng, F., Wang, K.L. et al. (2014). Electro- and magneto-modulated ion transport through graphene oxide membranes. *Sci. Rep.* 4 (1): 6798.
6 Prozorovska, L. and Kidambi, P.R. (2018). State-of-the-art and future prospects for atomically thin membranes from 2D materials. *Adv. Mater.* 30 (52), 1801179.
7 Fick, A. (1995). On liquid diffusion (reprinted from the London, Edinburgh, and Dublin Philosophical Magazine and Journal of Science, Vol 10, Pg 30, 1855). *J. Membr. Sci.* 100 (1): 33–38.
8 Malaeb, L. and Ayoub, G.M. (2011). Reverse osmosis technology for water treatment: state of the art review. *Desalination* 267 (1): 1–8.
9 Mohammad, A.W., Teow, Y.H., Ang, W.L. et al. (2015). Nanofiltration membranes review: recent advances and future prospects. *Desalination* 356: 226–254.
10 Marchetti, P., Solomon, M.F.J., Szekely, G., and Livingston, A.G. (2014). Molecular separation with organic solvent nanofiltration: a critical review. *Chem. Rev.* 114 (21): 10735–10806.
11 Gao, W., Liang, H., Ma, J. et al. (2011). Membrane fouling control in ultrafiltration technology for drinking water production: a review. *Desalination* 272 (1–3): 1–8.
12 Mehta, A. and Zydney, A.L. (2005). Permeability and selectivity analysis for ultrafiltration membranes. *J. Membr. Sci.* 249 (1–2): 245–249.
13 Anis, S.F., Hashaikeh, R., and Hilal, N. (2019). Microfiltration membrane processes: a review of research trends over the past decade. *J. Water Process Eng.* 32.
14 Yoo, M.J., Kim, K.H., Lee, J.H. et al. (2018). Ultrathin gutter layer for high-performance thin-film composite membranes for CO_2 separation. *J. Membr. Sci.* 566: 336–345.
15 Yoo, M.J., Lee, J.H., Yoo, S.Y. et al. (2018). Defect control for large-scale thin-film composite membrane and its bench-scale demonstration. *J. Membr. Sci.* 566: 374–382.
16 Robeson, L.M. (2008). The upper bound revisited. *J. Membr. Sci.* 320 (1–2): 390–400.
17 Sanders, D.E., Smith, Z.P., Guo, R.L. et al. (2013). Energy-efficient polymeric gas separation membranes for a sustainable future: a review. *Polymer* 54 (18): 4729–4761.

18 Lee, T.H., Jung, J.G., Kim, Y.J. et al. (2021). Defect engineering in metal-organic frameworks towards advanced mixed matrix membranes for efficient propylene/propane separation. *Angew. Chem. Int. Ed.* 60 (23): 13081–13088.

19 Park, H.B., Jung, C.H., Lee, Y.M. et al. (2007). Polymers with cavities tuned for fast selective transport of small molecules and ions. *Science* 318 (5848): 254–258.

20 Carta, M., Malpass-Evans, R., Croad, M. et al. (2013). An efficient polymer molecular sieve for membrane gas separations. *Science* 339 (6117): 303–307.

21 Rangnekar, N., Mittal, N., Elyassi, B. et al. (2015). Zeolite membranes – a review and comparison with MOFs. *Chem. Soc. Rev.* 44 (20): 7128–7154.

22 Ying, Y.P., Kamarudin, S.K., and Masdar, M.S. (2018). Silica-related membranes in fuel cell applications: an overview. *Int. J. Hydrogen Energy* 43 (33): 16068–16084.

23 Lee, T.H., Moghadam, F., Jung, J.G. et al. (2021). In situ derived hybrid carbon molecular sieve membranes with tailored ultramicroporosity for efficient gas separation. *Small* 17 (47), 2104698.

24 Koros, W.J. and Zhang, C. (2017). Materials for next-generation molecularly selective synthetic membranes. *Nat. Mater.* 16 (3): 289–297.

25 Qian, Q.H., Asinger, P.A., Lee, M.J. et al. (2020). MOF-based membranes for gas separations. *Chem. Rev.* 120 (16): 8161–8266.

26 Rashed, A.O., Merenda, A., Kondo, T. et al. (2021). Carbon nanotube membranes – strategies and challenges towards scalable manufacturing and practical separation applications. *Sep. Purif. Technol.* 257 (15), 117929.

27 Lee, T.H., Ozcan, A., Park, I. et al. (2021). Disclosing the role of defect-engineered metal-organic frameworks in mixed matrix membranes for efficient CO_2 separation: a joint experimental-computational exploration. *Adv. Funct. Mater.* 31 (38), 2103973.

28 Galizia, M., Chi, W.S., Smith, Z.P. et al. (2017). 50th anniversary perspective: polymers and mixed matrix membranes for gas and vapor separation: a review and prospective opportunities. *Macromolecules* 50 (20): 7809–7843.

29 Dubertret, B., Heine, T., and Terrones, M. (2015). The rise of two-dimensional materials. *Acc. Chem. Res.* 48 (1): 1–2.

30 Moghadam, F. and Park, H.B. (2019). 2D nanoporous materials: membrane platform for gas and liquid separations. *2D Mater.* 6 (4): 042002.

31 Moghadam, F. and Park, H.B. (2018). Two-dimensional materials: an emerging platform for gas separation membranes. *Curr. Opin. Chem. Eng.* 20: 28–38.

32 Nair, R.R., Wu, H.A., Jayaram, P.N. et al. (2012). Unimpeded permeation of water through helium-leak-tight graphene-based membranes. *Science* 335 (6067): 442–444.

33 Yoo, B.M., Shin, H.J., Yoon, H.W., and Park, H.B. (2014). Graphene and graphene oxide and their uses in barrier polymers. *J. Appl. Polym. Sci.* 131 (1): 39628.

34 Kim, H.W., Yoon, H.W., Yoon, S.M. et al. (2013). Selective gas transport through few-layered graphene and graphene oxide membranes. *Science* 342 (6154): 91–95.

35 Joshi, R.K., Carbone, P., Wang, F.C. et al. (2014). Precise and ultrafast molecular sieving through graphene oxide membranes. *Science* 343 (6172): 752–754.

36 Sun, P.Z., Zhu, M., Wang, K.L. et al. (2013). Selective ion penetration of graphene oxide membranes. *ACS Nano* 7 (1): 428–437.

37 Wang, S.F., Yang, L.X., He, G.W. et al. (2020). Two-dimensional nanochannel membranes for molecular and ionic separations. *Chem. Soc. Rev.* 49 (4): 1071–1089.

38 Li, C.Y., Meckler, S.M., Smith, Z.P. et al. (2018). Engineered transport in microporous materials and membranes for clean energy technologies. *Adv. Mater.* 30 (8).

39 Kavokine, N., Netz, R.R., and Bocquet, L. (2021). Fluids at the nanoscale: from continuum to subcontinuum transport. *Annu. Rev. Fluid Mech.* 53: 377–410.

40 Shen, J., Liu, G.P., Han, Y., and Jin, W.Q. (2021). Artificial channels for confined mass transport at the sub-nanometre scale. *Nat. Rev. Mater* 6 (4): 294–312.

41 Geise, G.M., Lee, H.S., Miller, D.J. et al. (2010). Water purification by membranes: the role of polymer science. *J. Polym. Sci. Pol. Phys.* 48 (15): 1685–1718.

42 Ren, J.Z., Li, Z.S., and Wong, F.S. (2006). A new method for the prediction of pore size distribution and MWCO of ultrafiltration membranes. *J. Membr. Sci.* 279 (1-2): 558–569.

43 Han, E.D., Kim, B.H., and Seo, Y.H. (2019). Experimental verification of Poiseuille flow in nanochannels. *Jpn. J. Appl. Phys.* 58 (6): 065001.

44 Osmanbeyoglu, H.U., Hur, T.B., and Kim, H.K. (2009). Thin alumina nanoporous membranes for similar size biomolecule separation. *J. Membr. Sci.* 343 (1-2): 1–6.

45 Tracey, E.M. and Davis, R.H. (1994). Protein fouling of track-etched polycarbonate microfiltration membranes. *J. Colloid Interface Sci.* 167 (1): 104–116.

46 Shen, W.J., Song, F.Q., Hu, X. et al. (2019). Experimental study on flow characteristics of gas transport in micro- and nanoscale pores. *Sci. Rep-Uk* 9 (1): 10196.

47 Liu, R.C., Jiang, Y.J., Li, B., and Yu, L.Y. (2016). Estimating permeability of porous media based on modified Hagen–Poiseuille flow in tortuous capillaries with variable lengths. *Microfluid. Nanofluid.* 20 (8): 120.

48 Loudon, C. and McCulloh, K. (1999). Application of the Hagen–Poiseuille equation to fluid feeding through short tubes. *Ann. Entomol. Soc. Am.* 92 (1): 153–158.

49 Gruener, S. and Huber, P. (2008). Knudsen diffusion in silicon nanochannels. *Phys. Rev. Lett.* 100 (6): 064502.

50 Roh, J.S., Lee, H., Lee, T.H. et al. (2020). Unprecedentedly low CO_2 transport through vertically aligned, conical silicon nanotube membranes. *Nano Lett.* 20 (7): 4754–4760.

51 Zhu, L.X., Yavari, M., Jia, W.G. et al. (2017). Geometric restriction of gas permeance in ultrathin film composite membranes evaluated using an integrated experimental and modeling approach. *Ind. Eng. Chem. Res.* 56 (1): 351–358.

52 Barrer, R.M. and Gabor, T. (1960). Sorption and diffusion of simple paraffins in silica-alumina cracking catalyst. *Proc. R. Soc. London Ser., A* 256 (1285): 267–290.

53 Shelekhin, A.B., Dixon, A.G., and Ma, Y.H. (1992). Adsorption, permeation, and diffusion of gases in microporous membranes .II. Permeation of gases in microporous glass membranes. *J. Membr. Sci.* 75 (3): 233–244.

54 Ash, R., Barrer, R.M., and Pope, C.G. (1963). Flow of adsorbable gases and vapours in a microporous medium. II. Binary mixtures. *Proc. R. Soc. London Ser., A* 271 (1344): 19–33.

55 Beniere, F. (2001). Diffusion in and through polymers. *Defect Diffus. Forum* 194-1 897–907.

56 Pinnau, I., Yampolskii, Y., and Freeman, B.D. (2006). *Materials Science of Membranes for Gas and Vapor Separation.* Wiley.

57 Wijmans, J.G. and Baker, R.W. (1995). The solution-diffusion model – a review. *J. Membr. Sci.* 107 (1-2): 1–21.

58 Lin, H.Q., Kai, T., Freeman, B.D. et al. (2005). The effect of cross-linking on gas permeability in cross-linked poly(ethylene glycol diacrylate). *Macromolecules* 38 (20): 8381–8393.

59 Wu, Y.H., Park, H.B., Kai, T. et al. (2010). Water uptake, transport and structure characterization in poly(ethylene glycol) diacrylate hydrogels. *J. Membr. Sci.* 347 (1-2): 197–208.

60 Jang, E.S., Kamcev, J., Kobayashi, K. et al. (2019). Influence of water content on alkali metal chloride transport in cross-linked Poly(ethylene glycol) diacrylate.1. Ion sorption. *Polymer* 178 (12): 121554.

61 Jang, E.S., Kamcev, J., Kobayashi, K. et al. (2020). Influence of water content on alkali metal chloride transport in cross-linked Poly(ethylene glycol) diacrylate.2. Ion diffusion. *Polymer* 192 (27): 122316.

62 Wind, J.D., Staudt-Bickel, C., Paul, D.R., and Koros, W.J. (2002). The effects of crosslinking chemistry on CO_2 plasticization of polyimide gas separation membranes. *Ind. Eng. Chem. Res.* 41 (24): 6139–6148.

63 Budd, P.M., Elabas, E.S., Ghanem, B.S. et al. (2004). Solution-processed, organophilic membrane derived from a polymer of intrinsic microporosity. *Adv. Mater.* 16 (5): 456–459.

64 Luo, T., Abdu, S., and Wessling, M. (2018). Selectivity of ion exchange membranes: a review. *J. Membr. Sci.* 555: 429–454.

65 Kamcev, J. and Freeman, B.D. (2016). Charged polymer membranes for environmental/energy applications. *Annu. Rev. Chem. Biomol.* 7: 111–133.

66 Park, S.C., Lee, T.H., Moon, G.H. et al. (2019). Sub-5 nm graphene oxide nanofilm with exceptionally high H^+/V selectivity for vanadium redox flow battery. *ACS Appl. Energy Mater.* 2 (7): 4590–4596.

67 Leong, J.X., Daud, W.R.W., Ghasemi, M. et al. (2013). Ion exchange membranes as separators in microbial fuel cells for bioenergy conversion: a comprehensive review. *Renewable Sustainable Energy Rev.* 28: 575–587.

68 Shin, D.W., Guiver, M.D., and Lee, Y.M. (2017). Hydrocarbon-based polymer electrolyte membranes: importance of morphology on ion transport and membrane stability. *Chem. Rev.* 117 (6): 4759–4805.

69 Donnan, F.G. (1995). Theory of membrane equilibria and membrane-potentials in the presence of non-dialyzing electrolytes – a contribution to physical-chemical physiology (reprinted from Zeitshrift Fur Elektrochemie Und Angewandte Physikalische Chemie, Vol 17, Pg 572, 1911). *J. Membr. Sci.* 100 (1): 45–55.

70 Kamcev, J., Paul, D.R., and Freeman, B.D. (2015). Ion activity coefficients in ion exchange polymers: applicability of manning's counterion condensation theory. *Macromolecules* 48 (21): 8011–8024.

71 Krishna, R. and Wesselingh, J.A. (1997). Review article number 50 – the Maxwell-Stefan approach to mass transfer. *Chem. Eng. Sci.* 52 (6): 861–911.

72 Geise, G.M., Paul, D.R., and Freeman, B.D. (2014). Fundamental water and salt transport properties of polymeric materials. *Prog. Polym. Sci.* 39 (1): 1–42.

73 Kamcev, J., Paul, D.R., Manning, G.S., and Freeman, B.D. (2017). Predicting salt permeability coefficients in highly swollen, highly charged ion exchange membranes. *ACS Appl. Mater. Interfaces* 9 (4): 4044–4056.

74 Geise, G.M., Falcon, L.P., Freeman, B.D., and Paul, D.R. (2012). Sodium chloride sorption in sulfonated polymers for membrane applications. *J. Membr. Sci.* 423: 195–208.

75 Miyake, T. and Rolandi, M. (2016). Grotthuss mechanisms: from proton transport in proton wires to bioprotonic devices. *J. Phys. Condens. Matter* 28 (2).

76 Sata, T. (1991). Ion-exchange membranes and separation processes with chemical-reactions. *J. Appl. Electrochem.* 21 (4): 283–294.

77 Tuckerman, M.E., Marx, D., and Parrinello, M. (2002). The nature and transport mechanism of hydrated hydroxide ions in aqueous solution. *Nature* 417 (6892): 925–929.

78 Robeson, L.M. (1991). Correlation of separation factor versus permeability for polymeric membranes. *J. Membr. Sci.* 62 (2): 165–185.

79 Freeman, B.D. (1999). Basis of permeability/selectivity tradeoff relations in polymeric gas separation membranes. *Macromolecules* 32 (2): 375–380.

80 Geise, G.M., Park, H.B., Sagle, A.C. et al. (2011). Water permeability and water/salt selectivity tradeoff in polymers for desalination. *J. Membr. Sci.* 369 (1-2): 130–138.

81 Robeson, L.M., Hwu, H.H., and McGrath, J.E. (2007). Upper bound relationship for proton exchange membranes: empirical relationship and relevance of phase separated blends. *J. Membr. Sci.* 302 (1-2): 70–77.

82 Beuscher, E.J.K.U. and Wijmans, J.G. (2022). Membrane research beyond materials science. *J. Membr. Sci.* 643 (1): 119902.

83 Park, H.G. and Jung, Y. (2014). Carbon nanofluidics of rapid water transport for energy applications. *Chem. Soc. Rev.* 43 (2): 565–576.

84 Faucher, S., Aluru, N., Bazant, M.Z. et al. (2019). Critical knowledge gaps in mass transport through single-digit nanopores: a review and perspective. *J. Phys. Chem. C* 123 (35): 21309–21326.

85 Tunuguntla, R.H., Henley, R.Y., Yao, Y.C. et al. (2017). Enhanced water permeability and tunable ion selectivity in subnanometer carbon nanotube porins. *Science* 357 (6353): 792–796.

86 Zhang, S., Hedtke, T., Zhou, X. et al. (2021). Environmental applications of engineered materials with nanoconfinement. *ACS ES&T Eng.* 1 (4): 706–724. https://doi.org/10.1021/acsestengg.1c00007.

87 Sparreboom, W., van den Berg, A., and Eijkel, J.C.T. (2009). Principles and applications of nanofluidic transport. *Nat. Nanotechnol.* 4 (11): 713–720.

88 Kannam, S.K., Todd, B.D., Hansen, J.S., and Daivis, P.J. (2013). How fast does water flow in carbon nanotubes? *J. Chem. Phys.* 138 (9).

89 Baddour, C. and Briens, C. (2005). Carbon nanotube synthesis: a review. *Int. J. Chem. Reactor Eng.* 3 (1): 1–20.

90 Das, R., Ali, M.E., Abd Hamid, S.B. et al. (2014). Carbon nanotube membranes for water purification: a bright future in water desalination. *Desalination* 336: 97–109.

91 Cheng, L., Liu, G., Zhao, J., and Jin, W. (2021). Two-dimensional-material membranes: manipulating the transport pathway for molecular separation. *Acc. Mater. Res.* 2 (2): 114–128. https://doi.org/10.1021/accountsmr.0c00092.

92 Yoon, H.W., Cho, Y.H., and Park, H.B. (2016). Graphene-based membranes: status and prospects. *Philos. Trans. R. Soc. A* 374 (2060), 20150024.

93 Cho, Y.H., Kim, H.W., Lee, H.D. et al. (2017). Water and ion sorption, diffusion, and transport in graphene oxide membranes revisited. *J. Membr. Sci.* 544: 425–435.

94 Abraham, J., Vasu, K.S., Williams, C.D. et al. (2017). Tunable sieving of ions using graphene oxide membranes. *Nat. Nanotechnol.* 12 (6): 546–550.

95 Bocquet, L. and Charlaix, E. (2010). Nanofluidics, from bulk to interfaces. *Chem. Soc. Rev.* 39 (3): 1073–1095.

96 James, C.E.W., Welty, R., Wilson, R.E., and Rorrer, G.L. (2008). *Fundamentals of Momentum, Heat, and Mass Transfer*, 5the. Wiley.

97 Sparreboom, W., van den Berg, A., and Eijkel, J.C.T. (2010). Transport in nanofluidic systems: a review of theory and applications. *New J. Phys.* 12 (1): 015004.

98 Kannam, S.K., Todd, B.D., Hansen, J.S., and Daivis, P.J. (2011). Slip flow in graphene nanochannels. *J. Chem. Phys.* 135 (14): 016313.

99 Schoch, R.B., Han, J.Y., and Renaud, P. (2008). Transport phenomena in nanofluidics. *Rev. Mod. Phys.* 80 (3): 839–883.

100 Wu, K.L., Chen, Z.X., Li, J. et al. (2017). Wettability effect on nanoconfined water flow. *Proc. Natl. Acad. Sci. U.S.A.* 114 (13): 3358–3363.

101 Thomas, J.A. and McGaughey, A.J.H. (2008). Reassessing fast water transport through carbon nanotubes. *Nano Lett.* 8 (9): 2788–2793.

102 Tao, J.B., Song, X.Y., Zhao, T. et al. (2018). Confinement effect on water transport in CNT membranes. *Chem. Eng. Sci.* 192: 1252–1259.

103 Keerthi, A., Goutham, S., You, Y. et al. (2021). Water friction in nanofluidic channels made from two-dimensional crystals. *Nat. Commun.* 12 (1): 3092.

104 Epsztein, R., DuChanois, R.M., Ritt, C.L. et al. (2020). Towards single-species selectivity of membranes with subnanometre pores. *Nat. Nanotechnol.* 15 (6): 426–436.

105 Cheng, C., Iyengar, S.A., and Karnik, R. (2021). Molecular size-dependent subcontinuum solvent permeation and ultrafast nanofiltration across nanoporous graphene membranes. *Nat. Nanotechnol.* 16 (9): 989–995.

106 Karnik, R., Fan, R., Yue, M. et al. (2005). Electrostatic control of ions and molecules in nanofluidic transistors. *Nano Lett.* 5 (5): 943–948.

107 Xiao, K., Jiang, L., and Antonietti, M. (2019). Ion transport in nanofluidic devices for energy harvesting. *Joule* 3 (10): 2364–2380.

108 Kohler, M.H. and da Silva, L.B. (2016). Size effects and the role of density on the viscosity of water confined in carbon nanotubes. *Chem. Phys. Lett.* 645: 38–41.

109 Falk, K., Sedlmeier, F., Joly, L. et al. (2010). Molecular origin of fast water transport in carbon nanotube membranes: superlubricity versus curvature dependent friction. *Nano Lett.* 10 (10): 4067–4073.

110 Lee, T.H., Oh, J.Y., Jang, J.K. et al. (2020). Elucidating the role of embedded metal-organic frameworks in water and ion transport properties in polymer nanocomposite membranes. *Chem. Mater.* 32 (23): 10165–10175.

111 Kim, J.Y., Balderas-Xicohtencatl, R., Zhang, L.D. et al. (2017). Exploiting diffusion barrier and chemical affinity of metal-organic frameworks for efficient hydrogen isotope separation. *J. Am. Chem. Soc.* 139 (42): 15135–15141.

112 Algara-Siller, G., Lehtinen, O., Wang, F.C. et al. (2015). Square ice in graphene nanocapillaries. *Nature* 519 (7544): 443–445.

113 Corry, B. (2018). The naked truth about K^+ selectivity. *Nat. Chem.* 10 (8): 799–800.

114 Sahu, S., Di Ventra, M., and Zwolak, M. (2017). Dehydration as a universal mechanism for ion selectivity in graphene and other atomically thin pores. *Nano Lett.* 17 (8): 4719–4724.

115 Gopinadhan, K., Hu, S., Esfandiar, A. et al. (2019). Complete steric exclusion of ions and proton transport through confined monolayer water. *Science* 363 (6423): 145–147.

116 Zhang, M.C., Guan, K.C., Ji, Y.F. et al. (2019). Controllable ion transport by surface-charged graphene oxide membrane. *Nat. Commun.* 10: 1253.

117 Liu, M., Zhang, L.D., Little, M.A. et al. (2019). Barely porous organic cages for hydrogen isotope separation. *Science* 366 (6465): 613–620.

118 Hu, S., Lozada-Hidalgo, M., Wang, F.C. et al. (2014). Proton transport through one-atom-thick crystals. *Nature* 516 (7530): 227-+.

119 Yoon, S.I., Seo, D.J., Kim, G. et al. (2018). AA'-stacked trilayer hexagonal boron nitride membrane for proton exchange membrane fuel cells. *ACS Nano* 12 (11): 10764–10771.

120 Boutilier, M.S.H., Jang, D.J., Idrobo, J.C. et al. (2017). Molecular sieving across centimeter-scale single-layer nanoporous graphene membranes. *ACS Nano* 11 (6): 5726–5736.

121 Jang, D., Idrobo, J.C., Laoui, T., and Karnik, R. (2017). Water and solute transport governed by tunable pore size distributions in nanoporous graphene membranes. *ACS Nano* 11 (10): 10042–10052.

122 Cheng, P.F., Kelly, M.M., Moehring, N.K. et al. (2020). Facile size-selective defect sealing in large-area atomically thin graphene membranes for sub-nanometer scale separations. *Nano Lett.* 20 (8): 5951–5959.

123 Fu, Y.J., Su, S.H., Zhang, N. et al. (2020). Dehydration-determined ion selectivity of graphene subnanopores. *ACS Appl. Mater. Interfaces* 12 (21): 24281–24288.

124 Jain, T., Rasera, B.C., Guerrero, R.J.S. et al. (2015). Heterogeneous sub-continuum ionic transport in statistically isolated graphene nanopores. *Nat. Nanotechnol.* 10 (12): 1053–1057.

125 Mogg, L., Zhang, S., Hao, G.P. et al. (2019). Perfect proton selectivity in ion transport through two-dimensional crystals. *Nat. Commun.* 10: 4243.

126 Gao, J., Feng, Y.P., Guo, W., and Jiang, L. (2017). Nanofluidics in two-dimensional layered materials: inspirations from nature. *Chem. Soc. Rev.* 46 (17): 5400–5424.

127 Xu, W.W.L., Fang, C., Zhou, F.L. et al. (2017). Self-assembly: a facile way of forming ultrathin, high-performance graphene oxide membranes for water purification. *Nano Lett.* 17 (5): 2928–2933.

128 Kim, S., Nham, J., Jeong, Y.S. et al. (2015). Biomimetic selective ion transport through graphene oxide membranes functionalized with ion recognizing peptides. *Chem. Mater.* 27 (4): 1255–1261.

129 Huang, L., Jia, W.G., and Lin, H.Q. (2020). Etching and acidifying graphene oxide membranes to increase gas permeance while retaining molecular sieving ability. *AIChE J.* 66 (12): e17022.

130 Chen, X.Y., Feng, Z.H., Gohil, J. et al. (2020). Reduced holey graphene oxide membranes for desalination with improved water permeance. *ACS Appl. Mater. Interfaces* 12 (1): 1387–1394.

131 Jang, J., Nam, Y.T., Kim, D. et al. (2020). Turbostratic nanoporous carbon sheet membrane for ultrafast and selective nanofiltration in viscous green solvents. *J. Mater. Chem. A* 8 (17): 8292–8299.

132 Kim, D.W., Kim, I., Jang, J. et al. (2017). One dimensional building blocks for molecular separation: laminated graphitic nanoribbons. *Nanoscale* 9 (48): 19114–19123.

133 Roh, J.S., Choi, T.H., Lee, T.H. et al. (2019). Understanding gas transport behavior through few-layer graphene oxide membranes controlled by tortuosity and interlayer spacing. *J. Phys. Chem. Lett.* 10 (24): 7725–7731.

134 Karunakaran, M., Shevate, R., Kumar, M., and Peinemann, K.V. (2015). CO_2-selective PEO-PBT (PolyActive (TM))/graphene oxide composite membranes. *Chem. Commun.* 51 (75): 14187–14190.

135 Ding, L., Wei, Y.Y., Li, L.B. et al. (2018). MXene molecular sieving membranes for highly efficient gas separation. *Nat. Commun.* 9: 155.

136 Karahan, H.E., Goh, K., Zhang, C.F. et al. (2020). MXene materials for designing advanced separation membranes. *Adv. Mater.* 32 (29): 1906697.

137 Wang, J., Zhang, Z.J., Zhu, J.N. et al. (2020). Ion sieving by a two-dimensional $Ti_3C_2T_x$ alginate lamellar membrane with stable interlayer spacing. *Nat. Commun.* 11 (1): 3540.

138 Chen, C., Wang, J.M., Liu, D. et al. (2018). Functionalized boron nitride membranes with ultrafast solvent transport performance for molecular separation. *Nat. Commun.* 9: 1902.

139 Wang, Z.Y., Tu, Q.S., Zheng, S.X. et al. (2017). Understanding the aqueous stability and filtration capability of MoS_2 membranes. *Nano Lett.* 17 (12): 7289–7298.

140 Jian, M.P., Qiu, R.S., Xia, Y. et al. (2020). Ultrathin water-stable metal-organic framework membranes for ion separation. *Sci. Adv.* 6 (23), eaay3998.

141 Peng, Y., Li, Y.S., Ban, Y.J. et al. (2014). Metal-organic framework nanosheets as building blocks for molecular sieving membranes. *Science* 346 (6215): 1356–1359.

142 Yang, H., Yang, L.X., Wang, H.J. et al. (2019). Covalent organic framework membranes through a mixed-dimensional assembly for molecular separations. *Nat. Commun.* 10: 2101.

143 Monsalve-Bravo, G.M. and Bhatia, S.K. (2018). Modeling permeation through mixed-matrix membranes: a review. *Processes* 6 (9): 172.

144 Hoang, V.T. and Kaliaguine, S. (2013). Predictive models for mixed-matrix membrane performance: a review. *Chem. Rev.* 113 (7): 4980–5028.

145 Bharadwaj, R.K. (2001). Modeling the barrier properties of polymer-layered silicate nanocomposites. *Macromolecules* 34 (26): 9189–9192.

12

Conclusions and Perspectives

Gongping Liu and Wanqin Jin

Nanjing Tech University, College of Chemical Engineering, State Key Laboratory of Materials-Oriented Chemical Engineering, 30 Puzhu South Road, Nanjing, P. R. China

In general, the emergence of various 2D nanosheets demonstrates great potential in developing ultrathin membranes for gas or liquid separation. Particularly, various attempts on tuning the size and microenvironment of subnanometer channels in 2D-material membranes have demonstrated several orders of magnitude higher permeance than conventional membranes, largely advancing membrane technology toward energy-efficient separation. Despite the significant progresses achieved during the past years, some critical issues on 2D-material membranes remain to be addressed from the following four aspects:

(1) **Materials preparation**. Up to now, only limited 2D materials have been applied for membrane separation and much more 2D nanosheets should be developed in the future. As the starting unit of the subsequent membranes, the acquisition of high-quality nanosheets is imperative yet challenging. Therefore, it is urgent to realize the reproducible fabrication of defect-free 2D materials in high yield. Advanced characterization techniques with atomic-scale resolution, such as positron annihilation lifetime spectra and *in situ* high-resolution transmission electron microscopy are required to monitor the synthesis process of nanosheets. Last but not the least, the large-scale fabrication of 2D nanosheets with micron lateral size and atomic thickness by greener and safer methods is necessary for future production of 2D-materials membranes.
(2) **Membrane fabrication**. Effective strategies for fabricating large-area and non-defective 2D-material membrane are of great significance. Usually, 2D-material membranes demonstrate *low pore density*, undesirable defects or nanochannels with high tortuosity. More efforts on tuning the size or chemically modifying the microenvironment of nanochannels should be made to construct nanochannels with molecular-sieving properties and enhanced affinity to target penetrants. Reliable and convenient fabrication methods of 2D-material

Two-Dimensional-Materials-Based Membranes: Preparation, Characterization, and Applications, First Edition. Edited by Gongping Liu and Wanqin Jin.
© 2022 WILEY-VCH GmbH. Published 2022 by WILEY-VCH GmbH.

membranes should be exploited. Spin coating and vacuum filtration are the most extensively adopted strategies for preparing 2D-material membranes, which are not beneficial for achieving large-area 2D-material membranes. Simple-printing and blade-coating technologies should be further explored to accelerate the scale-up production of 2D-material membranes. Besides, for 2D mixed-matrix membranes, issues, such as homogenous dispersion, at high loading and interfacial voids also need to be addressed.

(3) **Scale-up application**. We all know that membrane performance assessed in the laboratory scale under ideal feed streams differs significantly from that in the industry under rigorous conditions. Evaluating the separation performance under practical conditions is quite crucial to promote 2D membranes as the next-generation membrane. For industry applications, membranes of area >1000 m^2 and with a low concentration of defects are required (e.g. in gas separation, below about 1 cm^2 of defects for every 105 cm^2 of membrane surface area). These requirements for membrane-based separation, to some extent, are more stringent compared with other applications. High pressure, harsh pH, complex components, and water stability are common challenges for developing 2D-material membranes in liquid separation. In addition, designing large-area 2D-material membranes on hollow-fiber supports in a simple and scalable way is quite important for wide implementation due to the high packing efficiency. Besides, the thermal stability, plasticization, and aging problems of 2D-material membranes should also be considered in the future. The long-term stability of graphene oxide (GO) membrane in the time span of 10 000 hours in the working conditions needs to be evaluated to demonstrate the potential of 2D-material membranes in practical application.

(4) **Transport mechanism**. Up to now, transport mechanism of 2D-material membranes is limited in the study of the interlayer of nanosheets and the wrinkle morphology. Much more attention should be paid to investigate the effect of edge property and in-plane defect pore on the separation performance. Brand new theories, including facilitated transport theory and confined mass-transfer theory, should be constructed to describe transport behaviors in membranes and guide the design of 2D-material membranes. The relationship between experimental results and models should be established by combining simulation with advanced characterizations. For example, by optimizing the *in situ* characterization methods, real-time observation of formation of laminates, and the evolution of channel structures can be achieved. Monte Carlo, density functional theory (DFT) simulations, and related kinetic/thermodynamic calculations could be employed to study the adsorption, diffusion, and facilitated transport behaviors of molecules/ions in 2D confined subnanometer channels. In the future, mass-transfer mechanisms and models concerning 2D-material membranes for various separation processes should be revealed and established, which

would provide theoretical guidance for the scale-up fabrication of 2D-material membranes.

With the above issues addressed in the future, 2D-material membranes are believed to become a next-generation separation membrane and be extended to other fields such as catalysis and energy conversion.

Index

Two-Dimensional-Materials-Based Membranes: Preparation, Characterization, and Applications, First Edition. Edited by Gongping Liu and Wanqin Jin.
© 2022 WILEY-VCH GmbH. Published 2022 by WILEY-VCH GmbH.

d

h

i

q

r

s